Principles of

RESIDENTIAL CONSTRUCTION

ESSENTIAL SKILLS AND PRACTICES

R. Jack Jones

Associate Professor of Building Construction Technology
and Residential Construction Technology and Management
Pennsylvania College of Technology
Williamsport, PA

Publisher
The Goodheart-Willcox Company, Inc.
Tinley Park, IL
www.g-w.com

ISBN 979-8-89118-813-6

2 3 4 5 6 7 8 9 – 26 – 30 29 28 27 26 25

Image Credits. Front cover: Kirkam/Shutterstock.com

Preface

Congratulations—you are about to gain knowledge and skills that will serve you well throughout your life. You are preparing to explore the foundational principles of residential construction, and you will learn how a home is constructed from site preparation to the finishing touches and punch list. As you embark on this journey, you will develop the basic skills needed to build and maintain a structure, to analyze building blueprints, and to apply green construction innovations.

You have several choices of how you will apply this knowledge and skill in your future. The construction industry is a broad field, and career options are vast. You may choose to pursue a career in the construction trades, perhaps as a carpenter, plumber, electrician, or HVAC technician. Some tradespeople transition into fields that are related to construction but are not actively building. Many companies eagerly seek out people with knowledge in construction. Insurance adjusters, inspectors, safety professionals, trade journalists, and salespeople are just a few examples of careers that you can qualify for with experience and knowledge of the construction field. Even if you choose a career outside of construction, this knowledge is still valuable to you as a future homeowner.

Principles of Residential Construction is structured to follow the logical sequence of home construction that builders use to complete the various phases of home construction. This text has a unique approach that utilizes these phases as a framework on which to base associated content including relevant materials, tools, fasteners, and safety practices.

While traditional textbooks attempt to give a thorough study on a single topic for each chapter, *Principles of Residential Construction* builds on a broad array of topics through lessons across each chapter. Just as the construction process originates with permitting and concludes with the attainment of a certificate of occupancy, each chapter builds on previous content.

Utilizing this lesson structure allows educators the freedom and flexibility to teach horizontally by subject matter (Lesson 1.6, Lesson 2.6, Lesson 3.6, etc.) or vertically (Chapter 1, Chapter 2, Chapter 3, etc.), following content that complements the construction path. For instance, the construction path in Chapter 2 addresses excavation and foundation construction. Each lesson within the chapter covers different aspects during this phase, such as materials to create the foundation, tools to build the footings (concrete and CMU block), how to read foundation plans, and necessary safety precautions. The safety lessons of each chapter not only complement the construction path, but also follow the content required for an OSHA 10 course, allowing teachers to use this text to supplement and enhance a traditional or online OSHA class.

Principles of Residential Construction acknowledges the importance of the history of construction while looking toward the future of the industry, maintaining relevance regarding current industry practices and procedures, current building codes, legal requirements, and green building practices.

About the Author

R. Jack Jones is an Associate Professor of Building Construction Technology and Residential Construction Technology and Management at Pennsylvania College of Technology, a Pennsylvania State University affiliate. As an educator, he enjoys watching students' skills, understanding, and passion for the construction trade develop during their time in his program. As a professional, he has over 25 years of experience working in the construction industry as a carpenter, general contractor, and educator. He has been a member of the United Brotherhood of Carpenters and Joiners of America and has experience in both commercial and residential construction. Mr. Jones earned a Bachelor of Science degree in Business Administration through Mansfield University of Pennsylvania and serves as an Outreach Trainer for the Occupational Safety and Health Administration. He is an honorary member of The National Society of Leadership and Success, where he received the Excellence in Teaching Award. Additionally, Mr. Jones is a member of the Rochester Builder's Exchange and has been designated a Certified Installer by the Vinyl Siding Institute.

Reviewers

The author and publisher wish to thank the following industry and teaching professionals for their valuable input into the development of *Principles of Residential Construction: Essential Skills and Practices*.

Matthew Anastasi
Orleans/Niagara BOCES
Medina, NY

Knut Boersma
King Tech High School
Anchorage, AK

Hugh Creamer
Bell Senior High School
Bell, CA

Dominic Dipietro
Delsea Regional High School
Franklinville, NJ

Jerome Ebbs
Monroe 2-Orleans BOCES
Spencerport, NY

Arthur L. Erbe
City School District of Albany
Albany, NY

Leonard Friedman
Saugus High School
Santa Clarita, CA

Craig Griffie
Wauwatosa East High School
Wauwatosa, WI

Paul Hoch
Erie 2-Chautauqua-Cattaraugus BOCES
Wallace D. Ormsby Education Center
East Aurora, NY

Chad Hoobler
Venango Technical Center
Oil City, PA

Anthony R. Iannucci
Nassau BOCES
Joseph M. Barry Career & Technical Education Center
Westbury, NY

John W. Jackson
Skyline High School, Dallas ISD
Dallas, TX

David Jay
Waipahu High School
Waipahu, HI

Kent Kiess
Warren High School
Downey, CA

Eli Lihau Maioho
Waipahu High School
Waipahu, HI

Gerald Markel
Putnam/Northern Westchester BOCES
Yorktown Heights, NY

Arnold Mendez
Del Mar College
Corpus Christi, TX

Michael W. Mills
Northeast Technical High School
Watertown, SD

Shane Morrow
Bishop State Community College
Mobile, AL

Richard Nicholson
Society of Wood Manufacturing
Anaheim, CA

Oscar Ortiz
Orange Coast College
Costa Mesa, CA

Brian Skates
Trimble Technical High School, Fort Worth ISD
Fort Worth, TX

Mark Spoerk
Lynde and Harry Bradley Technology and Trade School
Milwaukee, WI

William Stobie
Wayne-Finger Lakes BOCES
Finger Lakes Technical and Career Center
Stanley, NY

Sean Strathearn
Genesee Valley BOCES
Mount Morris, NY

Rodney Dale Utt
Santa Barbara City College
Santa Barbara, CA

Peter Wachtel
Adolfo Camarillo High School
Camarillo, CA

Cameron Wambsganss
Erie 1 BOCES
Potter Career and Technical Center
West Seneca, NY

Thomas A. Witter, Jr.
Cattaraugus-Allegany BOCES
CTE Center at Ellicotville
Ellicotville, NY

Acknowledgments

The author and publisher would like to thank the following companies, organizations, and individuals for their contribution of resource material, images, or other support in the development of *Principles of Residential Construction: Essential Skills and Practices*.

A & C Paslode
Accufooting
Alfred State College
Alum-A Pole Corp
American Heart Association
American Olean Tile Co.
APA—The Engineered Wood Association
Arizona Tile
Ark Seal, Inc.
Arlington Industries
Bird Division, CertainTeed Corporation
Black & Decker
Bosch Tools
Brammer Mfg. Co.
BuildSMART, LLC
Canvas
Carly Coulson, AIA, LEED AP, CPHD
Carrier Corporation, subsidiary of United Technologies Group
CertainTeed Corporation
CPG Building Products—TimberTech ReliaBoard
Crown Aluminum Industries
David White
DeWalt
Dŏrken Systems Inc.
Dutcher Glass and Paint
Empire Level, Inc.
Enercept, Inc.
FBR Limited
Formica Group North America
Forney Industries, Inc.
Frank Paxton Lumber Co.
Greco Painting
Greenfiber Cellulose Insulation
Hilti Tools
Huber Engineered Woods
Independent Nail and Packing Co.
International Residential Code
James Hardie Building Products
Keuffel & Esser
Klein Tools
Kunkle Valve Co. Inc.
Lift-All Company, Inc.
LP Corp
Malco Tools, Inc.
MARSHALLTOWN
Marvin Windows
Masonite Corp.
Merillat Industries, Inc.
Miller and Richard Masonry
Milwaukee Tools
Monier Group
National Association of the Remodeling NOSHI
National Forest Products Assn.
National Gypsum Co.
National Woodwork Manufacturers Association
OSHA
Owens-Corning
Passive House Institute US, Inc.
Patent Scaffolding Co.
Pease Industries, Inc.
Pennsylvania College of Technology
PHIUS
Pierce Construction, Ltd
Plumwall Bracing Systems Inc.
Portland Cement Association
Reed Manufacturing Co.
RenewAire
Rev-A-Shelf, Inc.
Rheem Manufacturing Company
Rik Vandermeulen, Unalam
Robbins/Sykes, Tibbias Flooring Inc.
Robert Bosch Tool Corporation®
RotoZip, a division of Robert Bosch Tool Corporation
Simpson Strong-Tie Company, Inc.
Slant/Fin Corp.
Specify Concrete
Spectra-Physics Laserplane, Inc.
Stephen P. Kielar
Stinger
Sto Finish Systems Division, US Gypsum Co.
Superior Walls
The Energy Conservancy
The Garlinghouse Company
Trex Company
Trudeau Construction Co.
TruStile Doors
Typar
United States Steel
United Technologies Carrier
US Department of Energy
US Department of Labor
US Environmental Protection Agency
US Gypsum Co.
Village of Flossmoor, Illinois
Wellsboro Borough Zoning Map
Wermers Design and Architecture
Western Wood Products Association
Weyerhaeuser

Credentialing Partners and Support

Goodheart-Willcox appreciates the value of industry credentials, certifications, and accreditation. We are pleased to partner with leading organizations to support students and programs in achieving credentials. Integrating industry-recognized credentialing into a career and technical education (CTE) program provides many benefits for the student and for the institution. By achieving third-party certificates, students gain confidence, have proof of a measurable level of knowledge and skills, and earn a valuable achievement to include in their résumés. For educators and administrators, industry-recognized credentials and accreditation validate learning, enhance the credibility of programs, and provide valuable data to measure student performance and help guide continuous program improvement.

Principles of Residential Construction: Essential Skills and Practices is correlated to the construction credentials offered by the Home Builders Institute, NOCTI, and the Construction Trades Foundation YouScience® Industry Certification.

HBI Certifications

Home Builders Institute (HBI) is the nation's leading provider of trade skills training in residential construction. HBI provides pre-apprenticeship training, certification programs, and job placement services to secondary school students, veterans and transitioning military personnel, displaced workers, and at-risk and justice-involved youth and adults. HBI trains in carpentry, electrical, plumbing, building construction technology, HVAC, landscaping, masonry, and solar installation.

HBI is a nonprofit organization that collaborates with public, private, and nonprofit partners dedicated to workforce development. Working with its partners, HBI attracts and certifies new students through a combination of hands-on training, innovation, and technology to build an emerging generation of skilled construction workers. For more information about HBI, visit www.hbi.org.

NOCTI Certifications

Goodheart-Willcox is pleased to partner with NOCTI and Nocti Business Solutions (NBS), a leading provider of industry certification solutions for skill verification across the nation. With over 55 years of experience, NOCTI/NBS is a valuable partner both within the CTE community and industry to improve America's workforce. Goodheart-Willcox has created correlations between select products and the standards and competencies that are the basis for NOCTI/NBS credentials, to the benefit of states, instructors, and students working toward NOCTI/NBS credentials.

NOCTI/NBS certifications (knowledge-based and skill-based) are developed by national teams of subject matter experts as part of the process that meets personnel accrediting standards and requirements under ISO 17024, resulting in credentials measuring skills and competencies critical for learner success outside the classroom. NOCTI/NBS offers test development, test delivery, and psychometric services to support skill verification, using the latest tools and methods to provide relevant and helpful solutions for workforce development. For more information about NOCTI, visit www.nocti.org.

YouScience® Industry Certifications

Goodheart-Willcox is pleased to partner with YouScience®, correlating select Goodheart-Willcox titles to YouScience® Industry Certifications to the benefit of states, instructors, and students working to achieve them. YouScience® is the leading technology provider dedicated to solving the skills and exposure gap crisis for students and employers.

YouScience® Industry Certifications, part of the YouScience Brightpath academic and career guidance platform, offer students entry- to mid-level certifications that act as tangible proof of their skills and knowledge. For more information about YouScience, including a complete listing of their Industry Certifications, please visit https://www.youscience.com/certifications/.

To see how *Principles of Residential Construction: Essential Skills and Practices* correlates to credentialing and certification standards, visit the Correlations tab at www.g-w.com/principles-residential-construction-2026.

Guided Tour

The instructional design includes student-focused learning tools to help students succeed. This visual guide highlights the features designed for the textbook.

Chapter Outline offers a quick way to locate content within a chapter.

Introduction provides an overview and preview of the chapter content.

Learning Outcomes clearly identify the knowledge and skills to be obtained when the chapter is completed. These outcomes are labeled and correlated to related assessment elements to facilitate reporting and remediation.

Technical Terms list the key terms important to the content in the chapter.

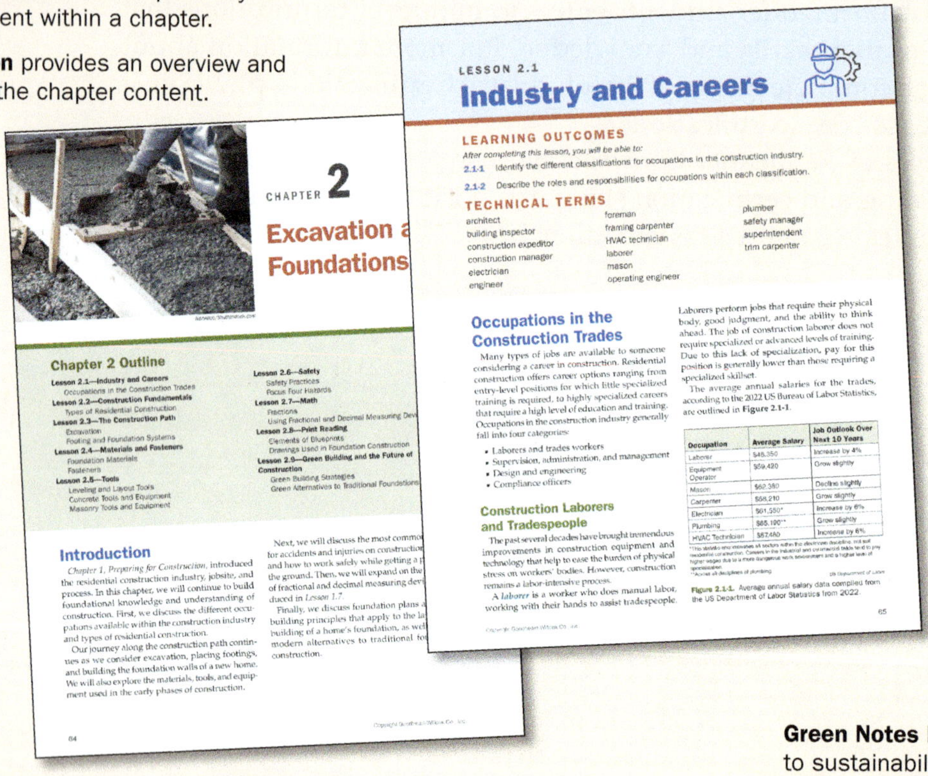

Green Notes highlight key items related to sustainability, energy efficiency, and environmental issues.

Safety Notes alert you to potentially dangerous materials and practices.

Code Notes point out specific items from environmental regulations and typical building codes.

Practice problems allow you opportunities to hone your skills around math concepts relevant to the construction industry.

Pro Tips provide you with advice and guidance that is especially applicable on the jobsite.

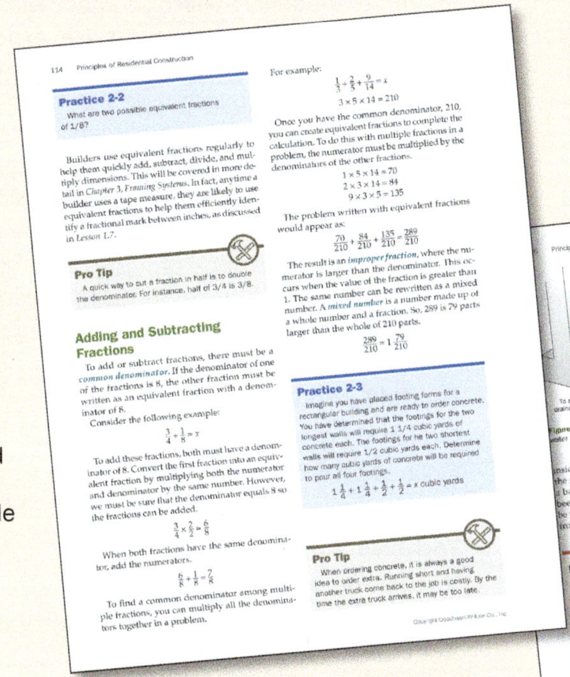

Illustrations have been designed to clearly and simply communicate the specific topic. Illustrations have been completely replaced and updated for this edition. Photographic images have been updated to show the latest equipment.

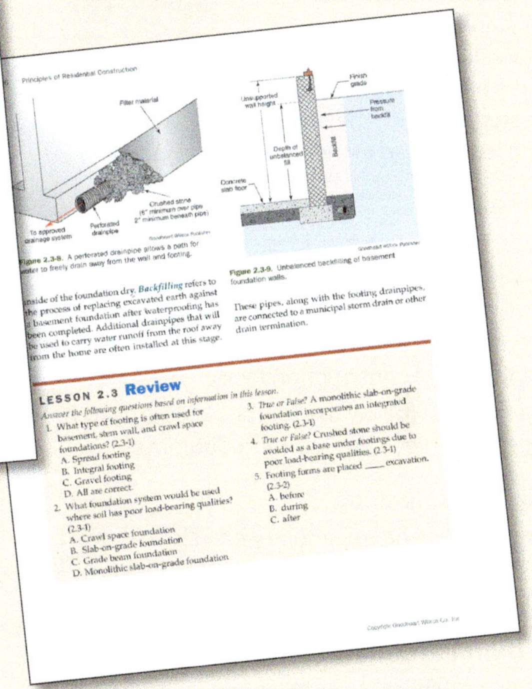

Lesson Review questions enable you to demonstrate knowledge, identification, and comprehension of lesson material.

Summary feature provides an additional review tool for you and reinforces key learning outcomes.

Apply and Analyze questions extend learning and develop your abilities to use learned material in new situations and to break down material into its component parts.

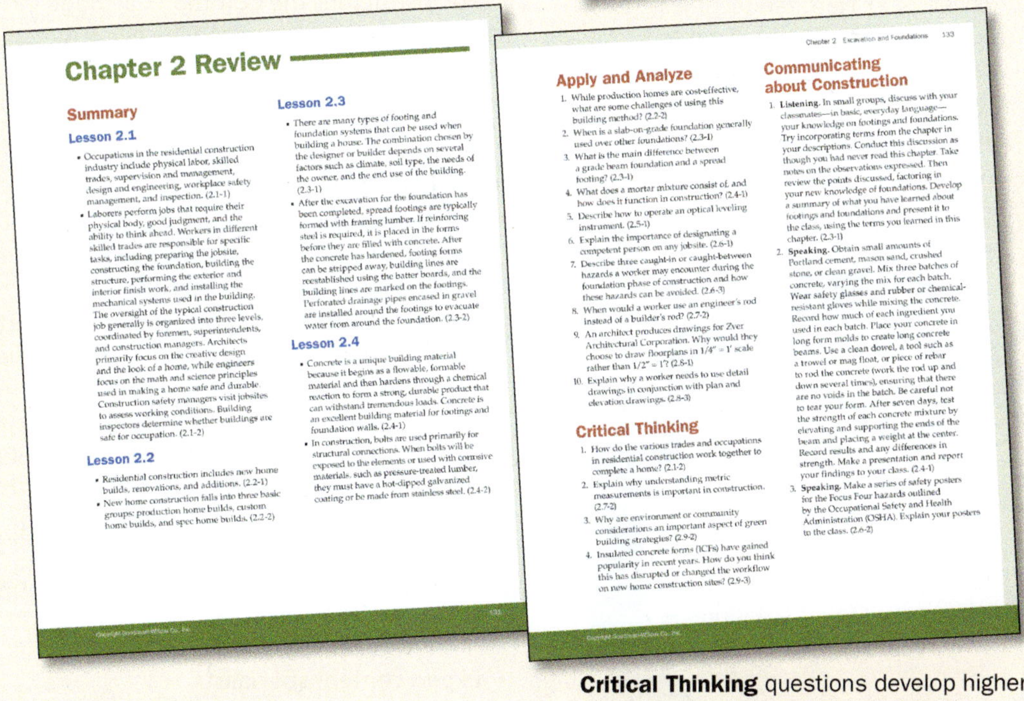

Critical Thinking questions develop higher-order thinking and problem-solving abilities.

Communicating about Construction activities extend your learning and help you apply knowledge in relevant, real-world scenarios.

TOOLS FOR STUDENT AND INSTRUCTOR SUCCESS

Student Tools

Student Text

Principles of Residential Construction is an introductory text that covers all the phases of construction using a student-friendly writing style that makes complex subjects easier to understand. Clear, easy-to-understand illustrations, along with real-world, hands-on activities, provide students with the necessary opportunities to build skills to succeed in the construction industry.

Lab Workbook

- Minds-on practice includes chapter review questions.
- Hands-on practice includes skill building activities and comprehensive projects for students to work on a variety of construction tasks.
- Organized to follow the textbook lessons to help students achieve key learning outcomes.

G-W Digital Companion

- E-flash cards and vocabulary exercises allow interaction with content to create opportunities to increase achievement.

Online Learning Suite

- Videos enrich learning: Narrated technical animations provide a unique visual aid to technical concepts discussed in the textbook.
- Provides easy-to-use access and navigation
- Includes accessible resources for all learners
- Encourages exploration and discovery (or practice and repetition)

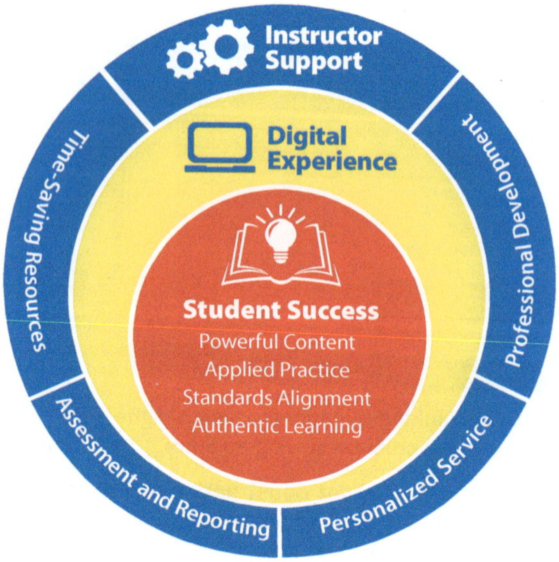

Instructor Tools

LMS Integration

Integrate Goodheart-Willcox content within your Learning Management System for a seamless user experience for both you and your students. LMS-Ready Content in Common Cartridge® format facilitates single sign-on integration and gives you control of student enrollment and data. With a Common Cartridge integration, you can access the LMS features and tools you are accustomed to using and G-W course resources in one convenient location—your LMS.

G-W Common Cartridge provides a complete learning package for you and your students. The included digital resources help your students remain engaged and learn effectively:

- **Digital Textbook**
- **Drill and Practice** vocabulary activities
- Online **Lab Workbook content**
- **Construction Videos**
- **Technical Animations**
- **Virtual Toolbox**
- **Understanding Measurement Tools Videos and Activities**

When you incorporate G-W content into your courses via Common Cartridge, you have the flexibility to customize and structure the content to meet the educational needs of your students. You may also choose to add your own content to the course.

For instructors, the Common Cartridge includes the Online Instructor Resources, which link directly to a cloud-based assessment platform that works alongside your LMS. Create, customize, edit, and print prebuilt questions and tests to meet your needs.

Online Instructor Resources

- The **Instructor Resources** provide instructors with time-saving preparation tools such as answer keys, editable lesson plans, and other teaching aids.
- **Instructor's Presentations for PowerPoint®** are fully customizable, richly illustrated slides that help you teach and visually reinforce the key concepts from each chapter.
- Easily create, customize, edit, and print assessments to meet your classroom needs using a cloud-based platform with hundreds of questions that assess student knowledge of the content in each chapter.

See www.g-w.com/principles-residential-construction-2026 for a list of all available formats.

Professional Development

- Expert content specialists
- Research-based pedagogy and instructional practices
- Options for virtual and in-person Professional Development

Brief Contents

Contents

CHAPTER 3
Framing Systems

CHAPTER 4
Drying In

Topic Contents

Materials and Fasteners

Tools

Safety

Math

Print Reading

Green Building and the Future of Construction

Oleg Zaslavsky/Shutterstock.com

CHAPTER 1

Preparing for Construction

Chapter 1 Outline

Introduction

Congratulations on your decision to explore the construction industry as a career option! Construction is a broad industry filled with exciting opportunities. This text primarily explores the field of residential construction, though the skills and applications learned here will apply to other fields of construction as well. Residential construction can be a tremendously personal and rewarding segment of the construction industry. It is not uncommon for workers to interact with the owners of the home they are building. It is often said that home ownership is a big part of the American Dream. In this context, those who work in the residential construction industry are actively building the American Dream. Trades workers are an essential fixture in the process of materializing the home ownership and renovation dreams of their clients.

LESSON 1.1

Industry and Careers

LEARNING OUTCOMES

After completing this lesson, you will be able to:

1.1-1 Explain how residential construction differs from other types of construction.

1.1-2 Describe the employment outlook in the construction trades.

1.1-3 Outline the history of the field of construction.

TECHNICAL TERMS

residential construction

Employment in the Construction Trades

Residential construction is the process of building single and multifamily homes, **Figure 1.1-1**. It also includes the addition to or renovation of these buildings. Beginning a career in the construction field can seem overwhelming to those who are not familiar with it. The tools and equipment used on jobsites can be dangerous. Construction workers often work from heights, performing tasks that can take a high degree of focus or strength. Construction drawings, product offerings, systems, assemblies, and even site conditions can be complex and difficult to understand, **Figure 1.1-2**. Furthermore, construction takes a high degree of planning and problem-solving ability to achieve good results.

Goodheart-Willcox Publisher

Figure 1.1-1. A foreman and worker discuss their daily plans on a new home build.

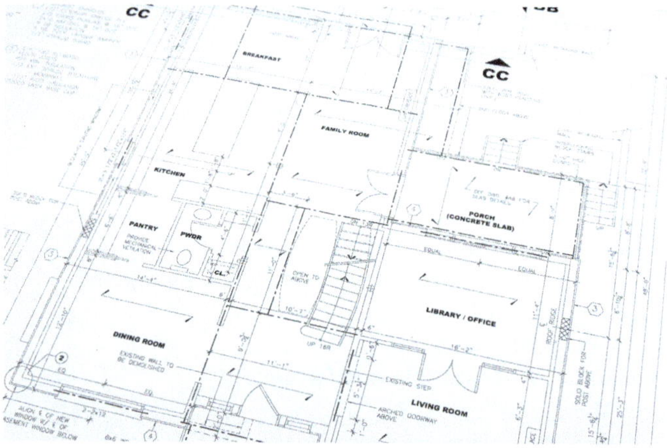

Branislav via Getty Images

Figure 1.1-2. This drawing illustrates the complexity of the information that a builder needs to process.

It is understandable that people may be reluctant or intimidated by the idea of jumping into this kind of work, especially if they have never been around it. If you find that you are experiencing these feelings, give yourself permission to relax. Construction is an occupation where even the best of the best are continually learning, developing their skills, and, believe it or not, making mistakes. Mastering construction skills and processes is a lifelong journey. As you progress, you will begin to build confidence in your abilities. When you practice the skills and techniques in this book, you will likely make plenty of mistakes. In these moments, give yourself grace. You may be disappointed, but making mistakes is simply part of the learning process.

Over time your skill and understanding will develop, **Figure 1.1-3**.

Construction work includes the building of new structures and additions, as well as alterations, maintenance, and repairs on existing structures. The industry encompasses not only residential structures where people live, but also commercial, agricultural, industrial, and civil projects. Any contracted building project from reroofing a shed to the design and assembly of huge factories, bridges, and airports would be classified as part of the construction industry. Consider all the employees it takes to build all such projects in the United States. According to a recent statistic by the US Bureau of Labor Statistics, more than 8.1 million people work in the construction sector in the United States. Even with this impressive statistic, the construction industry is suffering from a lack of skilled workers. New home construction has not kept pace with population growth over recent years. This, coupled with an aging workforce, has increased demand for construction workers. With construction skills and experience in high demand, wages in the construction industry have steadily increased at a rate greater than many other industries. Steadily increasing wages, a high demand for workers, a strong need for more housing, and rewarding work are some of the reasons that it is a great time to consider a career in the construction trades.

Within construction are a wide range of employment roles. Each of these roles can fall into specific categories of employment: labor, skilled trades, supervision and management, and design and engineering. These roles will be explored in more detail in *Chapter 2, Excavation and Foundations*.

History of Construction

As long as there has been a need for shelter, protection, and transportation, there has been construction. As we study ancient history, we see evidence that even cave dwellers made improvements to their natural environment, **Figure 1.1-4**. Some of the oldest known structures were built of stone, an incredibly durable building material taken directly from nature. However, because ancient humans also used sticks, mud, grass, and other materials that decomposed or eroded with time, we will never know exactly how the processes of construction developed. Over time, and with the advancement and discovery of tools and new ways of using materials, humans learned more advanced methods. Eventually, individual disciplines emerged as people started practicing their expertise

Goodheart-Willcox Publisher

Figure 1.1-3. Students in a heavy equipment operations program learn how to respond to an emergency water break.

Milosk50/Shutterstock.com

Figure 1.1-4. These cliff dwellings in Mesa Verde National Park illustrate the ingenuity of Native American architecture.

in a particular craft. Specializing in a particular set of skills led to the birth of the trades.

Some of the earliest evidence of large-scale construction comes from Mesopotamia, and the dwellings, palaces, and other public structures uncovered there, **Figure 1.1-5**. Other notable historic construction examples come from the Egyptian pyramids, the ancient Mayan ruins in Mexico, the Great Wall of China, and the remarkable architecture of ancient Greece and Rome. On a more contemporary timeline, we see examples of inspiring construction projects that required remarkable ingenuity and triumph over immense challenges. Projects such as the Hoover Dam, the Brooklyn Bridge, and more recently, the Scorpion Tower in Miami, **Figure 1.1-6**.

Peter Sobolev/Shutterstock.com

Figure 1.1-5. These ancient remnants of stone walls are some of the oldest known examples of construction in the history of organized civilization.

When we look at the architecture and construction of homes and buildings today compared to those built 100 years ago, we get a glimpse of how society has changed. We also notice differences from one geographic region to another, **Figure 1.1-7**. Construction has a way of expressing the values of a culture at large. Construction workers get to be a unique part of expressing and recording their culture throughout history.

Today, construction is complex. It has advanced to the point that there are specialties developed within each trade. In fact, some trades have

A *Benoit A/Shutterstock.com* B *TTstudio/Shutterstock.com* C *Felix Mizioznikov/Shutterstock.com*

Figure 1.1-6. Mankind has overcome great obstacles in construction projects. A—The Hoover Dam. B—The Brooklyn Bridge. C—The One Thousand Museum, also known as the Scorpion Tower, in Miami, Florida.

A *Ritu Manoj Jethani/Shutterstock.com* B *MaxFX/Shutterstock.com*

Figure 1.1-7. These two homes are examples of typical regional architectural styles found in the United States. A—New England. B—The Southwest.

specialties within specialties. For instance, some companies not only specialize in insulation, but some specialize in *only* spray foam insulation, or *only* blown-in cellulose insulation, **Figure 1.1-8**. Such a high level of specialization encourages further advancement within the industry. Often new techniques, tools, or products emerge because these companies have a strong focus and in-depth understanding of their segment of the industry.

Material science, building science, and digital technology continue to move the construction industry forward. Just a few decades ago, heating and cooling equipment was almost solely dependent on mechanical components to function. Today, along with the mechanical switches, valves, compressors, and thermocouples, this equipment incorporates digital interfaces and control modules, allowing for more efficient operation. We now use solar and geothermal resources to heat, cool, and electrify our homes in ways our ancestors never would have dreamed of. As technology continues to advance, construction will experience tremendous developments. Now is an exciting time to be in construction.

A *Cire notrevo/Shutterstock.com*

B *Greenfiber Cellulose Insulation*

Figure 1.1-8. Companies today specialize in specific types of insulation, working with a narrow range of materials. A—Spray foam insulation. B—Cellulose insulation.

LESSON 1.1 Review

Answer the following questions based on information in this lesson.

1. *True or False?* The construction of a duplex (a two-family home) is classified as residential construction. (1.1-1)
2. *True or False?* The construction of a school is classified as residential construction. (1.1-1)
3. *True or False?* New home construction has kept pace with population growth over recent years. (1.1-2)
4. Some early construction sites made of _____ are still intact for us to study today. (1.1-3)
   A. stone
   B. mud
   C. grass
   D. All are correct.

LESSON 1.2

Construction Fundamentals

LEARNING OUTCOMES

After completing this lesson, you will be able to:

1.2-1 Discuss skills needed by those in the carpentry field.

1.2-2 Discuss ethical practices in the workplace.

1.2-3 Describe the qualities employers value in an entry-level carpenter.

TECHNICAL TERMS

critical thinking
ethics
foreman
growth mindset
hard skills
problem solving
skill set
soft skills
work ethic

The Successful Construction Employee

Construction is rewarding work that can offer multiple pathways to success. Though the barriers to entry are relatively low in construction, it can be an intimidating field, particularly if you have no experience around a construction jobsite. Simply knowing what will be expected of you as an employee can be challenging. If you are embarking on a career in the construction industry, it will be useful to understand what qualities will help you thrive in a construction environment. In this chapter, you will be introduced to some of the character traits, ethical behaviors, and soft skills it takes to be successful in the industry. Many of these traits, behaviors, and skills are common to success in any career, regardless of the industry. In fact, it is likely that you practice them on some level already.

Your ***skill set*** is the range of abilities that help you perform your job. Your skill set can be grouped into two categories, *hard skills*, and *soft skills*, **Figure 1.2-1**. ***Hard skills*** are abilities you develop through the technical knowledge of your trade. Construction workers develop hard skills by practicing their trade and through on-the-job training, education, research, and experience. Hard skills range from basic tasks like learning to measure and cut materials accurately, to more complex and specialized tasks and problem-solving abilities. Your hard skills develop over time and improve throughout your career.

Hard skills are tremendously important in order to perform specific tasks correctly. However, even the most technically qualified and competent person may not be able to hold a job if they lack soft skills. ***Soft skills*** are competencies in productively interacting with others. They include a willingness to learn or challenge yourself, effective communication, listening and following directions, time management, and other related skills. Most construction work is performed in teams (sometimes called crews) with other people. It is important for anyone working in the trades to be able to work well with others. Soft skills can be gained through social interaction and life experiences; they can also be gained through working with others in a variety of circumstances. Developing your soft skills will improve your employability in any industry, not just construction.

A Darkroom Graphic/Shutterstock.com B RossHelen/Shutterstock.com

Figure 1.2-1. It is important to develop both hard and soft skills. A—Measuring lumber is a hard skill. B—Effective communication and working well with others are examples of soft skills.

Ethics

Ethics are moral principles that guide our decisions. Ethical behavior involves knowing and acknowledging right from wrong and basing your decisions on that knowledge. In construction, as in all careers, ethical behavior is key to long-term success. Unfortunately, when an employee practices unethical behavior, they may experience a short-term benefit. For instance, a lie told to cover up a mistake may get someone out of some immediate consequences; however, that lie is likely to be uncovered later, producing deeper consequences. Even if unethical behavior is not immediately found out, it is indicative of a pattern that will eventually emerge. Coworkers, bosses, and clients often observe small compromises in ethical behavior and make judgment calls based on the content of one's character. Even if specific examples and evidence cannot be produced against an employee, ethical compromises eventually catch up with that individual.

In construction it is beneficial to protect your reputation by proving your good work ethic. ***Work ethic*** is a desire and attitude that shows a determination to do what is right in the workplace. A good work ethic speaks to your character as a moral person and a valuable employee. Workers with a good work ethic generally rise to the top as candidates for promotions, raises, and special assignments. Some of the primary qualities associated with a good work ethic are honesty, integrity, respect, punctuality, discipline, responsibility, and accountability. See **Figure 1.2-2**.

Ethical Conduct in the Workplace	
Desired Quality	**Actions and Behaviors**
Honesty	Work with truth and transparency. Perform your job free of deception in word, thought, and deed.
Integrity	Act on what you know to be right, even when there will be no consequences for doing otherwise.
Respect	Honor the humanity of other people and yourself. Respect does not mean you have to agree with others, but it does mean that you have regard for their feelings, beliefs, and ambitions. Respect for others may mean you need to show greater awareness in how you dress, speak, or carry yourself on the job.
Punctuality	Rather than just making it to work on time, arrive early and be ready to work at the beginning of your shift. Do not abuse scheduled times for breaks or pack up early at quitting time.
Discipline	Follow established workplace rules and codes of conduct. Keep a positive attitude and do not complain.
Responsibility	Prove yourself to be reliable by doing what you say you will do. Unless you are legitimately sick or there is an emergency, do all you can to show up for work each and every day.
Accountability	Take responsibility for your actions. When you make a mistake, own up to it. When someone else deserves credit, acknowledge who is responsible and do not take credit away from the other employee.

Goodheart-Willcox Publisher

Figure 1.2-2. Employers expect their workers to have a good work ethic. Exhibiting the qualities listed will help you succeed and become a valued employee.

These qualities speak to the fact that a good work ethic requires an emotional investment toward the goals of the project. The benefits of maintaining a good work ethic may not seem worth it in a moment of trial. However, exercising these traits will earn admiration from coworkers and managers and will increase your status as a trustworthy and valuable employee or contractor.

Expectations

Good performance on the job begins with understanding what is expected of you. Construction is not a career where you should wait around until someone tells you what to do. A certain level of autonomy is expected on a jobsite. Rather than asking your supervisor what the next assignment is after completing each task, you should make a sincere effort to understand the construction process and know for yourself what comes next. Work is often fast paced on a construction site. Supervisors, usually called foremen, will be busy keeping things running smoothly and making decisions that affect the entire crew. The *foreman* is the person assigned to lead the crew; they have the required training, experience, and leadership ability necessary to supervise and oversee projects. They make decisions, secure required materials and equipment, manage jobsite logistics, and keep the crew safe, along with many other responsibilities. This job requires continual focus and forward thought. Part of a foreman's job is to keep the crew working steadily and maximize production. Depending on the number of crew members, supervision can become taxing and hinder the other responsibilities of a foreman's position.

Workers who can think ahead and determine the next meaningful steps to take have an edge on their peers. Doing so requires critical thinking and problem solving. These are among the most valuable skills you can develop as a tradesperson. *Critical thinking* involves analyzing an issue to form a judgment. An example that would require critical thinking in construction could involve studying a set of prints to properly lay out the components of a wall, **Figure 1.2-3**. The wall might intersect other walls; it is likely to contain windows or doors; it might require blocking to mount fixtures or cabinets. Each of these factors requires wall framing to accommodate these components. Thinking critically about how to properly lay out and build the wall beforehand will save time and money. *Problem solving* is the process of identifying an issue and finding a solution. Problem solving is often a reaction to an issue, where critical thinking is used to prevent issues from arising. These skills can prove to be invaluable in the construction industry. They can be developed by paying close attention to what is happening on the jobsite. Observing and reflecting on the who, what, where, why, and how of a job will help you to understand the overall construction process and enable you to perform at a higher level. Your foreman or supervisor will expect you to exercise some level of critical thinking and problem solving on the job.

PanuShot/Shutterstock.com

Figure 1.2-3. Good planning and execution of construction work requires critical thinking skills.

Learning these skills will enable you to stay busy and on task. Workers are expected to carry themselves in a productive manner. It is important to remain engaged in meaningful work, which involves more than just staying busy. It is expected that workers will contribute to the goals of the project. Consider what the foreman would have you work on next; then, if you are able, do that.

Additionally, as you are working, think about your posture toward your work. This includes both your physical posture and your attitude. When engaged, construction workers try to move quickly in order to maximize their time. As an example, they should not sit down and work. If working on something low, it is generally acceptable for workers to squat or kneel. Sitting may be more comfortable, but it reduces mobility and gives the impression that you value comfort over efficiency.

Sitting while working in construction is just one example of how to think about the impression your posture can have on others. You want to convey that you care about your work. Having this type of approach does not mean you can't enjoy work or converse with coworkers. In fact, these are great perks of the type of work construction workers do. The camaraderie that comes from building things with others is rewarding.

Having a growth mindset can increase your job satisfaction and contribute to your skill set and employability in the construction industry. A ***growth mindset*** means that you are willing to take on new challenges to improve your skills. Rather than complaining about an assignment or telling the foreman you don't know how to do a task, try seeing it as an opportunity to learn and improve your skills. This approach may mean an occasional mistake is made along the way. It may even mean embarrassment as you struggle to learn something new. However, your ambition and attitude will speak volumes to your coworkers and your boss. Learning from mistakes is part of having a growth mindset. Construction, like so many other fields of work, is a lifelong learning occupation.

Computer and technology skills are also important. Various computer applications are used in the construction industry, including the following:

- Drafters use CAD software to create building plans.
- Estimators use spreadsheet software and other programs to prepare estimates.
- Construction managers use scheduling software to keep track of progress on projects.
- Architects use word processing software to create specification documents.
- Carpenters use mobile apps for calculation tasks, such as measurement conversions, rafter layout, and stair layout.

Digital tools are constantly evolving. Throughout your working career, you will need to continue to learn about and apply the latest technology.

Critical thinking, problem solving, staying productive, and maintaining a good posture and attitude toward your work are just a few of the expectations that construction industry employers have for their employees. If you can maintain focus on these few things, you will be well on your way to a fulfilling construction career.

It is natural to be anxious about beginning a new career. However, it is important to begin with the end in mind. A career in construction can be both rewarding and lucrative, even if it starts out with humble beginnings. Any successful career will take hard work and determination. If you begin your career knowing that there will be challenging days ahead, it will help prepare you to persevere. Along with the challenging days come many days that are rewarding. In fact, the larger the challenge, the greater the sense of accomplishment you feel when a task has been completed.

LESSON 1.2 Review

Answer the following questions based on information in this lesson.

1. Which of the following is an example of a hard skill? (1.2-1)
   A. Effective communication
   B. Measuring materials accurately
   C. Time management
   D. All are correct.
2. Some of the primary qualities associated with a good work ethic are _____. (1.2-2)
   A. punctuality
   B. honesty
   C. respect for others
   D. All are correct.
3. *True or False?* After completing a task, it is best practice to sit and wait for the foreman to give you your next task. (1.2-3)

LESSON 1.3

The Construction Path

LEARNING OUTCOMES

After completing this lesson, you will be able to:

1.3-1 Discuss the administrative tasks involved in planning a construction project.

1.3-2 Outline the steps involved in site preparation.

TECHNICAL TERMS

batter board
benchmark
building codes
building line
building permit
excavation
site preparation
zoning laws

Construction Terminology and Building Projects

The Construction Path lesson in each chapter of this textbook will introduce you to the basic construction steps of building a new house. This section is designed to explain in general how a building project progresses from beginning to end. Many of the other sections of this chapter will complement *The Construction Path*, helping provide a greater understanding of the different components and terminology involved in each step of construction.

A challenge for many workers entering the construction field is simply understanding some of the language used throughout the building process. Many of the terms and expressions used in construction are foreign to someone not familiar with the industry or the process of building. Administrative contract language such as *RFI* (request for information) or *change order* can be confusing without context. Building code terminology such as *frost line*, *climate zone*, and *dead load* might seem difficult to understand without prior knowledge. Tools and building components can have strange names like *transit*, *bull float*, *transom*, or *astragal*. Even framing members can be unfamiliar—what is the difference between a *joist*, a *stud*, and a *rafter*? Even when you sort that out, there are variations of each type of member, such as a *hip* rafter, a *valley* rafter, a *common* rafter, and a *jack* rafter—you may wonder, how is that different from a *jack stud*? As these parts come together, they form assemblies that can also have quirky names, such as *cornices*, *mansard roofs*, *headers*, *footers*, and *girders*. To complicate matters further, you might hear a carpenter say they have to *drop a stringer* or cut the board seventy-five and five-eighths, *light*. There are also variations of similar building products or different names for the same product. For instance, perhaps someone is installing *MR board* in the bathroom or *type X* in the garage; these are both variations of *gypsum board*, which is also called *sheetrock* by some and *drywall* by others. Clearly, it takes time and experience to understand the language of construction. Even those who have been in the industry for decades can get thrown off when they work with someone new or talk with someone from a different part of the country or another part of the world.

Understanding the language of construction is something that one is not likely to accomplish by memorizing definitions. However, through careful study of this text and associated projects, you will gain a much broader comprehension and function more comfortably within the construction space. As you continue your education and gain more experience in construction, these terms will become second nature. Each chapter of this text will help establish a basic understanding of the terms and language used when working with residential construction methods and materials. Terms relating to a specific phase of construction are introduced and defined where appropriate. Become familiar with these terms and refer to the glossary at the end of this textbook for definitions when necessary.

It is worth noting that this text is not a comprehensive resource covering every aspect or definition of terms used in construction. In fact, even seasoned construction veterans learn new terminology and are exposed to new things now and then. It is enough to be aware of how experience and education play a role in developing your construction vocabulary.

Administration and Planning

Before starting a construction project, it is important to ensure that all parties are fully aware of the details of the construction. Because the home will be built inside the jurisdiction of a local government, officials must be informed and give their approval before the project begins. This is accomplished through permitting. To obtain a building permit, the builder or owner of the property must apply through their local building department. The ***building permit*** is the document authorizing the construction of the project. The building department will collect information on the project, such as the size of the building, where it will be located on the property, cost of the project, and a detailed set of plans. They can also determine if other special permits or procedures might be required for the project. Often builders will need to have approved plans or permits in place to explain how they will manage storm water runoff and erosion control on their site, where and how the driveway will be connected to the street, approval for municipal or on-lot sewer system, and how they will satisfy environmental and flood plain concerns. Many jurisdictions have ***zoning laws*** in place that designate and regulate the use of specific land for residential, commercial, industrial, agricultural, or mixed use.

In addition to approving initial construction plans, the building department will inspect the contractor's work at critical points in the process. These inspectors come to the jobsite to ensure the construction meets all applicable building code requirements. ***Building codes*** are the minimum construction standards that a building is required to meet. They ensure the health, safety, and general welfare of a building's occupants and users. The International Residential Code (IRC) is the most common building code used in residential construction in the United States.

Building codes and how to obtain a building permit will be covered in greater detail in *Chapter 3, Framing Systems.*

Site Preparation

Site preparation, or site prep, begins after all approvals have been given for construction. Site preparation work is generally conducted by an excavation contractor. Site prep includes any demolition work that needs to occur, clearing the land of obstacles or brush, stripping the top layer of vegetation and soil, and laying out where the building will be located. See **Figure 1.3-1**. During site prep, a benchmark is established. A ***benchmark*** is a reference point that is a fixed (will not change throughout the job) and known elevation. The benchmark is the point used to measure how much of the home's foundation will be underground and how much will be out of the ground when construction is complete. It is generally marked on the prints so any worker who needs to reference it can find the information required.

To establish where the building will be located, batter boards are constructed. A ***batter board*** is a wooden frame built in line with a perimeter wall of the building, **Figure 1.3-2**. It consists of two vertical stakes driven into the ground and a horizontal ledger for fastening a string line. Two batter boards are built per wall, and a string is stretched from one to the other. This string is positioned to align with the wall to be built. Batter boards can be established for each perimeter wall of the building but are sometimes only built to represent the longest walls. A minimum of two pairs of batter boards must be built to represent two perpendicular walls. Short walls can be measured from the strings representing the longer walls. Once situated, the strings on the batter

A B C Goodheart-Willcox Publisher

Figure 1.3-1. Site preparation prepares a plot of land for construction. A—Cutting down trees. B—Clearing obstacles. C—Removing the top layer of soil.

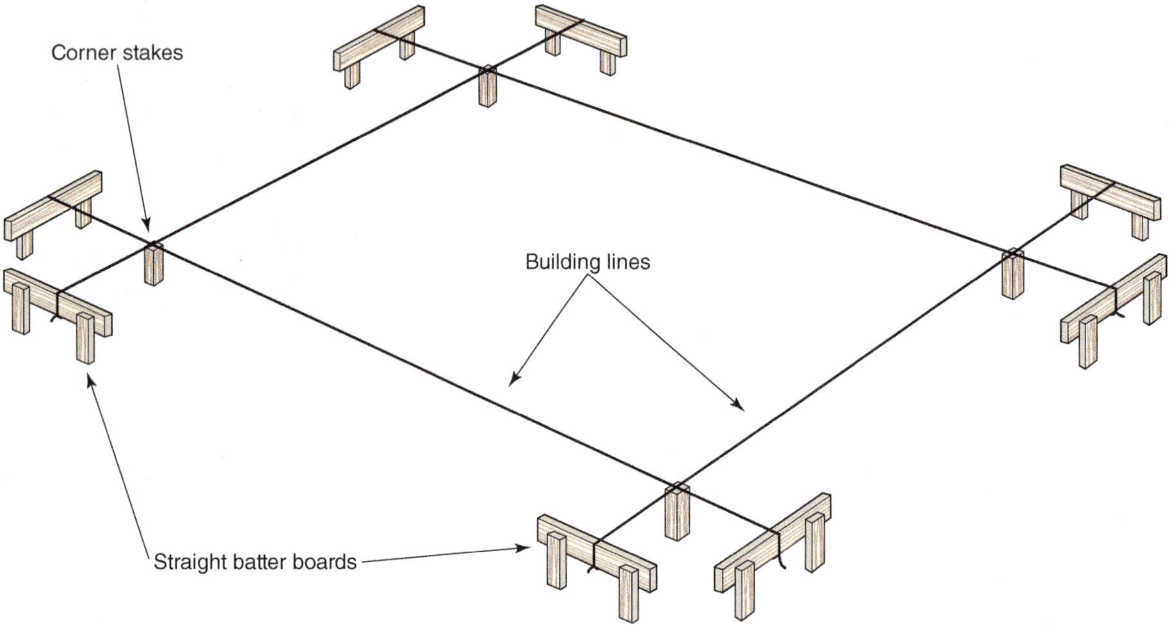

Goodheart-Willcox Publisher

Figure 1.3-2. Batter boards support strings that mark the building lines. Strings can be taken down while excavation is occurring, then put back up as needed.

boards represent the building line of the home. The ***building line*** marks the exterior perimeter of a building's foundation.

After the building line of a house has been established using batter boards and strings, the excavation can be marked using marking paint. The ***excavation*** is the depression or cavity in which the foundation will be built, **Figure 1.3-3**. With the excavation marked, the strings can be taken out of the way so the work of digging the foundation

Green Note

Whenever the ground surface is disturbed, the site should be protected from storm water eroding the soil. Doing so protects waterways and municipal stormwater systems from runoff polluted with silt or other contaminates.

Goodheart-Willcox Publisher

Figure 1.3-3. A fully excavated site for a home with a full basement.

can begin. At any time, the strings can easily be reinstalled on the batter boards to indicate the building lines, **Figure 1.3-4**.

Other site prep might include installing additional drainage measures, cutting or filling dirt in high or low areas, establishing a rough grade, and installing a driveway base. If soil erosion measures and stormwater runoff controls are needed, they would be established during the site prep stage of the project.

E-Iona/Shutterstock.com

Figure 1.3-4. The building line is reestablished using batter boards and strings after excavation.

LESSON 1.3 Review

Answer the following questions based on information in this lesson.

1. *True or False?* To obtain a building permit, the builder or owner of the property must apply through their local OSHA office. (1.3-1)
2. The minimum construction standards that a building is required to meet to ensure the health, safety, and welfare of a building's occupants and users are known as _____. (1.3-1)
   A. zoning laws
   B. building codes
   C. OSHA standards
   D. specifications
3. _____ are built to establish where the building will be located. (1.3-2)
   A. Benchmarks
   B. Drainage systems
   C. Batter boards
   D. All are correct.
4. *True or False?* The depression or cavity in which the foundation will be built is known as the excavation. (1.3-2)

LESSON 1.4

Materials and Fasteners

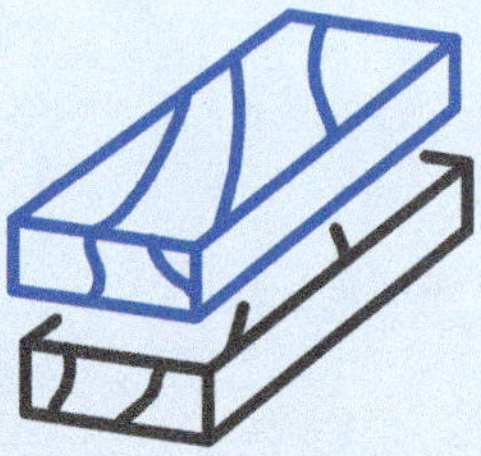

LEARNING OUTCOMES

After completing this lesson, you will be able to:

1.4-1 Understand the various soil considerations that must be considered in construction.

1.4-2 Express the dressed dimensions of various sizes of framing lumber, given their nominal dimensions.

1.4-3 Identify the types of screws used in construction and their applications.

TECHNICAL TERMS

counterbore	dressed dimension	nominal dimension
countersink	framing lumber	topsoil
crushed stone	frost line	

Soils

A wide variety of materials and building products is used in the construction of a home. Each chapter of this textbook will introduce construction materials and fasteners that are commonly used in the type of work featured in the chapter.

It is important to understand that soil composition and characteristics vary. To build a structure that is stable and will not settle, it cannot be built on topsoil. ***Topsoil*** is the upper layer of soil and contains organic matter. Organic matter will break down over time, creating voids. It also encourages further plant growth. Topsoil is stripped away from the building site and can be retained for later use.

It is common for a foundation to be excavated to a depth below the frost line. The ***frost line*** is the maximum depth to which soil is likely to freeze during winter months in a given climate. The frost line varies from one region to another. In cold-weather climates, moisture from rain and snow seeps into the ground and freezes during the winter months. Water expands when it freezes, pressing into the surrounding ground. It is important to protect the building from the forces imposed when this happens. If the foundation is constructed at a depth too shallow, moisture in the soil that freezes can force the building upward, causing major damage. Most buildings are designed to be built with their foundation below the frost line, protecting them from the effects of freeze-thaw cycles.

The soil upon which the building is built must also be strong enough to support the weight without concern over settling. Soil that has been disturbed will continue to settle for years, so it is important that the depth of the excavation is established and is not overdug.

Crushed stone is a building material that is used to promote stability and drainage under concrete slabs and against foundation walls, **Figure 1.4-1**. Crushed stone is rock that has been quarried and run through a rock crusher, then screened and sorted by size. Crushed stone is also used in construction for driveways, for septic systems, and as an ingredient in concrete.

ungvar/Shutterstock.com

Figure 1.4-1. Crushed stone is being installed as a stable base for this foundation.

Framing Lumber

Framing lumber, also known as dimensional lumber, is a common building material that gets its name from its primary use. It is the material that is most often used to build the frame, or skeleton, of a wood-framed home, **Figure 1.4-2**. Although framing lumber is used most often to build the frame of the home, it is also used during the layout of a home to build batter boards. It is common to use framing lumber to build footing forms, scaffolding, temporary shelter, sawhorses, and for many other temporary and permanent uses throughout the construction process. Framing lumber is convenient because of its modular sizing system. Sizes are consistent regardless of where the lumber was produced or sold.

Goodheart-Willcox Publisher

Figure 1.4-2. Framing lumber is dimensionally consistent, making it a convenient material for framing floors, walls, ceilings, roofs, and decks.

Standard framing lumber sizes are shown in **Figure 1.4-3**. Notice that sizes are given in nominal (named) dimensions. The *nominal dimensions* are its rough, unfinished dimensions (width and thickness). The actual dimension is its *dressed dimension*. The dressed dimensions are less than the nominal dimensions because the boards are surfaced after they are sawed into the nominal dimensions. Framing lumber is generally available in lengths from 8′ to 24′ in two-foot increments. However, many suppliers only stock lengths up to 16′.

1×, 2×, 4×, and 6× thicknesses are common in residential construction. 2× lumber is used more than the others for framing. While uses vary, 2×4 and 2×6 lumber is commonly utilized in wall framing. 2×8, 2×10, and 2×12 lumber is often used for ceiling, floor, and roof framing.

Framing lumber is sawed from a variety of species of softwood trees. It is kiln-dried for indoor applications or can be pressure treated with preservatives for outdoor, masonry, or ground-contact uses. Pressure-treated lumber has a green color and is not usually kiln-dried. Because moisture in the wood causes it to swell, its dimensions are slightly larger than kiln-dried lumber. See **Figure 1.4-3** for actual dressed dimensions.

Safety Note

Pressure-treated lumber can present a number of unforeseen hazards. Use the following safety rules when working with lumber.

- Wash any skin that is exposed while cutting or handling the lumber.
- Always wear eye protection and a dust mask when cutting pressure-treated lumber.
- Wash clothing covered in sawdust separately from other clothing.
- Do not burn pressure-treated lumber. The ash is a potential health hazard.

Associations of lumber producers develop and apply detailed rules for lumber grading.

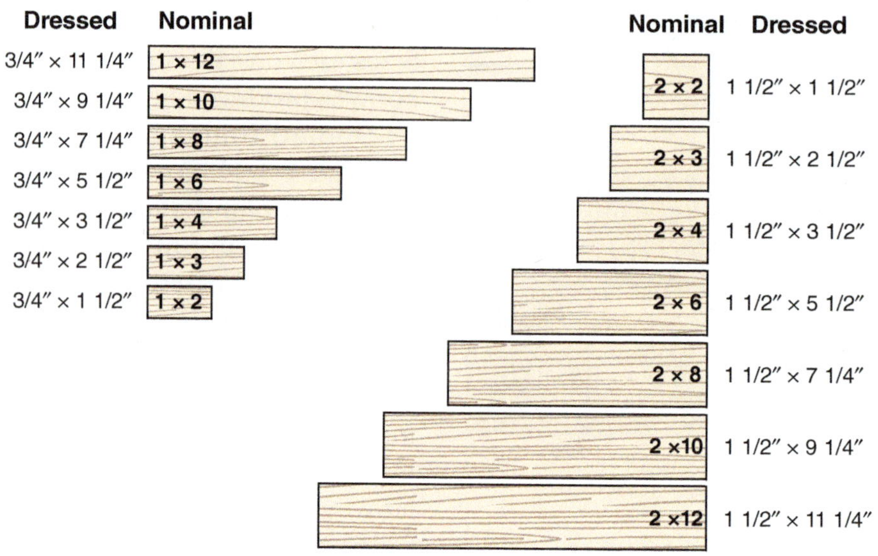

Goodheart-Willcox Publisher

Figure 1.4-3. Knowing the dressed dimensions of framing lumber is important for builders.

These agencies publish grading rules for the species of lumber produced in their regions. They also have qualified personnel who supervise grading standards at sawmills.

Basic classifications of softwood grading include boards, dimensions, and timbers. The grades within these classifications are shown in **Figure 1.4-4**. Another classification, called factory

Dimensions	
Light framing 2″ to 4″ thick 2″ to 4″ wide	Construction Standard Utility Economy
Studs 2″ to 4″ thick 2″ to 4″ wide 10′ or shorter	Stud Economy stud
Structural light framing 2″ to 4″ thick 2″ to 4″ wide	Select structural No. 1 No. 2 No. 3 Economy
Structural joists and planks 2″ to 4″ thick 6″ and wider	Select structural No. 1 No. 2 No. 3 Economy

Grade Markings

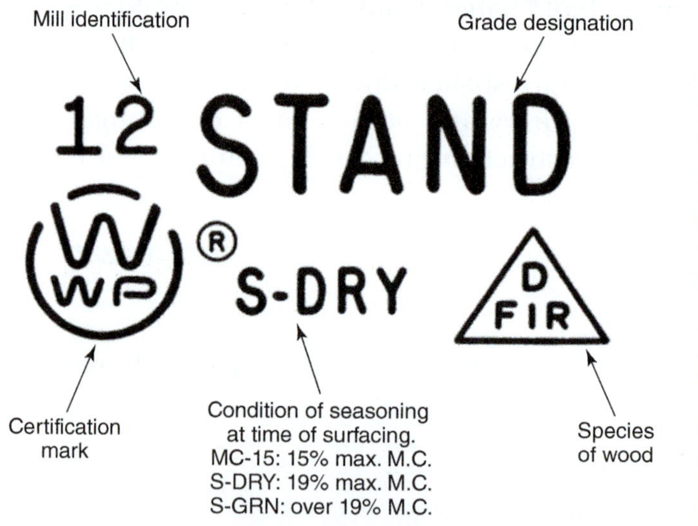

Timbers			
Beams and stringers	Select structural No. 1 No. 2 (No. 1 mining) No. 3 (No. 2 mining)	**Posts and timbers**	Select structural No. 1 No. 2 (No. 1 mining) No. 3 (No. 2 mining)

Goodheart-Willcox Publisher

Figure 1.4-4. Softwood lumber classifications and grades. Names of grades and their specifications will vary among lumber manufacturers' associations and among regions producing lumber. The grade listed on top is better than the grades below it for each classification.

and shop lumber, is graded primarily for remanufacturing purposes. Millwork plants use it to fabricate windows, doors, moldings, and other trim items.

The carpenter must understand that quality construction does not require that all lumber be of the highest grade. Today, lumber is graded for specific uses. In a given structure, several grades may be appropriate. The key to good economical construction is matching the proper grade with its function.

Screws

Screws are designed for fastening applications that require strong holding power and certain desirable properties, such as resistance to corrosion. Screws have a variety of applications in construction. This chapter introduces some of the most common types.

Construction Screws

Construction screws are available for a variety of applications. The most common construction screws have a bugle head. These screws have a coarse thread with an extra sharp point so they can be driven in with a screw gun. Phillips drive, torx drive, and square drive are available, **Figure 1.4-5**. Their aggressive thread pattern and thin shank allow them to be installed without drilling a pilot hole. Construction screws can be coated with a corrosion-resistant finish or can be made from stainless steel for outdoor applications. Drywall screws and deck screws are the most commonly used construction screws.

Goodheart-Willcox Publisher

Figure 1.4-5. A small sample of commonly used construction screws. Note the difference in the shape of the heads. Always use the correct drive tip or damage will result.

Wood Screws

Wood screws, like those shown in **Figure 1.4-6**, are generally installed by hand with a screwdriver. Unlike construction-grade screws, they do require a pilot hole and are generally countersunk or counterbored, **Figure 1.4-7**. A ***countersink*** pilot hole is

Goodheart-Willcox Publisher

Figure 1.4-6. Wood screws are often decorative and are installed with a screwdriver, rather than a screw gun.

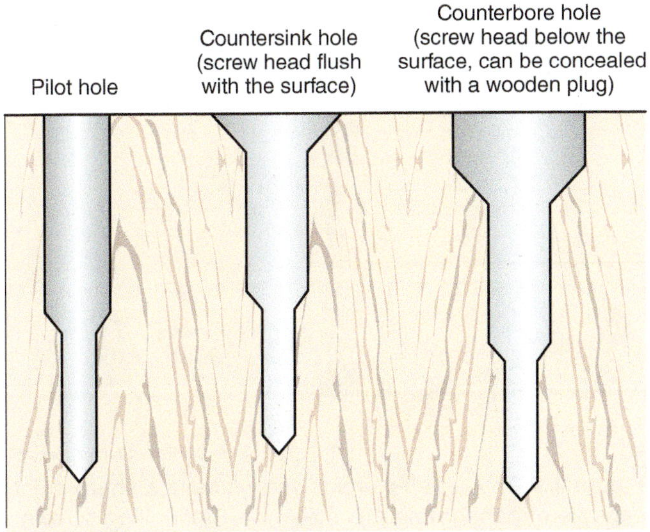

Goodheart-Willcox Publisher

Figure 1.4-7. Countersink and counterbore pilot holes allow wood screws to be installed flush or below the surface. Counterbored screws are sometimes covered with a wooden plug to hide the screw head.

angled at the top. A *counterbore* pilot hole has sides perpendicular to the top. This extra space below the surface creates room for the screw head to be level with or below the surface.

Wood screws are commonly used in applications that require the screw head to be exposed, such as on door hinges or wooden furniture.

Sheet Metal Screws

Sheet metal screws are often used to assemble light-gauge metal framing, ductwork, and other applications where thin metal is used. They are available in self-piercing and self-drilling varieties, **Figure 1.4-8**.

Lag Screws

Lag screws are large screws used to fasten load-bearing framing members together. They have a hex head and are categorized by their length and the diameter of their screw shaft, **Figure 1.4-9**. They are sometimes called lag bolts. Lag screws require a pilot hole to be drilled before installing them.

Specialty Screws

A wide variety of specialty screws are used in construction. Structural screws, **Figure 1.4-10A**, have more shear strength than other construction screws and can be used for structural applications according to manufacturer's recommendations. Roofing screws, **Figure 1.4-10B**, have a hex head with a rubber gasket that seals against metal roofing material to keep water out. Cabinet screws, **Figure 1.4-10C**, have a special head that is designed to improve its holding power on more delicate finish work. Their head remains above the surface of the material, rather than digging in like a bugle-shaped head. Concrete screws, **Figure 1.4-10D**, are made of hardened steel and have twin-lead threads that can cut and bind into masonry surfaces when the correct size pilot hole has been drilled. They are used to hold materials to concrete and masonry.

These examples are some of the more common specialty screws used in construction. Whatever the screw, ensure you are using it appropriately

A

amlet/Shutterstock.com

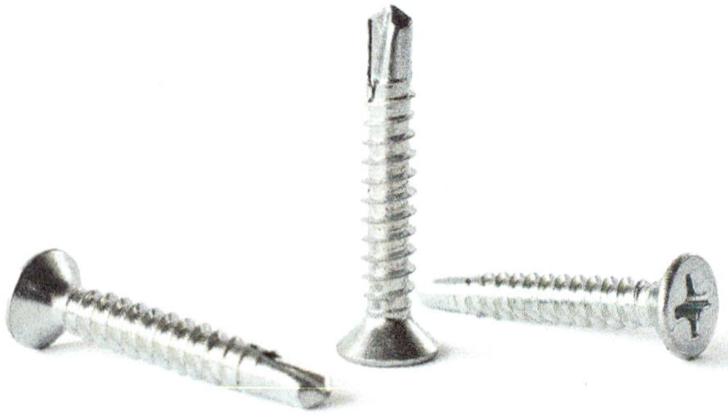

B

Iryna Bidovska/Shutterstock.com

Figure 1.4-8. A—Self-piercing screws have a sharp point capable of penetrating thin steel. B—Self-drilling screws are equipped with a drill point to cut through steel.

Goodheart-Willcox Publisher

Figure 1.4-9. Lag screws can be used for structural applications. The four bolts in the foreground are galvanized for exterior use with pressure-treated framing lumber. The one in the back is zinc plated for mild corrosion resistance.

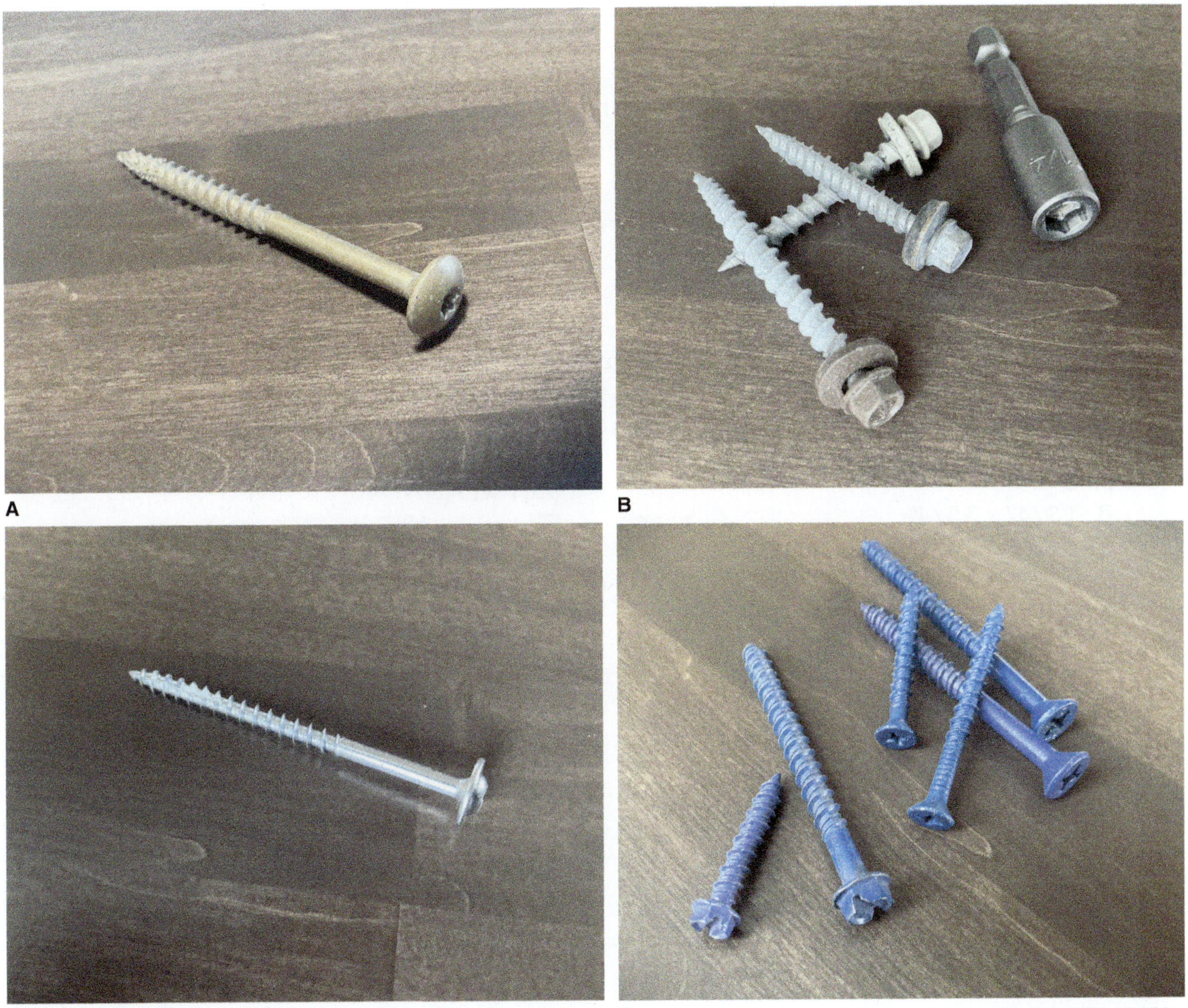

Goodheart-Willcox Publisher

Figure 1.4-10. A—Structural screws. B—Roofing screws. C—Cabinet screws. D—Concrete screws are used to hold materials to concrete and masonry.

and for its intended use. A broken or damaged screw can damage materials and do more harm than good. When installing screws, be sure to maintain good contact between the driver bit and the screw. If the bit slips, stop and reengage with better alignment and more pressure. If the bit continues to slip it will likely damage the screw head. Becoming proficient may take some time, but practicing correct methods will help speed the process.

LESSON 1.4 Review

Answer the following questions based on information in this lesson.

1. A foundation is generally excavated to a depth _____. (1.4-1)
   A. above the frost line
   B. below the frost line
   C. in the middle of the topsoil
   D. of three feet
2. *True or False?* Soil that has been disturbed will continue to settle for years. (1.4-1)
3. The dressed dimensions of a 2×4 are _____. (1.4-2)
   A. 2″ × 4″
   B. 1 3/4″ × 3 3/4″
   C. 1 1/2″ × 3 1/2″
   D. 1 1/2″ × 3 1/4″
4. The dressed dimensions of a 2×8 are _____. (1.4-2)
   A. 2″ × 8″
   B. 1 3/4″ × 7 3/4″
   C. 1 1/2″ × 7 1/2″
   D. 1 1/2″ × 7 1/4″
5. *True or False?* Construction screws require a pilot hole to be drilled before installation. (1.4-3)
6. *True or False?* A countersink hole has sides perpendicular to the top. (1.4-3)

LESSON 1.5

Tools

LEARNING OUTCOMES

After completing this lesson, you will be able to:

1.5-1 Use measuring, marking, and layout tools.

1.5-2 Identify various hand and power tools used in construction.

TECHNICAL TERMS

circular saw
level
measuring tape
plumb
sledgehammer
tape measure
utility knife

Construction Tools

Tools are an essential part of construction work. Nearly every task performed on a jobsite requires the use of tools. As you become familiar with the tools required for a task, it is important to understand the proper function of each.

Safety Note

Inspect tools for defects before using them. Frayed electrical cords, damaged blades, or loose parts can cause serious injury, or even death. Always respect tools and use them according to manufacturer recommendations.

This text will explore tools and their uses gradually, beginning with the tools that are more likely to be used during the early phases of construction. Each chapter will introduce additional hand and power tools that are commonly used on construction sites. This approach will expose you to tools with varying levels of complexity throughout the book.

Safety Note

Do not use a tool without a proper understanding of its purpose, how it works, and any required safety training.

Measuring Tools

In the United States, the US customary system of measurement is used for construction. This system employs yards, feet, and inches for lineal measurements. Standard construction tape measures are graduated in feet, inches, and fractions of an inch. Tape measures can be purchased with metric units, and some feature both systems of measurement. Surveyors often use feet in decimal form, rather than using inches. These dimensions are commonly found on site plans and should not be confused with the metric system.

Early phases of construction, such as site prep, involve the use of measuring and layout tools. The most common measuring tools are the tape measure and the measuring tape.

Tape measures, **Figure 1.5-1**, sometimes called pocket tapes, are available in a variety of lengths ranging from 6′ to 35′. Tape measures are spring loaded, so the steel tape automatically retracts back into the case when it is released. The hook on the end of the tape measure is designed to slide slightly so that measurements remain accurate whether the tape is butted to the inside or hooked on the outside of the object, **Figure 1.5-2**.

When longer measurements need to be made, such as when laying out a building, a measuring tape is used. ***Measuring tapes*** are available in lengths from 50′ to 300′. Measuring tapes are

sirtravelalot/Shutterstock.com

Figure 1.5-1. Tape measures are among the most widely used tools for shop work and on the jobsite.

bulkier than tape measures and have a crank to wind the tape back into the case, **Figure 1.5-3**. When using measuring tapes, be mindful that some are graduated in feet and inches (1/12 of a foot) while others, called engineer's measuring tapes, are graduated in feet and decimals (1/10 of a foot), **Figure 1.5-4**. The two can be easily confused, resulting in inaccuracies.

Pro Tip

It is easy for dirt and moisture to enter the tape case when retracting the tape. To prevent damage to the tape measure, wipe the tape with a cloth as it is retracted, especially if it is wet.

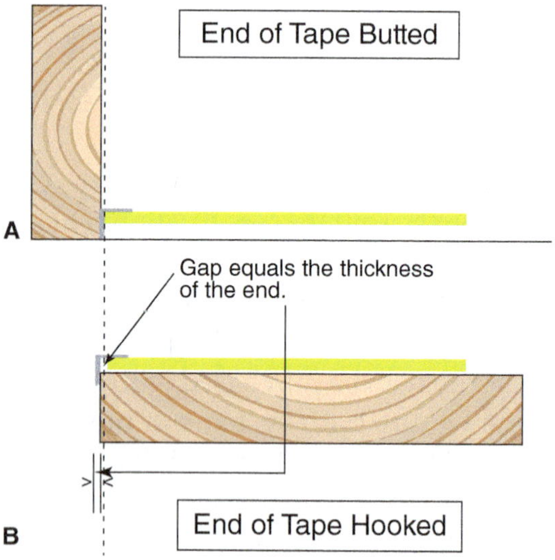

BW Folsom/Shutterstock.com

Figure 1.5-2. The end hook on the tape measure is designed to shift slightly so the thickness of the hook does not interfere with the accuracy of the reading for measurements that are either butted (A) or hooked (B).

Goodheart-Willcox Publisher

Figure 1.5-3. Measuring tapes are generally used for long measurements. They are not spring-loaded and must be cranked back into their case.

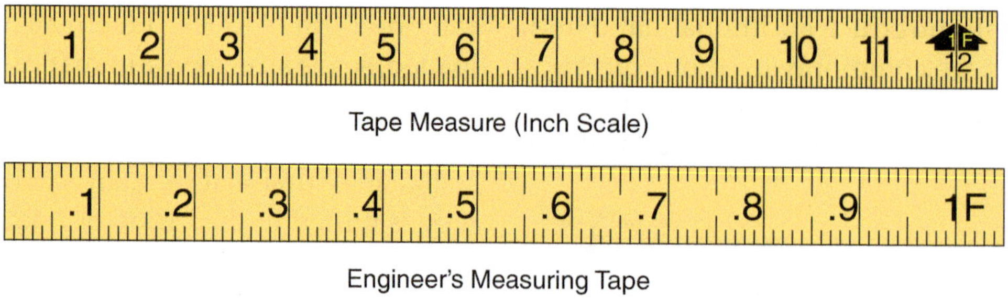

Goodheart-Willcox Publisher

Figure 1.5-4. Tape measures for general use in construction are graduated in feet, inches, and fractions of an inch. In surveying work, architects and engineers often use tenths and hundredths of a foot when referring to property lines and distances related to the site. Engineer's measuring tapes are graduated in feet, tenths, and hundredths of a foot. Note the difference between the two measurement systems.

Marking Tools

Construction requires marking a wide variety of materials ranging from the surface of the ground or rough concrete, to fine interior finish materials. At times wet or dirty materials must be marked. Often, straight reference lines are required for making long cuts or applying materials to a floor, wall, roof, or ceiling. Because there are so many conditions that require an accurate mark, construction workers use a variety of tools to do their marking.

The carpenter's pencil is the most common marking tool in construction, **Figure 1.5-5**. These pencils are larger and stronger than standard pencils. They are usually flat to keep them from rolling away. The flat edges of the carpenter's pencil can also be used as a spacer, shim, or measuring device since they are made to be exactly 1/4″ thick and 1/2″ wide.

Pro Tip

When making a rough mark with a carpenter's pencil, be sure that you distinctly communicate to others precisely what/where it is you are marking. It is common for newer carpenters to mark a small line or dash, **Figure 1.5-6A**. This can become confusing to others since they may not know which end of your mark is accurate. Instead, mark a point, **Figure 1.5-6B**. This type of mark is called a crow's foot. It clearly communicates where you intended your mark to land.

Goodheart-Willcox Publisher

Figure 1.5-5. Carpenter's pencils are used for marks on general materials. When marking finish materials, a traditional pencil is used.

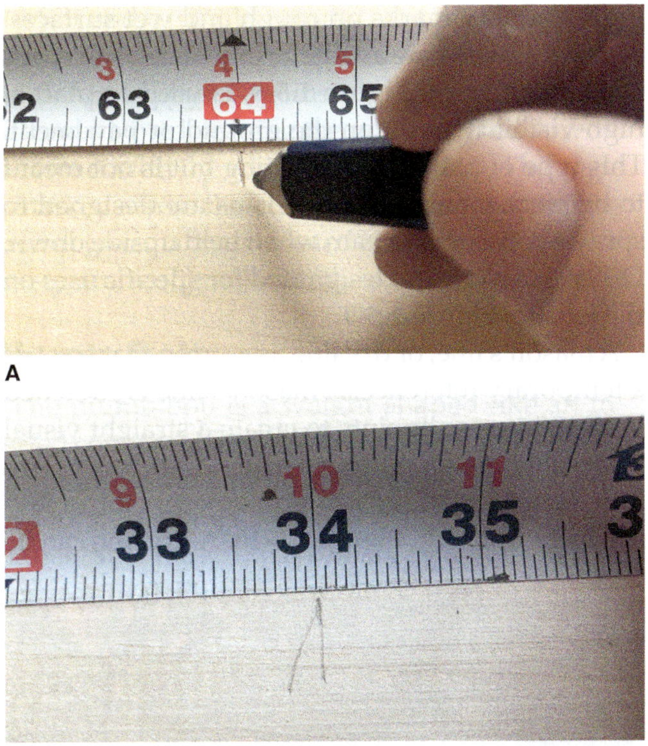

A

B

Goodheart-Willcox Publisher

Figure 1.5-6. Marking must be done precisely to avoid confusion. A—A mark such as this can be imprecise. When the tape measure is taken away, it can be difficult to determine exactly which end of the mark is accurate. B—The point of a crow's foot mark communicates exactly which part of the mark is accurate.

The lead is larger than a standard pencil for added strength and wear resistance. Carpenter's pencils are available in soft, medium, and hard lead for marking on a variety of materials.

A lumber crayon, **Figure 1.5-7**, is used to make coarse marks on surfaces that may not be easily marked with a pencil. They provide long-lasting,

BW Folsom/Shutterstock.com

Figure 1.5-7. Lumber crayons are used for rough, obvious marks.

Altrendo Images/Shutterstock.com

Figure 1.5-12. The plumb bob's weight and balance use gravity to create a plumb string or transfer a mark from the hanging point to the bottom tip of the tool.

BaLL LunLa/Shutterstock.com

Figure 1.5-13. The string of the plumb bob can be measured from various points to determine how plumb an object is.

Spirit Level

Another tool used to make something plumb is the spirit level, generally called a level. A level is an instrument designed with a vial filled with mineral spirits to create a bubble, **Figure 1.5-14**. Most levels have at least one vial indicating when the tool is plumb and at least one which indicates when the tool is ***level***, meaning it is perfectly horizontal, **Figure 1.5-15**. When the bubble is centered between the lines on the vial, the tool is level or plumb. Levels come in assorted sizes. Common lengths of levels are 2′, 4′, and 6′.

A level is delicately calibrated and should be treated with care. Dropping or abusing the tool can knock it out of calibration, leading to a false reading.

Hand Tools

In addition to the tools designed for measurement and layout work, there are many other types of hand tools used for work in construction. There are hand tools used for cutting, sawing, and fastening materials, as well as tools designed for more heavy work.

Tools don't have to be powered to be hazardous to work with. When handled or maintained incorrectly, hand tools have the potential to result in severe injury. To lessen or eliminate these risks, special attention must be paid to safety when using any tool. Always select the correct type and size of tool for your work. Be sure it is sharp and properly adjusted. Guard against using any tool if the handle is loose or in poor

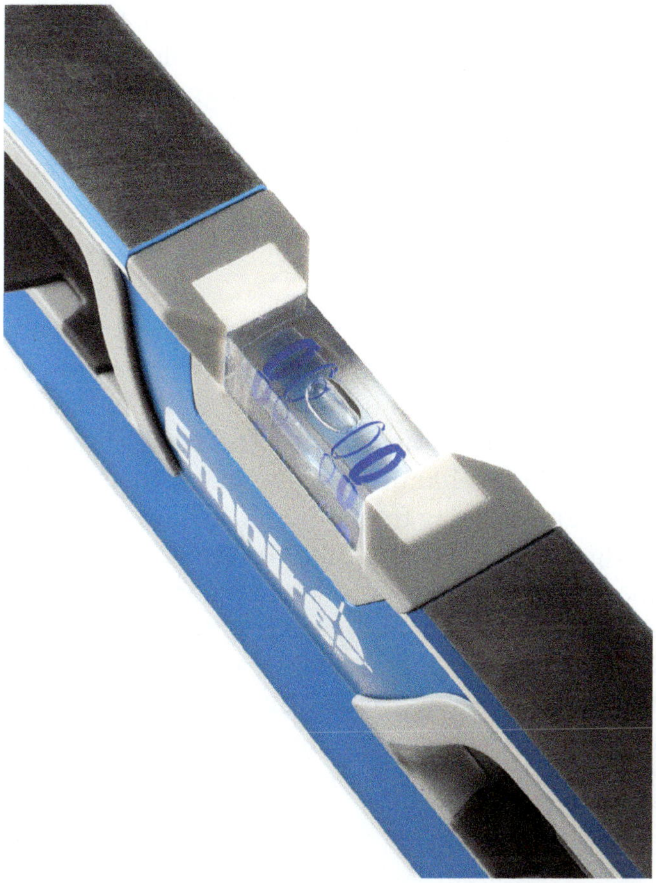

Empire Level, Inc.

Figure 1.5-14. Levels have one or more vials filled with mineral spirits. When the bubble is centered between the lines, the object is level or plumb.

Tyler Olson/Shutterstock.com

Figure 1.5-15. This carpenter is using a 4′ spirit level to ensure the framing is level.

condition. Dull tools are hazardous to use because additional force must be applied to make them cut. Oil or dirt on a tool may cause it to slip, resulting in injury.

Safety Note

Follow these safety rules when working with hand tools:

- Hold hand tools correctly. Most edge tools should be held in both hands with the cutting action away from your body.
- Be mindful when using your hand or fingers as a guide to start a cut with a handsaw.
- Keep edged and pointed tools turned downward.
- Carry only a few tools at one time unless they are mounted in a special holder.
- Do not carry sharp tools in pockets of your clothing.
- When not in use, tools should be kept in special boxes, chests, or cabinets.

Utility Knife

The ***utility knife*** is a basic cutting tool that is carried by nearly every tradesperson on the jobsite. Utility knives use sharp, replaceable blades for trimming and cutting a variety of materials, such as packaging, carpet, gypsum wallboard, and housewrap, **Figure 1.5-16**. Most modern utility knives feature a quick-change blade system and can store extra blades in the handle. Older and more basic models require a screw in the handle that must be removed to change the blade.

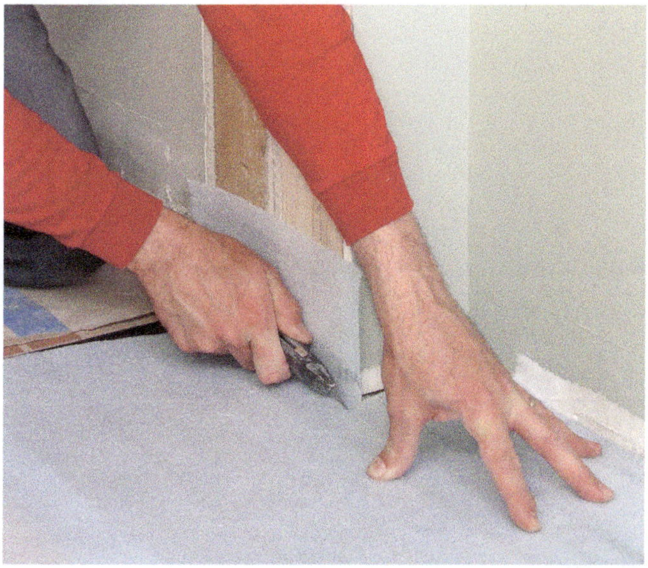

Christina Richards/Shutterstock.com

Figure 1.5-16. Utility knives are used frequently to trim, score, and cut materials.

Safety Note

Use caution and stay focused when cutting with a utility knife. It is always helpful to consider where the blade could go if you slip.

Handsaw

A handsaw can be used for rough cutting or finish work, depending on the design of the tool. Common types of handsaws used in construction are shown in **Figure 1.5-17**. Handsaws are available in several different lengths as well as various tooth sizes (given as teeth per inch or points per inch). Each type of saw has a specific purpose.

As the name indicates, a crosscut saw is designed to cut across the wood grain. It has teeth shaped so that they come to a point on alternating sides of the blade. A pull saw has a thin blade with fine teeth and is designed to make cuts as the tool is pulled toward the user. A coping saw has a thin, flexible blade and is designed to cut along curves. A drywall saw has large teeth designed for cutting through gypsum wallboard. A hacksaw is used to cut metal fasteners, such as bolts and nails, and metal trim.

Sledgehammer

A ***sledgehammer*** is a tool with a heavy head and a long handle to maximize driving force. Sledgehammers are used for demolition, breaking up materials, and for driving stakes into the ground. Sledgehammers are available in a variety of weights ranging from 2 lb to 20 lb heads. Smaller sledgehammers with heads between 2 lb and 4 lb that can be swung with one hand are sometimes called stake hammers, **Figure 1.5-18**. Larger sledgehammers have long handles and are designed to be swung using two hands. Caution must be exercised whenever using sledgehammers because of the tremendous forces they produce. It is common

A Nikita Rublev/Shutterstock.com B donatas1205/Shutterstock.com C Jeffrey B. Blanke/Shutterstock.com

D Stocksnapper/Shutterstock.com E RVillalon/Shutterstock.com

Figure 1.5-17. Handsaws used in construction. A—Crosscut saw. B—Pull saw. C—Coping saw. D—Drywall saw. E—Hacksaw.

Pamela Au/Shutterstock.com

Figure 1.5-18. A stake hammer is a small sledgehammer that can be swung with one hand.

for materials to fragment and fly apart when struck by a sledgehammer. See *Section 1.6* for proper safety practices.

Safety Note

Using any hammer introduces the potential of flying debris. Make sure you protect your eyes by wearing safety glasses.

Shovel and Pick

A shovel is a tool commonly used in construction for digging and moving materials like soil, gravel, sand, or concrete. It is essential for light work when excavating foundations, trenching for utilities, landscaping, and general site preparation. Shovels come in different shapes and sizes, each designed for specific purposes, such as spades for digging and square-point shovels for moving loose materials.

A pick is a tool used for tasks requiring breaking up hard, rocky, or compacted soil. Picks typically consist of a handle attached to a pointed metal head with a sharp end and a chisel end.

Power Tools

When used correctly and safely, power tools greatly reduce the time to perform work in construction. In addition, power tools enable work to be completed to a high level of accuracy. This chapter introduces power tools used in various phases of construction. Power tools used for specific operations are discussed throughout this text.

Before operating any power tool or machine, you must be thoroughly familiar with the way it works and the correct procedures to follow. Read the directions in the owner's manual if there is one. Make sure you understand them. Be alert and, above all, use common sense as you work. In general, when you learn to use equipment the correct way, you also learn to use it the safe way. If the tool you are using has a power cord, check it to make sure there are no breaks in the insulation. Power cords must *not* be taped or otherwise repaired. If the cord is damaged, it should be replaced.

While operating a power tool, do not allow yourself to be distracted. Do not distract the attention of others while they are operating power tools. Keep all safety guards in place and wear safety glasses. Never wear loose fitting clothing that could get caught in moving parts; dangling jewelry and earbuds should be removed when working with power tools. Tie back long hair before getting started.

Safety Note

Never carry a portable power tool by its cord or disconnect it from the receptacle by pulling on the cord. This can damage the insulation and electrical connections inside the plug or the tool, creating a potential shock hazard.

Circular Saw

The portable circular saw is one of the most used power tools on a construction site. ***Circular saws*** have a round blade and are used to cut framing lumber and other wood products. They are available in a variety of configurations, **Figure 1.5-19**. Corded circular saws are more powerful than battery-operated cordless saws. However, cordless circular saws offer the flexibility of working where a cord might present a safety hazard or an inconvenience. The drawback to working with cordless tools is that you have a limited power source. Making a high volume of cuts can deplete batteries quickly, leaving the user waiting for batteries to recharge.

Circular saws can be set up to cut different depths by adjusting the base plate. They are also able to

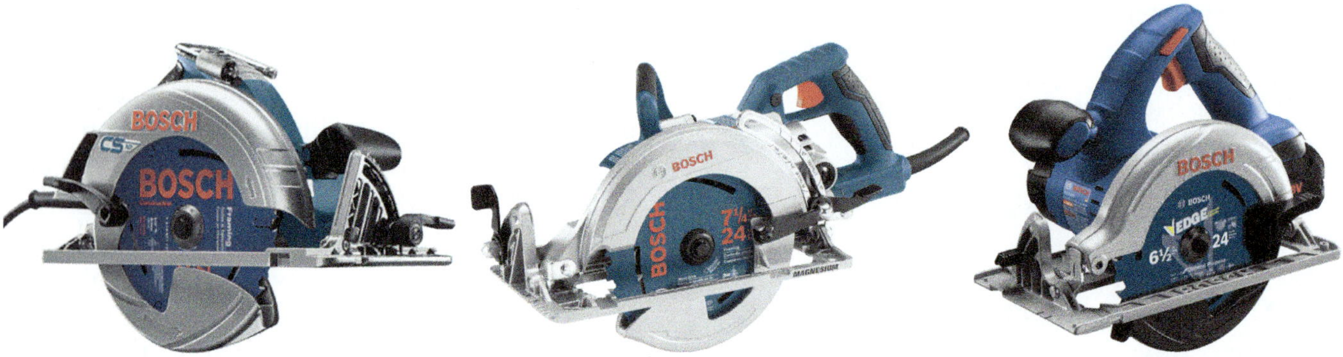

A B C

Photos courtesy of Robert Bosch Tool Corporation®

Figure 1.5-19. Circular saws can be purchased in multiple configurations. A—Corded, direct-drive with the blade on the right. B—Corded, worm-drive with the blade on the left. C—Cordless, direct-drive with the blade on the left.

cut bevel and angle cuts, **Figure 1.5-20**. To cut a bevel, the shoe tilts by loosening the bevel adjustment knob on the front of the saw.

Circular saws are sized according to their blade size. 7 1/4″ saws are the most common size on the jobsite; however, many cordless saws use a smaller 6 1/2″ blade. The larger the blade diameter is on a circular saw the deeper the cut it can produce. This capability can become important when cutting larger materials, or when cutting a bevel.

Various blades are available for different applications, **Figure 1.5-21**. The most commonly used is the framing or combination blade, which has relatively few teeth and an aggressive cut. These blades are ideal for rough framing and can be

Photo courtesy of Robert Bosch Tool Corporation®

Figure 1.5-20. The base shoe on circular saws can be angled to cut a bevel.

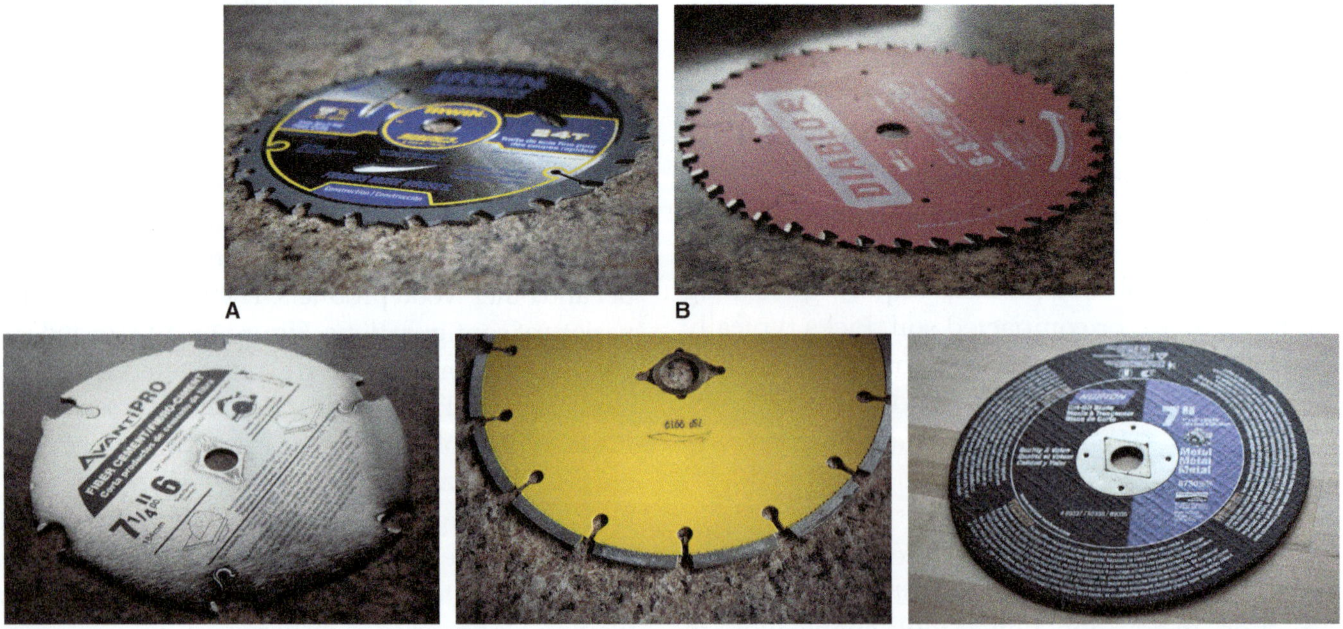

A B C D E

Goodheart-Willcox Publisher

Figure 1.5-21. These blades are designed to cut different materials. A—Combination (framing/general purpose). B—Finish/trim. C—Fiber cement. D—Masonry. E—Abrasive blade for cutting metal.

used for both rip cuts, along the grain, and cross-cuts, across the grain. Specialty blades are available for cutting plywood, steel, masonry, and fiber cement.

Safety Note

Circular saws can be dangerous when used improperly. Use the following safety rules when operating a circular saw.

- Wear safety glasses and hearing protection.
- Unplug the saw to replace a saw blade. Some saws have a control that locks the arbor during this operation, **Figure 1.5-22**.
- If the saw has two handles, keep both hands on them as much as possible during the cut. Start the cut using one hand to raise the guard (when needed, usually at the beginning of a compound cut). When the cut is completed, make sure the guard returns to the closed position.
- Allow cutoff waste to fall away from the cut when cutting framing lumber. Do not cut between two sawhorses.
- Before starting the saw, adjust the depth of cut to the thickness of the stock, plus about 1/8″.
- Check the base and angle adjustments to be sure they are tight.
- Plug the cord into a grounded outlet and be sure it will not become tangled in the work.
- Always place the saw base on the stock with the blade clear before turning on the switch.
- During the cut, stand to one side of the cutting line. Never reach under the workpiece during a cut.
- Never pull back on a running saw once the cut is started—to do so may cause a kickback.
- Keep hands clear of the cutting line.
- Always use a sharp blade of the proper type for the material and cut.
- Never force the saw through the material.

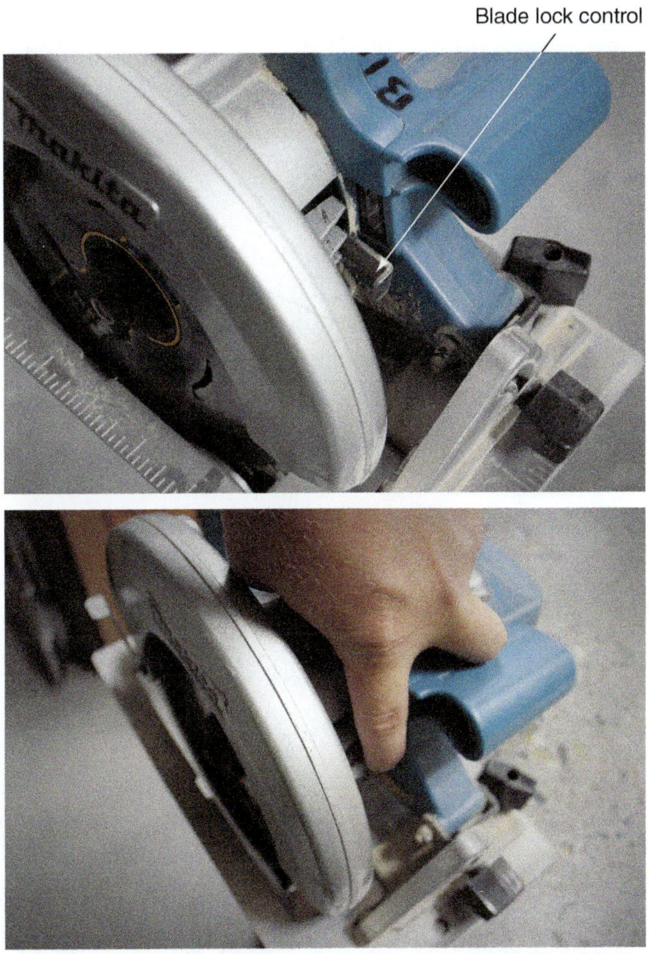

Black & Decker

Figure 1.5-22. Always disconnect the saw from the power source while adjusting or changing blades. Pressing a control locks the arbor during a blade change.

Cordless Drill

A cordless drill is a handheld power tool used for drilling holes, driving screws, and mixing materials. See **Figure 1.5-23**. These tools have a variety of speed and torque settings for various tasks. Many modern cordless drills also have a hammer setting for drilling small holes into masonry with a carbide bit. For drilling larger holes in masonry, a dedicated hammer drill is used, featured in

Photo courtesy of Robert Bosch Tool Corporation®

Figure 1.5-23. The cordless drill is a multipurpose tool that has become essential to construction workers.

Chapter 6, MEP Systems Rough-In. Bits designed for use in a portable drill are shown in **Figure 1.5-24**.

Green Note

While contributing significantly to increased production, the lithium-ion batteries used in cordless tools are toxic to the environment when disposed of improperly. Contact your local recycling facility to find out the best way to dispose of or recycle your faulty batteries.

Impact Driver

When it comes to installing screws, lags, and other mechanical fasteners, the impact driver, **Figure 1.5-25**, is the tool of choice. Although it operates differently, you can think of the impact driver as a cousin to the cordless drill. While the drill can be used to install screws and similar fasteners, it transfers the torque required to drive a fastener to the wrist of the user. Impact drivers have an internal hammering mechanism between the motor and chuck, ensuring better contact and less slip on the fastener. Unlike the drill, this mechanism allows

Goodheart-Willcox Publisher

Figure 1.5-24. Drill bits a construction worker might use. A—Spade bits for boring holes in wood, available in sizes from 3/8″ to 1 1/2″. B—Twist drill bits for drilling holes in metal, plastic, and wood, available in sizes from tiny to 1/2″. C—Auger bits for boring holes in wood, available in many sizes and types. D—Forstner bit for flat-bottom holes in wood, available in sizes from 1/4″ to 2 1/2″. E—Hole saws are preferred for cutting larger diameter holes up to 6″.

Goodheart-Willcox Publisher

Figure 1.5-25. Cordless impact drivers are used to drive screws and lags on the job.

the torque to be transmitted to the fastener without subjecting it to the user, resulting in less user fatigue. Impact drivers are more compact and lightweight than drills. They generally have a higher torque rating than comparable drills, providing better battery longevity and performance when driving screws. While the chuck on a drill accepts multiple sizes of drill bits, impact drivers have a chuck that will only accept a 1/4″ hexagonal shank, **Figure 1.5-26.**

Goodheart-Willcox Publisher

Figure 1.5-26. Impact drivers can only accept accessories with a 1/4″ hexagonal shank.

LESSON 1.5 Review

Answer the following questions based on information in this lesson.

1. *True or False?* A tape measure and a measuring tape are two names for the same tool. (1.5-1)
2. The most common marking tool in construction is the _____. (1.5-1)
   A. chalk box
   B. mason's line
   C. marking paint
   D. carpenter's pencil
3. A triangle-shaped tool that enables quick and easy marking across building materials is called a _____. (1.5-1)
   A. chalk box
   B. speed square
   C. marking triangle
   D. carpenter's pencil
4. *True or False?* Utility knives have replacement blades. (1.5-2)
5. *True or False?* A handsaw can be used for finishing work. (1.5-2)
6. Circular saws are sized according to their _____. (1.5-2)
   A. blade size
   B. amperage
   C. weight
   D. All are correct.
7. Impact drivers have a chuck that accepts a _____ hexagonal shank. (1.5-2)
   A. 3/16″
   B. 1/4″
   C. 3/8″
   D. 1/2″

LESSON 1.6

Safety

LEARNING OUTCOMES

After completing this lesson, you will be able to:

1.6-1 Understand basic workplace safety practices.

1.6-2 Discuss the role of OSHA in the construction industry.

1.6-3 Demonstrate basic construction tool safety.

TECHNICAL TERMS

general duty clause
Occupational Safety and Health Administration (OSHA)
OSHA standards
pneumatic
powder-actuated
safety mindset

Keeping the Work Environment Safe

Many workplace accidents can be prevented simply by using safer work practices. Often, we are concerned with getting a job done quickly and with excellence. However, part of doing a job with excellence is making every reasonable effort to keep yourself and others safe. This ***safety mindset***, or safety culture, is sincerely concerning yourself with not only your personal safety and welfare, but also the safety and welfare of others. One must always be alert and aware of dangers on the jobsite. Construction can be a dangerous occupation. The potential for accidents and injuries is always present on the job, even when everyone is at their best. The jobsite becomes even more dangerous when workers engage in recklessness, carelessness, or horseplay. It is critical to note that there is never a place for drugs or alcohol on a construction jobsite. Make it your mission to protect those around you by maintaining a strong safety mindset.

Be on the alert and work to recognize the hazards around you. Hazards can be caused by environment, people, materials, and various situations. A safety mindset involves being aware of your surroundings and identifying hazards before they become an issue.

Occupational Safety and Health Administration (OSHA)

Because construction can be a dangerous occupation, employers have a legal obligation to ensure a safe and healthful workplace. In 1971 the federal government created the ***Occupational Safety and Health Administration (OSHA)*** to improve worker safety and health protection. Since then, daily deaths on the job across all industries have declined from about 38 per day to approximately 15 per day in the United States. An impressive statistic, considering the number of people working in private industry doubled over the same period. Still, approximately 1 in 5 deaths on the job are in the construction sector. Even though these statistics show just how dangerous construction can be, the likelihood of a fatal accident is low if proper safety standards are followed.

Pro Tip

The mission of OSHA is "to ensure safe and healthful working conditions for workers by setting and enforcing standards and by providing training, outreach, education and assistance." If you work as an employee of a private (nongovernment) organization in the United States, you are protected by the authority and jurisdiction of OSHA. OSHA carries out its mission to ensure safe and healthful working conditions by enforcing something called the ***general duty clause***, a law requiring employers to take the primary responsibility regarding the safety and health of their employees. Some of the primary methods are:

- Developing job safety and health standards and enforcing them through worksite inspections.
- Providing training programs to increase knowledge about occupational safety and health.

It is helpful to consider OSHA as a partner in keeping workers safe, rather than an enforcement agency trying to catch someone violating a rule. The preference of OSHA is to encourage safety and health, rather than to enforce it.

Employee Rights and Responsibilities

OSHA has established a series of worker rights and responsibilities. As an employee, you have the right to:

- A safe and healthful workplace.
- Know about the hazardous chemicals associated with your work.
- Report an injury to your employer.
- Complain or request correction from your employer for hazards in the workplace.
- Safety and health training.
- Medical records indicating illness or injury on the job.
- File a complaint with OSHA.
- Participate in an OSHA inspection.
- Be free from retaliation for exercising safety and health rights.

Though employees are protected by OSHA's general duty clause, they also have a personal responsibility to uphold safe work practices and can be cited if they do not follow employer mandated safety rules or fail to take part in employer mandated training.

Employer Responsibilities

Employers are required to inform their employees of their rights by posting the *OSHA Job Safety and Health Poster: It's the Law,* **Figure 1.6-1**, prominently in a place frequented by all employees. Additionally, employers are responsible for the following:

- Provide a workplace free from recognized hazards and comply with OSHA standards.
- Provide training required by OSHA standards.
- Provide medical exams when required by OSHA standards.
- Provide and pay for most personal protective equipment (PPE), **Figure 1.6-2**.
- Maintain injury and illness records.
- Inform workers how to report an injury or illness to the employer.
- Make records available to workers and OSHA.
- Post annual summary of injuries and illnesses.

OSHA Standards

OSHA standards are rules that describe the methods employers must use to protect employees from health and safety risks. The OSHA standard that applies to the construction industry is broken down into subparts, each covering a different aspect of construction, **Figure 1.6-3**. OSHA publishes other specific standards that apply to industries with unique safety and health risks, such as agriculture and maritime. You can view all these standards digitally at OSHA.gov. This website provides a wealth of information regarding safety and health in the workplace.

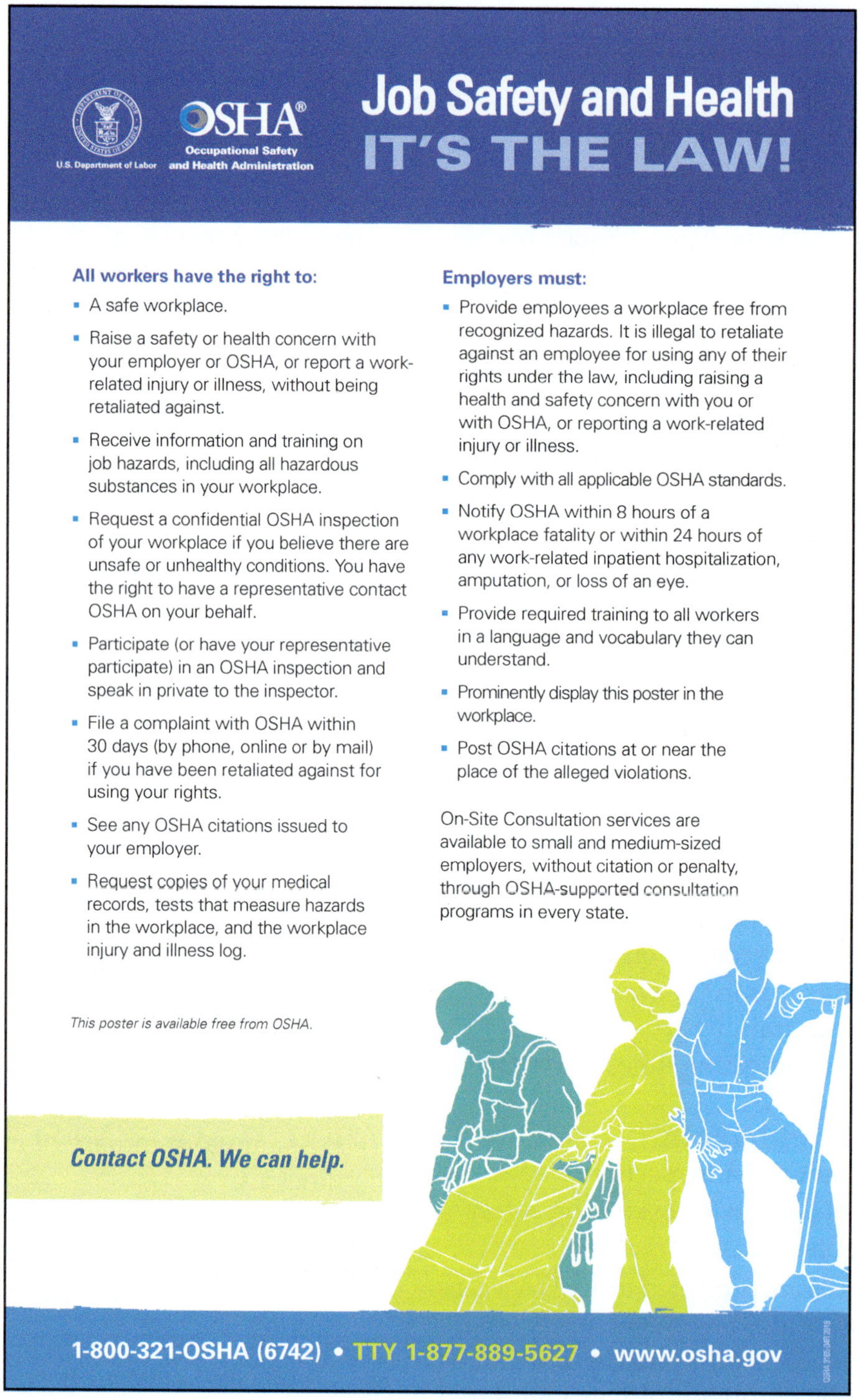

OSHA

Figure 1.6-1. OSHA requires employers to post this sign in an area where all employees can read it.

OSHA standards give guidance on how to best protect workers from hazards; ultimately, the best way to maintain a safe jobsite is to keep a safety mindset by maintaining a proper perspective and attitude toward safety. While it may take longer to complete a task in a safe manner, the risks of working unsafely do not outweigh the benefits. The results of unsafe work habits can be catastrophic. If an employer instructs you to perform a task without protection from known hazards, they are in

photka/Shutterstock.com

Figure 1.6-2. Personal protective equipment (PPE) provides protection against safety hazards on the construction site. The equipment to be worn is determined by the work being performed.

Part 1926	Safety and Health Regulations for Construction
Subpart A	General
Subpart B	General Interpretations
Subpart C	General Safety and Health Provisions
Subpart D	Occupational Health and Environmental Controls
Subpart E	Personal Protective and Life Saving Equipment
Subpart F	Fire Protection and Prevention
Subpart G	Signs, Signals, and Barricades
Subpart H	Materials Handling, Storage, Use, and Disposal
Subpart I	Tools - Hand and Power
Subpart J	Welding and Cutting
Subpart K	Electrical
Subpart L	Scaffolds
Subpart M	Fall Protection
Subpart N	Helicopters, Hoists, Elevators, and Conveyors
Subpart O	Motor Vehicles, Mechanized Equipment, and Marine Operations
Subpart P	Excavations
Subpart Q	Concrete and Masonry Construction
Subpart R	Steel Erection
Subpart S	Underground Construction, Caissons, Cofferdams and Compressed Air
Subpart T	Demolition
Subpart U	Blasting and the Use of Explosives
Subpart V	Electric Power Transmission and Distribution
Subpart W	Rollover Protective Structures; Overhead Protection
Subpart X	Stairways and Ladders
Subpart Y	Diving
Subpart Z	Toxic and Hazardous Substances
Subpart AA	Confined Spaces in Construction
Subpart CC	Cranes and Derricks in Construction

OSHA

Figure 1.6-3. This list shows how OSHA has organized the tasks of construction into subparts where safety hazards can be identified and safety standards applied.

LESSON 1.7

Math

LEARNING OUTCOMES

After completing this lesson, you will be able to:

1.7-1 Interpret markings on measuring devices.

1.7-2 Demonstrate math skills using whole numbers.

TECHNICAL TERMS

dividend	product	scale
divisor	quotient	whole number

Calculations in Construction

Basic math skills are important if you are going to consider a career in construction. Most of the math used on the construction jobsite is simple; however, math errors can result in bigger problems if they are not discovered and corrected. Therefore, it is vital to focus on an understanding of the math concepts being used. Math formulas are useful, but plugging numbers into a formula is not helpful if you do not understand the formula to begin with. Tasks like converting measurements from fractions to decimals, working with angles, and estimating costs or materials are vital to an efficient construction operation.

When performing any calculations in construction, it is important to see the big picture and ask if the solution you came up with is realistic. For example, let's say you are figuring out how much concrete to order for a garage floor. The garage measures 24′ by 24′. Concrete is delivered to the jobsite in a concrete mixing truck and is sold by volume in cubic yards. If you need to make the concrete floor 6″ thick, by your calculations, you incorrectly determine that you should order 288 cubic yards of concrete. Before ordering the concrete, you should ask yourself if your determination makes logical sense. It would be helpful to know how much concrete fits in a concrete truck. A quick online search reveals that a typical concrete truck can deliver about 10 cubic yards at a time, depending on the type of truck. According to your calculations, your garage floor would take nearly 29 truckloads of concrete.

Pro Tip

Always look at your answer and compare it with what might be a reasonable answer. This is the most important tip for calculator use. If the answer does not look reasonable, it is probably not correct.

Does this conclusion seem reasonable? That is a lot of concrete! After reviewing your calculations, you discover that you had figured the concrete in cubic feet, rather than cubic yards. You left out a step in the math. After correcting your calculations, you determine you will only need 10.7 cubic yards, just over 1 truckload of concrete. However, it is standard practice to order an extra 10%–15% to ensure you have enough to cover waste and other variables. You decide to order 12 cubic yards.

Stepping back to ask whether your estimate made logical sense saved a lot of money and embarrassment. This example illustrates how a simple math mistake can be discovered by using a little bit of critical thinking.

Reading a Ruler

Reading a measuring device is one of the most common tasks on the construction jobsite. It is critical that workers can read and communicate dimensions with complete clarity and accuracy. Doing so requires you to recognize the measuring system that is in use. Measuring devices such as rulers and tape measures may be marked for measuring feet, inches, and fractions of an inch; some may be marked with meters, centimeters, and millimeters; or you may see them marked with feet, inches, and tenths of an inch. The measuring system used to divide the spaces on a measuring device is called the *scale*. The most common linear (in a line) scale in construction uses yards, feet, inches, and fractions of an inch, although yards are seldom used. There are three feet in a yard, 12 inches in a foot, and the inches are divided fractionally. See **Figure 1.7-1**. The longest marks on the scale indicate inches. The inches on a measuring device may be divided into eighths, sixteenths, or even thirty-seconds. The second longest marks on the scale represent halves, the next longest represent fourths, and so on. The first step in reading the scale is to determine what the smallest marks on the scale represent. Count down from the whole inch to the halves, then the quarters, then eighths, sixteenths, and thirty-seconds, if they are used. You can count

Practice 1-1

What measurements are represented by letters A–E on **Figure 1.7-2?**

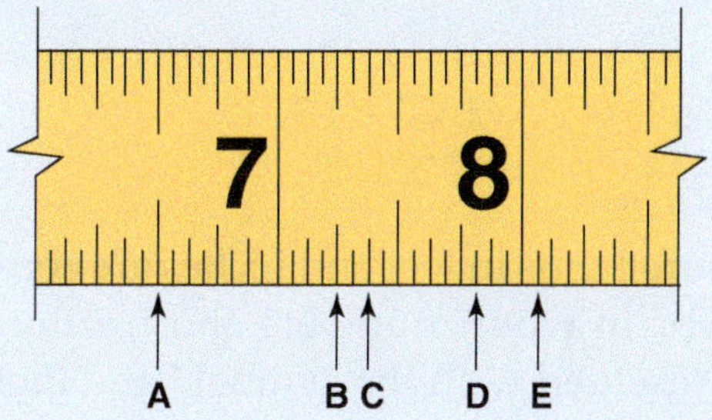

Goodheart-Willcox Publisher

Figure 1.7-2. Identify the measurements indicated on the ruler.

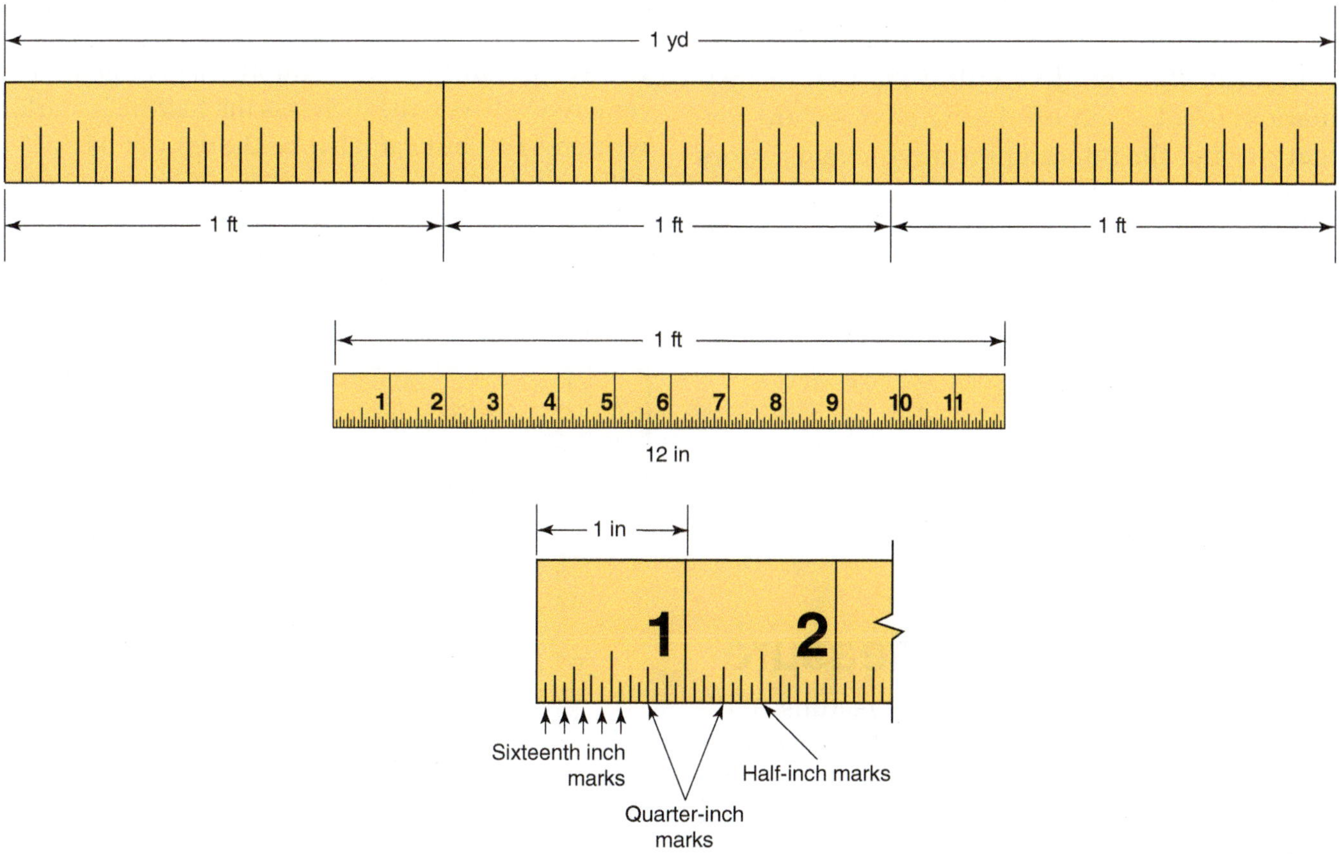

Goodheart-Willcox Publisher

Figure 1.7-1. In the US customary system of measurement, linear measurements are made in yards, feet, inches, and fractions of an inch.

the number of marks from the last inch mark to the mark you are reading, however, it is faster to learn marks by association, rather than counting them, as shown in the next session.

Often, the measurement you will take is several feet, a few more inches, and some fractions of an inch. For example, you might measure a room and find that it is 14 feet 4 and 3/4 inches wide. It is customary in construction to write this as 14′-4 3/4″. If it is necessary to express this as inches only, that can be done by multiplying the number of feet by 12 and adding the remaining inches.

$$\begin{aligned} 14'\text{-}4\ 3/4'' &= (14' \times 12\ \text{in}/1\ \text{ft}) + 4\ 3/4'' \\ &= 168'' + 4\ 3/4'' \\ &= 172\ 3/4'' \end{aligned}$$

Measurements in feet and inches are commonly expressed in whole numbers and fractions. Feet are represented with the symbol ′ and inches are represented with the symbol ″. A dash separates the number of feet from the number of inches. This format simplifies communication and helps prevent confusion. For example, if you were to write the measurement 14′-4 3/4″ as 14′4 3/4″, it could be mistaken for 144 3/4″. Similarly, miscommunication might result if the foot or inch mark symbol is left off. Using a consistent format clarifies the information and prevents errors in construction.

The reverse is also true. If you are given a dimension of 78 1/2″, it can be expressed as feet and inches. Divide the whole number by 12. Any remainder from the division will be left as inches.

$$\begin{array}{r} 6 \\ 12\overline{)78} \\ \underline{72} \\ 6 \end{array} \quad \text{inches left after division}$$

$$78\ 1/2 = 6'\text{-}6\ 1/2''$$

Reading a Tape Measure

Tape measures allow for an easy reading of dimensions in both the feet and inches format or the inches only format. For instance, **Figure 1.7-3** shows the portion of a tape measure that reads our 14′-4 3/4″ dimension from the previous example. Notice the 14′ mark is signified by the number 14 embedded in an arrow, centered on the line

Goodheart-Willcox Publisher

Figure 1.7-3. Notice the colors and shapes used in conjunction with the numbers on this tape measure to differentiate the feet and inches (top) from the continuous inches (bottom).

indicating 14′. Across the top of the tape measure, note the red numbers, beginning with 1 and counting to 11. These red numbers indicate the number of inches between each foot mark. Looking at the corresponding black numbers on the lower half of the tape measure, we can read the same dimension in inches only. So, we can read our dimension as either 14′-4 3/4″ or 172 3/4″.

Figure 1.7-4 shows a close view of the fractional scale between each inch mark. This tape measure is divided into 1/16 of an inch. Study the fractional parts shown along the bottom of the tape measure and the readings designating specific measurements along the top. The smallest markings represent 1/16 of an inch. The larger markings represent 1/8″, 1/4″, and 1/2″, each becoming slightly longer. These larger markings can be thought of as having multiple designations. The large mark centered between the inches is the half-inch mark, written 1/2″, but can also be thought of as 2/4″, 4/8″, or 8/16″.

Although construction workers do not say 4/16″ or 8/16″, these associations can be helpful to read

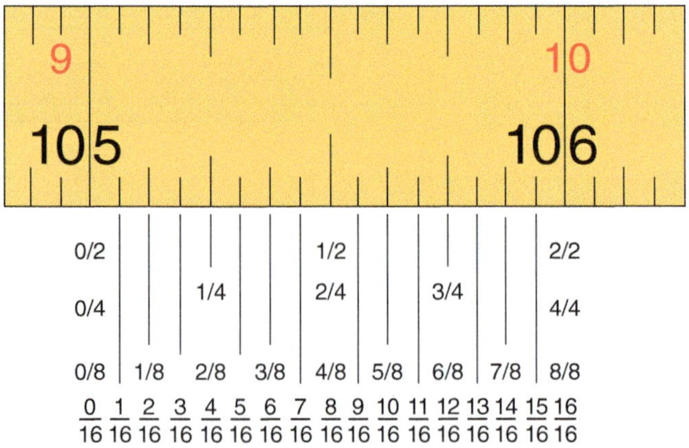

Goodheart-Willcox Publisher

Figure 1.7-4. To become more efficient in reading a tape measure, it is helpful to recognize how many sixteenths, eighths, and quarters each mark represents. On a tape measure, each 1″ division is typically divided into halves, fourths, eighths, and sixteenths. All the fractions on a tape measure can be expressed in sixteenths of an inch.

dimensions quickly and accurately to the nearest sixteenth of an inch. For instance, say you take a measurement and it is one mark beyond 105 3/4″. If you are communicating with a crewmember, it is not clear, nor is it helpful to say, "one-hundred and five and one little mark past three-quarters!" Alternatively, you could count the little marks and determine that this is the thirteenth, accurately concluding that the dimension is 105 13/16″. However, if you need to count each mark every time you read a dimension you will not be working efficiently. Working with measuring tools and dimensions regularly will help you to distinguish each mark quickly and accurately on a tape measure. Until you reach that point, it may be helpful to use the information in **Figure 1.7-4**. We know that there are four sixteenths in each quarter, so some quick skip counting will tell you that there are eight sixteenths in a half, and 12 sixteenths in three quarters. Add one to that, because the next mark is the one in question, and you have your dimension, 13/16″. This may seem clunky at first, but thinking about fractions in this way can help you identify each mark accurately. With time they will begin to come naturally. It can also help you to quickly convert and reduce fractions, which will help when you need to add, subtract, multiply, and divide dimensions.

Learning to read a tape measure quickly and accurately is one of the most important skills a construction worker can acquire. It is worth spending the time and effort to build your abilities early. Some mistakes made reading a tape measure can be identified and fixed easily. However, some errors may go unnoticed until it is too late, resulting in an expensive repair or a poor-quality product in the end. Avoid these mistakes by practicing with your tape measure every chance you get.

Practice 1-2

What measurements are represented by letters A–D on **Figure 1.7-5**?

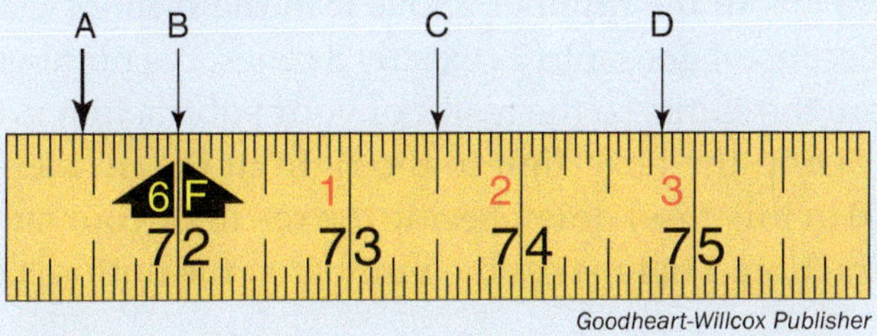

Figure 1.7-5. Identify the measurements indicated on the tape measure.

Whole Numbers

Whole numbers are simply numbers without fractions or decimal points, numbers such as 1, 2, 3, 4. Working with whole numbers is a building block to more complex math operations. Adding, subtracting, multiplying, and dividing whole numbers primarily requires memorizing a few math facts.

Adding and Subtracting Whole Numbers

Adding a column of whole numbers requires memorizing the sum of 3 + 5 and the sum of 8 + 2.

$$\begin{array}{r} 3 \\ 5 \\ +\ 2 \\ \hline 10 \end{array}$$

The same type of memorization of math facts is required for adding and subtracting larger whole numbers. When working with numbers whose values are ten or more (those with more than one digit), align the digits representing 0 through 9 (the 1s) in the right-hand column.

$$\begin{array}{r} \text{10s, 1s} \\ 37 \\ -\ 12 \\ \hline 25 \end{array}$$

We can easily subtract 12 from 37 and get 25 by looking at each column separately. In the 1s column, we know that 7 minus 2 is 5, and, in the 10s column, we know that 3 minus 1 is 2. The key to both addition and subtraction is to line up the columns of digits correctly.

In subtraction, if the number in a column being subtracted (the number on the bottom) is larger than the number it is being subtracted from (the number on the top), borrow 10 from the next digit to the left and add it to the one on the right. Write small numerals above the column to help you keep track.

$$\begin{array}{r} {\scriptstyle 2\ 16} \\ \not{3}\not{6} \\ -\ 19 \\ \hline 17 \end{array}$$

Practice 1-3

A general contractor needs to demolish an existing abandoned home and detached garage before starting construction on a new home. The contractor estimates that the demolition of the house will produce 46 cubic yards of debris. The garage will produce another 16 cubic yards of debris. Additionally, there is a deteriorated garden shed in the back of the lot that will produce another 3 cubic yards of debris.

1. How much debris will need to be disposed of on the job?
2. If dumpsters are available in 20, 30, and 40 cubic yard capacities, what combination of dumpster sizes could be used without overpaying for extra capacity?

Multiplying Whole Numbers

Multiplication of whole numbers requires memorization of a multiplication table. The only way to get $6 \times 5 = 30$ is to know that multiplication fact or to add $6 + 6 + 6 + 6 + 6$. That becomes way too tedious for bigger multiplication problems. To multiply numbers whose values are ten or more, align the numbers in the right-hand column. Then, multiply the top row by the 1s digit in the second row.

$$\begin{array}{rrr} & 3 & 1 \\ \times & 1 & 5 \\ \hline 1 & 5 & 5 \end{array}$$

Next, multiply the top row by the 10s digit in the second row. Because you multiplied by the 10s digit, the ***product*** (the result of multiplication) is written with its right-most digit in the 10s column.

$$\begin{array}{rrr} & 3 & 1 \\ \times & 1 & 5 \\ \hline 1 & 5 & 5 \\ 3 & 1 & \end{array}$$

If the problem has more digits in the second row, the above steps are repeated for each digit with the products being written in rows beneath one another, with the right-most digit in each row being written in the column for the place it represents.

When all the multiplication is complete, add the products just as you would for a simple addition problem. The result of this addition is the product (answer) of the multiplication problem.

$$\begin{array}{rrr} & 3 & 1 \\ \times & 1 & 5 \\ \hline 1 & 5 & 5 \\ 3 & 1 & \\ \hline 4 & 6 & 5 \end{array}$$

Dividing Whole Numbers

Division of whole numbers is the reverse of multiplication, but the problem must be set up differently. The ***dividend*** (the number being divided) is written inside the division symbol. The ***divisor*** (the number the dividend will be divided by) is written to the left of the symbol:

By knowing the multiplication table, we know that $7 \times 4 = 28$. So, if 28 is divided into 7 parts, each part will have 4, or $28 \div 7 = 4$. 4 is the ***quotient*** (the result of a division problem), and it is written above the division symbol and above the 1st place of the 28:

$$\begin{array}{rl} & \underline{\ 4\ } \leftarrow \text{Quotient} \\ \text{Divisor} \rightarrow 7 & \overline{)28} \leftarrow \text{Dividend} \end{array}$$

When the dividend is more than 9, the process is divided into steps. Starting at the largest value of the dividend, 3, determine how many times 4 can go into 3. 3 is too small, so we continue to the next digit. 4 goes into 32 8 times. Write the 8 above the 2 (the right column of the 32). Now multiply 4×8, which is 32. Write the 32 beneath the 32 in the division symbol.

$$\begin{array}{rl} & 08 \\ 4 & \overline{)320} \\ & \underline{32\ } \end{array}$$

Subtract the product of your multiplication (32) from the number above it in the symbol (32). Because 4 goes into 32 exactly 8 times, the numbers are the same, so the result of your subtraction is 0.

Drop the next digit to the right in the dividend (0 in this case) down beside the result of your subtraction. That makes the number at the bottom 00.

4 will not go into 0 (or 00), so the quotient of that step is 0. The quotient (answer) of the division problem is 80.

```
   080
4 |320
   32
   ---
    00
```

320 can be divided by 4, 80 times. If there are more places under the division symbol, just keep doing the same division, multiplication, subtraction, and drop down for each digit moving to the right.

If the last number produced by the drop-down cannot be divided evenly by the divisor, that number is called the remainder.

In construction, adding, subtracting, multiplying, and dividing are math functions that are used nearly every day. Tasks such as estimating and ordering materials, translating blueprints, and laying out various assemblies built when framing, plumbing, roofing, or excavating are just a few examples of when you will use math on the jobsite.

LESSON 1.7 Review

Answer the following questions based on information in this lesson.

1. Read in feet and inches, what is the dimension shown in **Figure 1-A**? (1.7-1)
   A. 16′-2 3/4″
   B. 16′-2 9/16″
   C. 16′-3 1/16″
   D. 17′-2 3/4″
2. Read only in inches, what is the dimension shown in **Figure 1-A**? (1.7-1)
   A. 194 9/16″
   B. 194 3/4″
   C. 195 1/16″
   D. 195 3/16″
3. What is the product of 8 and 6? (1.7-2)
   A. 24
   B. 32
   C. 48
   D. 96
4. What is 96 divided by 3? (1.7-2)
   A. 24
   B. 32
   C. 48
   D. 96

Figure 1-A.

LESSON 1.8

Print Reading

LEARNING OUTCOMES

After completing this lesson, you will be able to:

1.8-1 Define terms associated with construction drawings.

1.8-2 Differentiate plan view from elevation view in a set of construction drawings.

1.8-3 Outline the type of information typically included in a site plan.

TECHNICAL TERMS

construction drawing
elevation view
orthographic projection
plan view
print
reference symbol
set of plans
site plan
specs
view

Conventions and Terminology

Imagine trying to build a home using only carefully written instructions, with no pictures or illustrations. Even if the instructions were extraordinarily descriptive and well written, this task would be nearly impossible. Instead, builders use ***construction drawings***, also known as working drawings. Construction drawings are the language of the construction industry. Understanding this language enables architects, designers, and engineers to communicate what is to be built and how it is to be built, **Figure 1.8-1**. Builders must read and interpret the construction drawings accurately to transfer the design from paper or digital form into an actual building, often referred to as the built environment.

Copies of an architect's original drawings used on the jobsite are usually called ***prints***, or blueprints. At one time, all drawing copies were made on chemically treated paper that displayed white lines on a blue background, **Figure 1.8-2**. This is where the term blueprints originated. Today, prints are usually reproduced with dark lines on a white background. Although the term *blueprint* is still used, *print* is the preferred term.

Goodheart-Willcox Publisher

Figure 1.8-1. Prints are the copies of the construction or working drawings that are used on the jobsite.

Vast amounts of information are needed to build a house. A single drawing could not possibly hold all that information, so many drawings are needed. When printed and bound together, these sheets are

Mega Pixel/Shutterstock.com

Figure 1.8-2. True "blue" prints are no longer used now that plotters and printers can reproduce drawings.

known as a ***set of plans***. A complete set of plans is a comprehensive collection of drawings that communicates all necessary information to build a home.

A set of house plans usually includes the following drawings.

- Plans. Bird's-eye views that include plot plan, foundation or basement plan, and floor plans.
- Building elevations. Drawings that show the front, rear, and sides of a building or room.
- Section drawings. Drawings that show interior parts and structures as though the assembly had been cut through.
- Detail drawings. Enlarged drawings that show the details of how a particular aspect of the building is to be constructed.
- Drawings of the mechanical systems. Drawings that describe and give the location of electrical, plumbing, heating, and air conditioning components. These are often not included in drawing sets for relatively simple houses.

A complete set of plans also includes specifications and detailed notes, often referred to as ***specs***. The specs of a job supplement the drawings and are used by the architect to specify materials and define other details, such as the level of quality that must be met by the builder, **Figure 1.8-3**. For more information on drawing specifications, see *Chapter 7, Interior Insulation and Surfaces*.

FOUNDATIONS

1. Footings shall bear on firm, undisturbed soil a minimum of 24" below the final finished grade line for one and two story structures. However, should the frost line be locally deeper than 24", then the bottom of footings shall extend below this local frost line. Where sloping grades are involved, footing depths must always be measured from the point at the final finished grade line that is closest to the footing.

2. To ensure proper support for the footings and foundation walls, footings must always be poured with a flat bottom surface. Footings may be stepped where necessary to allow the footing to remain below frostline on a sloping grade, but the structure may not be built on a site where the finished grade line exceeds thirty degrees down from level. Sloping sites require special engineering for extended foundation walls and stepped footings.

3. A minimum 2000 psf soil bearing is required.

4. Drains shall be provided around all usable spaces located below ground. Drain tiles, perforated pipe, or other methods approved by local codes should be installed at or below the area to be protected, and must discharge by gravity or mechanical means to a drainage system. Top of drain pipe shall be level with top of footing.

CONCRETE

1. All concrete for walls, footings, and basement slabs shall develop and maintain a minimum compressive strength of 2500 psi at 28 days.
2. All concrete for garage slabs and porches shall develop and maintain a minimum compressive strength of 3500 psi at 28 day strength.
3. Concrete forms, shoring and pouring methods shall conform to all current practices endorsed by the American Concrete Institute.

Goodheart-Willcox Publisher

Figure 1.8-3. A short list of specs, such as these, can be printed on drawings. Larger jobs often have more comprehensive specs that are included as a spec book.

The Language of Print Reading

As previously mentioned, print reading is a method of communication for construction. It may be helpful to think of the set of plans as you might think of your favorite book. Taken as a whole, construction drawings give a complete picture of what is to be built with details telling the story of how each part goes together.

Sheets

Sticking with our analogy of your favorite book, each sheet in a set of plans is like a chapter of the book. If you were missing one sheet, you may have an idea of how the building will turn out, but some vital information would be omitted, just like if you were missing a chapter in a book.

Drawings

Individual drawings produced by the architect are detailed descriptions showing important aspects of what is to be built, **Figure 1.8-4**. These drawings may show large-scale general information about the project, or they may zoom in on minute details such as how the framing of the house is to be attached to the foundation, or how a staircase should be built. Individual drawings give dimensions, sizes, and locations of building components along with other important details. Consider individual drawings the paragraphs an author may use to describe a scene.

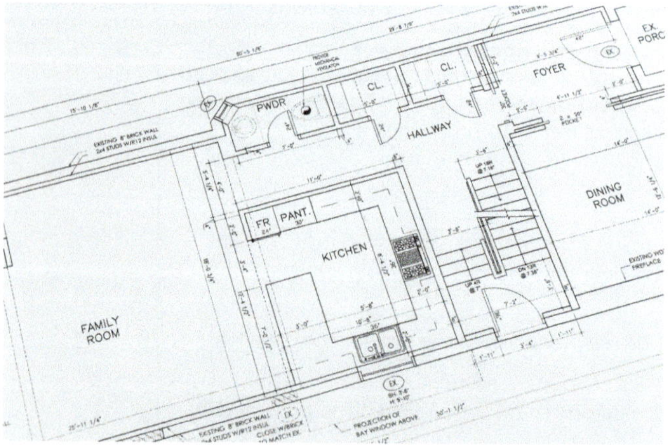

Branislav via Getty Images

Figure 1.8-4. Large drawings, such as a floor plan, often take up an entire sheet. Some sheets include multiple drawings.

Callouts and Notes

Architects also need to communicate information on the drawings that cannot be represented graphically; instead, it must be written with text. Callouts and notes, **Figure 1.8-5**, are used to communicate many of these details. It is important that builders pay close attention to callouts and notes to get every detail right on a job. These are like the sentences of your book. When you are reading, it may be easy to skip a word or a line of text without realizing it. In the same way, callouts and notes are easy to miss, but often communicate vital information. An example might be what size reinforcing steel is to be used in the footings or the basement floor, or where important fire blocking needs to be installed. Framing member sizing, insulation values, and the identification of materials used throughout a wall assembly are just a few other examples of where callouts and notes are used.

Symbols and Abbreviations

Symbols to an architect are like words to a writer. They are used to put a picture in the builder's mind that shows or gives a representation of a feature or type of material that is to be used. Many elements of a home cannot be drawn exactly as they look in real life. Imagine an architect drawing a light switch on a floor plan to let the electricians know where it needs to be. Doing so would not be a good communication strategy. Instead, the architect draws a symbol that represents a switch. Symbols are drawn to indicate materials like gravel under a concrete slab, or the concrete of the slab itself. Insulation, doors and windows, exterior finishes such as siding and roofing, plumbing and electrical fixtures, even assemblies like fireplaces are represented on a drawing using symbols. **Figure 1.8-6** features the use of symbols to represent the bathroom sinks, toilet, and tub/shower unit, as well as symbols to represent two different door types. Two of them swing on hinges, while one is a pocket door that slides into the wall when opened.

Additionally, ***reference symbols*** are used in conjunction with callouts to refer to another drawing in the set of plans. The reference symbols link the two drawings to show an alternative perspective,

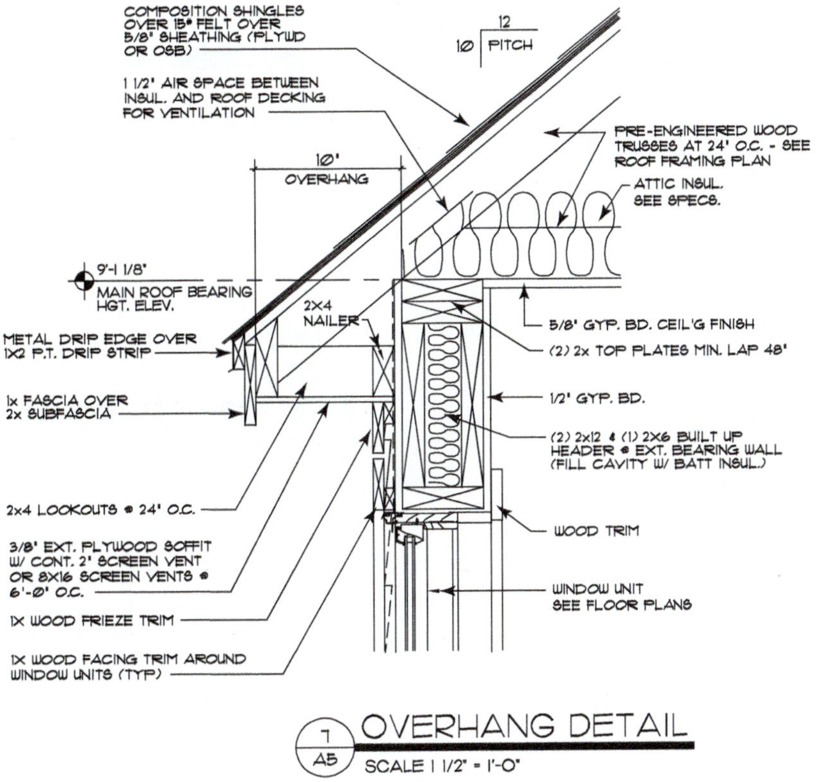

A

NOTES:

1. CONTRACTOR TO VERIFY ALL DIMENSIONS & EXISTING SITE CONDITIONS PRIOR TO CONSTRUCTION.
2. DOOR & WINDOW SIZES ARE GIVEN IN FEET & INCHES IN WIDTH & HEIGHT RESPECTIVELY.
3. DOOR EXAMPLE: 3068. 3′-0″ WIDE BY 6′-8″ TALL. WINDOW EXAMPLE: 2840. 2′-8″ WIDE BY 4′-0″ TALL.
4. CONTRACTOR SHALL VERIFY ALL INTERIOR FINISHES, FLOOR COVERINGS, AND TRIM SIZES & CONFIGURATIONS WITH OWNER PRIOR TO CONSTRUCTION.
5. FURTHERMORE, IT SHALL BE THE SOLE RESPONSIBILITY OF THE CONTRACTOR TO VERIFY CLEARANCES OF THE AFOREMENTIONED ITEMS PRIOR TO CONSTRUCTION.
6. REFER TO EXTERIOR ELEVATIONS FOR EXTERIOR WALL FINISHES.
7. REFER TO MANUFACTURER'S SPECS FOR ROUGH OPENING DIMENSIONS.
8. WINDOWS ARE SINGLE-HUNG VINYL, UNLESS OTHERWISE NOTED.
9. ALL BEAM SIZES TO BE VERIFIED BY STRUCTURAL ENGINEER.

B

Goodheart-Willcox Publisher

Figure 1.8-5. Notes may be used in conjunction with callouts on drawings or may be included in a list in the margin of a drawing or sheet. A—Callouts. B—Notes.

Figure 1.8-7, a cutaway view, or a close-up drawing (called a detail drawing).

Abbreviations are used extensively to consolidate callouts and notes on drawings. A list of abbreviations used in the prints is typically found on the title page.

Lines

Finally, think of lines as the alphabet the architect is using to build the words. On the surface, a line may just seem like a line. However, just as the alphabet is the building block of language,

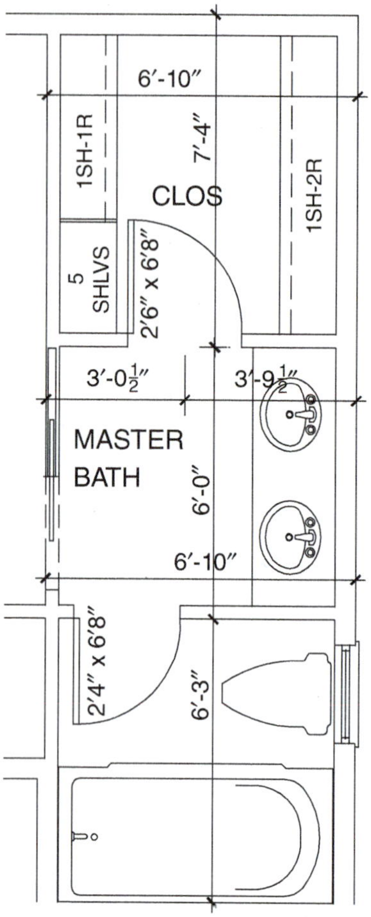

Goodheart-Willcox Publisher

Figure 1.8-6. Symbols are used extensively on construction drawings. What symbols can you identify in this bathroom drawing?

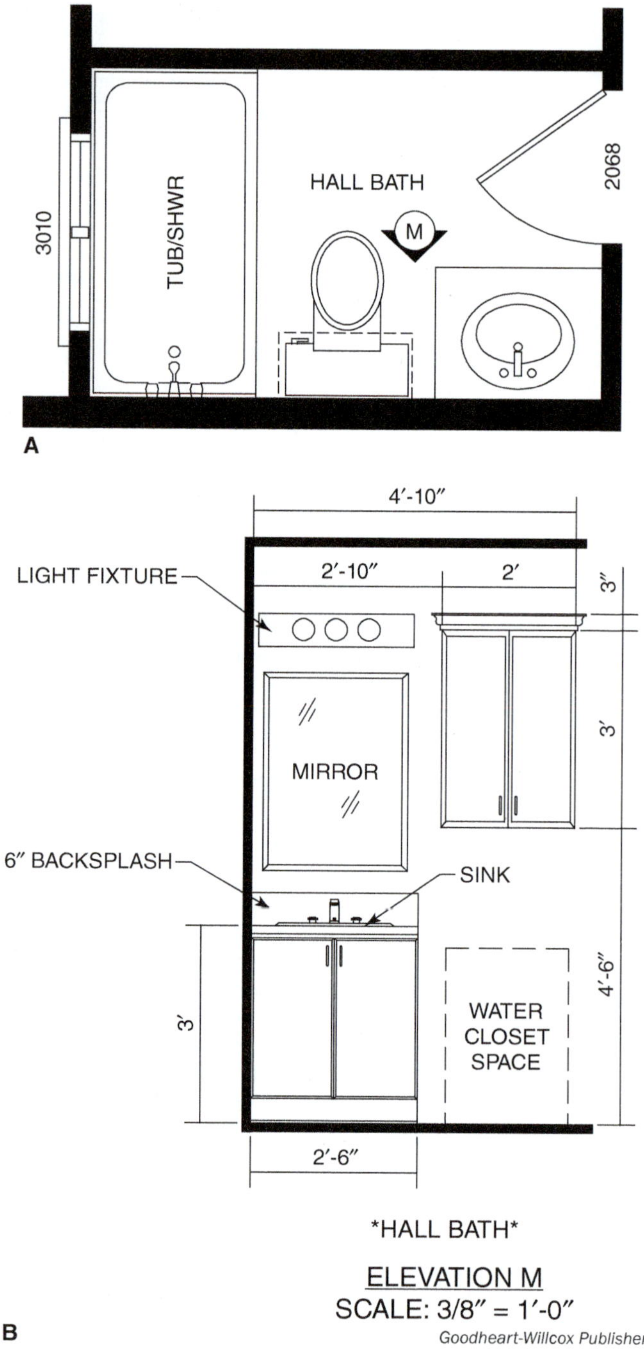

Goodheart-Willcox Publisher

Figure 1.8-7. A—The reference symbol "M" in the drawing above cross-references B—Elevation drawing "M" on a separate sheet of the set. Note the direction of the arrowhead on the reference symbol in the plan view indicates the direction of the perspective in the elevation drawing.

and individual letters have unique sounds and distinctions, so do line types. In fact, architects use a standard set of line types that is referred to as the "alphabet of lines." See **Figure 1.8-8**. Lines used in construction drawings communicate different meanings based on the thickness (also called weight), as well as the pattern of the line.

Views

When reading a print, it is important to understand the part of the building that is being shown and the type of information being communicated. Having this knowledge helps a builder visualize the building layout and identify where to locate information on other drawings. For instance, the foundation plan shows the information needed to build the foundation and includes references to other drawings for additional construction details.

The exterior elevation drawings show vertical height dimensions for the building and the types of construction materials used on the exterior. These drawings use different types of ***views***, or perspectives of a drawing, to establish the viewing direction for specific purposes.

Line Types Commonly Used on House Plans	
Line Name and Appearance	**Description**
Property line	Identifies the boundaries of a building lot.
Border line	Thick, continuous line used around the border of the drawing.
Object line	Thick, continuous line identifies objects such as walls, windows, doors, foundations, and other physical components of the building.
Hidden line	Thin, dashed line identifies the edge of an object located behind another object in the view shown in the drawing.
Center line	Thin line of alternating long and short dashes used to mark the locations of centers of circular features, column lines, and lines of symmetry.
Dimension line **Extension line**	Thin lines used with dimension values to show the "real life" distance between two points.
Section cutting line	Identifies the place in the drawing from which a section drawing is viewed, and identifies the location of the section view in the set of plans.
Break lines	Thick or thin lines identifying a break in an object.

Goodheart-Willcox Publisher

Figure 1.8-8. Construction drawings contain several types of lines. Each line type has a specific meaning and use.

Most views on construction drawings are created using a process called ***orthographic projection***, a method of showing a three-dimensional object in two dimensions by displaying various views. An orthographic view is created by defining one face, or side, of an object. To visualize this process, imagine a building placed inside a glass box. See **Figure 1.8-9**. When looking through one of the sides of the glass box, what you see is one face, or side, of the building. Each side of the glass box represents a projection plane. The front, top, and right-side views are projected. Next, imagine that the sides of the glass box are unfolded so that the top and right-side views are revolved into the same plane as the front view. The views are then arranged as shown in **Figure 1.8-10**. This is a standard arrangement of orthographic views. Note that the glass box has six sides, resulting in six possible views: front, top, right side, left side, rear, and bottom. The bottom view of a building is not used on construction drawings, but all the other views are used to show the height, width, and depth of the structure.

Views used on construction drawings include plan views, elevations, sections, and details. Plan views and elevations are discussed in this chapter. More information on section views and details can be found in *Chapter 4, Drying In*.

Plan Views

In the previous example, imagine hovering over the glass box and looking straight down. What would you see? This bird's-eye view is called a plan view in print reading. A ***plan view*** shows the horizontal layout of a part of the structure viewed from above. Site plans, foundation plans, and floor plans are drawn as plan views. See **Figure 1.8-11**.

A floor plan shows the layout of a designated level in the building, such as the first floor or second floor. It is created by making a theoretical section cut through the building 4′ above the floor.

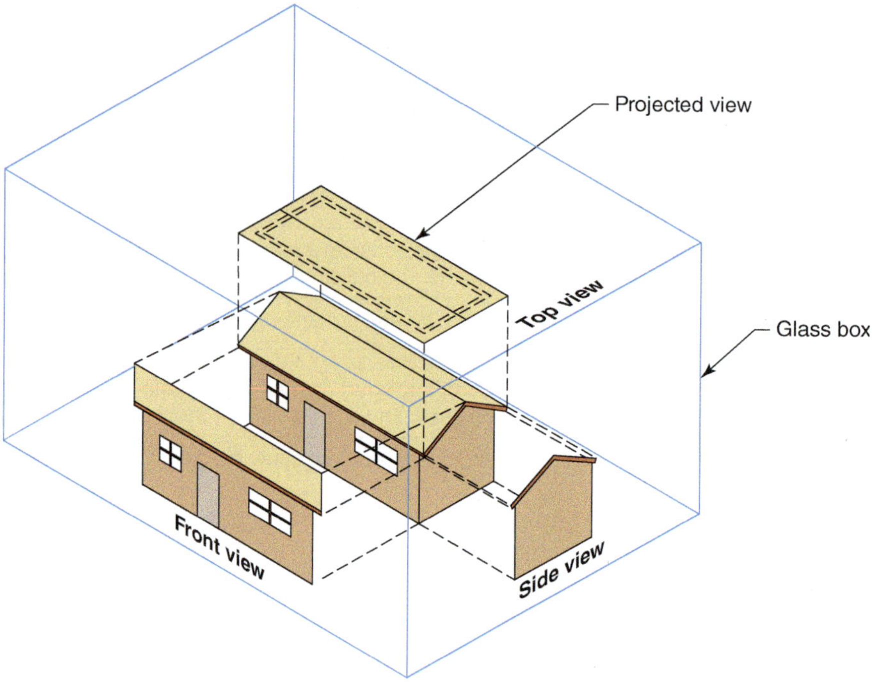

Goodheart-Willcox Publisher

Figure 1.8-9. Multiple views are needed to describe the features of a structure. On construction drawings, views are defined using orthographic projection. To visualize this process, imagine a building placed inside a glass box. Each view represents what you see when you look through one side of the glass box.

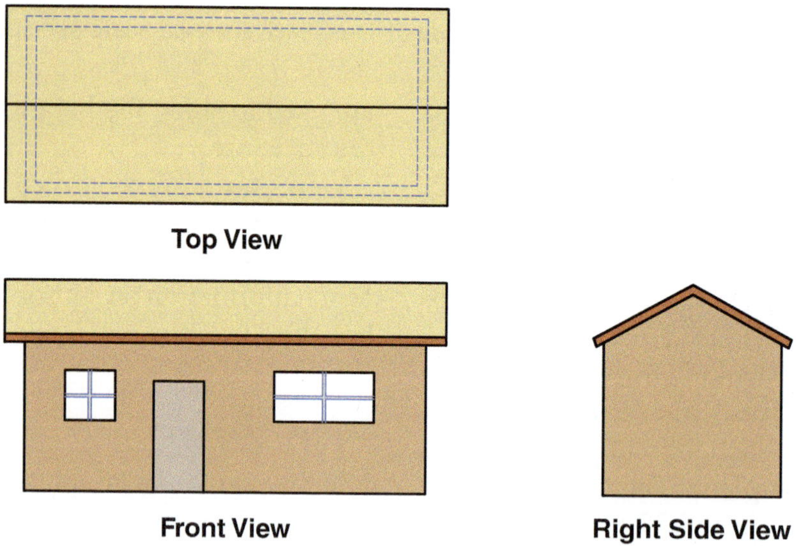

Goodheart-Willcox Publisher

Figure 1.8-10. A standard arrangement of views in orthographic projection. The views show the height, width, and depth of the building.

A floor plan serves a key purpose in a set of plans. It shows the general layout of the building, along with its dimensions, and is referenced in making other types of plan drawings, such as a framing plan, electrical plan, or plumbing plan. For more on floor plans, see *Chapter 3, Framing Systems*.

When using a plan view on a detail or section drawing, the drawing will be annotated with a note

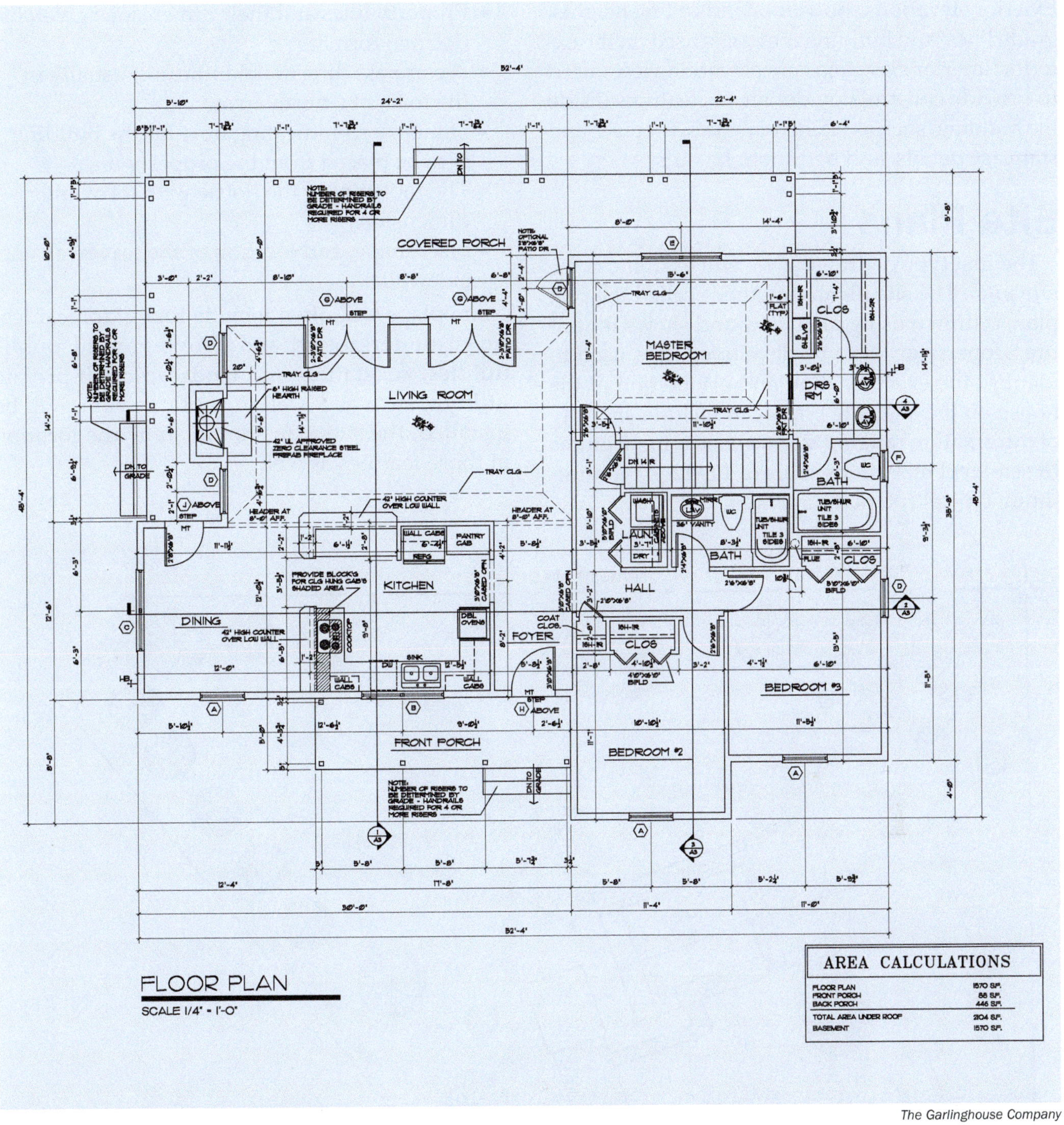

The Garlinghouse Company

Figure 1.8-11. Floor plans are drawn in plan view. In this plan view drawing, we see the house as if the roof has been removed. All floor plans are drawn in this way.

for clarity. However, most detail and section drawings are drawn using elevation view.

Elevation Views

Elevation views are profile views that show the front, rear, and sides of a building. Elevation views show the vertical layout of the building. Referring to the previous example, if you look directly through the front surface of the glass box, you see the front profile view of the building. If you look through the glass box from either side, you see a side profile view of the building. In print reading, these views are called the front elevation view and side elevation views. To see the building from the rear, you would need to go around to the rear side of the glass box and look at the back face.

Exterior elevations show floor and ceiling heights, grade lines, and building materials used on the exterior. Interior elevations are elevation views used to provide construction details for features inside the building, such as kitchen cabinets, fireplace and staircase details, and bathroom layouts.

Site Plans

The first type of drawing we will examine is the *site plan*. The site plan, sometimes called the plot plan, is drawn using plan view and shows the entire property or lot to be developed. It is used to identify the exact location and orientation of the house on the site. Site plans can include an array of information ranging from landscaping features to sea-level elevations, **Figure 1.8-12**. At a minimum, typical site plans include:

- Property lines and their dimensions given in decimal format
- A compass directional indicator, usually in the form of a north arrow
- Dimensions indicating how far the building is to be placed from the property lines
- The footprint of the house with its overall dimensions
- Size, shape, and location of the driveway and sidewalks

Site plans will often include topographical features, underground and above-ground public utilities, vegetation, and landscaping. If private utilities, such as a well or septic system, will be installed, the site plan should show the location of those features as well.

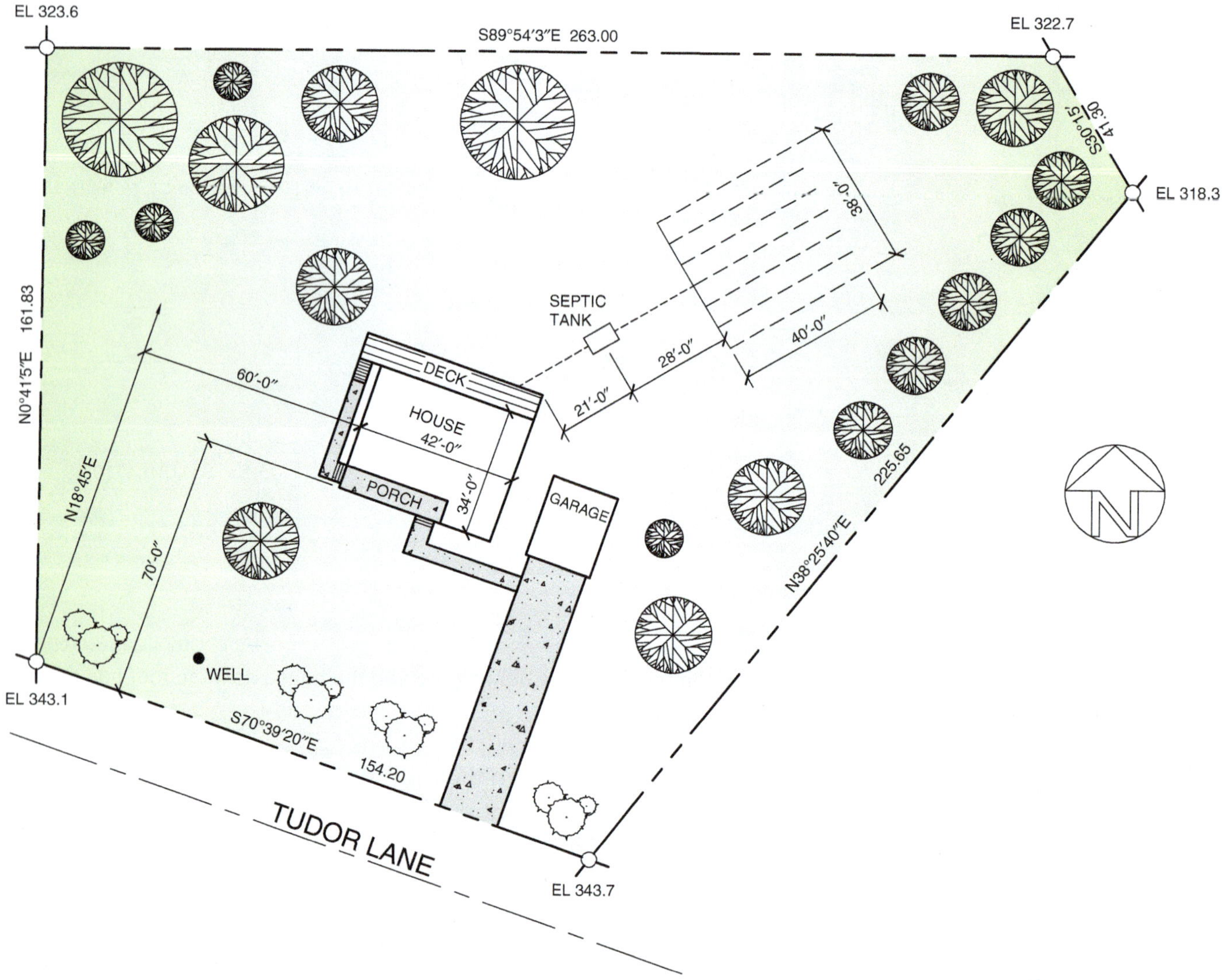

Goodheart-Willcox Publisher

Figure 1.8-12. The site plan shows important information regarding the property where the building is to be located.

LESSON 1.8 Review

Answer the following questions based on information in this lesson.

1. A series of blueprint sheets that coincide with each other is known as _____. (1.8-1)
   A. orthographic drawings
   B. specifications
   C. site plans
   D. a set of drawings
2. On a blueprint, _____ show a representation of a specific feature or material. (1.8-1)
   A. callouts
   B. symbols
   C. lines
   D. notes
3. A(n) _____ view is drawn from the front or side perspective. (1.8-2)
   A. elevation
   B. plan
   C. foundation
   D. front/side
4. A(n) _____ view is drawn as if looking down from above. (1.8-2)
   A. elevation
   B. plan
   C. top
   D. site
5. *True or False?* A site plan typically includes the driveway and sidewalks. (1.8-3)

LESSON 1.9

Green Building and the Future of Construction

LEARNING OUTCOMES

After completing this lesson, you will be able to:

1.9-1 Describe the basic principles of green building.

TECHNICAL TERMS

green building

Rethinking Home Design and Construction

Looking back through history, it is easy to recognize that construction has arisen through necessity. People need shelter. With the advent of tools, humans were able to build shelter that was more suitable for living than caves. As technology and knowledge developed, so did the quality of the buildings. Construction is on a path of continuous improvement. As buildings continue to improve, they must also improve regarding sustainability. To see where the construction industry is heading, we must consider how the limited resources of our planet can be conserved, which is the primary goal of green building. However, green building is not limited to building a home that operates with more efficiency.

Green building is a system of planning and construction where all parts of the home work together to lessen the impact on the environment, reduce energy consumption, and extend the usable life of the building with minimum maintenance, **Figure 1.9-1**. Green building must achieve these benchmarks while maintaining a comfortable and healthy living environment for a building's occupants. Standards are met by using a comprehensive approach to building design and construction. Simply increasing the insulation value or adding photovoltaic solar panels to a home will not make a home green. Increasingly, builders, developers, and regulators understand that the entire life cycle of a home, starting with the planning phase all the way through to the eventual demolition of a home, contributes to how green a home is. Within reason, the goal is to generate the lightest possible impact on natural resources to build and operate the home throughout its life cycle, while ensuring the home is built in a way that is durable and long-lasting. Doing so helps ensure that resources are not prematurely used to deconstruct it and build a replacement.

This all must be achieved while maintaining the design and function of the home in a way that is comfortable and enjoyable to live in, **Figure 1.9-2**. This is an important piece to the equation. Imagine building a concrete home deep underground. This home might cost little to heat and cool, making it more efficient to operate. With no exposure to the elements, it may even be cheap to maintain. After all, a normal home needs to have the roof and siding replaced over time. An underground home eliminates this expense. However, if the house is stuffy and damp with no natural light, it would not be a pleasant living environment. Although efficient and easy to maintain,

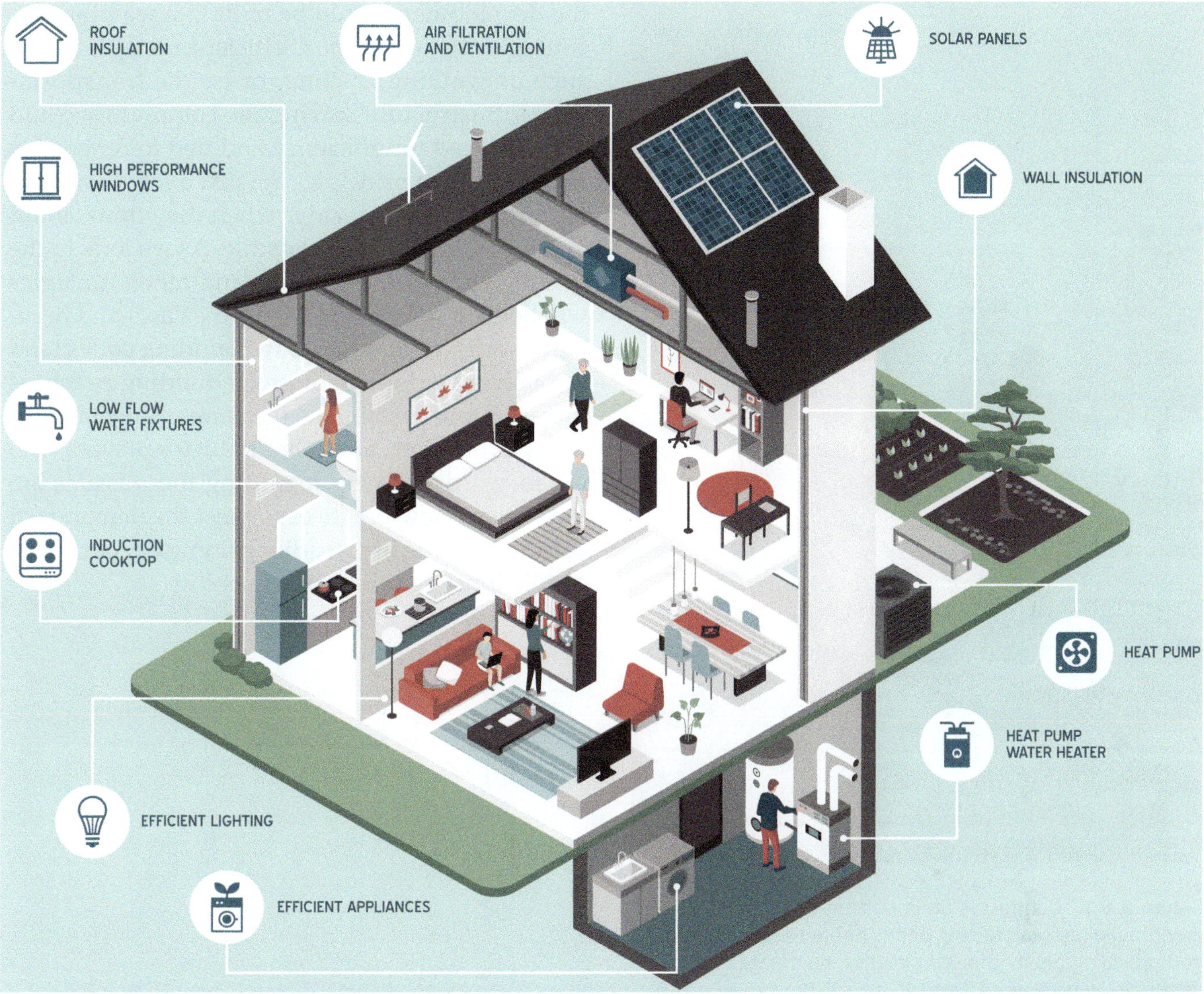

elenabsl/Shutterstock.com

Figure 1.9-1. Green building is a method of construction that considers the short- and long-term environmental impact of construction, maintenance, use, and deconstruction of a building.

it is not a very green design if no one wants to live there.

As we develop effective green building principles, it has become clear that we must continue to learn and grow. Resistance to change is always a factor when new methods or requirements are introduced. Often, seasoned builders prefer the old ways of building, rather than adapting and learning more environmentally friendly alternatives. Green building can include using new and unproven methods or products, which generates risk that many builders are not willing to accept. As more municipalities adopt green building codes, such as the International Green Construction Code (IgCC) and the International Energy Conservation Code (IECC), these practices will become more mainstream and accepted by the average builder.

When it comes to green building, small incremental changes can make a difference. However, to experience significant change, a paradigm shift needs to occur. Compare the impact of a single green-built home to an entire housing development designed and built to minimize its environmental impact. Not only would each home be built and operated efficiently, but the overall layout of

MBI/Shutterstock.com

Figure 1.9-2. Comfort is an important aspect of green building. Maintaining comfortable temperatures and providing plenty of natural light help create more desirable living spaces.

the development could be built in a way that encourages the use of more efficient transportation, such as walking, cycling, or public transportation. Infrastructure serving the community could be optimized for efficiency and tied to renewable sources. These principles are just a few that community planners consider when they think about green building on a larger scale. As we look to the future of construction, these and other strategies will begin to take form. Over time, laws and regulations such as updates to the building codes tend to push late adopters of green building practices toward more sustainable measures. The many architects, builders, and other industry players who are forward-thinking regarding these issues are developing new practices to meet the demands of consumers who are steadily gaining deeper social convictions regarding sustainability.

LESSON 1.9 Review

Answer the following questions based on information in this lesson.

1. The primary goal of green building is to _____. (1.9-1)
   A. build underground
   B. mass-produce homes
   C. design beautiful buildings
   D. conserve limited resources
2. *True or False?* The life cycle of a house includes the planning phase all the way through to its eventual demolition. (1.9-1)

Chapter 1 Review

Summary

Lesson 1.1

- Residential construction is the process of building single-family and multifamily homes. It also includes the addition to or renovation of these buildings. (1.1-1)
- New home construction and an aging workforce have led to an increased demand for construction workers. With construction skills and experience in high demand, wages in the construction industry have steadily increased at a rate higher than many other industries. (1.1-2)
- Early shelters were made of sticks, mud, grass, and stone. Over time, and with the discovery and advancement of tools and new ways of using materials, humans learned more advanced methods. (1.1-3)

Lesson 1.2

- Your skill set is the range of abilities you have that help you perform your job. Your skill set can be grouped into hard skills and soft skills. (1.2-1)
- Ethics are moral principles that guide our decisions. Ethical behavior involves knowing and acknowledging right from wrong and basing your decisions on that knowledge. (1.2-2)
- Critical thinking and problem-solving abilities are among the most valuable skills you can develop as a tradesperson. (1.2-3)

Lesson 1.3

- Administration, planning, and site preparation are the first steps when starting a new home building project. A building permit must be obtained and all zoning laws and building codes must be followed. (1.3-1)
- Site prep includes any demolition work that needs to occur. Site prep includes clearing the land of obstacles or brush, stripping the top layer of vegetation and soil, and laying out where the building will be located. (1.3-2)

Lesson 1.4

- Soil composition and characteristics vary. Topsoil contains organic matter and must be stripped away from the building site. Foundations are typically excavated to below the frost line, or the maximum depth to which soil is likely to freeze. (1.4-1)
- Framing lumber is available in a variety of lengths. Dressed size across width and thickness is smaller than nominal dimensions due to surfacing. (1.4-2)
- A variety of screw types are used in construction. It is important to know and understand their proper uses. Most screws are not rated for structural use. However, structural screws are available. (1.4-3)

Lesson 1.5

- Tape measures and measuring tapes are the most commonly used measuring tools in construction. The carpenter's pencil, lumber crayon, marking paint, and mason line are used for marking. The speed square, plumb bob, and spirit level are used for creating proper alignment. (1.5-1)
- Common hand tools include utility knives, handsaws, and sledgehammers. Common power tools include circular saws and cordless impact drivers. (1.5-2)

Lesson 1.6

- Many accidents and injuries on construction jobsites can be prevented by keeping a safety mindset. (1.6-1)

- OSHA is a federal agency that advocates for worker safety by setting and enforcing standards and by providing training, outreach, education, and assistance. (1.6-2)
- Tool safety should always be a priority. Both hand and power tools can be dangerous when not used properly. (1.6-3)

Lesson 1.7

- Reading fractional scales on measuring devices such as rulers or tape measures is an essential part of daily activity on a construction jobsite. (1.7-1)
- Whole numbers are numbers without fractions or decimal parts. Working with whole numbers is a building block to more complex math operations. (1.7-2)

Lesson 1.8

- Construction drawings are the language of the construction industry. Builders must read and interpret the construction drawings accurately to transfer the design from paper or digital form into an actual building. (1.8-1)
- Drawings use different types of views to establish the viewing direction for specific purposes. A plan view shows the horizontal layout of a part of the structure viewed from above. Elevation views are profile views that show the front, rear, and sides of a building. (1.8-2)
- The site plan shows the entire property or lot to be developed. Site plans can include an array of information ranging from landscaping features to sea-level elevations. (1.8-3)

Lesson 1.9

- Green building is a system of building in which all parts of the home work together to lessen the impact on the environment, reduce energy consumption, and extend the usable life of the building with minimum maintenance. (1.9-1)

Apply and Analyze

1. List three character traits you feel are essential in a construction career. How can these traits help you have success in a construction career? (1.2-3)
2. List one character trait you feel you should improve on. What is one step you can take today to help you grow in this area? (1.2-3)
3. Use materials available at your school or training center or visit a local lumber supplier to measure and list the dressed (actual) size for 2×4, 2×6, 2×8, 2×10, and 2×12. (1.4-2)
4. Using the OSHA website as a resource, list the costs a company might incur if there were a safety-related accident on the job. (1.6-2)
5. Explain why builders might be resistant to green building. (1.9-1)

Critical Thinking

1. Research three occupations in the construction field. List typical education and skill requirements associated with the occupations. How do these align with your skills and experience? Use the *Occupational Outlook Handbook* to find recent job outlooks and annual wages. (1.2-1)
2. Imagine you are the owner of a small construction company. What are the top five things you would look for in an employee? (1.2-3)
3. Consider working for a construction company that disregards safety. Contrast this with a company where safety is a core value and employees work with a safety mindset. From an employee perspective, what feelings do you think you would have toward your employer? How do you think these feelings might impact employee morale across the company? (1.6-1)

4. Research the production and delivery of framing lumber. List each step from tree harvesting to end use. What processes in this list require the consumption of energy? What steps could be taken to reduce this consumption of energy? How does this relate to green building? (1.9-1)

Communicating about Construction

1. **Speaking and Listening.** Interview someone who works in construction. Ask the person to describe a typical day at work. The following is a list of questions you might ask: (1.1-2)
   A. What is the work environment like?
   B. What are the job duties?
   C. What is your favorite part about being a carpenter? Least favorite?
   D. What kinds of injuries are you at risk for?
   E. What other types of professionals do you work with?
2. **Speaking.** Make a series of safety posters for the use of hand tools and power tools. Describe how each tool is used and basic safety practices, including how to clean and inspect the tool to determine if it is safe for use. Present your posters to the class. (1.6-3)
3. **Speaking and Listening.** Working with a partner, role-play the following scenario: you are a carpenter who has just observed a coworker under the influence of drugs or alcohol. One student acts as the carpenter and reports the behavior to their supervisor. The other student acts as the carpenter's supervisor. Role-play in front of the class and ask the class for feedback on the communication. (1.6-1)

Ivansabo/shutterstock.com

CHAPTER 2

Excavation and Foundations

Chapter 2 Outline

Introduction

Chapter 1, Preparing for Construction, introduced the residential construction industry, jobsite, and process. In this chapter, we will continue to build foundational knowledge and understanding of construction. First, we discuss the different occupations available within the construction industry and types of residential construction.

Our journey along the construction path continues as we consider excavation, placing footings, and building the foundation walls of a new home. We will also explore the materials, tools, and equipment used in the early phases of construction.

Next, we will discuss the most common reasons for accidents and injuries on construction jobsites, and how to work safely while getting a project off the ground. Then, we will expand on the coverage of fractional and decimal measuring devices introduced in *Lesson 1.7*.

Finally, we discuss foundation plans and green building principles that apply to the layout and building of a home's foundation, as well as some modern alternatives to traditional foundation construction.

LESSON 2.1

Industry and Careers

LEARNING OUTCOMES

After completing this lesson, you will be able to:

2.1-1 Identify the different classifications for occupations in the construction industry.

2.1-2 Describe the roles and responsibilities for occupations within each classification.

TECHNICAL TERMS

architect
building inspector
construction expeditor
construction manager
electrician
engineer
foreman
framing carpenter
HVAC technician
laborer
mason
operating engineer
plumber
safety manager
superintendent
trim carpenter

Occupations in the Construction Trades

Many types of jobs are available to someone considering a career in construction. Residential construction offers career options ranging from entry-level positions for which little specialized training is required, to highly specialized careers that require a high level of education and training. Occupations in the construction industry generally fall into four categories:

- Laborers and trades workers
- Supervision, administration, and management
- Design and engineering
- Compliance officers

Construction Laborers and Tradespeople

The past several decades have brought tremendous improvements in construction equipment and technology that help to ease the burden of physical stress on workers' bodies. However, construction remains a labor-intensive process.

A ***laborer*** is a worker who does manual labor, working with their hands to assist tradespeople. Laborers perform jobs that require their physical body, good judgment, and the ability to think ahead. The job of construction laborer does not require specialized or advanced levels of training. Due to this lack of specialization, pay for this position is generally lower than those requiring a specialized skillset.

The average annual salaries for the trades, according to the 2022 US Bureau of Labor Statistics, are outlined in **Figure 2.1-1**.

Occupation	Average Salary	Job Outlook Over Next 10 Years
Laborer	$46,350	Increase by 4%
Equipment Operator	$59,420	Grow slightly
Mason	$62,380	Decline slightly
Carpenter	$58,210	Grow slightly
Electrician	$61,550*	Increase by 6%
Plumbing	$65,190**	Grow slightly
HVAC Technician	$57,460	Increase by 6%

*This statistic encompasses all sectors within the electrician discipline, not just residential construction. Careers in the industrial and commercial fields tend to pay higher wages due to a more dangerous work environment and a higher level of specialization.

**Across all disciplines of plumbing.

US Department of Labor

Figure 2.1-1. Average annual salary data compiled from the US Department of Labor Statistics from 2022.

While some people may enjoy the physically demanding aspects of the position, starting a career as a construction laborer is one avenue where workers become more familiar with construction and explore options for a skilled trade. Tasks for a laborer can vary and often depend on the phase of construction, the size of the crew, or even the weather. Laborers generally work under the supervision of someone working in the skilled trades, **Figure 2.1-2**. Depending on the employer, a laborer may specialize and work with one trade as a concrete laborer, a carpenter's helper, or a mason's tender, for instance.

Pro Tip

A construction helper is similar to a construction laborer except that a helper typically focuses on laboring in a specific trade. This focus enables a higher level of training and specialization and can be a good path for gaining a better understanding and moving up as a skilled tradesperson.

Building construction involves the work of multiple trades. In residential construction, workers in different trades are responsible for specific tasks, including preparing the jobsite, constructing the foundation, building the structure, performing the exterior and interior finish work, and installing the mechanical systems used in the building. Each trade worker has specialized training and experience in a particular area to develop a certain set of skills. The following sections discuss the roles and responsibilities that exist within each construction trade.

Goodheart-Willcox Publisher

Figure 2.1-2. This laborer is helping pour concrete footings for a new home.

Equipment Operator

An *operating engineer* is a person who is qualified to operate different kinds of heavy equipment. Operators of heavy equipment on construction sites are responsible for such tasks as excavation and the movement of building materials around the site, **Figure 2.1-3**.

They operate equipment such as bulldozers, graders, lifts, loaders, excavators, and cranes. Heavy equipment is typically controlled through levers, foot pedals, switches, and dials. More complex equipment makes use of computerized controls. In addition to operating equipment, operating engineers are often responsible for setting up and inspecting equipment, maintaining and adjusting machines, and sometimes performing minor repairs.

Working conditions can be harsh; construction equipment operators are exposed to all types of weather conditions—heat, cold, rain, wind, snow, and sleet. Working conditions can also be noisy, and many types of equipment subject their operators to vibration and jarring impacts as they operate on rough ground. However, these conditions have improved over recent years and continue to do so as newer, better, and safer equipment becomes available.

Mason

A *mason* is a worker who builds walls, floors, walkways, or other structures from brick, concrete block, or stone, **Figure 2.1-4**. They often are referred

Goodheart-Willcox Publisher

Figure 2.1-3. An equipment operator installs the base material for a new driveway.

Goodheart-Willcox Publisher

Figure 2.1-4. Masons work to construct a concrete masonry unit (CMU) foundation wall for a new home.

to by the type of material they work with: brick masons, block masons, or stonemasons. Cement masons and concrete finishers work to prepare for and place concrete.

Stonemasons construct walls out of stone, set stone for veneer walls on buildings, and lay stone floors. They work with natural stone, such as marble, granite, and limestone, and artificial stone made with concrete, marble chips, and other masonry materials. Laborers may be employed to set up scaffolding and prepare and supply materials to the mason.

Masons must be able to accurately read and follow construction drawings. They make extensive use of measuring skills. Masons must be able to perform strenuous work that includes heavy lifting in a variety of weather conditions. Another required skill is efficient and safe use of a variety of hand and power tools.

Most masons acquire their skills by learning from experienced workers through informal on-the-job training. Some graduate from technical school programs. A small number—usually those working for large nonresidential contractors—complete a formal apprenticeship that is typically three years in length. Masons can advance to supervisory positions or start their own businesses. Opportunities also exist for masons in related fields, such as building inspection and construction management.

Carpenter

The scope of a carpenter's work can be vast and varied. Traditionally, carpenters have been known for specializing in working with wood. Today, carpenters work with many different types of materials and may specialize in any number of disciplines.

A ***framing carpenter*** is a carpenter who constructs the structure of a home, **Figure 2.1-5**. They fabricate the floor structures, wall sections, and roof framing of the building. This kind of work is rewarding for those who get a sense of achievement from seeing a physical product result from their efforts. Framing carpentry is strenuous work though, involving frequent lifting, carrying, climbing, standing, or kneeling, as well as working in situations where the danger of injury from falls or other accidents is present. Most of the work is done outdoors, sometimes in rainy, dusty, cold, or hot weather conditions. Wearing appropriate clothing and proper personal protective gear is important.

Trim carpenters perform most of their tasks indoors. A ***trim carpenter*** is responsible for the finishing touches on a job that may include building or installing fine cabinetry, trim, and paneling in a home, **Figure 2.1-6**. Trim carpentry is a trade that requires patience, accuracy, and a high attention to detail.

In residential construction, carpenters who do not specialize in a certain area must be familiar with the properties and installation of a broad range of materials. General contracting and remodeling companies employ carpenters to engage in nearly every phase of home construction. Aside from framing

TFoxFoto/Shutterstock.com

Figure 2.1-5. Framing carpenters build the frame of the house, which includes floor, wall, and roof systems. Here a carpenter is installing the top plate of a wall on a new home build.

even persuade them of the importance of safe work practices in their daily operations. Proficiency in writing and in presenting information are important skills for this job. Often, safety managers hold training sessions that range from toolbox talks to formal classroom training, such as OSHA 10-hour or OSHA 30-hour training. Safety managers also must exercise good judgment. Some hazard controls can be expensive and would cost their employer a fortune to enforce. Thinking creatively, a safety manager should look for alternative ways to maintain a safe work environment while keeping costs in check.

Degrees in health and safety are available; however, a degree is not required to enter the health and safety field. Safety managers need to understand the operations of their industry. For this reason, tradespeople are good candidates to transition into these roles. Many professional certifications are available through authorized OSHA training centers or other educational institutions.

The average annual pay for construction safety managers was $76,980 in 2022. Employment growth of construction safety managers is expected to be about the same as that of the skilled trades.

Inspector

Many different kinds of inspectors work in the construction industry. These professionals can be grouped under the general title of ***building inspector*** since they deal with some aspect of the building process, such as electrical, plumbing, or mechanical systems. Building inspectors usually have a background in the building trades. They determine whether buildings are safe for occupation, **Figure 2.1-16**. Building inspectors must have a thorough knowledge of all local building codes and regulations.

Approximately 50% of inspectors are employed by local governments. These inspectors perform the important task of making sure that a building and its systems conform to codes, standards, and local regulations. In addition to the inspectors employed by the government, a significant number of them are employed by engineering or architectural service firms. A growing number are self-employed, primarily in the field of home inspection where they are hired by a buyer or seller to examine a property being offered for sale.

KomootP/Shutterstock.com

Figure 2.1-16. When building a home, inspectors make multiple visits to ensure that construction practices and materials comply with the standards of the building code.

People employed as building inspectors and those working in the various specialties, such as electrical or plumbing inspection, should have a thorough knowledge of all codes and local regulations that the builder must meet. Many tradespeople "retire" to work as building inspectors. Some inspectors have education and training in construction technology or construction management. A degree in engineering or architecture is helpful in advancing to supervisory roles.

The average annual pay for a construction inspector was $70,130 in 2022. Employment growth of construction inspectors is expected to decline slightly over the next decade; however, over 15,000 openings are projected each year, resulting from workers retiring or transferring to other occupations.

LESSON 2.1 Review

Answer the following questions based on information in this lesson.

1. What occupation does *not* require specialized training or an advanced degree? (2.1-1)
   A. Inspector
   B. Laborer
   C. Skilled trades
   D. Architect
2. Who is often responsible for inspecting and operating heavy equipment? (2.1-2)
   A. Foreman
   B. Building inspector
   C. Operating engineer
   D. Superintendent
3. *True or False?* A framing carpenter builds walls, floors, walkways, or other structures from brick, concrete block, or stone. (2.1-2)
4. *True or False?* Plumbers are often among the best-paid members of the building trades. (2.1-2)
5. Who is responsible for writing contracts and purchase orders, preparing long-term schedules for jobs, and tracking overall costs on a jobsite? (2.1-2)
   A. Construction manager
   B. Foreman
   C. Operating engineer
   D. Superintendent
6. What is the superintendent's primary role? (2.1-2)
   A. Understand and demonstrate the principles of how a home is built
   B. Predict what issues are likely to arise in the project
   C. Plan how to overcome difficult and complex problems that arise
   D. Oversee project to ensure milestones and goals are met
7. Architects rely on _____ to determine the limits of specific materials. (2.1-2)
   A. carpenters
   B. construction managers
   C. foremen
   D. engineers

LESSON 2.2

Construction Fundamentals

LEARNING OUTCOMES

After completing this lesson, you will be able to:

2.2-1 Identify the primary types of residential construction.

2.2-2 Differentiate between the three groups of new home construction.

TECHNICAL TERMS

custom home
design-build firm
general contractor (GC)
production home
semi-custom home
spec home
stock plan
subcontractor

Types of Residential Construction

Residential construction includes new home builds, renovations, and additions. An addition is the merging of new construction onto an existing home, **Figure 2.2-1**.

Renovations, remodels, and additions to existing homes present unique challenges because there are no guarantees that the existing home was built correctly from the beginning. Additionally, the true history of a home can be a mystery. Going into a project, the contractor does not know what the home has experienced throughout its existence. Existing homes may have been subjected to a past flood or leak, or there could have been a fire or some other kind of traumatic event in the building's past. The home may conceal unseen termite damage or may have a structural problem that is not exposed until the job begins, **Figure 2.2-2**. Renovations are covered in more detail in *Chapter 9, Finish MEP Systems and Punchlist*.

Goodheart-Willcox Publisher

Figure 2.2-1. Builders engage in construction projects of all sizes. Here additional space is being added to an existing home.

Goodheart-Willcox Publisher

Figure 2.2-2. A remodeling contractor found floor joists of various widths shimmed up precariously, deep inside this crawlspace foundation.

These types of projects often require creative solutions and extensive expertise to determine the best path forward. Contractors who perform this type of work must remain flexible and should prepare their clients for changes in contract agreements, in case an unexpected complication surfaces.

New Home Construction

As with renovations and additions, new home construction can also be challenging for builders. However, this type of construction is somewhat more predictable.

New home construction falls into three basic groups:

- Production home builds
- Custom home builds
- Spec home builds

All new home builds follow a similar path, though there are slight differences between the three groups. Each build will have slightly different characteristics based on the builder.

Production Homes

A ***production home*** is a house that is built as cost-effectively as possible with highly specialized crews who focus on one aspect of the process, such as a crew for framing and a separate crew for roofing and/or siding, another crew for electrical systems, and so on. Production home builders produce a large volume of houses that are built close together in housing developments, **Figure 2.2-3**.

These crews can be employed by a general contractor. A ***general contractor (GC)*** oversees all aspects of construction and supervises the job. The GC is responsible for all the planning and much of the administrative work on a job. Most often, rather than directly employing each type of tradesperson, the GC hires crews of skilled tradespeople called subcontractors, usually referred to as *subs*. A ***subcontractor*** is an independent business hired to do a specific job such as masonry, framing, plumbing, or interior finish work on a project.

VCNW/Shutterstock.com

Figure 2.2-3. Production homes are built quickly and efficiently in housing developments.

Production homes within a development are typically limited to a small number of designs. Often there are multiple homes that are identical or nearly identical within the development. Homes like this are sometimes referred to as *tract homes*. Because of these similarities, builders can benefit from volume pricing discounts and reduced shipping costs when purchasing materials. When there are repeat builds of a design, material requirements are already known. This greatly reduces the cost of estimating quantities for the next build. Keeping construction of subsequent home builds in the same vicinity helps ensure crews can move quickly from one home to another without scheduling gaps between projects. Because crews have steady work in one area, they do not have costs associated with packing up their equipment and moving to another site. Additionally, because the homes are all owned by one developer, subs are able to utilize waste and extra materials on the next home build.

Production home building requires efficient and precise logistical control over all aspects of construction. Delay in material delivery or failure of a subcontractor to finish a job on time can cause major delays. A large production builder will schedule subcontractors to keep progress moving along as quickly as possible. This might mean several crews working on the home at the same time. In fact, it is common for crews to move from one house to the next, making their way through the housing development quickly and efficiently. The result is multiple homes being built on the same street, each in a slightly different phase of construction.

In summary, this type of home building keeps down costs and keeps up the volume of homes built by relying on strong management and logistical systems.

LESSON 2.3

The Construction Path

LEARNING OUTCOMES

After completing this lesson, you will be able to:

2.3-1 Describe the types of foundation systems used in residential construction.

2.3-2 Explain the sequence of construction when preparing and constructing the foundation of a home.

TECHNICAL TERMS

backfilling
crawl space foundation
excavation
footing
foundation
grade beam foundation
monolithic slab-on-grade foundation
slab-on-grade foundation
spread footing
stem wall foundation

Excavation

Building the footings and foundation of a home is an exciting time in a project. They are the first permanent pieces of the home to take shape. As these pieces develop, a sense of the size and shape of the building becomes more tangible, **Figure 2-3.1**. It becomes easier to grasp how the home will fit into the site and how it will interact with the landscape. However, before any of that can be accomplished, the site needs to be excavated. An ***excavation*** is any removal of earth resulting in a cavity, trench, or depression.

Goodheart-Willcox Publisher

Figure 2.3-1. With the footings formed and poured for a new addition to this home, a sense of space starts to develop.

Recall from *Chapter 1, Preparing for Construction*, that building sites with slopes or rugged terrain should be rough graded before the building is laid out. Topsoil should be removed and piled where it will not interfere with construction. This soil can be used to establish the finished grade after the building is completed.

Safety Note

When working around heavy equipment, never assume the operator knows where you are. Do not stand near the equipment or approach it from behind. Always make eye contact with the operator and wait for a signal from them before approaching a piece of equipment.

With batter boards in place, the outline of the excavation is marked with marking paint. The string lines are then removed from the batter boards so they will not interfere with the excavation.

For regular basement foundations, the excavation should extend at least 2′ beyond the building lines to allow clearance for formwork. Foundations for structures with a slab floor or crawl space need little excavating beyond the trench for footings and walls.

The depth of the excavation is calculated from a study of the blueprints. In cold climates, it is important to locate foundations below the frost line. If footings are set too shallow or are not insulated, moisture in the soil under the footing may freeze. This could force the foundation wall upward, a condition called *heaving*, resulting in cracks and serious damage. Local building codes usually cover these requirements.

Most architects and many local building codes require more than 4″ between the finished grade and the top of the foundation wall. At this height, wood framing and finish members are protected from moisture. The finished grade should slope away from all sides of the structure so surface water drains away from the foundation. These are all factors that help determine the height (or elevation) of the foundation.

Code Note

The *International Residential Code (IRC)* requires the foundation to extend at least 4″ above the finished grade where masonry veneer such as brick or stone is used, and at least 6″ on other wall types.

The depth of the excavation may also be affected by the elevation of the site. It may be higher or lower than the street or adjacent property. The level of sewer lines also has an effect. Normally, solving these problems is the responsibility of the architect. Information on grade, foundation, and floor levels is usually included in the construction drawings.

When the proper building elevations have been established, measurements are taken using a leveling instrument such as a builder's level or rotary laser to ensure the proper depth of the excavation. This process is covered in more detail in *Lesson 2.5*.

The method and extent of the excavation is determined by the type of foundation system being used. After the excavation is complete, footing forms are placed and construction can begin.

Safety Note

Follow these safety rules whenever the project involves foundation excavation for a house.

- Contact local utility companies to determine the locations of any underground utility lines. The phone number 811 has been established to check locations of utilities before excavation begins.
- If soil or moisture conditions make the sides of the excavation potentially unstable, provide adequate shoring. Monitor and assess soil conditions regularly.
- Pile all excavated material a distance at least equal to the height of the pile outside the excavation and be sure it rests at a stable angle.
- When necessary, pump water seepage or rainwater accumulation out of the excavation.
- Provide an adequate number of ladders or other means of safe exit from the excavation.

Footing and Foundation Systems

The ***foundation*** of a house is the part of the building that maintains contact with the ground, **Figure 2.3-2**. Some homes have nearly all their foundation underground while others have a foundation that is partially or mostly exposed. Foundation walls rest on a footing, **Figure 2.3-3**. The ***footing*** must support the weight of the entire structure. The

Goodheart-Willcox Publisher

Figure 2.3-2. This foundation is designed with a walk-out basement. The foundation walls will be fully exposed at the door opening. The remainder of the foundation will be partially exposed.

Goodheart-Willcox Publisher

Figure 2.3-3. The weight of the entire building is transmitted through the foundation walls to the footing.

foundation wall transmits its own weight and the weight of the rest of the home to the footing. The ground beneath the footing must be solid enough to resist compression and settling. Footings must be wide enough to spread the load over a sufficient area. Load-bearing capacities of different types of soils vary considerably. Soils with a lower load bearing capacity require wider and/or thicker footings.

There are many types of footing and foundation systems that can be used when building a house. The combination chosen by the designer or builder depends on several factors such as climate, soil type, the needs of the owner, and the end use of the building. Common foundation types include crawl space, basement, stem wall and slab-on-grade, monolithic slab-on-grade, and pier and grade beam, **Figure 2.3-4**.

Footing requirements vary from one foundation system to another. Some walls, like precast concrete walls, **Figure 2.3-5**, use an integrated footing as part of the wall system. These walls are placed directly on crushed stone that has been carefully leveled and graded. Precast walls are discussed in greater detail in *Lesson 2.9*.

For light construction, such as residential construction, a ***spread footing*** is a common type of footing that transmits the load from the walls, columns, or piers to the soil beneath. This type of footing is often used for basement, stem wall, and crawl space foundations.

A ***crawl space foundation*** is built like a basement foundation with the principal difference being that the foundation walls are not as tall, usually only a few feet high. This saves on construction and excavation costs, but results in less usable space for the home.

The ***slab-on-grade foundation***, also known as a *ground-supported slab*, is placed directly on the ground. The ***stem wall foundation*** is a type of slab-on-grade that also has a foundation wall and a footing placed below the frost line. The foundation wall transmits loads from the building to the footing. A stem wall foundation can be less expensive than a basement or crawl space foundation but eliminates access under the floor.

Monolithic slab-on-grade foundations are popular in warm climates because freezing of the ground beneath the footing is not a problem as it is in colder climates. A ***monolithic slab-on-grade foundation*** is a single, continuous concrete slab that serves as both the floor structure and the foundation of the building. Rather than using a typical spread footing, this foundation integrates the footing into the slab floor. However, building codes do make allowances for monolithic slab-on-grade foundations in cold climates when the footing is protected from freezing.

Code Note

According to the *IRC*, footings are able to be constructed above the frost line where protected from frost by foam insulation board according to requirements dictated by Section R403.3.

The ***grade beam foundation*** uses a concrete grade beam that spans a series of piers and serves as a footing above grade. The grade beam rests horizontally on concrete or wood piers extending into the soil, sometimes reaching down to bedrock. A machine-operated auger can be used to dig the

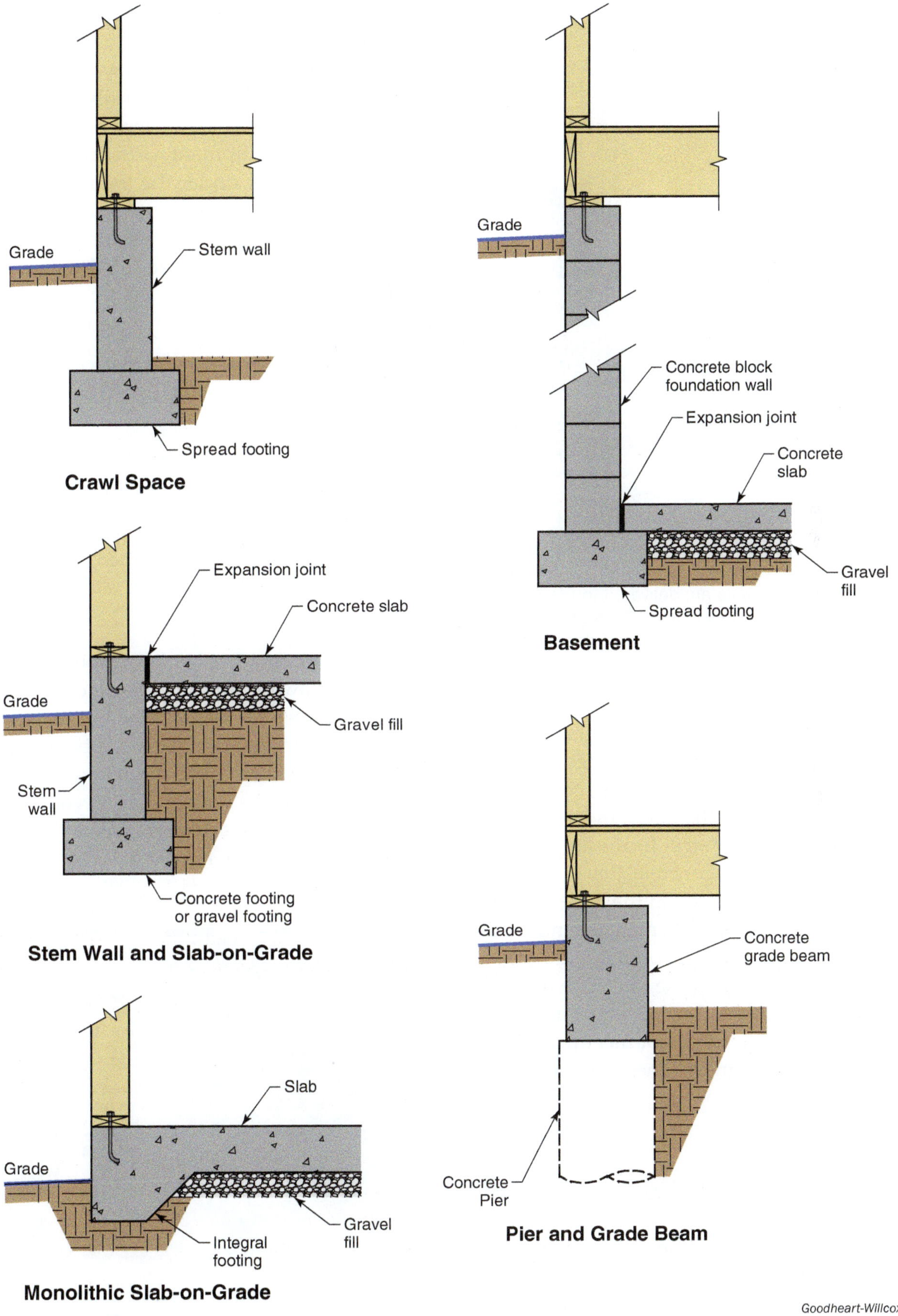

Goodheart-Willcox Publisher

Figure 2.3-4. Footing and foundation types vary by application.

Superior Walls

Figure 2.3-5. Precast walls are delivered to the site and placed on a compacted gravel base. The footings are integrated into the wall, and the weight of the building is displaced through the gravel.

holes for the piers. Alternatively, piers are sometimes driven into the soil by a large piece of equipment called a pile-driver. In firm soil, concrete can be poured directly into the holes. In less stable or soft soil, round fiber or steel forms are placed in the holes and filled with concrete. Rebar, discussed more in *Lesson 2.4*, is placed in the holes or forms before the pour. This reinforcement extends above the ground and ties into other rebar laid horizontally in the grade beam. These foundations can be expensive to construct and are generally used when other foundation types will not meet the requirements for the site conditions. These conditions may include flood risk, erosion, or soil with poor load-bearing qualities.

Building the Footings and Walls

After the excavation for the foundation has been completed, spread footings are typically formed with framing lumber. The footing forms are positioned relative to the walls or columns they will support. Builders use batter boards and strings to reset the building line. Recall from *Chapter 1, Preparing for Construction*, that the building line gives the builders a visual representation of where the foundation walls will be built. Batter boards should be carefully checked after excavation to ensure they were not disturbed by the excavation equipment. Crushed stone may be required as a base under the footings. The formwork is placed according to the construction drawings. If reinforcing steel is required, it is placed in the forms before they are filled with concrete, **Figure 2.3-6**.

Pro Tip

Building inspectors usually require the builder to make appointments for inspections at specified stages of construction. For example, most building permits require an inspection of the footing forms and reinforcement before any concrete is placed. However, inspectors may do unscheduled inspections as well and show up without notice.

After the concrete has hardened, footing forms can be stripped away, building lines are reestablished using the batter boards, and the building lines are marked on the footings. This can be done with a plumb bob or point laser, discussed in *Lesson 2.5*, and a chalk line. The color of chalk in your chalk line, or chalk box, denotes its permanency. When snapping lines on a footer, red chalk is recommended.

Accufooting

Figure 2.3-6. Specialty spreaders are used for this footing form. They hold the form boards the correct distance apart and support the reinforcing steel as the concrete is poured.

Foundation walls must be built strong and securely bonded to the footing to withstand natural forces caused by water, frost, and expansive soils. The building codes give guidance on how builders can meet minimum requirements for specific conditions such as differing climate or seismic zones. Foundation walls can be built from poured concrete, concrete masonry units (CMUs), insulated concrete forms (ICFs), precast concrete walls, or even treated wood, **Figure 2.3-7**. ICFs are discussed in greater detail in *Lesson 2.9*.

Perforated drainage pipes encased in gravel are installed around the footings to evacuate water from around the foundation, **Figure 2.3-8**. Before backfill (excavated soil or gravel) is placed around the outside of the foundation, walls are coated with a water-resistant layer to keep the

Code Note

Section R406 of the *IRC* specifies the conditions that require either dampproofing or waterproofing. Any concrete or masonry foundation walls "that retain earth and enclose interior spaces and floors below grade shall be dampproofed from the top of the footing to the finished grade." The *IRC* then provides a list of the permissible materials, which include bituminous coating and acrylic-modified cement. Waterproofing is only required by the *IRC* "in areas where a high water table or other severe soil-water conditions are known to exist."

Pro Tip

Unbalanced backfilling of foundation walls, such as occurs with a basement foundation, is not completed until the first-floor framing is finished. This is because the framing reinforces the top of the foundation wall. That reinforcement keeps the walls straight and able to resist unequal pressure created by the weight of the backfill, **Figure 2.3-9**.

A Northlight/Shutterstock.com B Miller and Richard Masonry C Goodheart-Willcox Publisher

D Superior Walls E Courtesy of the Southern Forest Products Association

Figure 2.3-7. Foundation walls can be built using several different methods. A—Poured concrete. B—Concrete masonry units (CMUs). C—Insulated concrete forms. D—Precast walls. E—Wooden foundation.

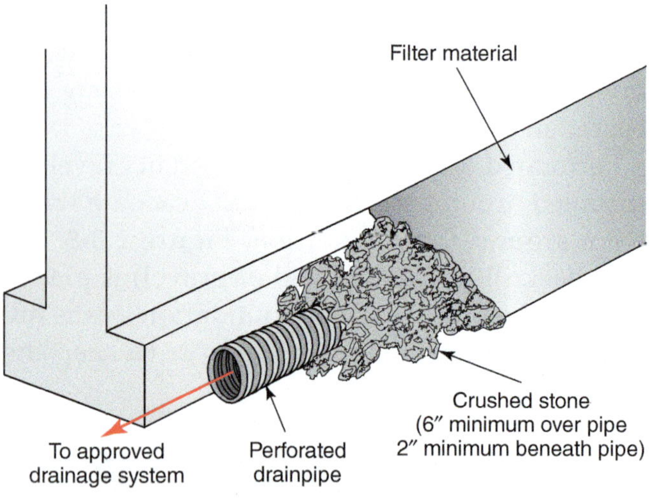

Figure 2.3-8. A perforated drainpipe allows a path for water to freely drain away from the wall and footing.

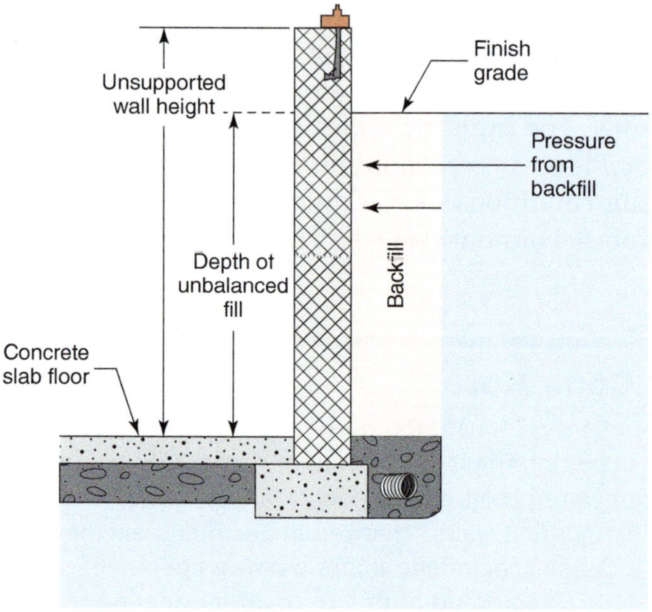

Figure 2.3-9. Unbalanced backfilling of basement foundation walls.

inside of the foundation dry. ***Backfilling*** refers to the process of replacing excavated earth against a basement foundation after waterproofing has been completed. Additional drainpipes that will be used to carry water runoff from the roof away from the home are often installed at this stage. These pipes, along with the footing drainpipes, are connected to a municipal storm drain or other drain termination.

LESSON 2.3 Review

Answer the following questions based on information in this lesson.

1. What type of footing is often used for basement, stem wall, and crawl space foundations? (2.3-1)
   A. Spread footing
   B. Integral footing
   C. Gravel footing
   D. All are correct.
2. What foundation system would be used where soil has poor load-bearing qualities? (2.3-1)
   A. Crawl space foundation
   B. Slab-on-grade foundation
   C. Grade beam foundation
   D. Monolithic slab-on-grade foundation
3. *True or False?* A monolithic slab-on-grade foundation incorporates an integrated footing. (2.3-1)
4. *True or False?* Crushed stone should be avoided as a base under footings due to poor load-bearing qualities. (2.3-1)
5. Footing forms are placed _____ excavation. (2.3-2)
   A. before
   B. during
   C. after

LESSON 2.4

Materials and Fasteners

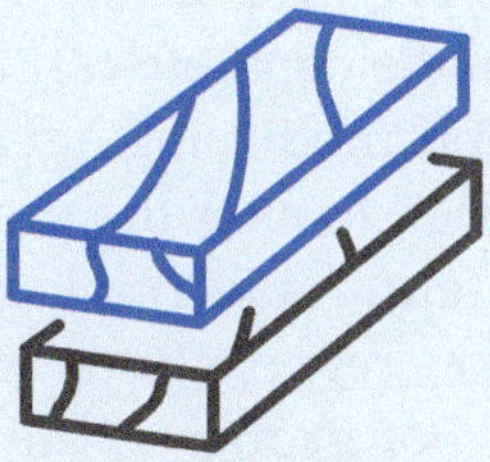

LEARNING OUTCOMES

After completing this lesson, you will be able to:

2.4-1 Identify materials used in the construction of a home's foundation.

2.4-2 Describe bolts used for structural connections when building a house.

TECHNICAL TERMS

admixture
aggregate
anchor bolt
compressive strength
concrete
concrete masonry unit (CMU)
mortar
rebar
tensile strength

Foundation Materials

This lesson focuses on construction materials and fasteners that relate to the foundation system on a new home build. In most cases, these materials and fasteners may also be used in other areas of home construction.

Concrete

Concrete is an excellent building material for footings and foundation walls. It is also used for basements, garage floors, sidewalks, and patios. Concrete is the most abundant building material on Earth because of its extensive use in commercial construction, especially civil construction. Bridges, dams, parking garages, airports, power plants, and roads are constructed using a tremendous amount of concrete.

Concrete is a unique building material because it begins as a flowable, formable material, **Figure 2.4-1**, then hardens through a chemical reaction to form a strong, durable product that can withstand tremendous loads. It is created by mixing measured amounts of cement, aggregate, and water together. *Aggregate* is a mixture of sand and gravel or crushed stone. The curing process for concrete can continue for years. When cured properly, concrete reaches approximately 70% of its full strength after 7 days. For construction purposes, concrete is considered fully cured after 28 days.

Specify Concrete

Figure 2.4-1. After it is mixed, concrete is poured into formwork and finished with a smooth or decorative finish.

Safety Note

The chemical reaction that causes concrete to cure generates heat. If concrete is left in contact with bare skin for an extended time, it can cause serious chemical burns. Therefore, always wear safety glasses, rubber boots, and gloves when working with concrete.

The proportions of the raw ingredients used to make concrete can impact its strength. Additives and modifying agents, called ***admixtures***, can be used to alter the properties or appearance of concrete. As technology improves, applications for concrete continue to advance.

For example, buildings are being constructed with lightweight concrete, and there are even concrete blocks and panels that allow light to pass through, **Figure 2.4-2A**. To avoid or minimize flooding, parking lots can be made of pervious concrete, which allows storm water to pass through instead of pooling at the surface, **Figure 2.4-2B**. Decorative concrete is used to build countertops, tables, and benches. Techniques are used to expose decorative aggregates or acid-stain the surface, **Figure 2.4-2C**. It is common to see patios, porches, and driveways made from concrete that has been dyed and stamped to resemble wood, stone, or brick, **Figure 2.4-2D**.

Green Note

Concrete is a versatile material that continues to prove its usefulness in construction. However, the amount of energy used to manufacture cement, and the binding material used in producing concrete, is a concern. In fact, other than water, concrete is the most utilized material on Earth. Cement production accounts for as much as 7%–8% of global emissions. Great efforts are underway to find materials and methods that make the production and use of concrete more carbon neutral.

A Specify Concrete B shaylahwood/Shutterstock.com

C Goodheart-Willcox Publisher D Miller and Richard Masonry

Figure 2.4-2. Concrete can be both functional and beautiful. A—New developments in concrete allow for applications where light can filter through the material. B—Water can run through this pervious concrete freely, helping to manage storm water runoff for parking lots and roads. C—This countertop was polished to expose the aggregate, then acid-stained. D—This stamped concrete patio was colored to look more like natural stone.

Concrete Reinforcement

By itself, concrete is rather brittle as a building material. Imagine a plank of concrete shaped like an 8-foot-long 2×4. This plank would break easily compared to a wooden 2×4, because concrete has lower tensile strength than wood. ***Tensile strength*** is a material's ability to resist fracture when being pulled apart. However, concrete has high compressive strength. ***Compressive strength*** is a material's ability to resist deforming when subjected to forces of compression. Because of its high compressive strength concrete is can withstand a tremendous amount of weight, which is a primary reason it is such a great material for footings and foundation walls.

Green Note

Fly ash is produced when coal is burned to produce electricity. This by-product can be used in concrete as a replacement for part of Portland cement. Environmental benefits of reusing fly ash to replace Portland cement include conservation of resources and energy, reduction of greenhouse gas emissions, and reduced costs. In addition, fly ash improves the workability of the fresh concrete; requires less water, which reduces cracking; and makes the final product more durable because permeability is reduced.

To counter concrete's low tensile strength, builders add a reinforcing material that has high tensile strength. For instance, when placing footings, the strength characteristics of concrete can be improved by installing rebar in the forms before concrete is poured, **Figure 2.4-3**. ***Rebar***, short for "reinforcing bar," is a steel bar commonly used to reinforce concrete against tensile forces. The size of rebar is designated by the bar number. For numbers 3–8, each number represents 1/8″ in diameter. So, a No. 3 bar is 3/8″ and a No. 8 bar is 1″. Bars that are larger than 1″ do not follow the same sizing system. It is rare to see anything larger than a No. 5 bar used on residential jobs. When concrete must resist compression forces only, reinforcement is usually not necessary.

Goodheart-Willcox Publisher

Figure 2.4-3. Rebar is tied together with steel tie wire. It should be held off the earth at the bottom by rebar chairs.

Slabs for floors or similar applications are often reinforced with welded wire fabric, **Figure 2.4-4**. The concrete can also be reinforced with high-strength fibers added to the mix, **Figure 2.4-5**. These fibers yield nearly as much tensile strength as welded wire fabric and allow builders to save time and money.

Safety Note

To avoid possible injury caused by a worker falling onto vertical rebar, plastic or wooden caps should be placed on the ends. An alternative is to bend the end at a 90° angle.

Code Note

Section R404 of the *IRC* requires vertical reinforcement in most CMU walls. Common practice is to fill cores with masonry grout and rebar every 48″, 56″, or 72″, depending on the width of the wall and the height of the backfill. Some walls require even more reinforcement. Always check local codes to ensure compliance.

Goodheart-Willcox Publisher

Figure 2.4-4. Welded wire fabric helps keep concrete slabs from cracking. It is pulled up toward the center of the slab as the concrete is placed.

Goodheart-Willcox Publisher

Figure 2.4-5. Fibers to reinforce concrete may be glass, polypropylene, or nylon.

Concrete Masonry Units (CMU)

A *concrete masonry unit (CMU)*, also known as a *CMU block* or *block*, is often used to build foundation walls. The standard CMU is made from a mixture of Portland cement and aggregates such as sand, fine gravel, or crushed stone.

CMUs are classified as solid or hollow core. Hollow core blocks are lighter and easier to handle than solid blocks. The core also provides a cavity inside the wall that can be filled with concrete grout and rebar to provide additional strength where required, **Figure 2.4-6**.

Goodheart-Willcox Publisher

Figure 2.4-6. Block cores allow space for grout and steel reinforcement.

Blocks are usually available in 4″, 6″, 8″, 10″, and 12″ widths and 4″ and 8″ heights. **Figure 2.4-7** shows some of the sizes and shapes.

To build a wall, blocks are bonded one to another with mortar, **Figure 2.4-8**. *Mortar* is a mixture of Portland cement, lime, and masonry sand. Water is mixed with these materials to give the mortar its strength, durability, and bonding capabilities. Mortar is also used in bricklaying in the same fashion.

Pro Tip

Block sizes are actually 3/8″ shorter than their nominal (named) dimensions to allow for the mortar joint. For example, an 8×8×16 block is actually 7 5/8″ × 7 5/8″ × 15 5/8″. With a standard 3/8″ mortar joint, the laid-in-the-wall height is 8″ and the length 16″.

Safety Note

CMU blocks are heavy and can pinch and smash fingers when you set them down or move them carelessly. It is a good idea to wear leather work gloves if handling them frequently. Be careful not to overload scaffolding with too much weight from CMU blocks. Use caution when stacking or pulling apart cubes of CMUs, as they can become unstable and fall over.

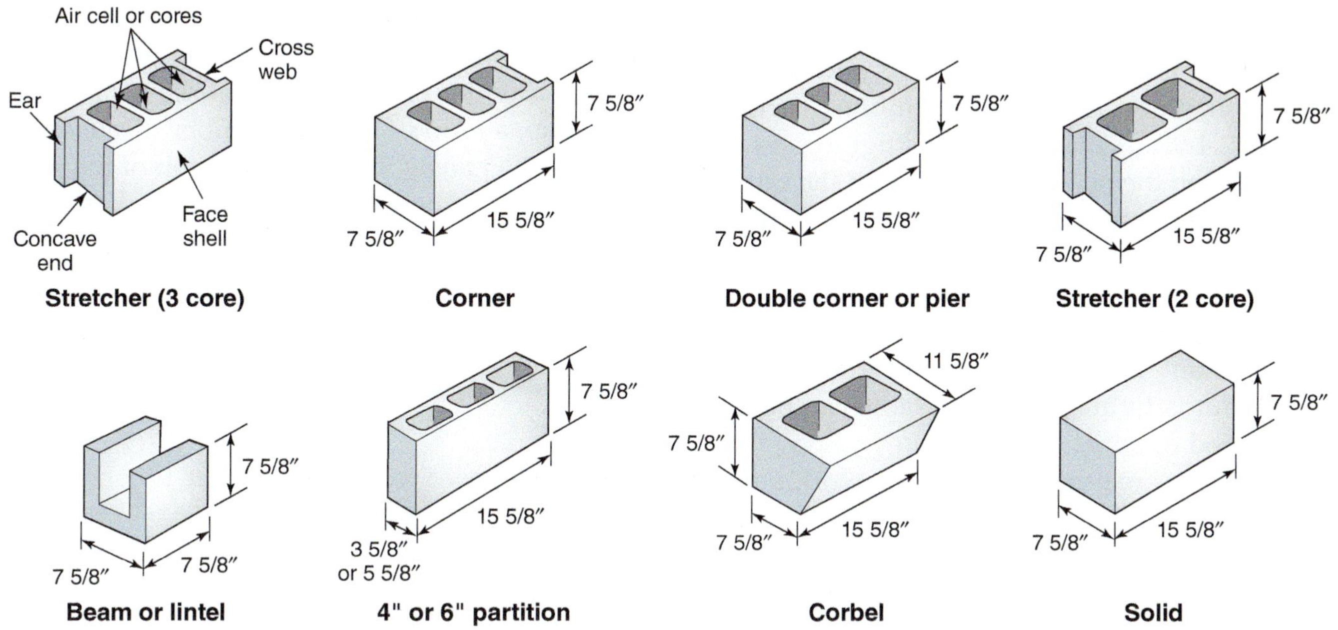

Portland Cement Association

Figure 2.4-7. Concrete blocks are manufactured for many different purposes. These shapes are typical. Sizes are specified in this order: width, height, and length. Sizes are nominal, allowing for mortar joints.

A B Portland Cement Association

Figure 2.4-8. Laying block. A—Mortar is placed on the head joints before the blocks are laid. B—Blocks should be lightly bumped against the previously laid block for a good joint.

Fasteners

In construction, bolts are used primarily for structural connections. Bolts have a head on one end and are threaded to accept a nut on the other end. Flat washers are used between the head or the nut and the material to distribute pressure over a larger area. Bolts have high holding strength and are available in a variety of diameters and lengths. Diameters ranging from 1/4″ to 1″ are commonly used in residential construction. When bolts will be exposed to the elements or used with corrosive materials, such as pressure-treated lumber, they must have a hot-dipped galvanized coating or be made from stainless steel.

While the types of bolts featured in this chapter are not used to construct footings and foundations, with the exception of anchor bolts, it is important to have a general understanding of what types of bolts are available and used for various purposes throughout the construction process.

Types of Bolts

Machine bolts have a hex head and are used to fasten wood or metal assemblies together, **Figure 2.4-9A**. A machine bolt is tightened using a wrench on the head and another on the nut.

The carriage bolt is unique because it only requires one wrench to tighten it. When carriage bolts are installed, the head looks like a dome, **Figure 2.4-9B**. Just under this domed head is a square shank that keeps the head from spinning as the nut is tightened. These bolts can be used with wood or metal.

Stove bolts are used for light-duty work, such as securing cabinets to walls or mounting small appliances. They have threads along the entire length of the shaft, allowing the head and the nut to clinch thin materials together. These bolts are available in slotted, square, or Phillips configurations and have flat or round-shaped heads, **Figure 2.4-9C**.

An ***anchor bolt*** is a special bolt used to fasten the sill plate, a wood framing member discussed in *Chapter 3, Framing Systems*, to the top of the foundation wall. Anchor bolts have a 90° bend or a J-bend on one end and are threaded on the other, **Figure 2.4-10**. The bent end is embedded into wet

Code Note

Building codes require anchor bolts to be at least 1/2″ in diameter, and they must be galvanized to prevent corrosion. They must also be embedded in a minimum of 8″ of concrete and be spaced no more than 6′ apart.

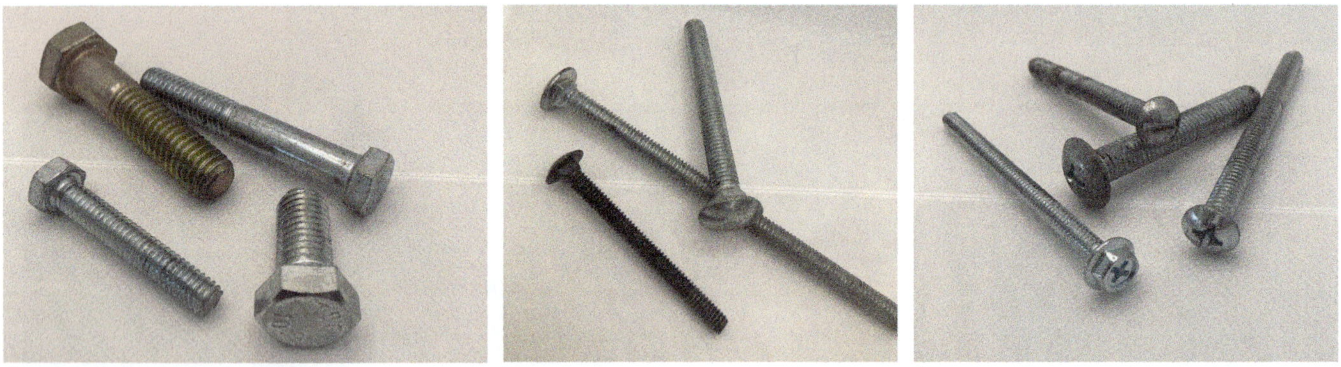

A B C Goodheart-Willcox Publisher

Figure 2.4-9. A—Machine bolts are used in conjunction with washers and nuts to fasten wood or metal pieces together. B—When installed, only the domed head of a carriage bolt remains visible, offering a low-profile alternative to machine bolts. C—Stove bolts are used for light-duty work.

A B Goodheart-Willcox Publisher

Figure 2.4-10. A—Anchor bolts or a suitable alternative are required at the top of the foundation wall to secure the sill plate to the foundation. B—The bend on an anchor bolt must be embedded in concrete or grout.

concrete or concrete grout at the top of the wall, leaving the threaded end sticking out of the top of the wall. When the wood framing begins, the sill plate is predrilled so the bolts can pass through it. Washers and nuts are added to secure the sill plate to the wall.

LESSON 2.4 Review

Answer the following questions based on information in this lesson.

1. For construction purposes, how long does it take for concrete to be considered fully cured? (2.4-1)
   A. 7 days
   B. 14 days
   C. 21 days
   D. 28 days
2. *True or False?* Wood is the most abundant building material on Earth. (2.4-1)
3. *True or False?* Compression strength is a material's ability to resist fracture when being pulled apart. (2.4-1)
4. To build a foundation wall, blocks are bonded one to another with _____. (2.4-1)
   A. admixture
   B. mortar
   C. anchor bolts
   D. machine bolts
5. What bolts are used to fasten a wooden framing member to the foundation wall? (2.4-2)
   A. Machine bolts
   B. Carriage bolts
   C. Stove bolts
   D. Anchor bolts

LESSON 2.5

Tools

LEARNING OUTCOMES

After completing this lesson, you will be able to:

2.5-1 Identify leveling, layout, concrete, and masonry tools and equipment used during the construction of a home's foundation.

TECHNICAL TERMS

angle grinder	concrete placer	line of sight
auto level	concrete saw	mason's trowel
builder's level	finish trowel	mortar hoe
bull float	jointer	point laser
chalk line	laser level	screed
concrete broom	leveling instrument	subgrade
concrete edger	leveling rod	wheelbarrow
concrete float	line laser	

Leveling and Layout Tools

Tools introduced in this lesson are commonly used when excavating the foundation, building forms for footings, pouring concrete, and building foundation walls.

A ***leveling instrument***, such as a builder's level, auto level, or laser level, is an instrument used for determining and identifying differences in elevation. In construction work, these instruments are used extensively when establishing critical elevations for the foundation, such as the depth of the excavation and the top of the foundation wall. Regardless of the type of leveling instrument, it is important to handle the tool with care. Levels are sensitive pieces of equipment that are highly calibrated. A tool knocked out of calibration will result in inaccurate readings, which could lead to major issues with construction of the building.

Pro Tip

Take your time when setting up and using leveling instruments. A mistake made establishing a permanent elevation of the structure or finish grade could result in a serious error that could impact the rest of the job.

Builder's and Auto Levels

The ***builder's level***, also called a *dumpy level* or *optical level*, consists of an accurate spirit level and a telescope assembly attached to a circular base that swivels 360°, **Figure 2.5-1**. Leveling screws are

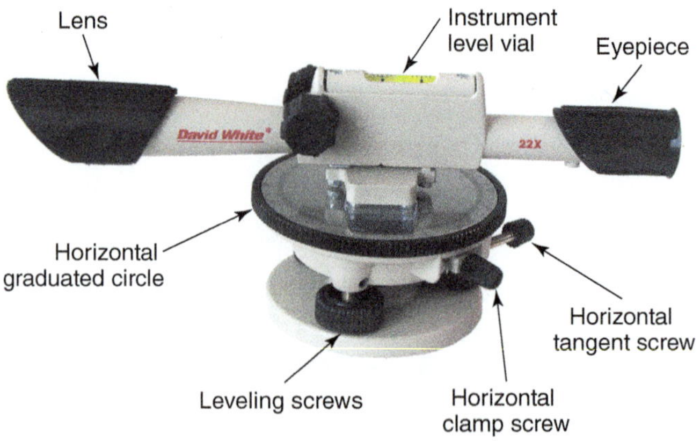

David White

Figure 2.5-1. A builder's level is used to measure elevations.

used to adjust the base after the instrument has been mounted on a tripod. The telescope is fixed so that it does not move up or down, but it rotates on the base.

The ***auto level***, **Figure 2.5-2**, is similar to the builder's level, except that it contains an internal compensator that eliminates any minor variant from level. Most auto levels use a circular bull's-eye level, **Figure 2.5-3**, rather than a cylindrical leveling vial. The bull's-eye level is not as precise as the cylindrical level. However, the internal compensator ensures accuracy. This feature makes auto levels quicker and easier to set up than builder's levels. Auto levels are also less sensitive to light bumps than builder's levels. For these reasons, auto levels have mostly replaced builder's levels on construction sites.

Builder's levels and auto levels are telescopes that are mounted on a tripod and leveled on a base that can be rotated. These tools take two people to operate, one to look through the scope to take a reading, and another to hold a measuring device. The optical device of these instruments operates on the basic principle that a ***line of sight*** is a straight line that does not dip, sag, or curve, **Figure 2.5-4**. When the line of sight is level, any point along that line of sight will be at the same height, regardless of how far it is away from the instrument. The rotating base allows the user to view objects through the scope, along the line of sight for a full 360°.

Using these leveling instruments along with a measuring device such as a leveling rod, the builder can determine a difference in elevation of two objects that are far from each other. This capability is useful for determining critical grades, such as the depth the equipment operator must dig to place the footings, or when grading the gravel base, called the ***subgrade***, for the concrete floor in the basement. The builder is likely to use a leveling instrument to establish the correct height for the footing forms. A mason uses a leveling instrument to ensure the block wall remains level as it is being built.

Photos courtesy of Robert Bosch Tool Corporations®

Figure 2.5-2. The auto level performs the same functions as a builder's level, but it is faster and easier to set up.

Goodheart-Willcox Publisher

Figure 2.5-3. A—Auto levels have a circular bull's-eye leveling vial, rather than a cylindrical leveling vial. During setup, the user centers the bubble in the ring by adjusting the leveling screws. B—Most auto levels are equipped with a mirror to easily check the instrument for level from the normal user position.

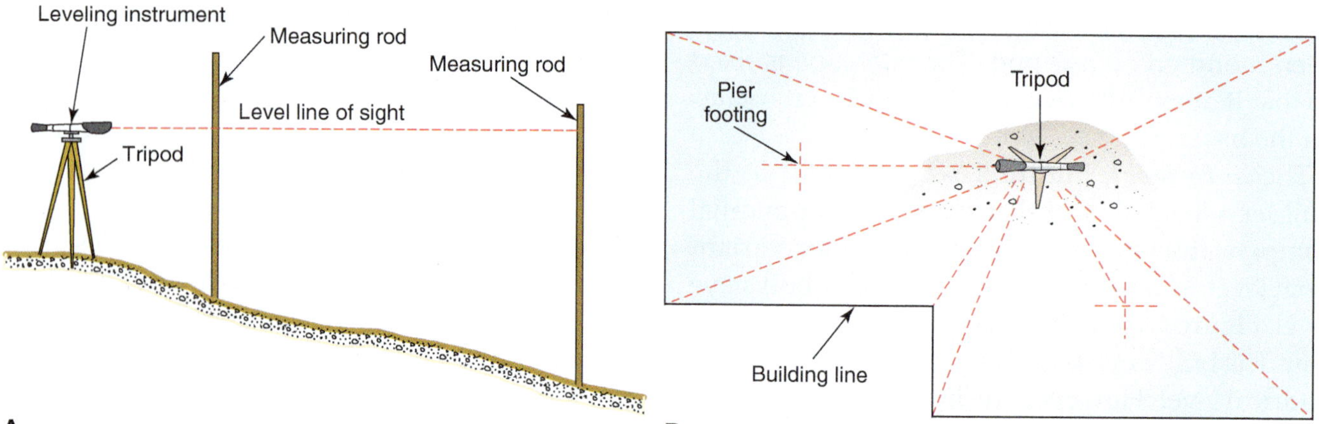

A B

Goodheart-Willcox Publisher

Figure 2.5-4. Elevation changes can be measured with a leveling instrument. A—The line of sight is level over uneven ground levels. B—Regardless of which direction the instrument is pointed, the line of sight remains level.

Leveling Rod

A ***leveling rod*** is used to measure vertical distances from a horizontal sight line when operating a leveling instrument like a laser level, auto level, or builder's level. It is a long rod marked with numbered graduations. It allows differences in elevation between the instrument height of the level and various positions where the rod is held to be easily read. The person operating the level can make the readings, or the target can be adjusted up and down to the line of sight and then the person holding the rod (rod holder) can make the reading, **Figure 2.5-5**. The reading denotes the difference between the horizontal line of sight projected by the leveling instrument (instrument height) and the object on which the leveling rod rests.

A leveling rod with graduations in feet and inches is used for regular surveying work. Rods are also available with graduations in feet and decimal parts of a foot. The process of reading a leveling rod is discussed in *Lesson 2.7*.

A B

David White

Figure 2.5-5. Leveling rod with target. The target can be attached to the rod and moved up or down to match the line of sight of the instrument.

Laser Levels

A laser level is a popular choice for leveling jobs, because one worker acting alone can perform most leveling and plumbing operations with this instrument. A laser transmitter and its receiver are shown in **Figure 2.5-6**. The ***laser level*** emits a beam of light in a 360° circle. It can be adjusted so the projected circle is perfectly level or plumb (perfectly vertical). Some lasers are also capable of projecting on a defined slope. A separate laser detector or target signals when the laser beam strikes it. A laser level

Photos courtesy of Robert Bosch Tool Corporation®

Figure 2.5-6. Laser level transmitters project a beam 360°. The beam can be set to level, plumb, or slope on many lasers. The laser receiver detects the beam and indicates whether it is higher or lower than the mark.

functions much like a builder's level for establishing a level line. During use, the transmitter may be left unattended. The receiver can be tended by a rod holder. Heavy equipment can also be outfitted with laser receivers that can be read in the cab of the equipment. Operators use this information to tell how deep they have dug or how close to the final grade their blade is.

Another type of laser level used in construction is a ***point laser***, **Figure 2.5-7**. Point lasers project stationary beams that can be used for reference. These lasers are available in a variety of configurations. The most basic point lasers project two beams vertically: one beam projects a dot, or a *point*, below the laser, and another projects a dot above. These two dots are perfectly plumb in reference to one another, **Figure 2.5-8**. A 3-beam point laser also projects a beam out the front as a level reference.

Pro Tip

Unlike a plumb bob, a point laser is not impacted by the wind.

A ***line laser*** works much like a point laser but projects lines, rather than points, **Figure 2.5-9**. A line laser is used when a line is preferred to an exact point. This can be helpful when installing products such as ceiling, floor, or wall tile; kitchen cabinetry; and wall paneling.

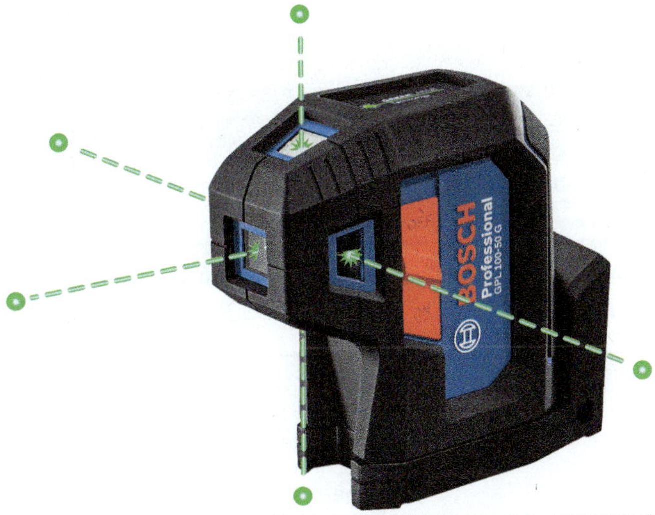

Photos courtesy of Robert Bosch Tool Corporation®

Figure 2.5-7. Point lasers are helpful tools that project a laser point in multiple directions.

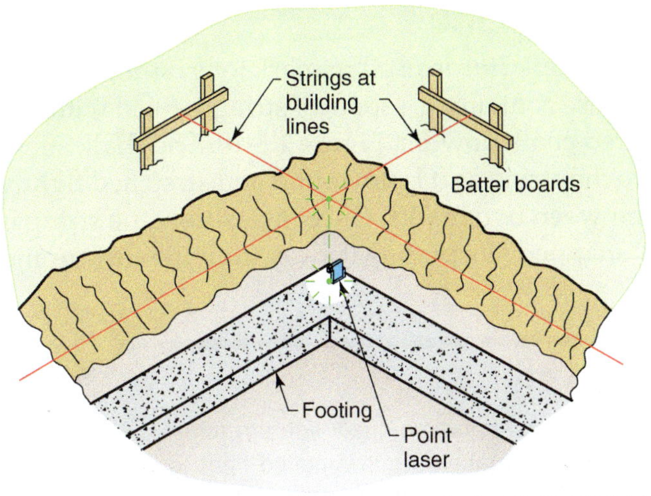

Goodheart-Willcox Publisher

Figure 2.5-8. A point laser is a convenient way to transfer the building lines to the footing. It is not affected by the wind like a plumb bob.

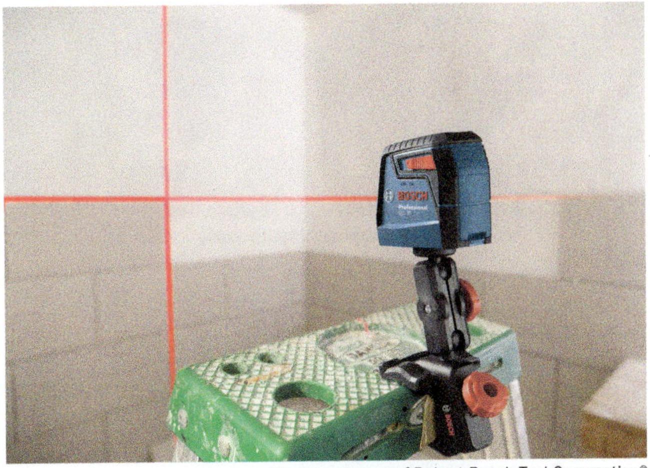

Photos courtesy of Robert Bosch Tool Corporation®

Figure 2.5-9. A line laser projects multiple lines for plumb, level, and square reference marks.

Each of the beams or lines on point and line lasers have a 90° orientation to one another, so they can be used in a variety of ways to plumb, level, and square building assemblies and finishes.

Safety Note

Never look directly into the beam of a laser. For safety, there are regulations limiting power output of a laser. Federal regulations limit Class II construction lasers to 1 milliwatt and Class IIIa units to less than 5 milliwatts. These levels are safe for human eyes as long as the user does not stare into the beam. User directions and cautions provided by the manufacturer should be followed carefully.

Chalk Line

A *chalk line* is used to mark long, straight lines. A case containing a spool of string is filled with colored chalk powder, **Figure 2.5-10**. The chalk sticks to the string as it is unwound and stretched tightly between two marks. Once in position, the string is "snapped" by pulling it away from the surface and releasing it. Snapping the string leaves behind a straight line of chalk on the surface. Chalk is available in an assortment of colors, each with a different level of permanence and visibility, **Figure 2.5-11**.

Pro Tip

When snapping a chalk line stretched over a long distance, make sure it is pulled tight to keep the line from sagging. It is helpful to hold it against the surface at the midpoint, then snap it on each side. This results in a more accurate line.

TFoxFoto/Shutterstock.com

Figure 2.5-10. A chalk line case contains a spool of string and is filled with colored chalk. The line is stretched tight between two marks. When snapped, it leaves a straight line of colored chalk.

Concrete Tools and Equipment

Placing and finishing concrete is a job that requires practice and experience for a quality finished product. A good finish starts with accurate placement, honing an ability to recognize high and low spots in a slab or form before moving on to the next area or form. It takes experience to get the timing right on how quickly or how much concrete is moving down the chute of a concrete truck or coming out of a concrete pump hose, **Figure 2.5-12**. Concrete workers need to understand the characteristics of how concrete moves and flows and how it reacts to various tools and techniques that are applied. Concrete is a time-sensitive material, so decisions must be made quickly. It is critical to finish tasks with proper timing as the chemical reactions take place.

Concrete tools are used in foundation work to place concrete in the footing forms and to finish the surface of concrete slabs, such as basement floors. CMU block walls, formed concrete walls, and insulated concrete forms all require full or partial filling of concrete or masonry grout. Often this work is done from scaffolding, using a concrete pump or conveyor truck, **Figure 2.5-13**. Common tools and equipment used for these processes are

Marking Chalk Colors							
Types							
Standard		**Temporary**	**High Visibility**			**Permanent**	
Colors							
Blue	White	Violet	Orange	Green	Yellow	Black	Red
Uses							
Interior/Exterior		Interior	Exterior			Exterior	
Applications							
Interior layout Roofing Siding		Interior layout Interior trim Cabinet layout	Corrections Plumbing Electrical Roofing Masonry			Exterior layout Concrete layout Roofing Sheet metal fabrication	

Goodheart-Willcox Publisher

Figure 2.5-11. Assorted colors of chalk are used for different applications.

Ungvar/Shutterstock.com

Figure 2.5-12. A concrete pour is a busy time on a jobsite. For a successful pour, workers need to stay aware of hazards and work efficiently as a team.

Plumwall Bracing Systems Inc.

Figure 2.5-13. A site prepared for a concrete pour with a concrete pump truck.

discussed in the following sections. Note that this lesson covers typical tools used on a concrete pour. Additional concrete tools are available but are not discussed in detail here.

Wheelbarrow

A *wheelbarrow* is used to move materials such as sand, gravel, concrete, and mortar around the jobsite, **Figure 2.5-14**. Wheelbarrows can come in a variety of configurations. Most wheelbarrows on construction sites have one wheel attached to two wooden handles with a steel or polymer tub mounted on them. The handles have metal frames that sit on the ground when the wheelbarrow is at rest. When in use, the worker lifts the handles, splitting the weight of the load with the wheel. In this way, workers are able to push heavy loads around the site.

ungvar/Shutterstock.com

Figure 2.5-14. Workers often use wheelbarrows to move concrete and other materials around the jobsite.

Safety Note

It may be tempting to fill a wheelbarrow to the top when moving concrete. Doing so is likely to result in spillage or, worse, an overturned load. Only fill a wheelbarrow with as much as you can safely manage.

Concrete Placer

A *concrete placer* is used to spread and move wet concrete as it is being poured, **Figure 2.5-15**. This tool is similar to a wide hoe, or a rake with no tines. Placers can be made of steel but are also available in lightweight aluminum and stainless steel.

Screed

A *screed* is a straight board used for striking off concrete to the proper and consistent level after it is placed, **Figure 2.5-16**. For instance, a short screed could be used after concrete is poured into footing forms to strike off the concrete evenly with the top of the forms, producing a consistently flat and level surface across the footer. A longer screed would be used to do the same when pouring concrete for a sidewalk. Long screeds are used for floor surfaces. The screed board can be made from a straight piece of lumber, or it can be made of aluminum or magnesium.

MARSHALLTOWN

Figure 2.5-15. Concrete placers are used to spread and move wet concrete.

Goodheart-Willcox Publisher

Figure 2.5-16. A group of students learn how to pour a sidewalk using concrete placers and a screed. The screed is shuffled back and forth across the top of the sidewalk form boards to produce a flat and consistent surface.

Pro Tip

It takes time to become proficient at using concrete tools. When working, it is worth the effort to observe and analyze how the concrete behaves under different techniques. Pushing or pulling the concrete with the screed can leave voids and pits in the surface, while shaking it back and forth leaves a more fluid surface. However, the shaking action also has a leveling effect, which may not be helpful when pouring on a slope.

Concrete Float

A ***concrete float*** is used to help level and smooth concrete surfaces for finish troweling, **Figure 2.5-17**. They range in size from 10″ to 18″ long. The most popular concrete floats are made from magnesium and are often called *mag floats*.

Bull Float

A ***bull float*** is a wide, flat finishing tool that is used to smooth ridges and fill voids left from screeding. The bull float is connected to a long handle that is used to push and pull the tool across a slab of concrete, **Figure 2.5-18**. Bull floats can be outfitted with a swivel bracket that enables the operator to change the pitch of the float by twisting the handle. This helps to keep the float from digging in to the wet concrete as it transitions from the push stroke to the pull stroke, or vice versa. Bull floats are usually 8″–12″ wide. Common lengths are 48″ and 60″.

Concrete Edger

Leaving a rough edge on concrete is acceptable for some work, such as concrete footings. However, when pouring concrete slabs like driveways, sidewalks, and patios, the perimeter of the slab must be shaped to avoid rough, jagged edges. A ***concrete edger*** is a tool used to produce a smooth and consistent radius edge on concrete, **Figure 2.5-19**. The tool is slid against the form along the edge of the slab while the concrete is still workable.

Finish Trowel

A concrete ***finish trowel*** is used to produce a smooth finish on concrete. Finish trowels have a

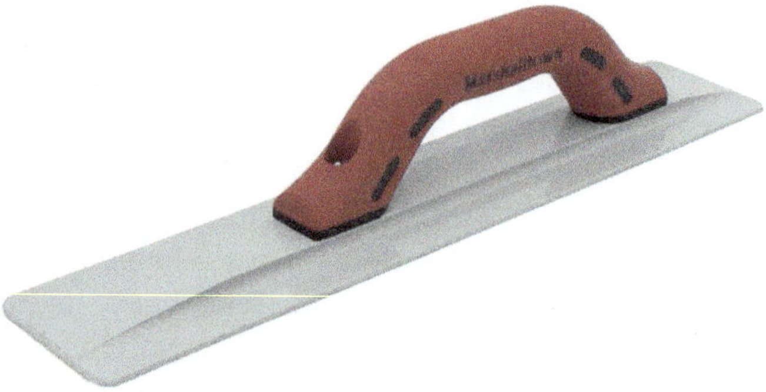

MARSHALLTOWN

Figure 2.5-17. A concrete float is used for the initial smoothing of concrete.

Goodheart-Willcox Publisher

Figure 2.5-18. The bull float smooths irregularities in the concrete after it is poured.

MARSHALLTOWN

Figure 2.5-19. Edgers smooth and shape the perimeter of a concrete slab.

thin, rigid steel blade with a steel shank attached in the center where the handle connects, **Figure 2.5-20.** The shank adds strength and rigidity so the tool can keep consistent pressure across its length. This tool is fanned across the surface of a concrete slab as the last step to achieve a smooth finish.

Pro Tip

When working with concrete, keep tools and formwork as clean as possible. Residual concrete loses moisture and hardens quickly, making it a hassle to deal with. This is particularly true when using an edger. If the formwork has crusted residual concrete all over it, producing a clean edge is more difficult.

Concrete Broom

A smooth surface is not a desirable finish for all concrete slabs. Outdoor surfaces like patios, steps, sidewalks, and driveways can require a more slip-resistant texture on their surface. A *concrete broom* is used to create a rough texture on the surface of concrete, **Figure 2.5-21**. A concrete broom can be attached to a standard broom handle or a bull float handle. It is pulled over the surface of concrete before it hardens in order to produce a consistent and attractive broomed surface texture.

Concrete Saw

A *concrete saw*, also called a *cutoff saw* or *demolition saw*, **Figure 2.5-22**, uses a large circular blade to cut through concrete, CMU block, or steel rebar. These saws are most often powered by a gasoline internal combustion engine but are available in electric, battery, and hydraulically powered models. Gas-powered saws are commonly equipped with a 14" diameter blade.

Saws use blades made for specific purposes. Be sure to use the appropriate blade for the job. Concrete cutting blades, **Figure 2.5-23A**, are rimmed with industrial diamonds that can quickly cut through hard masonry materials. To cut steel, an

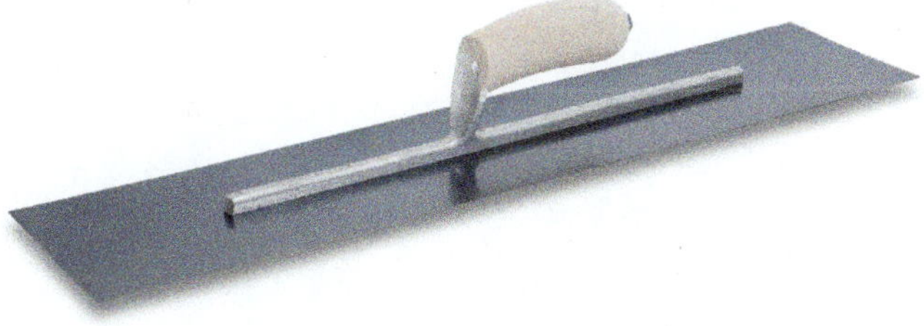

MARSHALLTOWN

Figure 2.5-20. Finish trowels are used for finishing a concrete slab to a smooth surface.

A MARSHALLTOWN

B ungvar/Shutterstock.com

Figure 2.5-21. A—Concrete brooms are available in different widths and bristle stiffnesses. B—Brooming the slab at the right time to produces a rough, lined texture.

abrasive blade is used, **Figure 2.5-23B**. Abrasive blades reduce in size as they are used and can chip easily if mishandled or misused.

Cutting concrete and masonry produces harmful dust, called *silica*. Some concrete saws are

a_v_d/Shutterstock.com

Figure 2.5-22. This demolition saw is equipped with a wet sawing feature that helps keep dust down when cutting concrete.

Safety Note

When cutting concrete or stone materials with a saw, wear an approved respirator or use controlled wet-sawing practices to prevent breathing crystalline silica dust. Excessive exposure to the dust can cause silicosis, a serious lung disease. Sawing should always be done outdoors so that air movement can carry the dust away from the person doing the cutting.

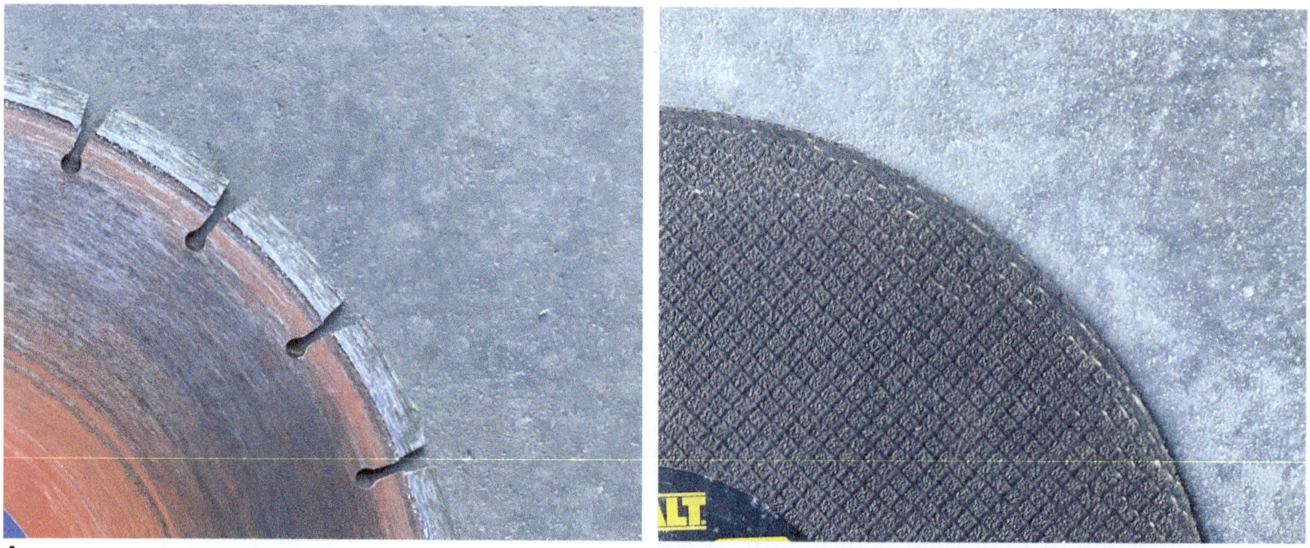

A B Goodheart-Willcox Publisher

Figure 2.5-23. A—The outer ring of a diamond-tipped blade wears away with use. B—Abrasive blades become smaller with use. This blade is for cutting steel. Abrasive blades are also available for cutting masonry.

equipped with wet-sawing or dust collection systems designed to protect the user from silica dust. These tools must be operated according to the manufacturer's instructions to work properly. Other protective measures may be required as well.

Masonry Tools and Equipment

A block mason is responsible for building CMU walls. Masonry tools are simple but require a high degree of skill and physical strength to use correctly.

Mortar Hoe

A *mortar hoe* is similar to a garden hoe but has two holes in the blade, **Figure 2.5-24**, and is used to mix mortar in a wheelbarrow or mortar box. The holes allow some material to move through the blade to allow for better mixing of the materials. They also help the user to judge the consistency of the mortar as it squeezes through the holes.

Mason's Trowel

A *mason's trowel* is used to spread and shape mortar for brick and block laying. Its tapered blade and shape allow experienced masons to apply mortar quickly and accurately, **Figure 2.5-25**.

Jointers

A *jointer* is a tool used to shape mortar joints in brick or block work before the mortar hardens, **Figure 2.5-26**. These tools give the joint a smooth, concave shape. Sled jointers can be used for long, horizontal joints.

MARSHALLTOWN

Figure 2.5-24. The holes in a mortar hoe allow for faster mixing of mortar.

MARSHALLTOWN

Figure 2.5-25. Mason's trowels are used for brick and block laying.

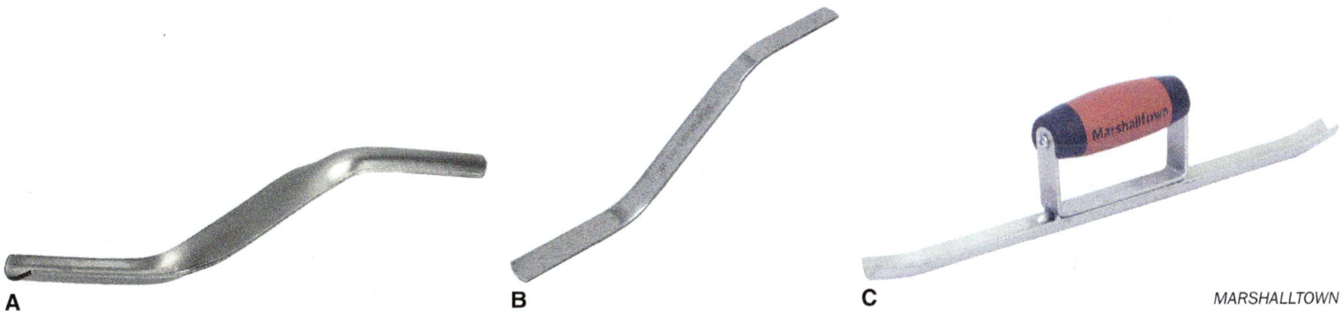

A B C

MARSHALLTOWN

Figure 2.5-26. Various examples of jointers used to finish mortar joints. A—A brick jointer creates a concave joint profile. B—A slicker is used to smooth and pack the joints flat. C—A sled runner is used on long horizontal joints for consistency and efficiency.

Angle Grinder

An *angle grinder* is a handheld power tool used for cutting and grinding steel and masonry, **Figure 2.5-27**. Grinders use abrasive discs or cutting wheels that are mounted at a right angle to the shaft. The disc spins at a high rate of speed in revolutions per minute (rpm) to cut through or grind away material. A handle is fitted to the side of the tool for two-handed operation.

Angle grinders produce sparks, dust, and flying debris when used and may generate toxic fumes. Use appropriate control measures to protect yourself and those around you! Appropriate personal protective equipment (PPE) should always be worn.

Lonely Walker/Shutterstock.com

Figure 2.5-27. This angle grinder is equipped with a diamond blade for cutting masonry materials.

Safety Note

Use the following safety rules when operating an angle grinder.

- Use the correct size and type of disc for the grinder and the material being cut.
  - Cutting discs must not be used to grind materials.
  - Do not use the outside edge of a grinding disc for cutting or grinding.
- Inspect discs for cracks, chips, excessive wear, and other defects. Do not use damaged cutting or grinding discs.
- Unplug the tool before changing discs.
- Ensure the guard is adjusted so it is positioned between the tool and the user to protect the user from flying debris.
- Eye and hearing protection should always be worn when using an angle grinder. Leather gloves, natural fiber clothing, an apron, or respiratory protection may also be needed.
- Keep hands, loose jewelry, clothing, and hair away from the blade at all times.
- Allow the grinder to come up to speed before making contact with the surface.
- Maintain full control of the tool with two hands while cutting or grinding.
- Cut and grind with the tool so that forces produced are pulling the tool away from your body, not pushing it toward you.
- Maintain good footing, balance, and a comfortable body position.
- Be mindful of pressures being applied to the disc. A pinched cutting disc or an unexpected glance against the material can produce a violent kickback of the tool. It can also damage the disc.
- Allow the tool to come to a complete stop before setting it down.

LESSON 2.5 Review

Answer the following questions based on information in this lesson.

1. What leveling device can be operated by one person? (2.5-1)
   A. Auto level
   B. Builder's level
   C. Laser level
   D. All are correct.
2. In which of these situations would a leveling instrument be used? (2.5-1)
   A. To ensure the block wall remains level as it is being built
   B. To determine the depth an equipment operator must dig to place the footings
   C. To establish the correct height for the footing forms
   D. All are correct.
3. *True or False?* With regard to leveling instruments, the cylindrical level is more precise than the bull's-eye level. (2.5-1)

Matching: Match each concrete tool with its correct function.

A. Concrete broom
B. Concrete edger
C. Concrete float
D. Concrete placer
E. Finish trowel

4. Tool used to create a rough texture on the surface of concrete (2.5-1)
5. Tool used to help level and smooth concrete surfaces (2.5-1)
6. Tool used to produce a smooth finish on concrete (2.5-1)
7. Tool used to produce a smooth and consistent radius edge on concrete (2.5-1)
8. Tool used to spread and move wet concrete as it is being poured (2.5-1)

LESSON 2.6

Safety

LEARNING OUTCOMES

After completing this lesson, you will be able to:

2.6-1 Explain the importance of designating a competent person on a jobsite.

2.6-2 Explain each of the "Focus Four" hazards as outlined by the Occupational Safety and Health Administration (OSHA).

2.6-3 Identify common caught-in or caught-between hazards prevalent during the construction of a home's foundation.

TECHNICAL TERMS

competent person

Focus Four

trench

Safety Practices

Recall from *Chapter 1, Preparing for Construction,* that construction workers are regularly exposed to jobsite hazards and hazardous conditions. An awareness of possible hazards and how to manage them will go a long way in maintaining personal and jobsite safety. However, awareness on its own will not protect workers. This lesson explores safety standards established by the Occupational Safety and Health Administration (OSHA) and practical ways to apply safety in residential construction.

Designating a Competent Person

Recall that OSHA requires employers to follow and enforce safety standards on their construction jobs. These standards are extensive and can be complex, especially to workers who are new to the construction industry. Employers must train employees on proper safety protocol for their duties and then designate a competent person to each job. OSHA defines the ***competent person*** as "one who is capable of identifying existing and predictable hazards in the surroundings or working conditions which are unsanitary, hazardous, or dangerous to employees, and who has authorization to take prompt corrective measures to eliminate them."

The competent person is on-site, making sure work is proceeding in a way that complies with the OSHA standards. In a small company, the owner of the company might be the competent person. If the company is a little bigger, it might be the foreman. The competent person is not necessarily the boss but must have the authority to take corrective action if needed, as stated by OSHA's definition.

Pro Tip

Many employers participate in OSHA Outreach Training Programs that provide workers with basic (OSHA 10-hour) and more advanced (OSHA 30-hour) training about common safety and health hazards on the job.

Knowing the common construction accidents that can occur will help you recognize moments or tasks that require a higher level of safety awareness and take the necessary precautions to do the job safely. Most construction accidents are preventable if we recognize the hazard and control the risk by following the proper safety guidelines.

Focus Four Hazards

The majority of construction accidents can be grouped into four categories, referred to by OSHA as the ***Focus Four***, formerly known as the *Fatal Four*: struck-by, falls, electrocution, and caught-in or caught-between hazards, **Figure 2.6-1**.

According to the National Institute for Occupational Safety and Health (NIOSH), 991 work-related fatalities occurred in the construction industry in 2019. Of those fatalities, 64.4% were attributed to one of the Focus Four.

Struck-By Hazards

Construction jobsites are often busy places with a lot of activity occurring up high and down low. A tape measure falling from a few stories up has the potential energy to do tremendous harm if it comes into contact with a person working below. Nail guns, saws, hammers, and other tools can put materials into motion unintentionally, and sometimes unpredictably. Materials that are not stacked, stored, or transported properly can fall or get blown over by a strong gust of wind. Vehicles and equipment like backhoes, skid steers, delivery trucks, and the like are often operating and moving around on the jobsite. When a worker is struck by an object that is falling or moving, there is potential for injury or death. *Chapter 3, Framing Systems*, addresses some ways that workers can protect themselves and others from struck-by hazards.

Safety Note

It is best to take time before a task to assess any identifiable hazards and consider what controls need to be applied before beginning.

Goodheart-Willcox Publisher

Figure 2.6-1. OSHA has identified the four leading categories of construction accidents.

Fall Hazards

Consider how many construction tasks must be accomplished from ladders, aerial lifts, or scaffolding, **Figure 2.6-2**. Also, think about how often construction workers need to open a manhole cover, dig a trench, or work from elevated platforms such as roofs, or the second floor of a home that is being framed. These are just some of the many possible scenarios on a jobsite where there is a potential for falls to occur. Even an icy stairway or a messy jobsite full of tripping hazards can be dangerous.

Falls are the leading cause of construction injuries and fatalities, with more than 33% of construction fatalities attributable to falls. OSHA requires employees to be protected from falls over 6′ high on working surfaces such as roofs or into holes, or over 10′ high when working from scaffolding. This can be done by using guard rails, harnesses, and safety nets, in conjunction with training. *Chapter 4, Drying In*, provides more information on managing fall hazards. *Chapter 5, Exterior Finishes*, provides more information on working from heights.

Electrocution

Electricity is an important part of our daily lives, and we often do not consider its potential to harm us. On the construction site, this potential is increased because we not only use electricity, but we are also actively working on electrical systems and near sources of electricity that are not protected in

Klopping/Shutterstock.com

Figure 2.6-2. Many construction tasks require working from heights.

the same way a finished home's systems are protected. There are many ways construction workers can come into contact with electricity on a jobsite, **Figure 2.6-3**. From accidentally striking a buried power line with digging equipment, to a faulty extension cord, to installing siding or roofing near a service entrance cable, the opportunities for electrical shock are seemingly everywhere. In 2019, 79 construction workers died from electrocution on the job, representing nearly half of all work-related fatalities from electrocution across all occupations. *Chapter 6, MEP Systems Rough-In*, provides more information on electrical safety and managing electrocution hazards.

Caught-In or Caught-Between Hazards

Concrete formwork, masonry construction, and foundation work are all physically demanding and can involve situations where you may encounter caught-in or caught-between hazards. Tools and equipment like mortar mixers and power trowels, **Figure 2.6-4**, have moving parts that can easily pull in part of your body or your entire body or pin you to a wall unexpectedly. Any time there is potential for a worker's body or part of the body to get squeezed, caught, crushed, pinched, or compressed between objects or parts of an object, a caught-in or caught-between hazard is present. Any tool or equipment with rotating parts, gears, or belts must be used properly with caution and respect. Always use the guards on these tools and maintain a high level of awareness when working with or around them.

On the jobsite it may seem reasonable to work close to operating equipment such as a hole-boring auger, **Figure 2.6-5**. It may seem routine to jump in a freshly dug trench to glue up a pipe connection.

SAJE/Shutterstock.com

Figure 2.6-3. Always ensure an electrical system is properly deenergized before performing any work on it.

MARSHALLTOWN

Figure 2.6-4. Jobsite equipment used during foundation construction poses a risk of caught-in or caught-between hazards. A—Mortar mixer. B—Power trowel.

Keith Homan/Shutterstock.com

Figure 2.6-5. Hole-boring augers have no guard for protection from injury. Caution must be exercised around rotating equipment.

However, these situations can be potentially dangerous. The auger could snag a loose bootlace or pant leg and wrap you up. The trench may be unstable and could collapse, causing heavy soil to trap or bury you.

Although caught-in or caught-between hazards can exist in any phase of construction, these hazards are especially prevalent in the early stages of a project when sitework, excavation, and foundation construction are happening. Identifying these hazards takes a broad understanding and an analytical perspective of what could go wrong in each individual circumstance. In addition to caught-in or caught-between hazards, other hazards exist in the early phases of construction.

The following sections discuss common hazards you may encounter during the sitework and foundation phase of construction. This is not a comprehensive list of all hazards, nor is it a comprehensive list of all caught-in or caught-between hazards. The objective is to highlight common safety hazards and help you develop a habit of using a safety mindset to assess all conditions. Awareness is key for maintaining a safe jobsite. Think through your work, assess the jobsite, and pay attention. Consider what could go wrong and manage those hazards accordingly to protect yourself and others.

Working with Heavy Equipment

Recognizing an equipment hazard may not be the first thing on your mind when you are focused on completing a task. Although it may look to you like an equipment operator is in full control and has full awareness of the surroundings, you should know that many machines, such as excavators and skid steers, restrict the operator's visibility dramatically, **Figure 2.6-6**.

Operators can also make mistakes. It is critical that you do not put yourself between any movable part of that equipment, such as a digging bucket

Safety Note

High visibility clothing such as the vest worn by the worker in **Figure 2.6-7** can help equipment operators see you on the jobsite. Many construction companies require all workers to wear high visibility clothing.

Klimek Pavol/Shutterstock.com

Figure 2.6-6. Working around heavy equipment requires a heightened awareness of your surroundings. Machines can move quickly and abruptly. Never assume the operator is aware of you.

TFoxFoto/Shutterstock.com

Figure 2.6-7. Bright, reflective clothing can keep workers more visible to others.

Superior Walls

Figure 2.6-8. When a crane is on site, pay special attention to suspended loads. Always keep your distance from the crane itself when it is in operation.

or the rotating part of a crane or excavator, and a stationary object such as a wall. Do not enter a trench or excavation while a machine is digging or dumping material. Do not stand or walk between moving equipment and another object. Stay out from under loads that are being moved by forklifts, cranes, or other equipment as they could shift and fall.

When operating equipment, always wear a seatbelt to avoid being thrown from the machine in the event of a rollover or other unexpected event. When moving a load with a piece of equipment, always ensure it is stable and secure.

When inspecting or repairing equipment, it is important that you have been trained on proper procedures. Vehicles must be turned off to perform maintenance and repair work. Proper lockout/tagout processes need to be followed to protect yourself and others from vehicle movement. This might include blocking the wheels to keep equipment from rolling, or lowering or locking out blades, buckets, forks, dump mechanisms, and similar equipment.

When precast concrete construction is used for foundations, the wall sections are set in place by a crane and bolted together. When cranes are on site, stay out of the swing radius of the machine and stay out from under suspended loads. Also, never place yourself or any part of your body between wall sections that are being placed, **Figure 2.6-8**.

Working in Excavations

Recall that an excavation is any removal of earth resulting in a cavity, trench, or depression. A ***trench*** is an excavation that is less than 15′ in width at its bottom and is deeper than it is wide. Trenches can be dangerous for workers to enter, **Figure 2.6-9**. Cave-ins pose the greatest risk and are a caught-in hazard that can result in worker fatalities. Other potential hazards include workers being struck by falling objects or loads from above, low oxygen levels or high levels of other gases causing a hazardous atmosphere in the trench, and struck-by and caught-between hazards associated with heavy equipment. Workers can also be subject to

Bannafarsai_Stock/Shutterstock.com

Figure 2.6-9. This shallow trench poses little threat to these workers. When working in trenches more than four feet in depth, extra safety measures must be taken.

drowning, electrocution, burns, or explosions if they are in a trench when an unexpected utility strike occurs. In fact, workers facing excavation hazards such as these can be more than twice as likely to suffer a fatality compared to other construction work.

Of the hazards associated with trench work, cave-ins are the most likely. Cave-ins are dangerous because of the weight of the soil and the unpredictability with which they can occur. Some types of soil can weigh as much as 3,000 lb per cubic yard. That is as much as a small car! Never enter an excavation without the permission of the competent person.

Measures that can be taken to protect workers entering a trench include:

- A competent person must determine the soil type and establish applicable protective systems before any worker enters the excavation.
- Air quality must be tested for oxygen deficiency, flammable gases or vapors, and toxic gases or vapors.
- Any trench more than 4′ deep must have a ladder, stairway, or ramp provided for worker access. Access must be within 25′ of each worker at all times.
- Excavations deeper than 5′ are not to be entered without a protective system.

When working in a trench, workers must stay within the confines of the protective system. The following methods can be used to establish a protective system:

- Sloping or benching the soil according to OSHA standards, **Figure 2.6-10**.
- Shielding the worker from possible cave-in through the use of a trench box, **Figure 2.6-11**.
- Using sheeting or shoring systems to retain the sides of an excavation. These systems are costly and are usually reserved for commercial and industrial construction.

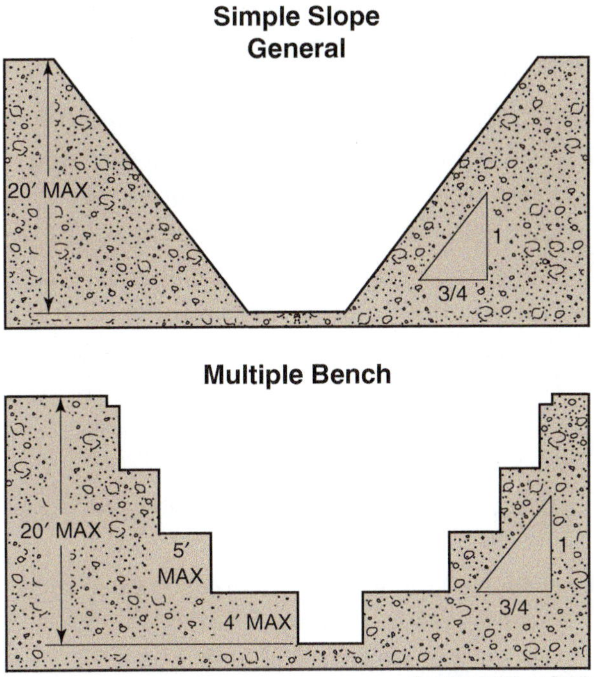

Goodheart-Willcox Publisher

Figure 2.6-10. Sloping and benching requirements vary by soil type.

Alfred State College

Figure 2.6-11. A trench box is lowered into place to provide these workers safe access to a fire hydrant tap connection. Once inside, the box will shield them from possible trench collapse.

LESSON 2.6 Review

Answer the following questions based on information in this lesson.

1. *True or False?* Someone who is designated as the competent person has *no* authority on a jobsite to eliminate risks or hazards. (2.6-1)
2. What hazard is the leading cause of construction injuries and fatalities? (2.6-2)
   A. Struck-by
   B. Falls
   C. Electrocution
   D. Caught-in or between
3. *True or False?* A faulty extension cord is a potential electrocution hazard. (2.6-2)
4. Which of the following hazards pose the *greatest* risk to a worker's safety when working in trenches? (2.6-3)
   A. Being struck by falling objects or loads from above
   B. High levels of gases causing a hazardous atmosphere
   C. Hazards associated with heavy equipment
   D. Caught-between hazards resulting from cave-ins
5. Which method can establish a protective system for excavations that are deeper than 5′? (2.6-3)
   A. Sloping or benching the soil
   B. Using a trench box
   C. Using shoring systems to retain the sides of an excavation
   D. All are correct.

LESSON 2.7

Math

LEARNING OUTCOMES

After completing this lesson, you will be able to:

2.7-1 Add and subtract fractions.

2.7-2 Read measurements using fractional and decimal foot measuring tapes and leveling rods.

TECHNICAL TERMS

builder's rod
common denominator
decimal foot system
denominator
engineer's rod
equivalent fraction
fraction
improper fraction
mixed number
numerator

Fractions

Fractions are used regularly in construction. A *fraction* is a part of something larger. For example, if there are three anchor bolts in a pound, each anchor bolt weighs one-third of one pound. One third can be written as:

$$\frac{1}{3}\begin{array}{l}\leftarrow \text{numerator}\\ \leftarrow \text{denominator}\end{array}$$

The *numerator* is the top number of the fraction, the number that appears above the fraction bar. It indicates how many parts are in the fraction, in this case one anchor bolt. The *denominator* is the number below the fraction bar. The denominator indicates how many parts are in the whole.

Practice 2-1

If there are 50 anchor bolts in a carton and we take seven of them out, how many have we taken out? Write the answer as a fraction to illustrate the parts (anchor bolts) of the whole (carton).

A strong understanding of fractions; how they are used; and how to add, subtract, multiply, and divide with them is important to success in the construction field. In this chapter and the next, you will learn and practice using fractions in ways that will help you to function better when working on the jobsite.

For example, when pouring footings or a basement floor, concrete is ordered by the cubic yard. A cubic yard is a significant amount of material. For this reason, it is common practice to order in increments of 1/2 or 1/4 of a cubic yard. CMU blocks come in both full and 1/2 blocks. Also, anytime a builder pulls out their tape measure, they are using fractions. In fact, an inch is just a fraction (1/12) of a foot.

Equivalent Fractions

Equivalent fractions are needed to perform certain calculations with fractions. *Equivalent fractions* are two fractions that represent the same value.

For example:

$$\frac{1}{3} = \frac{2}{6}$$

These two fractions are equivalent fractions because they both represent one-third of the whole. To create an equivalent fraction, both the numerator and the denominator of the fraction must be multiplied by the same number.

For example:

$$\frac{1 \times 2}{3 \times 2} = \frac{2}{6}$$

Practice 2-2

What are two possible equivalent fractions of 1/8?

Builders use equivalent fractions regularly to help them quickly add, subtract, divide, and multiply dimensions. This will be covered in more detail in *Chapter 3, Framing Systems*. In fact, anytime a builder uses a tape measure, they are likely to use equivalent fractions to help them efficiently identify a fractional mark between inches, as discussed in *Lesson 1.7*.

Pro Tip

A quick way to cut a fraction in half is to double the denominator. For instance, half of 3/4 is 3/8.

Adding and Subtracting Fractions

To add or subtract fractions, there must be a ***common denominator***. If the denominator of one of the fractions is 8, the other fraction must be written as an equivalent fraction with a denominator of 8.

Consider the following example:

$$\frac{3}{4}+\frac{1}{8}=x$$

To add these fractions, both must have a denominator of 8. Convert the first fraction into an equivalent fraction by multiplying both the numerator and denominator by the same number. However, we must be sure that the denominator equals 8 so the fractions can be added.

$$\frac{3}{4}\times\frac{2}{2}=\frac{6}{8}$$

When both fractions have the same denominator, add the numerators.

$$\frac{6}{8}+\frac{1}{8}=\frac{7}{8}$$

To find a common denominator among multiple fractions, you can multiply all the denominators together in a problem.

For example:

$$\frac{1}{3}+\frac{2}{5}+\frac{9}{14}=x$$

$$3\times5\times14=210$$

Once you have the common denominator, 210, you can create equivalent fractions to complete the calculation. To do this with multiple fractions in a problem, the numerator must be multiplied by the denominators of the other fractions.

$$1\times5\times14=70$$
$$2\times3\times14=84$$
$$9\times3\times5=135$$

The problem written with equivalent fractions would appear as:

$$\frac{70}{210}+\frac{84}{210}+\frac{135}{210}=\frac{289}{210}$$

The result is an ***improper fraction***, where the numerator is larger than the denominator. This occurs when the value of the fraction is greater than 1. The same number can be rewritten as a mixed number. A ***mixed number*** is a number made up of a whole number and a fraction. So, 289 is 79 parts larger than the whole of 210 parts.

$$\frac{289}{210}=1\frac{79}{210}$$

Practice 2-3

Imagine you have placed footing forms for a rectangular building and are ready to order concrete. You have determined that the footings for the two longest walls will require 1 1/4 cubic yards of concrete each. The footings for the two shortest walls will require 1/2 cubic yards each. Determine how many cubic yards of concrete will be required to pour all four footings.

$$1\frac{1}{4}+1\frac{1}{4}+\frac{1}{2}+\frac{1}{2}=x \text{ cubic yards}$$

Pro Tip

When ordering concrete, it is always a good idea to order extra. Running short and having another truck come back to the job is costly. By the time the extra truck arrives, it may be too late.

Subtracting fractions is similar to adding them. Both fractions must have a common denominator, then the numerators are subtracted just like whole numbers.

For example:

$$\frac{2}{3} - \frac{1}{4} = x$$

Find a common denominator and subtract the numerators.

$$\frac{8}{12} - \frac{3}{12} = \frac{5}{12}$$

Practice 2-4

A CMU block must be cut to finish a course in a foundation wall. If 7 1/4″ is cut from the 15 3/8″ long block, then how much remains?

$$15\frac{3}{8}'' - 7\frac{1}{4}'' = x$$

Using Fractional and Decimal Measuring Devices

Measuring devices such as tape measures, measuring tapes, and leveling rods may be used to make measurements in feet and inches, decimal feet, or metric units. Metric measuring devices are not commonly used on construction sites in the United States, but they are common in other parts of the world. It is essential that you identify which unit format and measurement system is being used before reading, recording, or communicating dimensions.

When performing site and foundation work, it is common to use measuring devices based on the decimal foot system. The ***decimal foot system*** uses dimensions that are graduated in feet, tenths of a foot, and hundredths of a foot rather than feet and inches. Inches are not used in the decimal foot system, so 6″ is equivalent to 0.5′, and 3″ equals 0.25′ in the decimal foot system.

The site plan shown in **Figure 2.7-1** has dimensions in feet and inches to indicate distances from the building lines to the property lines. However, there are dimensions in decimal format indicating floor elevations and elevations at the corners of the property. In surveying, elevations are height measurements referenced to a known point. On site plans, elevations and lengths of property lines are commonly dimensioned in decimal feet.

Dimensions expressed in decimal format allow for simplicity when adding and subtracting dimensions on the job. Rather than converting a fraction to the lowest common denominator, a decimal form of each number exists and is simply calculated without needing to make conversions. For instance, using the dimensions in the following example, we can see that multiple steps are required using feet, inches, and fractions of an inch:

$$17'\text{-}5\ 3/8'' + 14'\text{-}8\ 3/4'' = x$$

Finding the common denominator…

$$17'\text{-}5\ 3/8'' + 14'\text{-}8\ 6/8'' = x$$

Then adding like units…

$$17'\text{-}5\ 3/8'' + 14'\text{-}8\ 6/8'' = 31'\text{-}13\ 9/8''$$

Then reducing fractions and converting mixed numbers…

$$31'\text{-}13\ 9/8'' = 31'\text{-}14\ 1/8''$$

Then converting the inches to feet…

$$31'\text{-}14\ 1/8'' = 32'\text{-}2\ 1/8''$$

However, the equivalent numbers can be added quickly and easily using the decimal foot system:

$$17'\text{-}5\ 3/8'' + 14'\text{-}8\ 3/4'' = 17.45' + 14.73' = 32.18'$$

Some construction drawings and measuring devices regularly feature the decimal foot system for ease of use in construction.

In **Figure 2.7-1**, notice the floor elevation for the main level of the house is labeled as 137.0′ and the elevation of the terrace behind the house is labeled as 136.5′. These elevations are height measurements in reference to sea level. This indicates a 0.5′ (or 6″) step down from the floor of the home to the terrace when stepping out the back door onto the terrace. This calculation would be just as simple if feet and inches were used.

Next, look at the garage. In this case, there is a room above the garage that is used for living space. The two elevations given specify the finished floor elevation for the garage (131.5′, or 131′-6″) and for the room above the garage (140.4′, or 140′-4 13/16″).

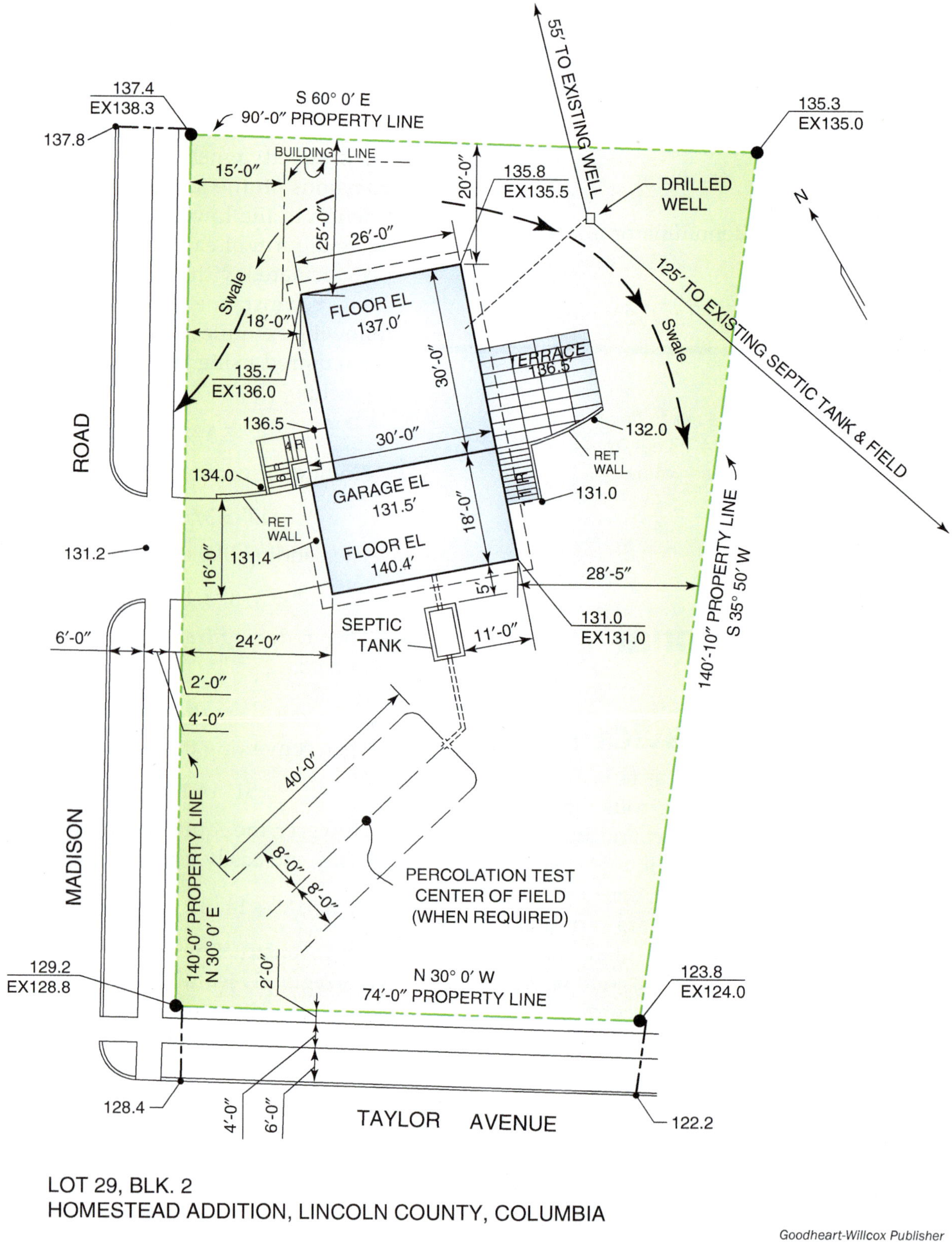

Goodheart-Willcox Publisher

Figure 2.7-1. It is common to see dimensions on a site plan given in decimal form. The elevations on this site plan are dimensioned in decimal feet.

To find the difference in elevation from one floor system to the next, we only need to do some quick subtraction.

$$140.4' - 131.5' = 8.9'$$

There is an 8.9′ (8′-10 13/16″) difference in elevation from one floor system to the next. If we were to calculate this using feet, inches, and fractions of an inch, we would end up with the same result; however, the additional steps increase the risk of error. *Chapter 4, Drying In*, covers how to convert from decimal feet to feet and inches, and vice versa.

Practice 2-5

Use **Figure 2.7-1** to determine the difference in floor elevation from the garage (131.5′) to the main floor (137.0′). What does your answer tell you about the relationship between the garage floor and the main floor of the house?

$$137.0' - 131.5' = x$$

Can you picture this scenario? Have you ever seen a home where this is the configuration? What did it look like? Asking questions such as these are good mental exercises that can help you understand how a project needs to be built.

Using Measuring Tapes

Measurements and layouts involving long distances can be made using measuring tapes. Measuring tapes are available in lengths from 50 feet to 300 feet. There are various types with different graduations. Surveyors use a measuring tape graduated in feet and decimal parts of a foot. On an engineer's measuring tape, **Figure 2.7-2A**, each major division numbered on the tape before "F1" (the mark representing one foot) is a tenth of a foot, written 0.1′. A carpenter, on the other hand, will usually select one that is marked in feet, inches, and eighths of an inch, **Figure 2.7-2B**. This measuring tape has seven marks between each inch. Each of these marks represents an eighth of an inch (1/8″).

Note that tenths of a foot are larger than inches. Recall that an inch is one-twelfth of a foot, equal to splitting one foot up into 12 equal parts. Splitting one foot up into tenths creates 10 equal parts.

An engineer's measuring tape is not used to read in fractions. For instance, we do not call the division in the middle of two tenths a half (1/2) of a tenth. These divisions represent hundredths of a foot. Therefore, the division halfway between 0.7′ and 0.8′ is 0.75′. Notice that this division aligns with 9″ on the fractional measuring tape. Just as 9″ is three-quarters of a foot, 0.75′ is three quarters of a foot. Each division between the tenths on the engineer's measuring tape represents a hundredth of a foot.

Using Leveling Rods

Leveling rods, introduced in *Lesson 2.5*, are used alongside leveling instruments such as builder's levels and rotating lasers to take readings when measuring differences in elevation. Different leveling rods are available graduated in feet and inches, or in decimal feet. An ***engineer's rod*** is a leveling

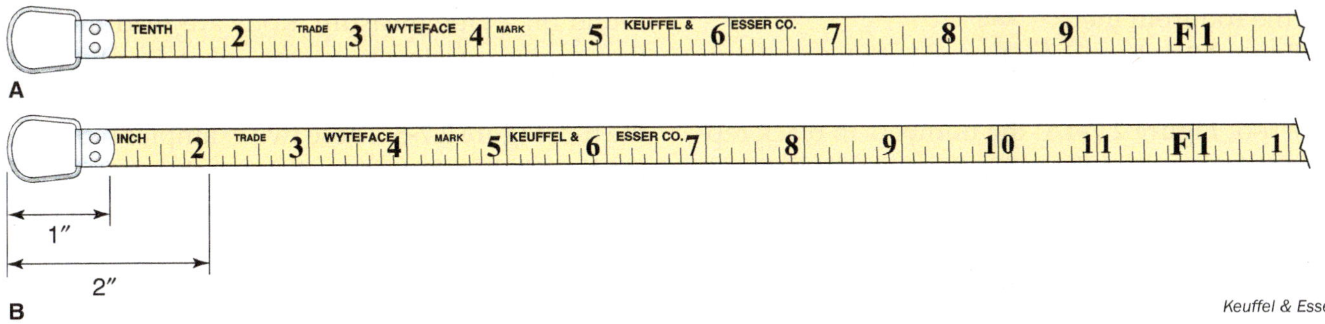

Figure 2.7-2. Measuring tape. Long tapes for layout use are made in lengths of up to 300′. Measuring tapes are made with different unit formats and graduations. A—Feet and decimal graduations. B—Feet, inch, and eighth-inch graduations.

Pro Tip

The term *scale* is also used to refer to the measuring device architects use to measure items on a scaled drawing. **Figure 2.8-1** shows an architect's scale and a partial list of conventional inch-foot scales.

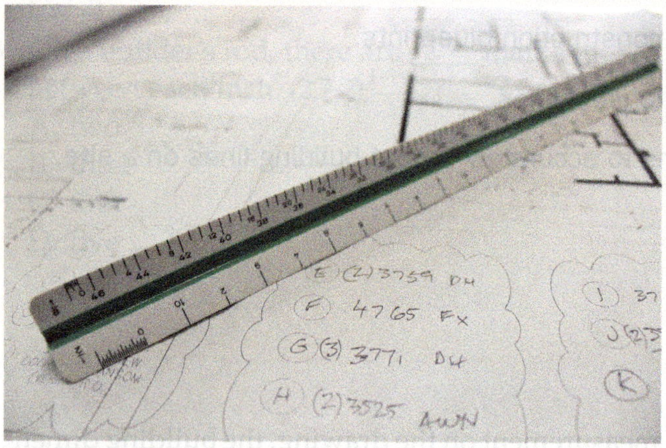

A

Common-Inch Foot Scales			
Scale	**Where Used**	**Ratio**	**Relationship to Actual Size**
1/32″	Site plans	1:384	1/32″ = 1′-0″
1/16″	Plot plan	1:192	1/16″ = 1′-0″
1/8″	Plot plan	1:96	1/8″ = 1′-0″
1/4″	Elevation plans	1:48	1/4″ = 1′-0″
3/8″	Construction details	1:32	3/8″ = 1′-0″
1/2″	Construction details	1:24	1/2″ = 1′-0″
3/4″	Construction details	1:16	3/4″ = 1′-0″
1″	Construction details	1:12	1″ = 1′-0″
1 1/2″	Construction details	1:8	1 1/2″ = 1′-0″
3″	Construction details	1:4	3″ = 1′-0″

B

Goodheart-Willcox Publisher

Figure 2.8-1. A—Architect's scale. On the side of the tool shown, scales of 1/8″ = 1′-0″ and 1/2″ = 1′-0″ can be seen starting at the left end; the 1/4″ = 1′-0″ and 1″ = 1′-0″ scales share this side of the tool but originate from the right end of the scale. Graduations indicating parts of a foot can be seen to the left of the 0 mark. Each graduation represents 2″ on the 1/8″ = 1′-0″ scale and 1/2″ on the 1/2″ = 1′-0″ scale. B—Common inch-foot scales used on residential drawings.

Dimensions are communicated using dimension lines and extension lines, **Figure 2.8-2**. A ***dimension line*** runs the length of the object being measured and terminates with endpoints. An ***endpoint*** is a tick mark, dot, or arrowhead at the ends of dimension lines indicating the extent of the dimension noted. On construction drawings, the dimension text indicating the distance measurement is usually placed above the dimension line. An ***extension line*** establishes the extents of a dimension. Extension lines extend away from the elements of the drawing to which the dimensions apply and are drawn perpendicular to the dimension line. In some cases, a dimension line and extension line will intersect without an endpoint, **Figure 2.8-3**. This indicates that the dimension is not measured to this point but continues to the next endpoint.

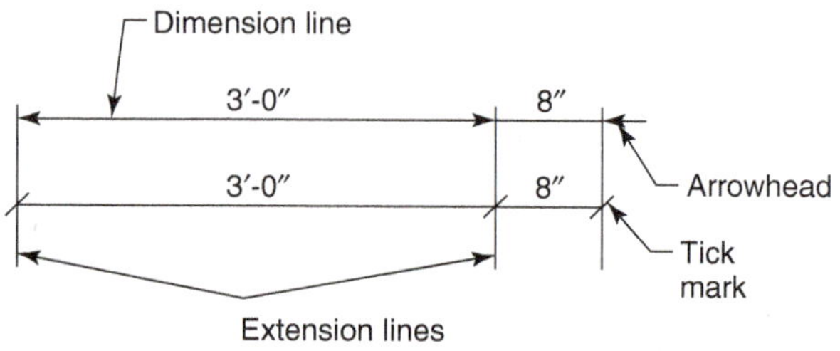

Goodheart-Willcox Publisher

Figure 2.8-2. Dimension lines and extension lines are used to define object dimensions. Dimension lines have terminating points (tick marks) to indicate the beginning and end of the dimension. Extension lines are drawn perpendicular to the dimension line and establish the extents of the dimension.

Drawings Used in Foundation Construction

Construction drawings work together to communicate the design of a home and provide key details needed to lay out and build all phases of the project. After using the site plan to identify where the building is to be located on the site, the builders shift their attention to the foundation plan to identify dimensions of and lay out the building lines on the site.

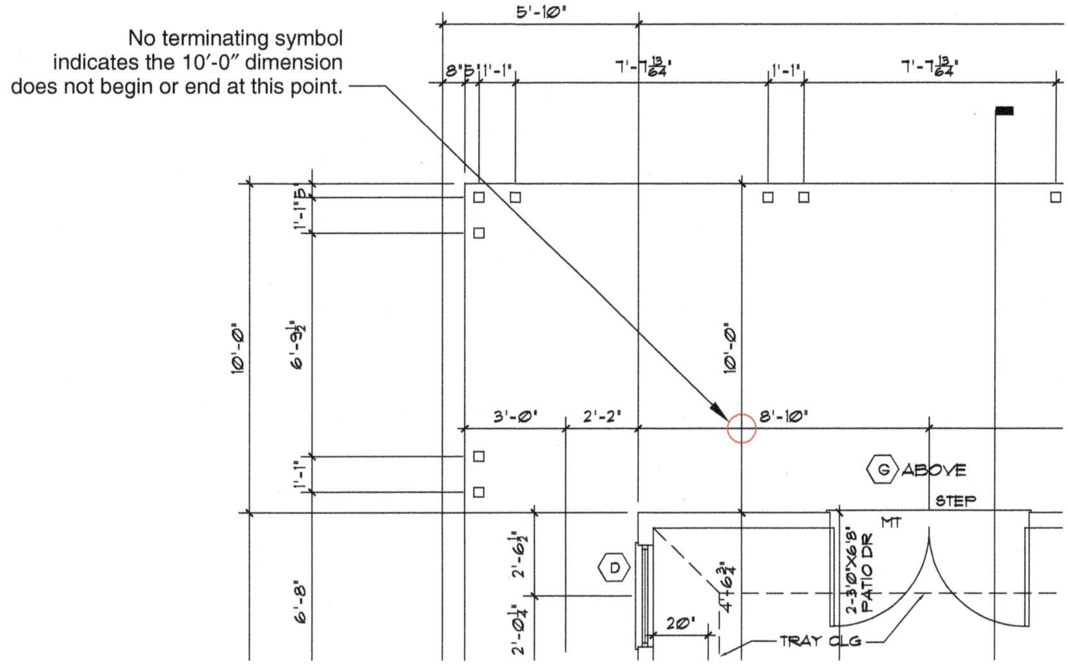

Goodheart-Willcox Publisher

Figure 2.8-3. Terminating symbols establish where a dimension starts and ends. Pay close attention to intersections such as these.

The following section discusses the information found on the foundation plan, which a builder uses in constructing the footings and foundation for the home.

Foundation Plan

When building a home, the foundation is the first permanent component that is built, starting with the footings. A *foundation plan* shows construction details for the foundation of a building. A foundation plan typically uses a 1/4″ = 1′-0″ scale and, as the name suggests, is drawn in plan view. The foundation plan is used to determine sizes and locations of footings, walls, beams, columns, windows, doors, and other elements of the foundation. The foundation plan also references section and detail drawings that show additional information needed to build the foundation according to the intended design.

After the excavation is completed, the builder references the foundation plan to establish the size and location of each footing based on the location of the foundation walls. On the foundation plan, footings are shown with hidden lines (thin dashed lines) on each side of the foundation wall.

A home may use footings that step up because of the slope of the grade or for a crawlspace, garage, or porch. These details will be found on the elevation drawings, **Figure 2.8-4**. Foundation walls, footings, and other features below grade are shown with hidden lines. A foundation wall section, **Figure 2.8-5**, indicates the materials used to construct the footings and walls. A drawing such as this is not part of the foundation plan but will be referenced by the foundation plan. Builders will consult the notes and specifications for further detail and clarification regarding specific requirements for materials.

Goodheart-Willcox Publisher

Figure 2.8-4. Elevation view of a residence showing the foundation wall and footing lines as hidden lines.

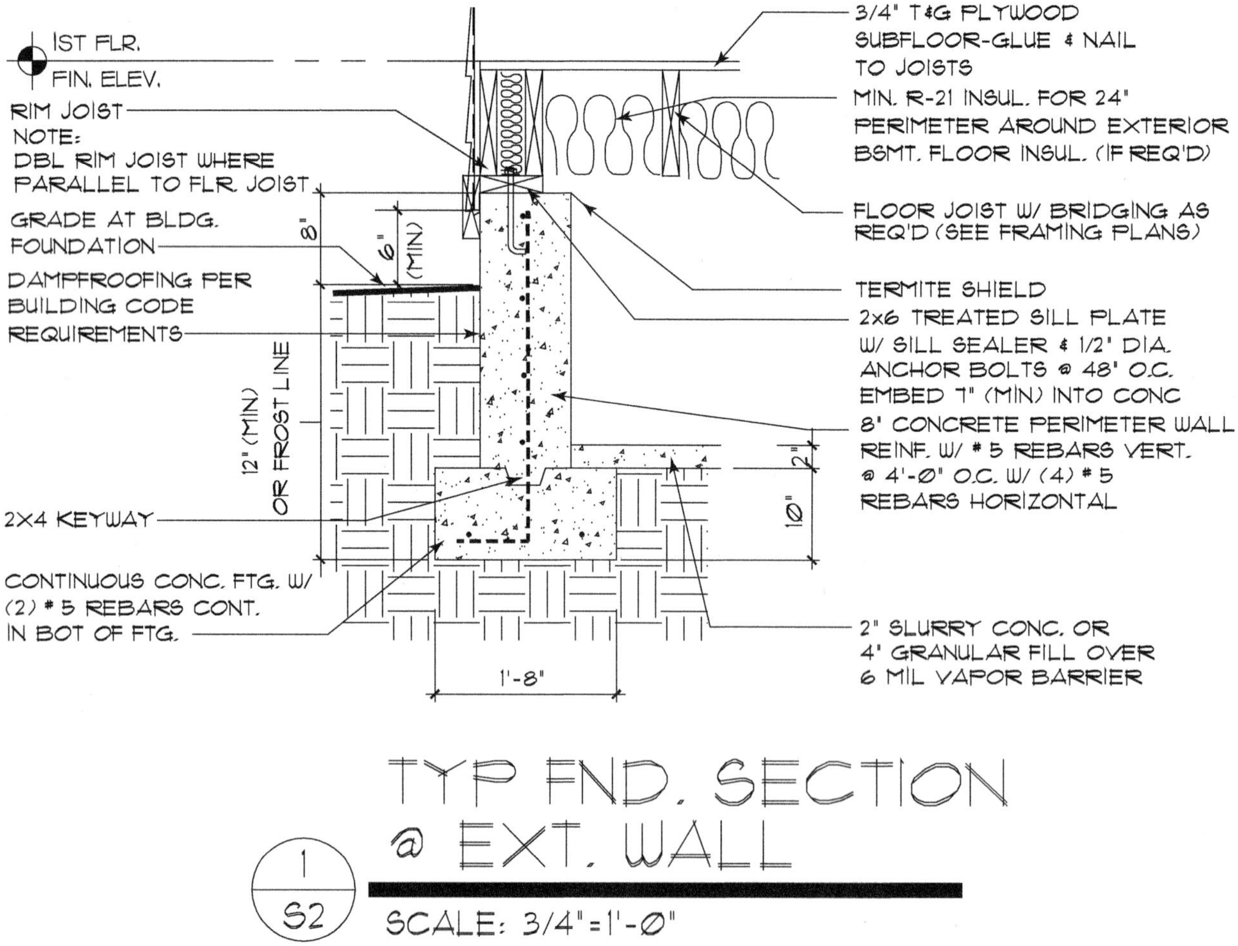

Goodheart-Willcox Publisher

Figure 2.8-5. A foundation wall section shows details of the footing and wall construction.

LESSON 2.8 Review

Answer the following questions based on information in this lesson.

1. *True or False?* A drawing that is drawn to scale is in exact proportion to the actual size. (2.8-1)
2. *True or False?* On construction drawings, all dimensions greater than 1′ are written in inches. (2.8-1)
3. Residential foundation plans are generally drawn to _____ scale. (2.8-1)
   A. 1/16″
   B. 1/8″
   C. 1/4″
   D. 1/3″
4. *True or False?* An extension line runs the length of the object being measured. (2.8-2)
5. Which of the following indicates the materials used to construct the footings and walls? (2.8-3)
   A. Foundation plan
   B. Site plan
   C. Framing plan
   D. Foundation wall section

LESSON 2.9

Green Building and the Future of Construction

LEARNING OUTCOMES

After completing this lesson, you will be able to:

2.9-1 Describe key considerations for green building in residential construction.

2.9-2 Explain how the design and orientation of a home impacts energy efficiency.

2.9-3 Discuss two alternatives to traditional foundations.

TECHNICAL TERMS

insulated concrete form (ICF)

precast concrete wall

Green Building Strategies

At the center of green building is energy conservation. In addition to cost savings to the owner in the way of lower utility bills, conservation efforts include slowing the consumption of finite natural resources and reducing atmospheric pollution before, during, and after construction. Making the home more durable so products and systems last a long time helps to achieve conservation goals that extend beyond lowering utility bills. These strategies should be considered with any green building project.

A lot of thought should go into the overall vision of a green building project. A successful green building project requires that the design team, the construction team, and the owner all work together from project conception through the completion of construction. After construction, the owners and occupants of the building also need to carry out the proper operation and maintenance of the home to ensure it makes the most of its potential. Design choices, starting with the selection of the lot and the placement and orientation of the home, can impact how the building and the users of that building consume resources.

Key considerations in any green building project include:

- Overall site development
- Controlling water runoff during and after construction
- Water use and efficiency for the home
- Energy use and efficiency for the home
- Energy sources and generation
- Material selection, sourcing, use, and waste disposal
- Indoor environmental control and air quality
- Ongoing maintenance and use of the occupied home

Teamwork is necessary to ensure all the components and systems are designed and installed properly. No one trade can complete its work without involving the work of other trades. Everyone on the jobsite must cooperate, working together toward overall project goals to ensure the house is built to the highest standard and achieves its green objectives.

Environmental Considerations

Part of green building includes a consideration of where the house is built and how the community surrounding the house is designed. Green building needs to include a visionary approach with consideration for how the building fits into the surrounding environment—whether that environment is a pristine, secluded woodland lot, a suburban housing development, or an urban setting.

Some of the environmental factors to be considered include:

- **General orientation**—A home with its longest axis oriented in an east-west direction will absorb more winter sun on the south side (in the northern hemisphere), **Figure 2.9-1**.
- **South-facing windows**—A home with more glass on the south-facing side and less on the north-facing side will absorb more heat from the winter sun and will give occupants more natural light and exposure to sunlight during the cold winter months, **Figure 2.9-2**.
- **Hilly terrain**—South-facing slopes are exposed to far more winter sun than north-facing slopes. Deep valleys and ravines can shade winter sun, causing a higher heating load on the home.
- **Deciduous trees**—Trees that lose their leaves in the winter (deciduous trees) can provide shade in the summer while allowing winter sun to come through. Established trees on a site should be incorporated into the design of a green-built home whenever possible.
- **Prevailing winds**—Positioning and designing a home so that it is shielded from harsh winter winds can improve both comfort and energy performance.

The objective when planning, selecting a lot, and conducting sitework for a green-built house is to minimize its impact on the site and the environment. Before construction begins, a plan is made outlining how the sitework and construction will be conducted. The plan addresses how to protect existing trees and shrubs, manage water runoff, control erosion, and minimize overall disturbance to the site.

Community Considerations

The design should also account for how the project interacts within the social context. This consideration can have a lot to do with reducing the impact on the environment and improving the lives of those living in the house. For example, a house built near public transportation may reduce the

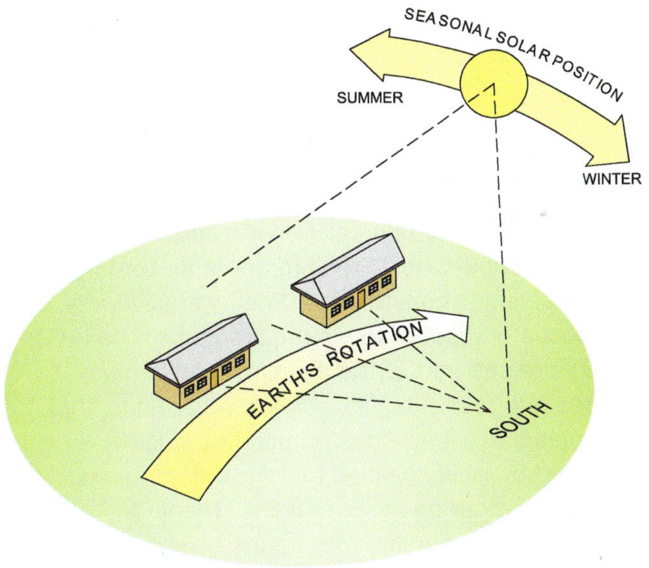

Goodheart-Willcox Publisher

Figure 2.9-1. Orienting the long side of the home toward the south maximizes its exposure to passive heating from the winter sun.

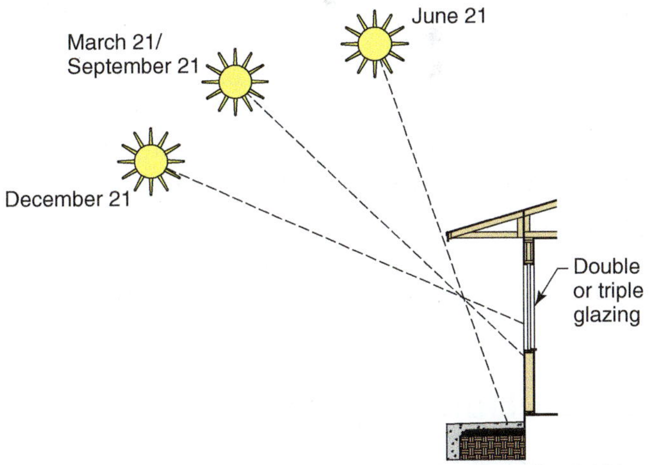

Goodheart-Willcox Publisher

Figure 2.9-2. The size of the roof overhang and amount of glass on the south-facing side of a home can control solar heating. In the winter, when the sun remains closer to the horizon, the windows allow for passive solar heating. In the summer, the sun rises higher in the sky, so the roof overhang shades the windows, reducing the solar heat gain in the building.

miles traveled in personal automobiles. A development located within walking distance of shopping will reduce vehicle traffic even more.

When a developer is planning a new housing development, the questions to consider might include the following:

- In what ways can the natural ecology of the site be preserved or restored?
- Does the layout of the streets and sidewalks promote efficient and safe travel?
- Is the design of the development contributing to an increase or decrease in vehicular traffic?
- Is existing public infrastructure capable of supporting the additional loads created by the increased population density?
- Does the design encourage the best use of green space, pedestrian traffic, and an active lifestyle?
- Are there charging accommodations to promote the use of alternative forms of transportation such as electric cars, public transport, and bicycles?
- Have considerations been made regarding light and sound pollution?

Paying attention to questions such as these in the planning stage can positively affect the entire community. Projects that are built with sustainability in mind should not only save natural resources but should also enhance the quality of life for the occupants. Encouraging transportation methods such as walking or biking promotes a healthy lifestyle while reducing carbon emissions, noise, and light pollution. Providing easy access to public transportation and carpooling accommodations also plays a role, **Figure 2.9-3**.

Goncharovaia/Shutterstock.com

Figure 2.9-3. Walkable communities promote a healthy lifestyle and reduce reliance on automobiles.

Green Building Programs

Various green building programs exist to set standards and to measure the design and performance of green-built homes. The National Green Building Standard (NGBS) is one of the leading green building programs used nationwide. Leadership in Energy and Environmental Design (LEED®) is the green building program of the US Green Building Council (USGBC®). These programs provide builders and developers with established structures that include rating systems for validating various levels of green certification.

Widely used model building codes such as the *International Residential Code (IRC)* continue to require higher levels of compliance with regard to building performance. Some areas of the country have adopted higher-level green building codes, such as the International Green Construction Code (IgCC) and the International Energy Conservation Code (IECC), to mandate a minimum level of performance regarding design, construction, and energy performance.

Code Note

The International Energy Conservation Code (IECC) defines specific requirements for commercial and residential construction. Energy efficiency is the focus of most of these requirements. The IECC is adopted as part of the building code in many states and municipalities. On the other hand, the International Green Construction Code (IgCC) is adopted by municipalities that have determined to raise the bar on green building. The IECC establishes minimums for energy conservation, whereas the IgCC establishes minimums for green building. It is up to each municipality to adopt all or parts of the code that they wish to enforce.

Green Alternatives to Traditional Foundations

Traditionally, homes have been built with poured concrete or CMU block foundations. Before that, many homes were built on stone foundations. Through new products, technology, and techniques, other cost-effective, energy-efficient, and more environmentally friendly alternatives have gained popularity.

Insulated Concrete Forms (ICFs)

Insulated concrete forms (ICFs) are a lightweight alternative to conventional concrete forms. An ***insulated concrete form (ICF)*** is made of light polystyrene foam or other types of rigid insulating material. ICFs are used for both foundation walls and the exterior walls of a building. Once forms are erected and braced, concrete is poured into them, **Figure 2.9-4**. The forms are left in place, resulting in a structural wall that is insulated on both sides. Depending on the specific system, an ICF wall system also provides surfaces and fastening points for a carpenter to install both exterior and interior wall coverings.

ICF foundations and even ICF above-grade walls are popular for green homes. Compared with conventional walls, ICF walls provide greater resistance to heat transfer and are airtight, which helps to eliminate drafts. Thus, the use of ICFs reduces the amount of energy required for heating and cooling throughout the life of the structure. In addition, ICF construction produces less waste than that produced by conventional form construction.

Precast Concrete Walls

Imagine arriving at a jobsite one morning where the excavation has been completed and prepared for a home foundation. Now imagine the same site with the foundation walls in place by lunchtime on the same day. This scenario is possible if a home builder decides to use precast concrete walls. ***Precast concrete walls*** are panels that are formed and poured in a factory, delivered to the construction site, and placed with a crane, **Figure 2.9-5**. The panels are built to the size and specifications required by the individual job. Precast walls have a continuous layer of insulation built in. Their design also leaves space to conveniently install more insulation after the panels are installed, giving these walls much better thermal resistance than traditional foundation walls. Precast concrete walls also reduce waste and the quantity of materials needed to build a foundation from traditional materials.

Plumwall Bracing Systems Inc.

Figure 2.9-4. Once fully prepared, ICFs are filled with concrete. The forms stay in place after placing the concrete.

Superior Walls

Figure 2.9-5. Precast concrete walls are moved into place using cranes.

LESSON 2.9 Review

Answer the following questions based on information in this lesson.

1. *True or False?* Green building methods consider how the building fits into the surrounding environment. (2.9-1)
2. *True or False?* Green building methods do *not* include ongoing maintenance and use within the occupied home. (2.9-1)
3. The objective when planning, selecting a lot, and conducting sitework for a green-built house is to ____. (2.9-2)
   A. protect existing trees and shrubs
   B. manage water runoff and control erosion
   C. select, source, use, and ethically dispose of materials
   D. All are correct.
4. A key community consideration for green building includes which of the following? (2.9-2)
   A. South-facing windows
   B. Terrain of hills and valleys
   C. Layout of sidewalks in the area
   D. All are correct.
5. *True or False?* Precast concrete walls have better thermal resistance than traditional foundation walls. (2.9-3)

Chapter 2 Review

Summary

Lesson 2.1

- Occupations in the residential construction industry include physical labor, skilled trades, supervision and management, design and engineering, workplace safety management, and inspection. (2.1-1)
- Laborers perform jobs that require their physical body, good judgment, and the ability to think ahead. Workers in different skilled trades are responsible for specific tasks, including preparing the jobsite, constructing the foundation, building the structure, performing the exterior and interior finish work, and installing the mechanical systems used in the building. The oversight of the typical construction job generally is organized into three levels, coordinated by foremen, superintendents, and construction managers. Architects primarily focus on the creative design and the look of a home, while engineers focus on the math and science principles used in making a home safe and durable. Construction safety managers visit jobsites to assess working conditions. Building inspectors determine whether buildings are safe for occupation. (2.1-2)

Lesson 2.2

- Residential construction includes new home builds, renovations, and additions. (2.2-1)
- New home construction falls into three basic groups: production home builds, custom home builds, and spec home builds. (2.2-2)

Lesson 2.3

- There are many types of footing and foundation systems that can be used when building a house. The combination chosen by the designer or builder depends on several factors such as climate, soil type, the needs of the owner, and the end use of the building. (2.3-1)
- After the excavation for the foundation has been completed, spread footings are typically formed with framing lumber. If reinforcing steel is required, it is placed in the forms before they are filled with concrete. After the concrete has hardened, footing forms can be stripped away, building lines are reestablished using the batter boards, and the building lines are marked on the footings. Perforated drainage pipes encased in gravel are installed around the footings to evacuate water from around the foundation. (2.3-2)

Lesson 2.4

- Concrete is a unique building material because it begins as a flowable, formable material and then hardens through a chemical reaction to form a strong, durable product that can withstand tremendous loads. Concrete is an excellent building material for footings and foundation walls. (2.4-1)
- In construction, bolts are used primarily for structural connections. When bolts will be exposed to the elements or used with corrosive materials, such as pressure-treated lumber, they must have a hot-dipped galvanized coating or be made from stainless steel. (2.4-2)

Lesson 2.5

- Tools commonly used when building forms for footings, pouring concrete, and building foundation walls fall into the following categories: leveling and layout tools, concrete tools and equipment, and masonry tools and equipment. (2.5-1)

Lesson 2.6

- The competent person is on-site, making sure work is proceeding in a way that complies with OSHA standards. (2.6-1)
- The majority of construction accidents can be grouped into four categories: fall, struck-by, electrocution, and caught-in or caught-between. OSHA refers to these types of hazards as the Focus Four or Fatal Four. (2.6-2)
- Any time there is potential for a worker's body or part of the body to get squeezed, caught, crushed, pinched, or compressed between objects or parts of an object, a caught-in or caught-between hazard is present. (2.6-3)

Lesson 2.7

- To add or subtract fractions, there must be a common denominator. Adding and subtracting fractions is common when working with construction materials and drawings. (2.7-1)
- Leveling rods use thick marks that are easily read through the scope of a leveling instrument. Rods measure from the bottom up. The bottom of the rod represents 0′, and measurements progress up the rod from there. Red numbers represent feet, and black numbers represent either inches or tenths of a foot. (2.7-2)

Lesson 2.8

- Drawings used in construction are made to a reduced scale so that the parts of the building or site fit on the drawing sheet. A drawing that is drawn to scale is in exact proportion to the actual size. (2.8-1)
- Dimensions are communicated using dimension lines and extension lines. A dimension line runs the object's length being measured. Extension lines are drawn perpendicular to the dimension line. (2.8-2)
- Foundation plans are drawings that show construction details for the foundation of a building. Like all construction drawings, interpreting them accurately requires a good understanding of the use of scale and dimensions. (2.8-3)

Lesson 2.9

- Key considerations for green building projects include overall site development; controlling water runoff during and after construction; water and energy use and efficiency for the home; energy sources and generation; material selection, sourcing, use, and waste disposal; indoor environmental control and air quality; and the ongoing maintenance and use of the occupied home. (2.9-1)
- Design and orientation can impact energy efficiency by positioning the house in a direction that can absorb the most sunlight, designing windows facing the direction that will allow the most natural light, and considering how landscape can block wind or provide shade in the summer. (2.9-2)
- Alternatives to traditional foundation materials, such as ICFs and precast concrete walls, have gained popularity and are becoming mainstream choices for construction. These products reduce waste and improve the energy performance of a home. (2.9-3)

Apply and Analyze

1. While production homes are cost-effective, what are some challenges of using this building method? (2.2-2)
2. When is a slab-on-grade foundation generally used over other foundations? (2.3-1)
3. What is the main difference between a grade beam foundation and a spread footing? (2.3-1)
4. What does a mortar mixture consist of, and how does it function in construction? (2.4-1)
5. Describe how to operate an optical leveling instrument. (2.5-1)
6. Explain the importance of designating a competent person on any jobsite. (2.6-1)
7. Describe three caught-in or caught-between hazards a worker may encounter during the foundation phase of construction and how these hazards can be avoided. (2.6-3)
8. When would a worker use an engineer's rod instead of a builder's rod? (2.7-2)
9. An architect produces drawings for Zver Architectural Corporation. Why would they choose to draw floorplans in 1/4″ = 1′ scale rather than 1/2″ = 1′? (2.8-1)
10. Explain why a worker needs to use detail drawings in conjunction with plan and elevation drawings. (2.8-3)

Critical Thinking

1. How do the various trades and occupations in residential construction work together to complete a home? (2.1-2)
2. Explain why understanding metric measurements is important in construction. (2.7-2)
3. Why are environment or community considerations an important aspect of green building strategies? (2.9-2)
4. Insulated concrete forms (ICFs) have gained popularity in recent years. How do you think this has disrupted or changed the workflow on new home construction sites? (2.9-3)

Communicating about Construction

1. **Listening.** In small groups, discuss with your classmates—in basic, everyday language—your knowledge on footings and foundations. Try incorporating terms from the chapter in your descriptions. Conduct this discussion as though you had never read this chapter. Take notes on the observations expressed. Then review the points discussed, factoring in your new knowledge of foundations. Develop a summary of what you have learned about footings and foundations and present it to the class, using the terms you learned in this chapter. (2.3-1)
2. **Speaking.** Obtain small amounts of Portland cement, mason sand, crushed stone, or clean gravel. Mix three batches of concrete, varying the mix for each batch. Wear safety glasses and rubber or chemical-resistant gloves while mixing the concrete. Record how much of each ingredient you used in each batch. Place your concrete in long form molds to create long concrete beams. Use a clean dowel, a tool such as a trowel or mag float, or piece of rebar to rod the concrete (work the rod up and down several times), ensuring that there are no voids in the batch. Be careful not to tear your form. After seven days, test the strength of each concrete mixture by elevating and supporting the ends of the beam and placing a weight at the center. Record results and any differences in strength. Make a presentation and report your findings to your class. (2.4-1)
3. **Speaking.** Make a series of safety posters for the Focus Four hazards outlined by the Occupational Safety and Health Administration (OSHA). Explain your posters to the class. (2.6-2)

Maudib/iStock/Thinkstock

CHAPTER 3

Framing Systems

Chapter 3 Outline

Introduction

So far, we have covered how the construction of a new home begins. After the foundation of a home is complete, the job seems to move faster as the house's frame takes shape. In this chapter, we will outline types of business structures, review the principles of framing, and explore some more materials, tools, and fasteners used when framing a home. To continue building your knowledge of print reading, you will be introduced to floor plans. We will also explore key safety concerns that might be encountered while framing and discuss methods for reducing safety risks in general. Finally, we will explore alternative methods of framing that improve energy efficiency and reduce waste.

LESSON 3.1

Industry and Careers

LEARNING OUTCOMES

After completing this lesson, you will be able to:

3.1-1 Summarize four types of business structures that are used in the United States.

TECHNICAL TERMS

corporation
limited liability company (LLC)
partnership
sole proprietorship

Types of Business Structures

In *Chapter 2, Excavation and Foundations*, we explored common occupations in residential construction. In this lesson, we will discuss common structures used to organize a business and their legal status in the United States. Many of these structures lend themselves to an individual with an entrepreneurial spirit who aspires to operate their own business, **Figure 3.1-1**. To operate legally, a business must be categorized as a sole proprietorship, a partnership, a limited liability company (LLC), or a corporation.

shisukaa/Shutterstock.com

Figure 3.1-1. Home builders and other construction business owners must understand business management as well as construction.

Sole Proprietor

A ***sole proprietorship*** is a type of business structure with only one owner who is responsible for all aspects of the business. Sole proprietors are self-employed and do business under their given name or can create a business name. The owner has complete control over the business, but also bears all the financial risk and legal liability. As a sole proprietor, the business and the owner are considered the same legal entity; the owner is personally responsible for all debts and obligations of the business. Sole proprietors are also responsible for ensuring compliance with safety regulations and standards to protect both their workers and clients. For instance, if an accident or property damage were to occur, a sole proprietor could be exposed to significant financial risk.

Sole proprietorships are easy to set up and operate. In fact, if you do business without setting

Pro Tip

Although the terms may sound similar, there is a big difference between a sole proprietor and an independent contractor. The term independent contractor describes a business relationship between companies. A subcontractor, for instance, may do work under contract for a general contractor. In this case, the subcontractor's business is independent of the general contractor's. However, the business structure of that subcontractor could be any of the four discussed in this chapter.

up a legal structure, then you are automatically categorized as a sole proprietor. As a sole proprietor's business grows or risk increases, they may opt to move to a different business structure that offers more protection. Many of the small residential construction companies operating in towns and cities across the United States operate as sole proprietorships.

Partnership

A *partnership* is a type of business structure in which two or more people own a business together. Each partner contributes to the business. Contributions might be through financial investment, skills, or labor. Partners share the profits and losses according to a legal partnership agreement. The partners also share responsibility for the liabilities, debts, and obligations of the business. Management of a partnership may be split between partners or delegated mostly to one partner. Sometimes a partner may simply be a financial investor who expects a share of the profit but is not involved in the operation or management of the business.

Limited Liability Company (LLC)

A *limited liability company (LLC)* is a type of business structure that offers the owner(s) some level of protection from the debts and liabilities of the business. With an LLC, the owner and their personal assets such as vehicles, bank accounts, and home are protected if the company gets sued or defaults on a debt. LLCs offer a higher level of separation and protection for a business owner than a sole proprietorship. Yet, when it comes to taxation, the government views an LLC similar to a sole proprietor. However, starting this type of business involves costs and legal considerations. An LLC typically requires more paperwork and higher initial costs compared to a sole proprietorship. You may need to register with the state, draft an operating agreement, and comply with ongoing regulatory requirements.

Pro Tip

For anyone who considers starting a construction business, it is advisable to consult with a business attorney or professional advisor. A professional who is familiar with the specific legal and regulatory requirements for construction businesses can ensure compliance before forming an LLC or corporation.

Corporation

A *corporation* is a type of business structure in which the business and its liabilities are separate from the owner or owners of a company. The corporation offers the best protection to the business owner. Legally, corporations are viewed as their own entity, separate and distinct from the owner(s). In fact, corporations generate profits that are taxed independently.

When a person incorporates a business, the ownership of that entity is allocated through shares. Shares of the corporation can be sold to generate funds for the business to grow. Shareholders own a stake in the corporation but are not actively involved in managing the operations of the company.

Corporations require extensive recordkeeping and reporting procedures. The cost to form a corporation is higher than that of other business structures.

LESSON 3.1 Review

Answer the following questions based on information in this lesson.

1. *True or False?* Sole proprietors are self-employed. (3.1-1)
2. In a partnership, what responsibilities do the owners share? (3.1-1)
   A. Financial investment
   B. Skills and labor
   C. Daily operations
   D. All are correct.
3. What business structure offers a level of protection for the owner's personal assets? (3.1-1)
   A. A sole proprietorship
   B. A partnership
   C. A limited liability company (LLC)
   D. None are correct.
4. *True or False?* Shareholders own a stake in an LLC but are not actively involved in managing the operations. (3.1-1)
5. The cost to form a _____ is higher than that of other business structures. (3.1-1)
   A. sole proprietorship
   B. partnership
   C. limited liability company (LLC)
   D. corporation

LESSON 3.2

Construction Fundamentals

LEARNING OUTCOMES

After completing this lesson, you will be able to:

3.2-1 Explain the purpose of building codes in the construction industry.

3.2-2 Outline the steps to obtaining a building permit.

3.2-3 Describe how inspections are conducted during a construction project.

TECHNICAL TERMS

authority having jurisdiction (AHJ)
building code
model code

Building Codes

Imagine living in a home built by someone who had a limited understanding of or experience in construction. Also, imagine if companies regularly built homes using cheap materials that were not safe to use, or assembled in a way that would not withstand a mild windstorm or minor earthquake. What problems might arise? A home may appear to be well built and yet fail to withstand daily use or protect you during a major weather event.

Materials and how they are used in home construction should be considered carefully. Some products used to build a home may be toxic or harmful, such as lead paint or toxic fumes from solvents or glues. How someone moves through the home should also be considered. Imagine how dangerous it would be in the event of a fire if there was only one way to exit a home, or if that exit was blocked. Even window sizes need to account for a first responder potentially needing access into the home to rescue the occupants.

Building codes are essential for many reasons. What issues could arise if a structural support beam was not strong enough, not installed properly, or accidentally removed in an open floor plan home renovation? Many serious safety considerations are part of building a home. These safety concerns are managed, avoided, or minimized through the regulation of building codes.

Code Note

In doing any kind of construction work, the importance of closely following all building codes cannot be overemphasized. A well-trained builder is expected to know the requirements of local building codes. These codes dictate where materials may be used and how a structure is to be built. Failure to follow the codes can prove costly, since inspectors may require work that does not meet code requirements to be redone. This means more time and money spent on materials and labor.

Building codes are a collection of laws that apply to a given state or municipality. A ***building code*** gives the minimum requirements to which a new building must be constructed and also covers important aspects of the alteration, repair, and demolition of existing buildings.

Building codes exist to provide for the health, safety, and general welfare of building occupants and other people in the community. Many cities and towns have their own local building codes. If they do not, the state building code applies. State codes usually require that local codes be at least as strict as the state code. A local government body, identified as the ***authority having jurisdiction (AHJ)***, regulates building construction in a particular area. Often, the AHJ is a city's building department.

Model Building Codes

Continual improvements take place in construction methods and materials, which can make the updating of local building codes cumbersome and costly. For this reason, many communities have adopted model codes. A ***model code*** is a set of building codes generally accepted across the industry and can apply to any geographic location. One of the first model codes was introduced by the American Insurance Association (successor to the National Board of Fire Underwriters). This publication was known as the *National Building Code*.

The International Code Council introduced the *International Residential Code for One- and Two-Family Dwellings (IRC)* in 2000. The *IRC* is a model code, meaning that it is available for authorities having jurisdiction to use as the basis for their local codes. The International Code Council publishes an updated version of the *IRC* every three years to keep the codes up to date. A sample of the language used in the *IRC* regarding ceiling height is shown in **Figure 3.2-1**.

SECTION R305
CEILING HEIGHT

R305.1 Minimum height.
Habitable space, hallways and portions of *basements* containing these spaces shall have a ceiling height of not less than 7 feet (2134 mm).

Bathrooms, toilet rooms and laundry rooms shall have a ceiling height of not less than 6 feet 8 inches (2032 mm).

Exceptions:

1. For rooms with sloped ceilings, the required floor area of the room shall have a ceiling height of not less than 5 feet (1524 mm) and not less than 50 percent of the required floor area shall have a ceiling height of not less than 7 feet (2134 mm).
2. The ceiling height above bathroom and toilet room fixtures shall be such that the fixture is capable of being used for its intended purpose. A shower or tub equipped with a showerhead shall have a ceiling height of not less than 6 feet 8 inches (2032 mm) above an area of not less than 30 inches (762 mm) by 30 inches (762 mm) at the showerhead.
3. Beams, girders, ducts or other obstructions in *basements* containing *habitable space* shall be permitted to project to within 6 feet 4 inches (1931 mm) of the finished floor.

R305.1.1 Basements.
Portions of *basements* that do not contain *habitable space* or hallways shall have a ceiling height of not less than 6 feet 8 inches (2032 mm).

Exception: At beams, girders, ducts or other obstructions, the ceiling height shall be no less than 6 feet 4 inches (1931 mm) from the finish floor.

Figure 3.2-1. This is the section of the *IRC* that regulates ceiling heights. Most sections of the code are much longer, and many have several pages of tables and illustrations.

Balloon framing, **Figure 3.3-4**, was once very popular, but it is rarely used in modern construction. Builders should still be familiar with this framing method as they will likely encounter it while remodeling or repairing older homes. ***Balloon framing*** is a framing system where the studs are continuous from the sill to the rafter plate. Balloon framing requires longer studs and the addition of a firestop at each floor level to meet modern building code requirements. It also does not provide the platform (floor) on which to assemble the wall framing, making it more difficult to build than platform framing. However, many balloon-framed homes still exist and builders may encounter this design in remodeling or repair projects.

Post-and-Beam Framing

Post-and-beam framing, also called *plank-and-beam framing*, differs from platform and balloon framing due to the use of heavy structural members that are at least 4″ thick. Because of their great strength, these members may be spaced farther apart than conventional framing members, **Figure 3.3-5**. Many old homes and agricultural buildings were built with post-and-beam framing. With modern requirements for mechanical systems and insulation, this type of construction is more costly than conventional stick-built framing. Today post-and-beam framing is primarily used for architectural appeal.

Common Structural Members Used in Conventional Framing

Understanding the name and function of each framing member is the first step in grasping the overall concept of modern home building. The frame is the structural core of a home and plays the primary role in providing a medium for all other aspects of construction. Each component of framing has a role to play within the structure. Without structural support and proper load distribution, a home could not withstand natural forces. This lesson outlines the basic structural members used in platform framing.

First-Story Framing Members

The first-story framing rests on top of the sill. The ***span*** is the distance between the sills of opposing walls or other structural supports. Usually, the span between sills on the foundation walls is

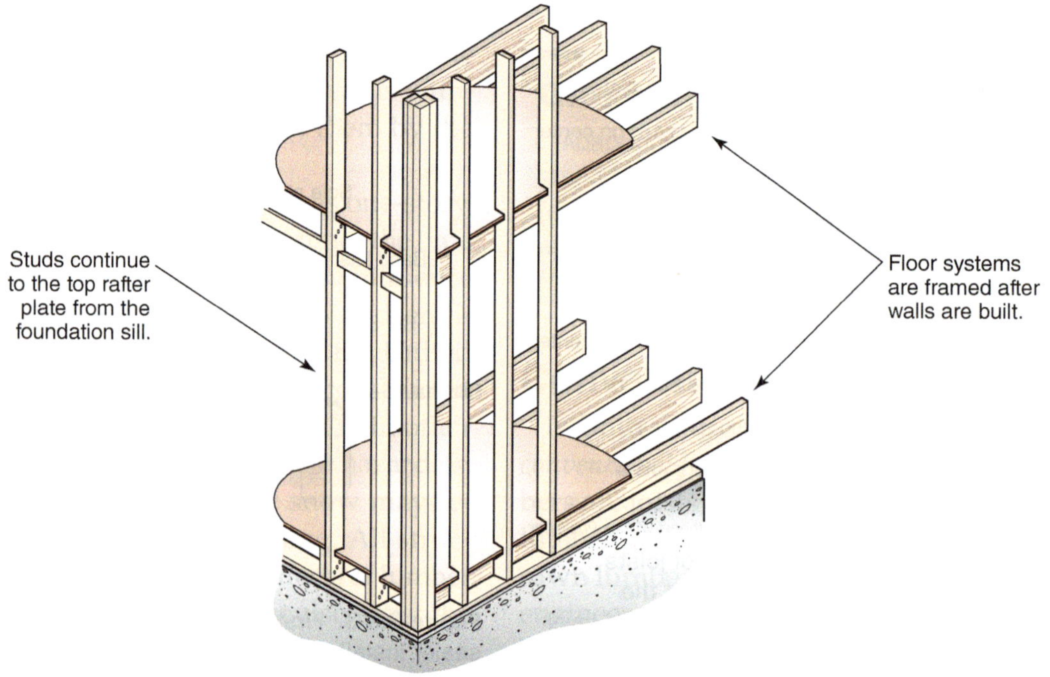

Figure 3.3-4. In balloon framing, the studs are continuous from the sill to the rafter plate.

FOTOGRIN/Shutterstock.com

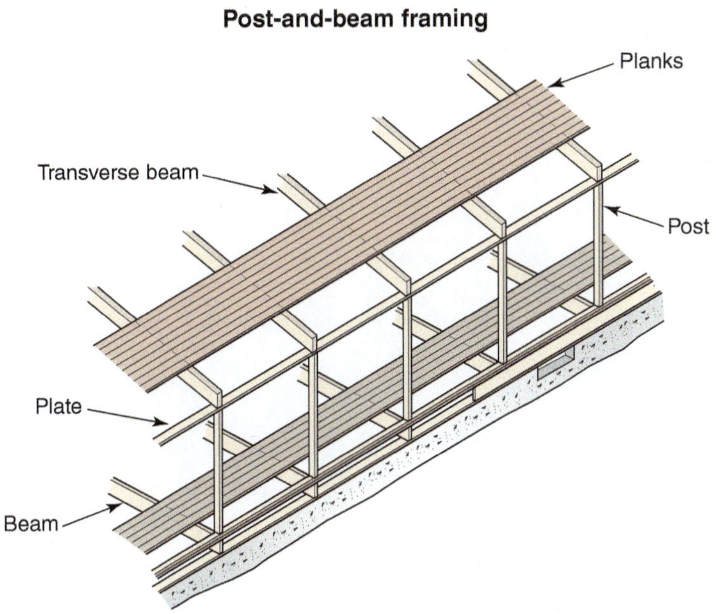

National Forest Products Assn.

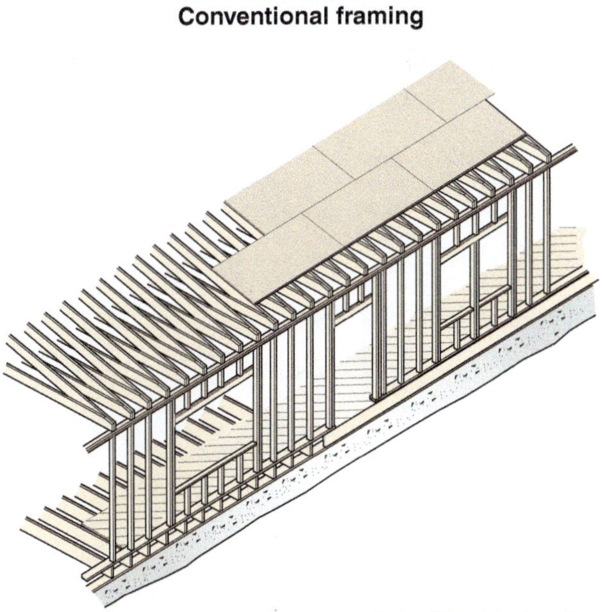

National Forest Products Assn.

Figure 3.3-5. Post-and-beam framing uses large framing members that are spaced further apart than conventional framing members. Posts and beams are often left exposed on the inside as an architectural feature.

so great that additional support must be provided between them. A ***girder***, also called a *beam*, rests on the foundation walls and on posts or columns. Girders support the floor system where the span is too great, **Figure 3.3-6**. Girders may be solid timbers, built-up lumber, engineered lumber, or steel beams. Sometimes, a loadbearing partition replaces a girder or beam. Specifications such as sizing and material for girders are typically found on the foundation plan of a home.

Built-up girders can be made of three or four pieces of 2″ lumber nailed together with 10d nails, **Figure 3.3-7**. Modern engineered lumber, such as laminated beams, may be used for girders. These are factory-made by gluing together thin strips of wood or veneer. Such beams can be manufactured in many different sizes and lengths. We will review more on engineered lumber in *Lesson 3.4*.

A ***joist*** is the horizontal member of the frame. Floor joists support the floor and transmit loads

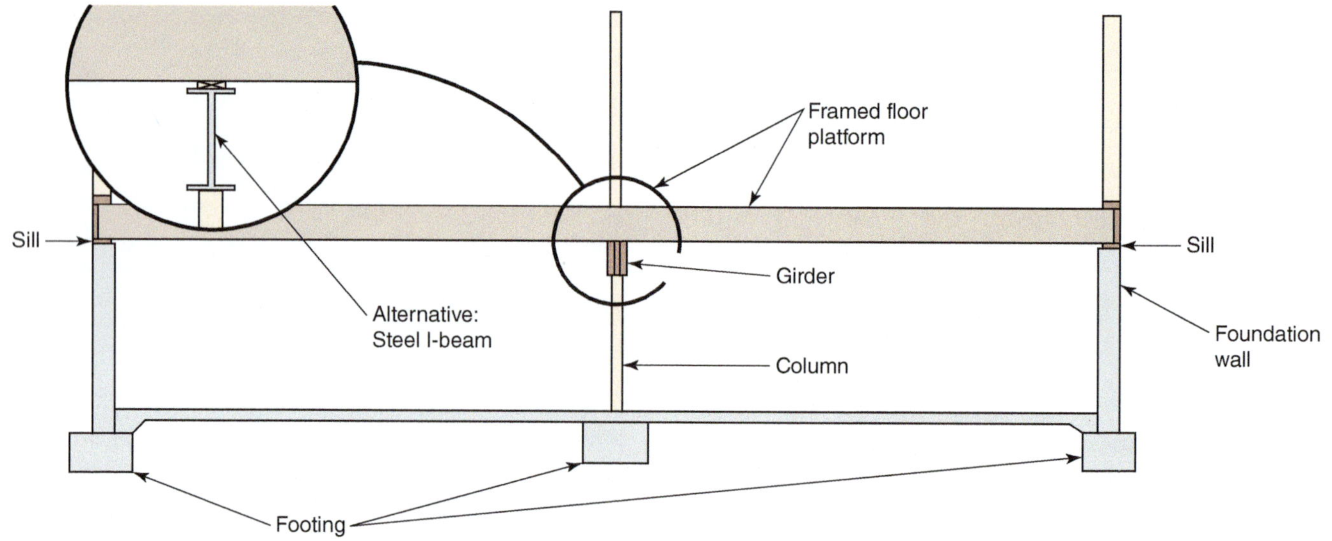

Goodheart-Willcox Publisher

Figure 3.3-6. Girders support the floor system where the span between foundation walls is too large. This drawing shows an option to use built-up lumber or a steel I-beam.

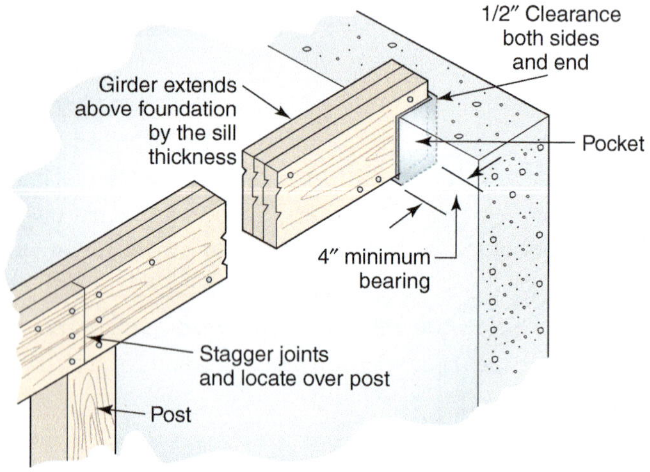

Goodheart-Willcox Publisher

Figure 3.3-7. When built-up wood girders are fabricated on the job, nails should be spaced no farther apart than 32″ along the top and bottom edges. A metal bearing plate should be placed under the girder at the foundation wall. Clearance in the pocket allows air circulation and prevents rot from moisture.

Goodheart-Willcox Publisher

Figure 3.3-8. Floor joists are being installed for a new home. Note the steel girders.

to the sills and girders, **Figure 3.3-8**. Outside joists, called *band joists* or rim joists, run along the perimeter of the floor system. Common floor joists span from the sill to the girder and provide a structure for the rest of the floor.

Traditionally, framing lumber such as 2×10s and 2×12s were used for floor joists. These are still used today, but it has become more common to use engineered lumber such as wood I-joists or floor trusses instead. These engineered products are available in longer lengths, are stronger, and are more stable than sawn lumber.

The most common spacing of wood joists is 16″ on center (OC). However, 12″, 19.2″, and 24″ OC

Code Note

Building codes typically specify allowable span distances for floor joists based on the size of the joist and the species and grade of wood. Always verify that the joist sizes for the project are acceptable to the applicable building code.

spacings are also used. Joists must not only be strong enough to carry the load that rests on them, but they must also be stiff enough to prevent undue bending or vibration.

Subfloors

The laying of the subfloor, or rough flooring, is the final step in completing the floor frame. Panel materials are used for this purpose, **Figure 3.3-9**. Oriented strand board (OSB) and plywood are the materials of choice. These products are discussed in *Lesson 3.4* of this chapter.

The subfloor serves three purposes:

- Adds rigidity to the structure
- Provides a base for finish flooring material
- Furnishes a work surface where a carpenter can lay out and construct wall framing

Wall-Framing Members

The wall-framing members used in conventional platform construction include sole plates, top plates, studs, headers, and sheathing. A ***stud*** is the vertical member of the wall frame. Except where interrupted by openings for windows and doors, they run full length from sole plate to top plate. The ***sole plate*** is the horizontal member at the bottom of the wall frame, supported by a wood subfloor, concrete slab, or other closed surface. The ***top plate*** is the horizontal member at the top of a wall frame.

Pro Tip

Wall studs can be purchased precut to 92 5/8″ for a standard wall height of just over 8′. Buying precut studs saves cutting time on the job.

A short stud, known as a ***cripple stud***, is used above and below wall openings. Cripple studs are placed on the same on-center layout as the common studs. Refer to **Figure 3.3-10**. A ***trimmer stud***, or *jack stud*, is a shortened stud that stiffens the sides of rough openings. Trimmer studs bear the direct load of a header and mate to the king stud. A ***king stud*** is a full-length stud at the sides of rough openings, which provides support to the trimmer stud. Headers are installed between the king studs on either side of the opening. Extra studs are used at corners, at the sides of the rough openings for doors and windows, and where an interior wall meets an outside wall. Studs and plates are made from 2×4 or 2×6 lumber. Headers usually require heavier material.

All these components are covered on the exterior by ***sheathing***, which are boards or prefabricated

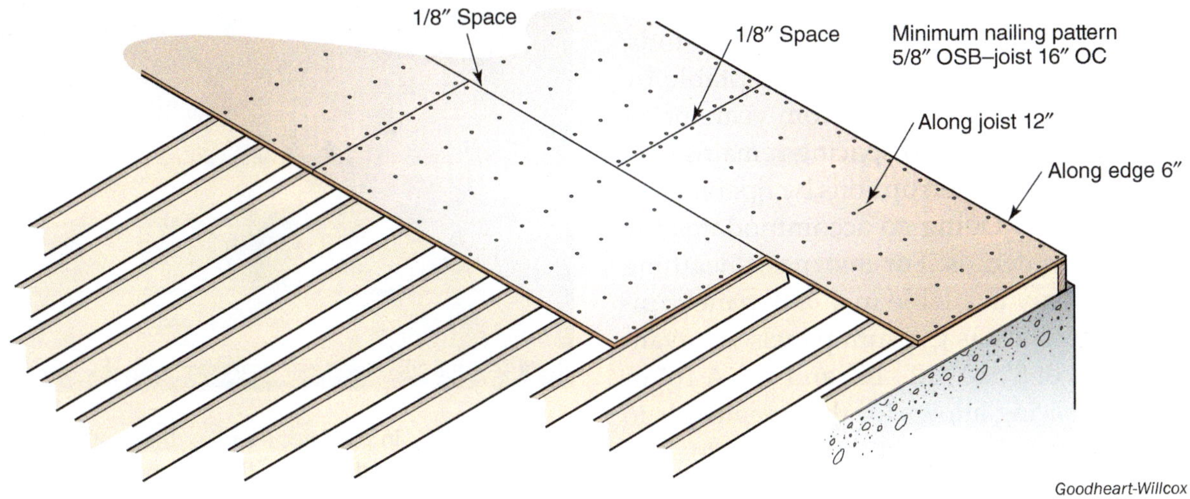

Goodheart-Willcox Publisher

Figure 3.3-9. Properly installing OSB subfloor. Note the nailing pattern and how joints are staggered for added strength.

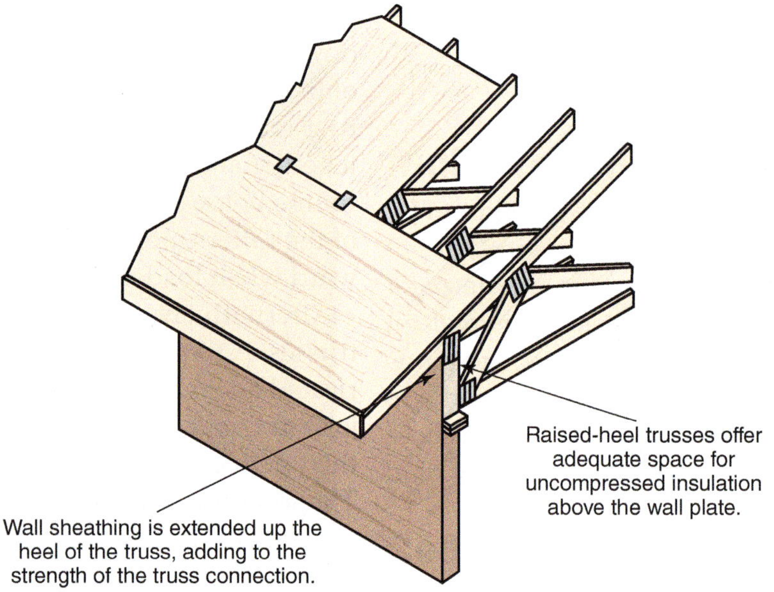

Figure 3.3-14. Raised-heel trusses give more space for insulation in the attic at the eaves.

Roof Sheathing

Sheathing, also called *decking*, provides a nailing base for the roof covering and adds strength and rigidity to the roof frame. The most common sheathing material is structural panels of plywood or oriented strand board (OSB). The required thickness of the panels varies with the roofing material and rafter spacing.

Green Note

Although not considered framework, structural insulated panels (SIPs) can also be used to construct a roof structure. SIP systems are engineered for the individual project. Consult the working drawings for assembly requirements. *Lesson 3.9* provides more information on SIPs.

LESSON 3.3 Review

Answer the following questions based on information in this lesson.

1. What is the most common framing method used in residential construction in the United States? (3.3-1)
   A. Platform framing
   B. Balloon framing
   C. Post-and-beam framing
2. What type of framing uses heavy structural members that are at least 4″ thick? (3.3-1)
   A. Platform framing
   B. Balloon framing
   C. Post-and-beam framing

Matching: Match each framing member with its function.

A. Girder
B. Joist
C. Rafter
D. Sole plate
E. Stud

3. Horizontal framing member of the floor (3.3-2)
4. Horizontal member at the bottom of the wall frame (3.3-2)
5. Beam that supports the floor system (3.3-2)
6. Vertical member of the wall frame (3.3-2)
7. Sloped framing member that runs downward from the peak of the roof (3.3-2)

LESSON 3.4

Materials and Fasteners

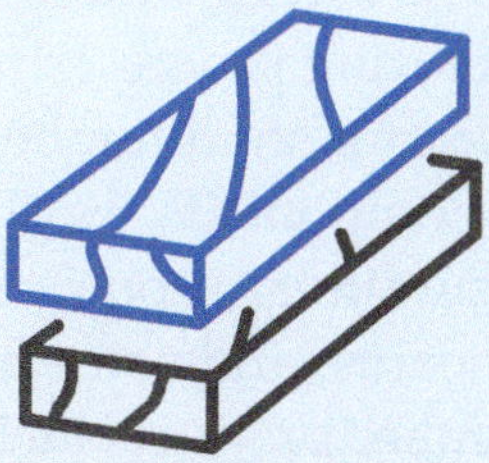

LEARNING OUTCOMES

After completing this lesson, you will be able to:

3.4-1 Describe commonly used engineered lumber products in framing.

3.4-2 Identify metal connectors, bonding agents, and fasteners used in framing.

3.4-3 Explain the difference between framing nails and trim nails.

TECHNICAL TERMS

engineered lumber product (ELP)
finish nail
floor truss
glue-laminated beam (glulam)
I-joist
laminated-strand lumber (LSL)
laminated-veneer lumber (LVL)
metal framing connector
oriented strand board (OSB)
parallel-strand lumber (PSL)
plywood
trim nail

Engineered Lumber Products

Chapter 1, Preparing for Construction, introduced dimensional framing lumber. Builders have long relied on traditional wood products such as sawn framing lumber for residential and light commercial framing. Modern tree-farming technology has ensured that this material continues to be available, keeping up with the increased demand for wood structures. It once took more than 80 years to grow a tree to sawmill size; tree farming has shortened the span to 20–30 years. In the process, unfortunately, the tree's annular rings, which determine its strength, are much farther apart, resulting in weaker dimension lumber. The industry has found a way to strengthen lumber and use more of the tree through the development of engineered lumber products.

An ***engineered lumber product (ELP)*** includes those wood structural units that have been altered through manufacturing processes to make them stronger, straighter, and more dimensionally stable than sawn lumber. These processes produce more lumber products from less timber and can utilize smaller trees that are unsuitable as sawn lumber. Where sawn lumber typically uses only 40% of a log, engineered materials use 75% of the log

Green Note

Some trees can be genetically modified to grow more quickly. These trees can be farmed on large tracts of land and are planted and harvested on a continuous cycle. These practices help to make construction lumber a sustainable material.

Green Note

Precut, custom, and fabricated components can be cut to specification at the lumber mill and shipped to the jobsite. This reduces jobsite waste due to material misuse, poor storage, carpenter error, and other mishaps.

for structural lumber. At the same time, the manufacturing processes require less energy than those used for solid lumber.

Components are glued together in different configurations—some are solid, some are shaped like I-beams, and some are in truss form. Engineered lumber products include the following:

- Plywood
- Oriented strand board (OSB)
- Laminated-veneer lumber (LVL)
- Glue-laminated beams (glulams)
- I-joists
- Open-web trusses
- Laminated-strand lumber (LSL)
- Parallel-strand lumber (PSL)

Plywood

Plywood is an engineered wood panel constructed by gluing together a number of thin layers (plies) of wood with the grain direction turned at right angles in each successive layer, **Figure 3.4-1**. An odd number (3, 5, 7) of plies is used so they are balanced on either side of a center core and so the grain of the outside layers runs in the same direction.

The two basic types of plywood are exterior and interior plywood. Exterior plywood is bonded with waterproof glues. It can be used for siding, concrete forms, and other construction where it will be exposed to the weather or excessive moisture. Interior plywood may be manufactured with either interior or exterior glue. Exterior glue is generally used. The grade of the panel is mainly determined by the quality of the layers.

Akmenra/shutterstock.com

Figure 3.4-1. Sheets of plywood on a construction site. Notice the layers that make up each sheet.

Plywood is made in thicknesses from 1/8″ to more than 1″. Common sizes are 1/4″, 3/8″, 1/2″, and 3/4″. Thicknesses of 15/32″, 19/32″, and 23/32″ are also available. A standard panel size is 4′ wide by 8′ long. Plywood can be used for sheathing on framed walls and roofs, but more often OSB is used as a less expensive alternative.

Safety Note

A common mistake for beginners is to place a piece of lumber or sheathing on a pair of sawhorses, then try to make a cut between the sawhorses. As the material begins to fall, it pinches the saw blade. The result can be a violent kickback of the saw and a poor cut. Always allow the waste to fall away freely or remain fully supported.

Plywood Grades

Structural plywood panels are quality graded in two ways:

- A lettering system indicates the veneer quality for the face and back of the panel, **Figure 3.4-2**.
- A name indicates the intended use or "performance rating."

Veneer Grades for Softwood Plywood	
A	Smooth paintable. Not more than 18 neatly made repairs, boat, sled, or router type, and parallel to grain, permitted. May be used for natural finish in less demanding application. Synthetic or wood repairs permitted.
B	Solid surface. Shims, sled, or router repairs, and tight knots up to 1″ across grain permitted. Wood or synthetic repairs permitted. Some minor splits permitted.
C	Tight knots to 1 1/2″. Knot holes to 1″ across grain and some to 1 1/2″ if total width of knots and knotholes is within specified limits. Synthetic or wood repairs permitted. Discoloration and sanding defects that do not impair strength permitted. Limited splits allowed. Stitching permitted.
C (plugged)	Improved C veneer with splits limited to 1/8″ width and knotholes or other open defects limited to 1/4″ by 1/2″. Wood or synthetic repairs permitted. Some broken grain permitted.
D	Knots and knotholes to 2 1/2″ width across grain and 1/2″ larger within specified limits. Limited splits are permitted. Stitching splits are permitted. Limited to Exposure 1 or interior panels.

APA—The Engineered Wood Association

Figure 3.4-2. Description of veneer grades for softwood plywood.

APA—The Engineered Wood Association has a rigid testing program based upon standard PS 1-95. Mills that are members of the association may use the official grade-trademark. It is stamped on each piece of plywood. **Figure 3.4-3** explains the parts of the APA trademark and grade stamp.

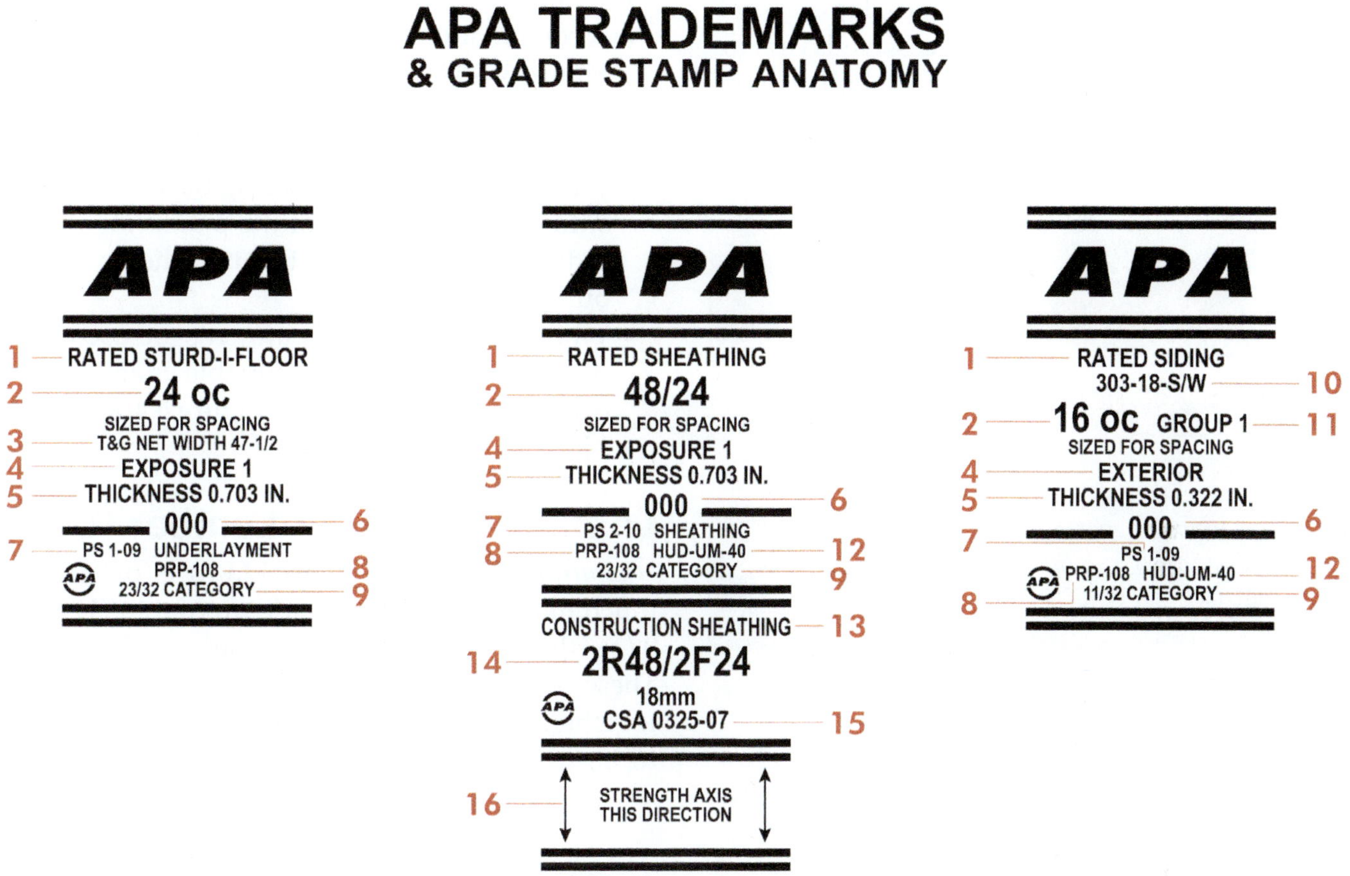

1. **Panel grade** — the term "grade" may refer to panel grade or veneer grade. Panel grades are generally identified in terms of the veneer grade used on the face and back of the panel (e.g., A-B, B-C, etc.) or by a name suggesting the panel's intended end use (e.g., APA Rated Sheathing, Underlayment, etc.).
2. **Span Rating** — Two numbers separated by a slash. The left-hand number is the maximum recommended center-to-center spacing for supports in inches when the panel is used for roof sheathing with long dimensions across supports. The right-hand number is the maximum center-to-center spacing of supports in inches when the panel is used for subflooring with the long dimension across supports. The Span Rating on APA RATED STURD-I-FLOOR and APA RATED SIDING panels appears as a single number. The Span Ratings for APA RATED STURD-I-FLOOR panels, like those for APA RATED SHEATHING, are based on application of the panel with the long dimension or strength axis across three or more supports.
3. **Tongue-and-groove** — A system of jointing in which the rib or tongue of one member fits into the groove of another. In Sturd-I-Floor panels, tongue-and-groove edges eliminate the need for blocking beneath panel edges for support.
4. **Bond Classification** — Exposure ratings for APA wood structural panels may be Exterior or Exposure 1. The classifications are based on the severity and duration of weather and moisture exposure the panels are designed to withstand, and the wood materials and adhesives used in manufacture.
5. **Mill thickness declaration** — Designated thickness subject to tolerances specified in standard.
6. **Mill number** — Manufacturing mill identification number.
7. **Product Standard** — An industry product manufacturing or performance specification. Voluntary Product Standard PS 1-09 for Construction and Industrial Plywood was developed cooperatively by the plywood industry and the U.S. Department of Commerce. Voluntary Product Standard PS 2-10, Performance Standard for Wood-Based Structural Use Panels, establishes performance criteria for specific designated construction applications.
8. **APA's Performance Rated Panel Standard** — A standard applying to panels such as APA RATED SHEATHING, APA RATED STURD-I-FLOOR, and APA RATED SIDING. Panels manufactured to meet APA performance standards must satisfy rigorous, exacting performance criteria for the intended applications.
9. **Performance Category** — A panel designation related to the panel thickness range that is linked to the nominal panel thickness designations used in the International Building Code (IBC) and the International Residential Code (IRC).
10. **Siding face grade** — Grade identification of siding face, based on number of repairs and appearance characteristics.
11. **Species group number** — Classified according to strength and stiffness under manufacturing standard PS 1 into Groups 1 through 5. Group 1 woods are the strongest, Group 2 seconded strongest, and so on.
12. **HUD recognition.**
13. **Panel grade, Canadian Standard** — Some APA panels are manufactured to both U.S. and Canadian standards, and carry a dual mark, as shown in the Rated Sheathing mark above. Other panels may carry a single mark for either the U.S. or Canadian standard.
14. **Panel mark** — Rating and end-use designation, Canadian standard. In the Canadian span marking, the "R" signifies roofs and the "F" represents subfloors.
15. **Canadian performance-rated panel standard.**
16. **Panel face orientation indicator.**

APA—The Engineered Wood Association

Figure 3.4-3. Explanation of the elements of the APA trademark and grade stamp found on all rated panels.

Oriented Strand Board (OSB)

Oriented strand board (OSB) is an engineered wood panel made of individual flakes of wood that are laid down like a blanket and adhered to each other with suitable resins and glues, **Figure 3.4-4**. The fibers are put down in successive layers arranged at right angles to one another. It is generally manufactured from aspen poplar in the northern part of North America and from southern yellow pine in the South. The panels are bonded under heat and pressure with thermosetting binders. Panels range in thickness from 1/4″ to 1 1/8″. There are many applications of OSB, including the following:

- A combination subfloor and underlayment, **Figure 3.4-5A**.
- Sheathing for roofs and exterior walls, **Figure 3.4-5B**. Oriented strand board provides an excellent nailing base for siding. The large size of the panels also reduces air infiltration through the building's walls.
- When special surface treatments are used, oriented strand board is suitable as an exterior finish as a siding, **Figure 3.4-5C**.
- I-joists, **Figure 3.4-5D**.

hodim/Shutterstock.com

Figure 3.4-4. Oriented strand board (OSB) has many applications in construction ranging from structural sheathing to exterior siding products.

Laminated-Veneer Lumber (LVL)

Laminated-veneer lumber (LVL) is produced much like plywood and made up of thin (1/10″–3/16″ thick) sheets of wood that are 27″–54″ wide. These veneer sheets are glued, stacked, and pressed together, then cut into various lengths and widths to be used as structural framing members like

A LP Corp

B blbimages/Shutterstock.com

Figure 3.4-5. A—OSB subflooring is designed to resist water damage from weather during the framing process. It is sanded smooth on the surface to accommodate finish flooring. B—Oriented strand board has become popular as sheathing material. The textured side should be installed facing up to provide better footing. (*Continued*).

C LP Corp D Weyerhaeuser

Figure 3.4-5. (*Continued.*) C—LP Smartside is an exterior-grade OSB siding that looks like real wood siding. D—I-joists supported by a glulam girder. I-joists can be used in place of sawn lumber for floor joists and rafters.

girders and headers, **Figure 3.4-6**. It is produced under various trade names, such as Microlam™, Parallam™, and Gang-Lam™.

LVL products are dimensionally stable, uniform in size throughout, and predictable in performance and can be nailed, drilled, and cut with ordinary construction tools. LVL products easily take stain and can be left exposed, where they become a natural part of the interior décor.

Although a cross section of LVL looks somewhat like plywood, there is an important difference—grain orientation. In conventional plywood, each veneer layer is placed at a 90° angle to the previous layer. In LVL, all the veneer panels have their grain running in the same direction. This provides greater strength, allowing the member to carry the heavy loads required of beams, headers, and truss components. Standard laminated beams are 1 3/4″ thick with depths of 5 1/2″, 7 1/4″, 9 1/2″, 11 7/8″, 14″, 16″, and 18″.

Goodheart-Willcox Publisher

Figure 3.4-6. LVL is available in thicknesses of 1 1/4″, 1 3/4″, and 3 1/2″, widths from 7 1/4″ to 24″, and any length needed.

Pro Tip

Because moisture content changes the dimension of lumber, it should be accounted for in the building process. Engineered lumber is stable and will remain more consistent in size than conventional lumber. When using both together, moisture content and shrinkage should be considered. If not adjusted for, irregularities in the finished product may result, **Figure 3.4-7**.

Glue-Laminated Beams (Glulams)

A ***glue-laminated beam (glulam)*** is made by gluing and then applying heavy pressure to a stack of four or more layers of 1 1/2″ thick stock. The result is a beam that can be up to 30″ deep. The layers, called *lams*, are arranged to produce maximum strength.

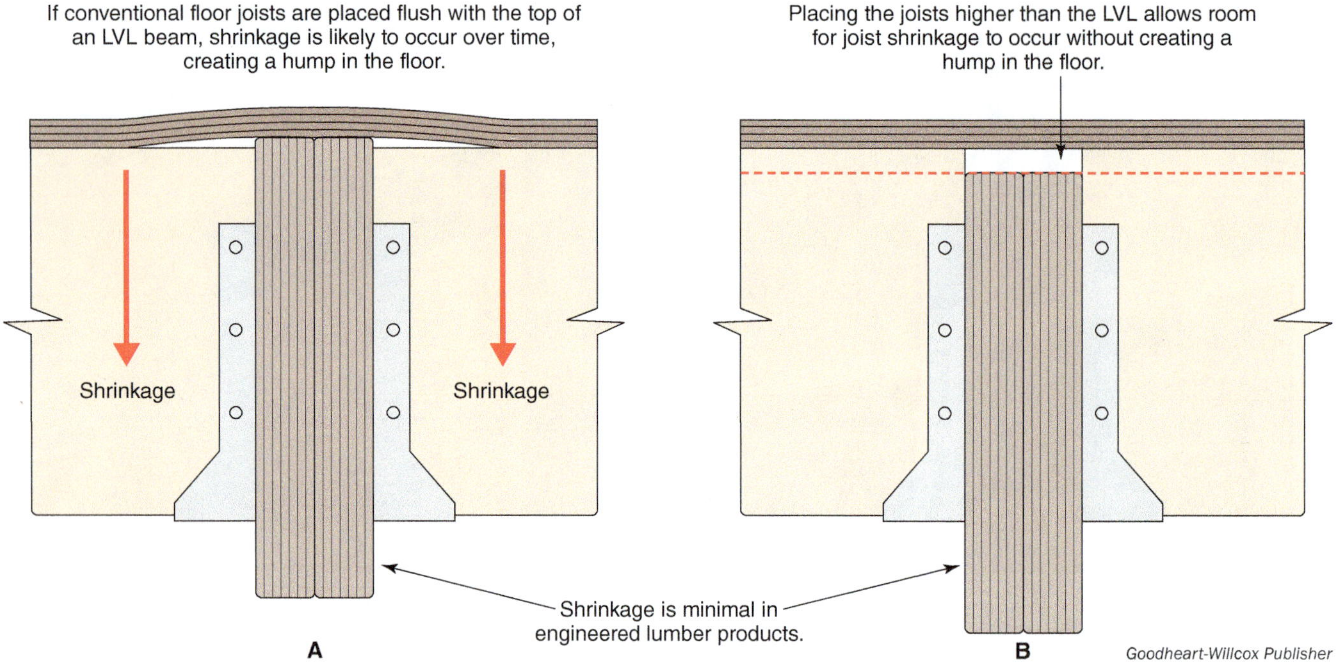

Goodheart-Willcox Publisher

Figure 3.4-7. Moisture content changes the dimension of lumber. Different lumbers will be affected differently. A—When abutting conventional lumber to an engineered beam, a hump in the floor can form if shrinkage is not accounted for. B—Conventional lumber should be held above the LVL during construction. The moisture content should be measured to determine how much shrinkage to expect.

Glulams are normally used (either exposed or hidden) for rafters (post-and-beam framing), floor-support beams, and stair stringers. Manufacturers also produce glulams in curved shapes for supporting arched roofs such as those found in churches or other public buildings, **Figure 3.4-8.**

The appearance grade of a glulam has no bearing on its strength. Three grades of appearance are available.

- **Industrial appearance**—Designed for use where appearance is not important
- **Architectural appearance**—Where appearance is important
- **Premium appearance**—When appearance is critical

Fire resistance is a significant advantage of glue-laminated beams. They are slow to ignite, and they burn slowly.

Sizes

Glulams are sold in both custom and stock sizes. Manufacturers produce to customer specifications for long spans, unusually heavy loads, or other conditions. Stock beams are manufactured in commonly used dimensions. Distributors and dealers cut them to order for customers. Typical stock widths are 3 1/8″, 3 1/2″, 5 1/8″, 5 1/2″, and 6 3/4″. These meet the requirements of most residential applications.

I-Joists

An ***I-joist*** has, as the name suggests, an "I" shape, **Figure 3.4-9.** They consist of flanges (also called *chords*) of structural composite lumber, such as laminated-veneer lumber or sawn lumber.

Rik Vandermeulen, Unalam

Figure 3.4-8. Glulams can be manufactured with curves and great depths. They are often an architectural highlight of a building.

Goodheart-Willcox Publisher

Figure 3.4-9. I-joist with top and bottom flanges consisting of either laminated-veneer lumber or solid finger-jointed lumber and OSB web. To avoid damage, always store I-joists with web upright.

I-joists also consist of a vertical web of oriented strand board that is 3/8″ or 7/16″ thick. The web is fastened into grooves in the flanges using waterproof glue.

Code Note

For fire protection, *IRC R501.3* requires that manufactured I-joists be covered with 1/2″ gypsum wallboard or equivalent, unless protected by an automatic sprinkler system.

I-joists are light, straight, and strong. They are designed to span long distances. Though mostly used as joists and rafters, I-joists may be used as headers, too. Most mills deliver them in lengths up to 60′. Distributors and dealers cut them to frequently used lengths (16′–36′) specified by builders. Most common depths are 9 1/2″, 11 1/2″, 14″, and 16″.

Open-Web Trusses

Open-web floor trusses are sometimes used in place of floor joists, **Figure 3.4-10**. This is especially true when dealing with long spans. Open-web *floor trusses* are fabricated in factories from solid 2×4 lumber. Sometimes, steel webs are used instead of wood. Depths of 14″ and 16″ are most common. The open webs of the trusses allow installation of pipes, wiring, drains, and other mechanical systems without having to cut openings, as in other types of joists.

Safety Note

Use great care in working with I-joists. They are unstable until properly braced. Failing to observe the following precautions and practices could result in accidents and damage:

- Never allow workers to walk on joists until the joists are braced.
- Do not stack building materials on unsheathed joists. Rather, place them only over beams or walls.
- Subflooring must be completely attached to each joist before additional loading of the floor.
- Install and nail all blocking, hangers, and rim boards at the ends of the joists.
- To temporarily brace joists, provide a rigid structure such as a braced wall or permanent deck (sheathing) nailed to the first 4′ of the joists.
- Attach 1×4 struts from the braced area across the tops of the rest of the joists. Unless this is done, sideways buckling or a rollover is likely.
- When joists are cantilevered, struts must be attached to both the top and bottom flanges.
- Keep flanges straight with a tolerance of no more than 1/2″ of true alignment.

Laminated-Strand Lumber (LSL)

Laminated-strand lumber (LSL) is laid up from 1/32″ × 1″ × 12″ strands of wood bonded with polyurethane adhesive. It is available in two thicknesses: 1 3/4″ and 3 1/2″. Depths vary up to a maximum

herreid/Shutterstock.com

Figure 3.4-10. Open-web floor trusses are assembled at a factory with metal plate connectors. Note the ribbon at the top of the outside perimeter. This ribbon holds the trusses upright.

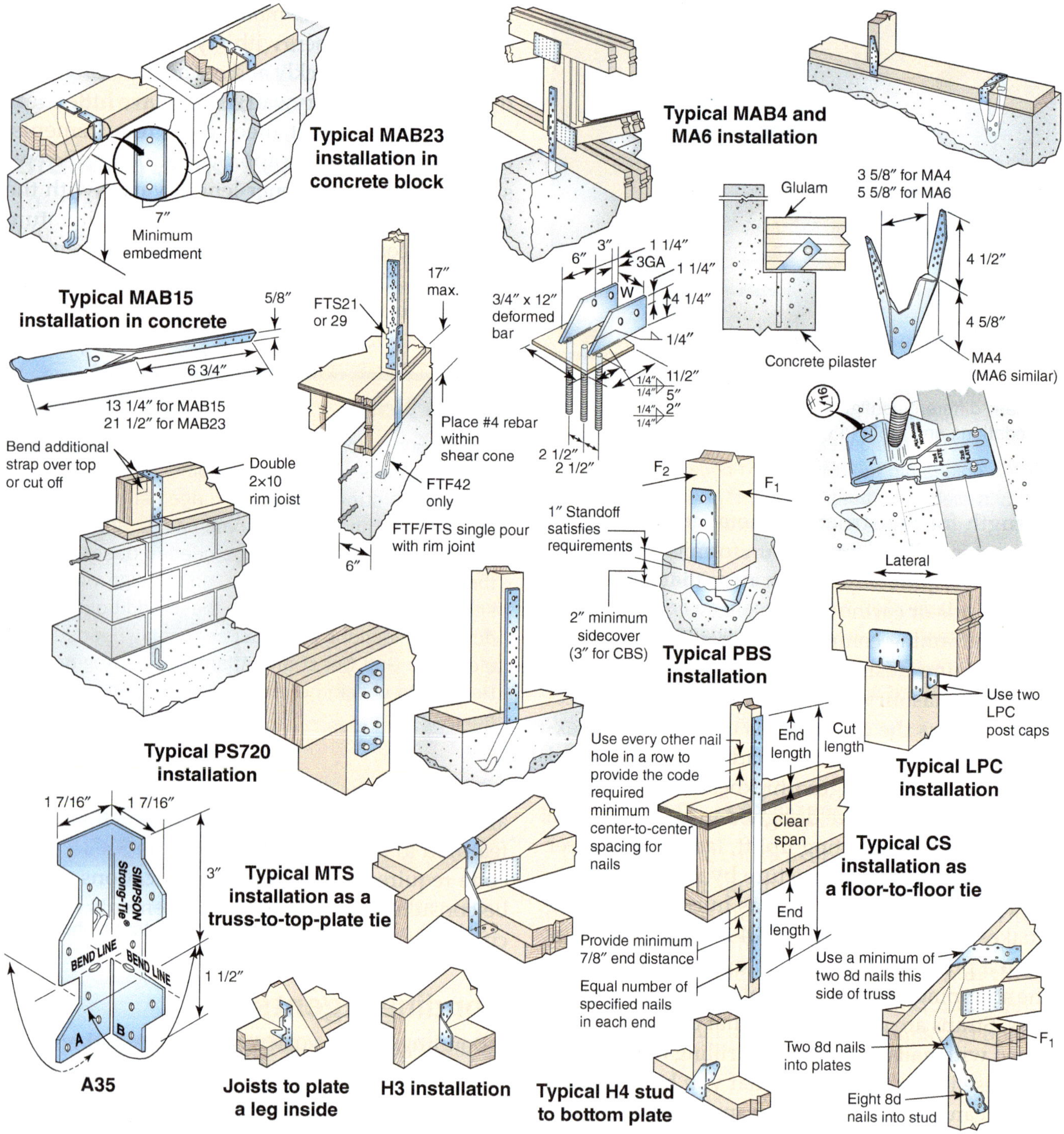

©Simpson Strong-Tie Company, Inc.

Figure 3.4-14. Straps are framing connectors that are used to resist forces caused by high winds, earthquakes, and floods.

Research and development have created many new products in this area. Some are highly specialized and designed for use with a specific material or application.

Bonding agents are available in many forms depending on their type and the application for which they are being used. For instance, subfloor sheathing is glued to the floor joists using adhesives applied with a caulk gun or a foam gun. Mastics used to glue down flooring come in a bucket and are applied with a notched trowel. Some contact cements are applied with a roller, while others

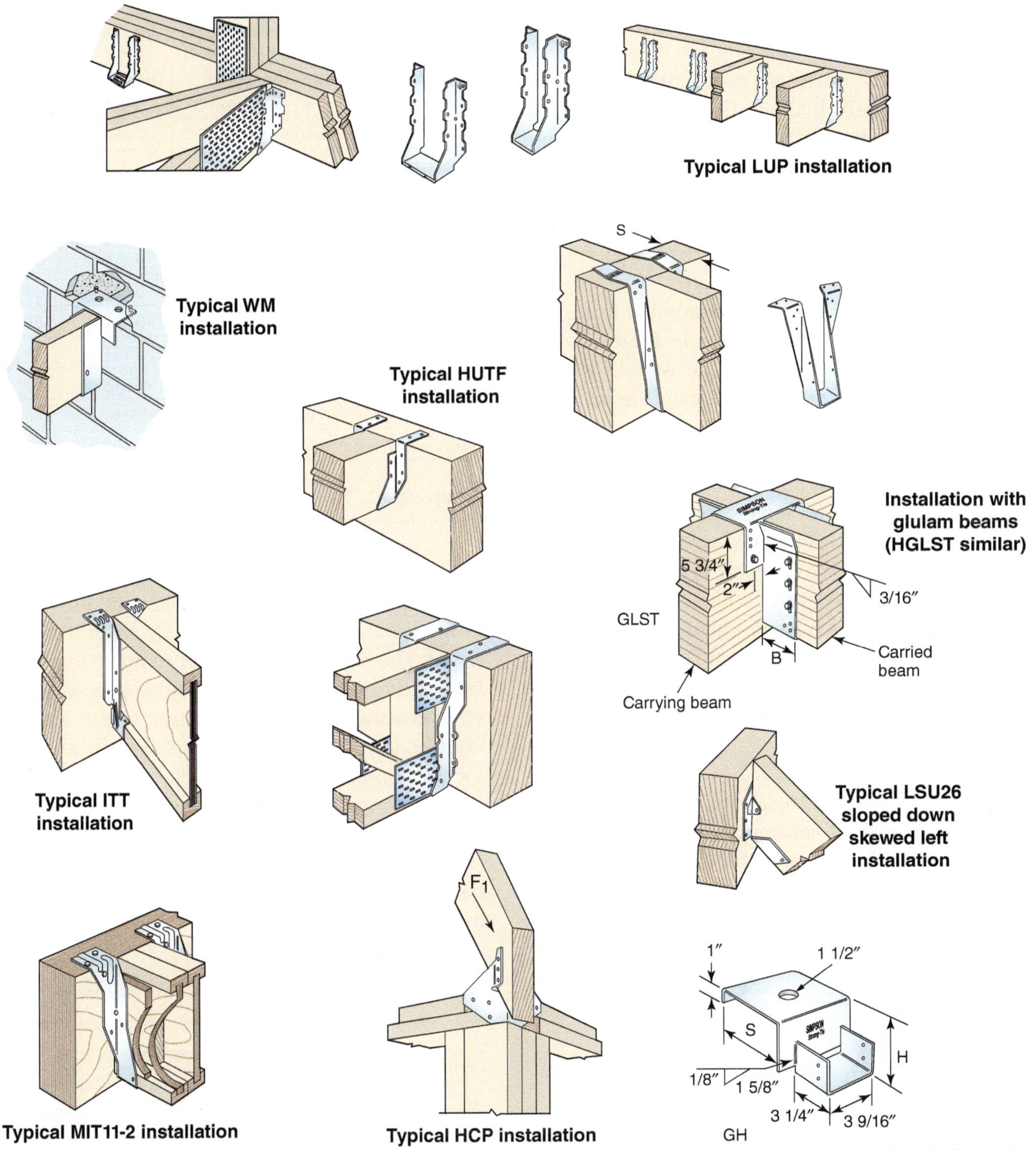

©Simpson Strong-Tie Company, Inc.

Figure 3.4-15. Wood-framing hangers can be used at various framing joints in residential construction to provide support for loads and stresses placed on joists and beams.

come in a spray can. Epoxies come in two parts—a resin and a hardener—and must be mixed just before use. Each of these examples may have multiple uses, or a product might require a specific bonding agent. Always read the manufacturer's instructions for particular applications. For more on bonding agents and their application, refer to *Chapter 7, Interior Insulation and Surfaces.*

Nails

Nails are a type of framing fastener that have been used for centuries. However, it is not as common to drive nails by hand these days. Nail guns have changed the way nails are manufactured and purchased.

Nails are the primary fastener used when framing a home. Hand nails are still driven in with a hammer for some applications on the job, though nail guns are used far more often.

Nails are available in a wide range of types and sizes. Basic sizes are illustrated in **Figure 3.4-16**. Nails for use in nail guns are designed for automatic feeding through the tool. They are provided in flat blanks called clips or strips, and in round coils. Nail gun nails are sized by their gauge (thickness or diameter) and their length in inches.

The nail size unit for hand nails is called a penny and is abbreviated with the lowercase letter d. It indicates the length of the nail. A 2d (2 penny) nail is 1″ long. A 6d (6 penny) nail is 2″ long. Refer to **Figure 3.4-17**. This measurement applies to common, box, casing, and finish nails. Brads and small box nails are specified by their actual length and gauge number.

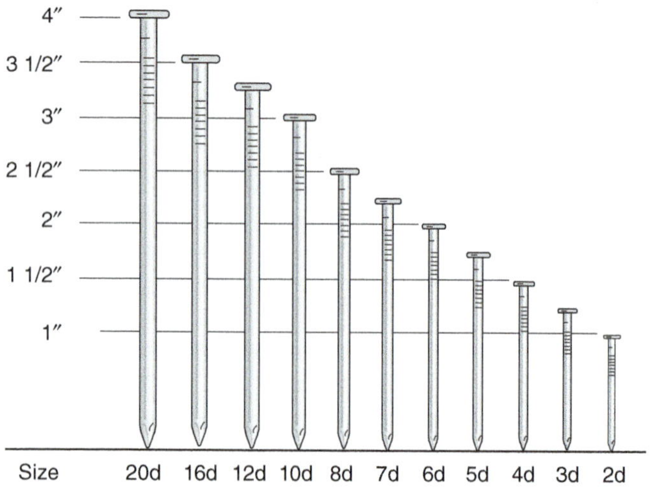

United States Steel

Figure 3.4-16. Nail sizes are given in a unit called the "penny." It is written as d.

Size	Length (in)	Common		Box	
		Diameter (in)	Number per Pound	Diameter (in)	Number per Pound
4d	1 1/2	0.102	316	0.083	473
5d	1 3/4	0.102	271	0.083	406
6d	2	0.115	181	0.102	236
7d	2 1/4	0.115	161	0.102	210
8d	2 1/2	0.131	106	0.115	145
10d	3	0.148	69	0.127	94
12d	3 1/4	0.148	63	0.127	88
16d	3 1/2	0.165	49	0.134	71
20d	4	0.203	31	0.148	52
30d	4 1/2	0.220	24	0.148	46
40d	5	0.238	18	0.165	35

Goodheart-Willcox Publisher

Figure 3.4-17. Nail sizes and approximate number in a pound are shown on this chart.

Framing Nails

The three main classifications of hand nails used for framing include common nails, box nails, commonly called *sinkers*, and duplex nails. The common nail has a heavy cross section and is designed for rough framing and general nailing. Sinkers are a thinner version of common nails and are used where splitting may be a problem. Sinkers have a thinner shank to help keep them from splitting the wood. The head is also thinner to make them easier to drive flush. Sinkers are often coated with a vinyl coating to make them easier to drive and to hold better, **Figure 3.4-18**. Duplex nails are used for temporary attachment when building concrete formwork, scaffolding, temporary bracing, or other assemblies that will be dismantled. These nails have a double head design which allows them to be easily pulled out using a pry bar or claw hammer, **Figure 3.4-19**.

A

Anton Starikov/Shutterstock.com

B

Panupong786/Shutterstock.com

Figure 3.4-18. A—Common nails are used for general framing. B—Sinkers are a thinner version of common nails. They are coated for lubrication and holding power.

Shutterstock-Pixelsquid/Shutterstock.com

Figure 3.4-19. The double head of a duplex nail allows it to be easily removed. They are used for temporary connections.

Trim Nails

A ***trim nail*** has a smaller head than a framing nail so the head can be set below the surface and filled with putty or caulk. Trim nails are used where nail heads should be hidden to improve the appearance of the work, **Figure 3.4-20**.

The casing nail is the same weight as the common nail and is typically used to attach exterior door and window casings and other wood trim.

Finishing nails and brads are quite similar and have the thinnest cross section and smallest head. A ***finish nail*** is the nail most often used to fasten interior trim. Brads are used for light-duty, more delicate work.

A few of the many nails designed for special purposes are shown in **Figure 3.4-21**. Annular or spiral threads greatly increase holding power.

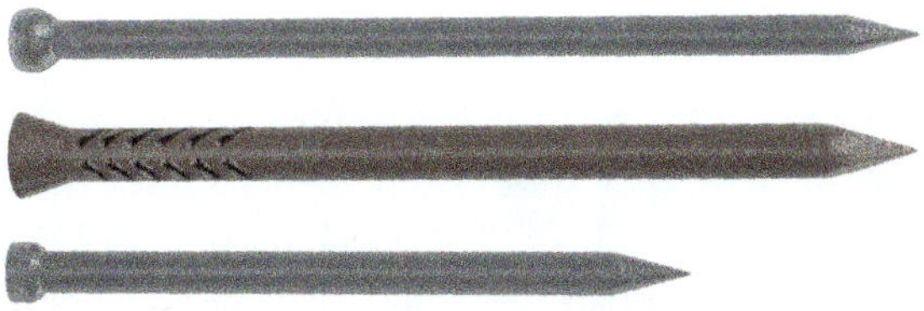

3drenderings/Shutterstock.com

Figure 3.4-20. Trim nails can be classified in three categories: finishing nails (top), casing nails (middle), and brads (bottom). Casing and finish nails are classified by the penny system, while brads are classified by length and thickness or diameter (gauge).

Some nails have special coatings of zinc, cement, or resin. Coating or threading increases nail-holding power. Nails are made from materials such as iron, steel, copper, bronze, aluminum, and stainless steel.

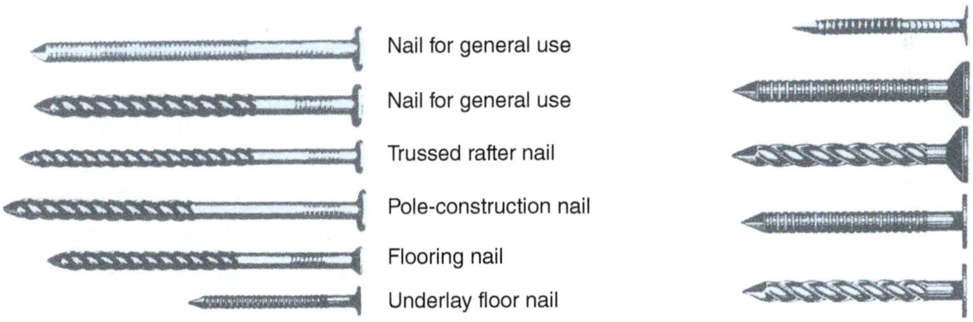

Independent Nail and Packing Co.

Figure 3.4-21. Annular- and spiral-threaded nails are designed for special purposes.

LESSON 3.4 Review

Answer the following questions based on information in this lesson.

1. *True or False?* Exterior glue is generally used on interior plywood. (3.4-1)
2. An engineered wood panel made of individual flakes of wood that are laid down like a blanket and adhered to each other with suitable resins and glues is called ____. (3.4-1)
   A. plywood
   B. oriented strand board (OSB)
   C. laminated-veneer lumber (LVL)
   D. parallel-strand lumber (PSL)
3. Which of the following is *not* an application of OSB? (3.4-1)
   A. Subfloor
   B. I-joists
   C. Studs
   D. Exterior siding
4. What bonding agent is developed from natural materials? (3.4-2)
   A. Glues
   B. Adhesives
   C. Cements
   D. Mastics
5. *True or False?* Trim nails have a smaller head than framing nails. (3.4-3)

LESSON 3.5

Tools

LEARNING OUTCOMES

After completing this lesson, you will be able to:

3.5-1 List hand and power tools commonly used during the framing of a house.

TECHNICAL TERMS

cat's paw	framing square	pry bar
chop saw	nail gun	reciprocating saw
finish hammer	power miter saw	table saw
flat bar	power stapler	wrecking bar

Framing Tools

Tools used in the framing process range from basic hand tools and measuring tools to an assortment of power saws and nail guns. Circular saws, featured in *Chapter 1, Preparing for Construction*, are used extensively for framing. This lesson covers some of the other commonly used tools during the framing process.

Hammers and Pry Bars

The claw hammer is the most often used fastening tool. Parts of a hammer are labeled in **Figure 3.5-1**. The face can be either flat or slightly rounded, called a *bell face*. The bell face is used most often because it drives nails flush with the surface without leaving hammer marks on the wood.

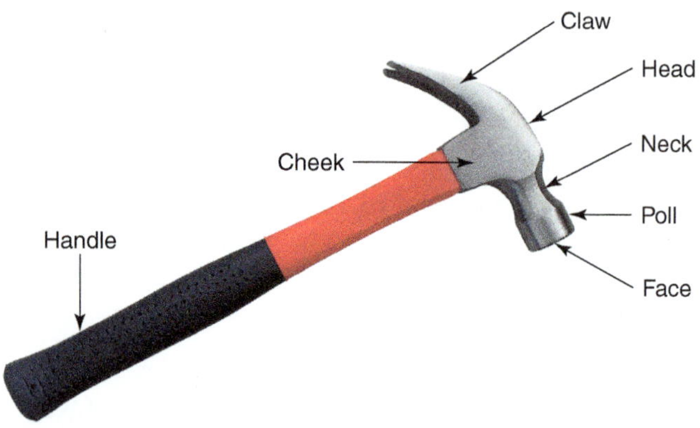

Figure 3.5-1. Parts of a hammer.

The hammer head is forged from high-quality steel and is heat treated to give the poll and face extra hardness.

Two common shapes of hammer heads are the curved claw and ripping (straight) claw. The curved claw is best suited for pulling nails. The ripping claw, which may be driven between fastened pieces, is used somewhat like a chisel to pry them apart. Straight claw hammers are also used for framing.

Safety Note

It is important to keep the handle of a hammer tight and the face clean. If a wooden handle becomes loose, it can be tightened by driving the wedges deeper or by installing a new handle. A hammer with a loose head can be a dangerous hazard on the jobsite.

The size of a hammer is determined by the weight of its head. They are available in sizes from 7 oz to 22 oz. Carpenters generally use either the 16 oz, 20 oz, or 22 oz, depending on the work to be done. Framing hammers are generally 20 oz or 22 oz with longer handles and straight claws. The face of a framing hammer may have a textured surface to help prevent the hammer from glancing off the nail, **Figure 3.5-2**. A ***finish hammer***, also known as a *trim hammer*, is used for tasks that require precision

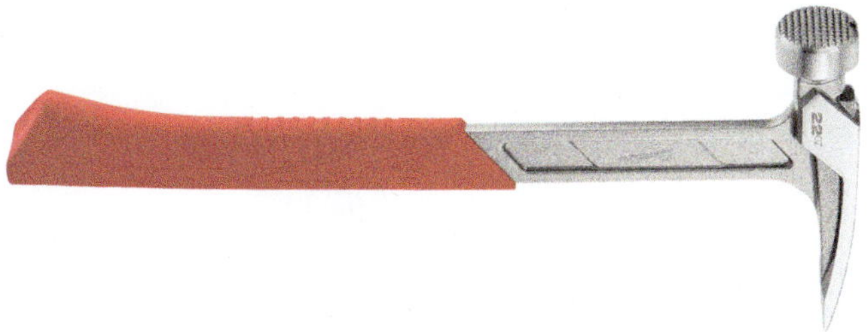

Milwaukee Tools

Figure 3.5-2. The textured head of a framing hammer.

and finesse. Unlike larger framing hammers, finish hammers have a shorter handle and a smaller head with a bell face.

Sometimes, it is necessary to take apart nailed structural members. The same is true when forms or temporary braces are removed from a building under construction. **Figure 3.5-3** illustrates the procedure for pulling nails to relieve strain on the hammer handle and protect the work. A ***pry bar*** is a tool that is specifically designed to pull nails and dismantle assemblies. The most common pry bars are wrecking bars, flat bars, and cat's paw pry bars, **Figure 3.5-4**.

A ***wrecking bar*** can vary in length from 12″ to 36″. They are used to strip concrete forms, to disassemble scaffolding, and for other rough work involving prying, scraping, and nail pulling. A ***flat bar*** is designed for removing larger nails or spikes from lumber. Most flat bars are 13″–15″ long. The smaller ones, around 7″, are handy because they fit in a carpenter's tool belt. A ***cat's paw*** is used to initially pull a nail head back above the surface of the wood. A hammer is used to drive the sharpened claw under the nail head, raising it sufficiently to be grasped by the hammer or a pry bar.

Safety Note

When framing, always remove nails from lumber scraps before discarding them. Failure to do so could result in injury from a puncture and lead to lost time on the job.

Framing Square

The ***framing square***, also called a *rafter square*, is a tool specially designed for a carpenter, **Figure 3.5-5**. The blade, or *body*, is 2″ wide and 24″ long. The tongue is at a right angle to the blade. It is 1 1/2″ wide and 16″ long. The corner of the square is called the heel. One side of the square, with the tongue to the right and toward you, is called the face.

A B C

Goodheart-Willcox Publisher

Figure 3.5-3. Proper methods for pulling nails and separating two wood pieces. A—Place a putty knife under the hammer to protect the wood surface. B—Use a block of wood to increase leverage and protect the surface. C—To separate two nailed surfaces, a ripping claw hammer works best. The ripping claw can also be used to split short boards.

Goodheart-Willcox Publisher

Figure 3.5-4. Pry bars are used for separating, disassembling, and aligning heavy building components. The most common pry bars are shown here.

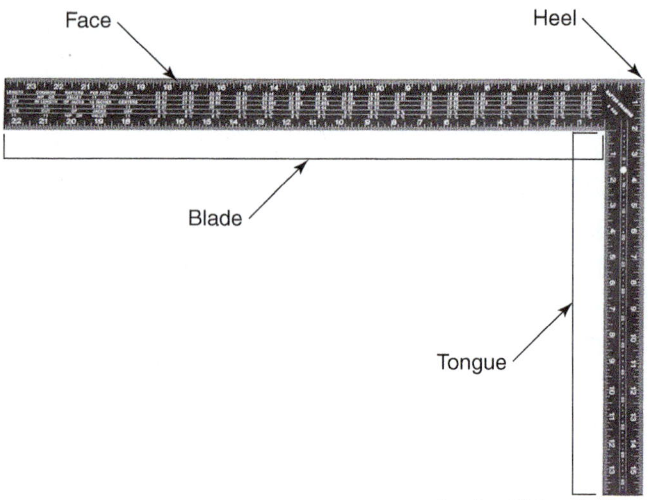

Goodheart-Willcox Publisher

Figure 3.5-5. The parts of a framing square are the blade, tongue, heel, and face.

Framing squares have a number of tables printed on them. The face has rafter tables printed on the body. These can be used in determining the length of rafters. The tongue carries the octagon scale. This scale is used to lay out the angles of an octagon. The opposite side of the square may have the Essex Board measure table and the brace measure table. These tables are used to help calculate board footage and lengths for 45-degree bracing.

The outside edge of the face is divided into 1/16″ graduations, while the inside edge has 1/8″ graduations. The outside edge of the opposite side of the square has each inch graduated in twelfths of an inch. This is useful in making a scale drawing. Each inch represents a foot, and each 1/12″ represents an inch. The inside edge of the tongue is graduated in tenths of an inch.

Like the speed square, framing squares can be used to mark various angles across framing lumber and other surfaces. However, in the hands of a skilled carpenter, the framing square is a tool with many other functions. The following are just a few tasks a carpenter can perform using a framing square:

- Check or square lines across boards
- Check the squareness of corners
- Make a 45° angle across wide boards
- Find the length and angles of rafters
- Lay out stairs and rafters

Nail Guns and Power Staplers

A wide variety of nail guns and power staplers is available. A ***nail gun*** is a power tool used for driving nails, **Figure 3.5-6**. Likewise, a ***power stapler*** is a power tool used for driving staples, **Figure 3.5-7**. Most of these tools are pneumatic (air-powered), but battery-operated guns continue to gain popularity and usability. Gas-operated models are also available.

Safety Note

Always wear safety glasses when working with pneumatic tools. Never direct an air stream at yourself or another person. Do not use compressed air to blow dust off your skin or clothes. Pneumatic tools should be equipped with quick-disconnect fittings, **Figure 3.5-8**. These fittings allow for fast and easy disconnection of the air supply.

A Paslode B DeWalt C Paslode

Figure 3.5-6. There are several types of nail guns. A—Pneumatic. B—Battery-operated. C—Gas-operated.

sergey0506/Shutterstock.com

Figure 3.5-7. This lightweight cabinet backing is being installed using a pneumatic stapler.

The following are two examples of nail gun types: strip-fed or coil-fed, **Figure 3.5-9**. Strip-fed nail guns hold a strip of nails in a long magazine located under the handle. Most framing nail guns and finish nail guns are strip-fed. Some use strips at a 20°–22° angle, others use straight strips. Some use clipped head nails; others use round head nails. Be sure you use the right type for your nail gun. Coil-fed nail guns carry nails in a round canister, usually located ahead of a trigger handle. Most roofing nail guns are coil-fed.

Power staplers are used for fastening materials where a staple serves the purpose better than a nail. Nails generally provide greater holding power than staples, especially in shear and tension. Therefore, nail guns are preferred for structural applications where stronger connections are required, such as framing a house or building. Where materials are soft and can tend to tear around a fastener, such as with shingles, fabrics, and thin sheathings, a staple may be a better choice. A staple may also be used when securing

Goodheart-Willcox Publisher

Figure 3.5-8. Pneumatic tools should always be equipped with a quick-disconnect coupling.

certain objects, such as flexible pipe. However, these are specific applications that require specialized staplers and fasteners.

Operating Nail Guns and Power Staplers

Nail guns and power staplers have two main controls: a trigger and a safety tip at the nose. The firing mechanism can work differently depending

A wingedwolf/Shutterstock.com

B Irra/Shutterstock.com

Figure 3.5-9. A—Clips for power nail guns are available in angled or straight, and round head or clipped. Each gun requires a specific type of clip. B—Coils are used in some nail guns.

on the type of gun. For example, guns that are designed for speed, such as roofing and framing nail guns let you shoot nails by holding down the trigger and bumping the tip against the material you are fastening, called *bump firing*. Each time the tip is bumped, a nail is fired.

Nail and staple guns with sequential trip firing mechanisms require you to release and press the trigger for each nail, called *sequential firing*. With this trigger, you have to push the safety tip onto the surface first, then squeeze the trigger to shoot a nail. Both the safety tip and the trigger must be released before shooting another nail. Sequential trip firing mechanisms cannot be bump fired.

Some nail guns have a switch that lets you choose which trigger type to use while others are factory-set to one or the other. The safest and most precise method uses sequential firing.

Safety Note

Use extreme caution when using bump-fired nail guns. A slight miscalculation can result in a nail turning into a projectile. And never attempt to sequential fire a nail gun designed for bump firing. The recoil on these nail guns causes repeated firing in rapid succession, damaging materials and creating a safety hazard.

Some nail gun manufacturers have stopped producing bump-fired nail guns with a contact trip mechanism. Sequential trip nail guns are considered to be a safer option. When purchasing a nail gun, look for nail guns with a sequential trip mechanism.

Using nail guns and power staplers speeds up work and allows a carpenter to work in tight places where swinging a hammer might be hard or impossible. Staplers offer an alternative to nails, especially for securing sheathing and shingles.

To operate a sequential nail gun, first press the nose against the work to disarm the safety device that keeps it from driving the fastener, and then press the trigger to drive the nail, **Figure 3.5-10**.

Most air compressors used on a jobsite are portable, **Figure 3.5-11**. They may be driven by either electric motors or gas engines. Manufacturers of pneumatic tools can assist carpenters in selecting a compressor large enough to handle air needs on the job.

ungvar/Shutterstock.com

Figure 3.5-10. Pneumatic nail guns speed work and can be used in close quarters where using a hammer may be difficult.

Goodheart-Willcox Publisher

Figure 3.5-11. An air compressor suitable for operating pneumatic tools must deliver up to 110 pounds per square inch (psi) of pressure.

Safety Note

Nail guns are used every day on construction sites. They speed up work and improve quality, but they are dangerous. A study on nail gun safety published by OSHA found that two out of five carpentry apprentices experience a nail gun accident sometime during their four-year apprenticeship. One out of five was injured twice.

Follow these safety rules when working with nail guns and power staplers:

- Wear safety glasses and hearing protection.
- Study the manufacturer's operating directions and carefully follow them. Always ensure the nail gun is in proper working order before use.
- Use the correct type and size of fastener recommended by the manufacturer.
- For air-powered nail guns, always use the correct pressure (seldom over 90 psi).
- Always keep the nose of the stapler or nail gun pointed toward the work. Never aim it toward yourself or others.
- Keep hands at least 12″ away from the nailing point at all times.
- Always disconnect the tool from the air or electrical power supply when it is not being used or before you attempt to clear a jam.
- Remove your finger from the trigger when carrying a nail gun or stapler. An accidental discharge could cause injury.
- Never hoist a nail gun onto a ladder or scaffold by the hose. Doing so risks accidental firing and damage to the gun and hose.

Saws

A wide variety of saws is used throughout the construction of a home. This lesson primarily focuses on the power saws used for cutting wood and associated products.

Reciprocating Saw

Reciprocating saws are useful under conditions where a circular saw is not safe or practical. A *reciprocating saw* uses a back-and-forth, linear cutting motion that allows for cutting through a variety of materials quickly and efficiently. These saws are used in construction, demolition, plumbing, and electrical work and extensively in remodeling work where sections of framing, sheathing, or inside walls must be removed. Blades are available to cut through wood, plaster, fiberglass, and all kinds of metals, **Figure 3.5-12**.

Safety Note

Follow these safety rules when working with saber saws and reciprocating saws:

- Wear safety glasses when using any power saw.
- Select the correct blade for your work and be sure it is properly mounted in the saw.
- Disconnect the saw from power to change blades or to make adjustments.
- Make certain that corded saws are properly grounded through the electrical cord.
- Firmly place the base of the saw on the stock before starting the cut. The blade should be clear of the wood before the tool is up to cutting speed.
- Turn the motor on before the blade contacts the work.
- Do not attempt to cut curves so sharp that the blade is twisted.
- Make certain the work is well supported. Check for clearance so that you do not cut into sawhorses or other supports.

Power Miter Saw

Other than portable circular saws, a *power miter saw* is used most often by carpenters for sawing operations. A sliding compound miter saw, **Figure 3.5-13**, is a particularly versatile tool. It has a base with a fence in the back to hold the work in position. A portion of the base rotates left and

Photos courtesy of Robert Bosch Tool Corporation®

Figure 3.5-12. A reciprocating saw. A—This type of saw has a blade that moves forward and back (reciprocates) through a stroke distance of about 1 1/8″. It can be fitted with a variety of blades to suit all purposes. B—This set of all-purpose blades shows the variety of reciprocating blades available.

right along with the sliding arm, motor, blade, and control handle. A 10″ or 12″ blade is common. The blade is fitted with a guard that retracts as the saw is lowered into the cut. The sliding arms allow the saw motor and blade to move from front to back while cutting wider work pieces.

The chop saw, **Figure 3.5-14**, is a variation of the power miter saw. A ***chop saw*** does not tilt for bevels, and it does not have the sliding mechanism

Goodheart-Willcox Publisher

Figure 3.5-13. Power miter saws are used extensively on the jobsite.

Goodheart-Willcox Publisher

Figure 3.5-14. A metal cutting chop saw is almost a necessity when working with light gauge steel framing.

for moving through wide stock. Some chop saws have a vise to hold metal in position to be cut. Chop saws can be set for miters or square cutoffs only. Most of the chop saws available today are used for cutting metal; however, some builders still use wood-cutting chop saws. Abrasive blades for cutting metal and special carbide-tipped metal cutting blades are available for cutting all types of light-gauge metal. The most common blade size for chop saws is 14″.

Safety Note

Follow these safety rules when working with power miter saws and chop saws:

- Wear safety glasses and hearing protection.
- Attach the saw securely to a platform, bench, or saw stand.
- Inspect the saw before using to ensure that all guards are in place and functioning properly.
- Avoid wearing loose clothing, as it can be caught by a saw blade. Gloves are safe when using an abrasive blade to cut metal, but never when using a toothed blade.
- Unplug the saw before changing blades.
- Clear the area around the saw of wood scraps.
- Keep your hands clear of the path of the blade. Use one hand to hold the stock against the fence and the other hand to operate the saw. Never cross your arms to reach across the saw.
- Always slide the saw from front to back (away from the operator) when cutting wide pieces.

Table Saw

The ***table saw*** is a tool mostly used for ripping stock to width and cutting it to length, but it also cuts bevels, chamfers, and tapers.

There are two categories of table saws: contractor saws and cabinet saws, **Figure 3.5-15**. Contractor saws are lighter and more portable for use on the jobsite. Cabinet saws are set up in a shop environment and are used mostly for finish work. As with all power tools, be sure to understand and follow the safety rules for operation. Table saws are dangerous machines.

The size of the saw is determined by the largest blade it will take. Stock to be ripped must have at least one flat face to rest on the table and one straight edge to run along the fence, thus serving as a guide for the cut. **Figure 3.5-16** shows the correct

Goodheart-Willcox Publisher

Figure 3.5-16. Ripping stock with a table saw. The guard should always be in place for safe operation.

Contractor saw **Cabinet saw**

Goodheart-Willcox Publisher

Figure 3.5-15. Table saws are versatile tools. They are most commonly used by carpenters for ripping boards to width.

Safety Note

Warning! Using the fence while attempting a crosscut on the table saw can cause violent kickback and result in serious injury. Never use a table saw unless properly trained in all safe cutting operations.

Follow these safety rules when working with table saws:

- Wear safety glasses.
- Be certain the blade is sharp and right for the job at hand.
- Make sure the saw is equipped with a guard and use it.
- Set the blade so it extends about 1/4″ above the stock to be cut.
- Stand to one side of the operating blade and do not reach across it.
- Maintain a 4″ margin of safety. Do not let your hands come closer than 4″ to the operating blade, even though the guard is in position.
- Stock should be surfaced and at least one edge jointed before the piece is cut on the saw.
- Use the fence or miter gauge to control the stock; never use both. Do not cut stock freehand.
- Always use push sticks when ripping short, narrow pieces.
- Stop the saw before making adjustments.
- Do not let small scrap cuttings accumulate around the saw blade. Use a push stick to move them away.
- Setups must be carefully made and checked before the power is turned on.
- Workers helping to "tail-off" the saw should not push or pull on the stock, only support it. The operator must control the feed and direction of the cut.
- Once work is complete, turn off the machine and remain in place until the blade has stopped. Clear the saw table and place waste in a scrap box.

procedure for making a rip cut. When crosscutting on a table saw, the fence is not used. **Figure 3.5-17** shows a standard crosscutting operation. A miter gauge is used to stabilize the material and set the angle of the cut. It is used to feed the stock into the blade.

A Goodheart-Willcox Publisher B Goodheart-Willcox Publisher

Figure 3.5-17. Crosscutting on the table saw allows you to square off stock. A—Squaring the stock. Securely hold the stock against the miter gauge. B—Guide the stock through the cut by pushing the miter gauge. One side of the guard has been lifted out of the way to show operation.

LESSON 3.5 Review

Answer the following questions based on information in this lesson.

1. What type of hammer is best suited for pulling nails? (3.5-1)
   A. Curved claw hammer
   B. Ripping claw hammer
   C. Straight claw hammer
   D. All are correct.
2. What type of pry bar is used to initially pull a nail head back above the surface of the wood? (3.5-1)
   A. Cat's paw
   B. Chisel
   C. Flat bar
   D. Wrecking bar
3. *True or False?* Most framing nail guns are coil-fed. (3.5-1)
4. *True or False?* Nail guns with a sequential-fire trigger are considered safer than those with a contact trip. (3.5-1)

Matching: Match each saw with its function.

A. Cabinet saw
B. Chop saw
C. Contractor saw
D. Sliding compound miter saw
E. Reciprocating saw

5. Saw with a linear cutting motion, which can be used to remove framing (3.5-1)
6. Portable table saw for use on the jobsite (3.5-1)
7. Saw set up in a shop environment and used mostly for finish work (3.5-1)
8. Saw usually used for cutting metal (3.5-1)
9. Saw with a rotating base and a fence to hold the work in position (3.5-1)

ungvar/Shutterstock.com

Figure 3.6-2. Lifting or storing materials at height produces a struck-by hazard.

in any stage of construction. However, these hazards can be elevated during the framing stages. The Center for Construction Research and Training reports that struck-by accidents are the largest cause of nonfatal injuries and the second leading cause of death for construction workers.

Falling objects are one type of struck-by hazard. The potential for falling objects is high when framing. Workers operating from elevated platforms, ladders, or higher levels of the structure itself are busy placing and cutting materials and using all kinds of tools. It is not uncommon to lose a grip on a hammer or have a tape measure fall out of a tool belt. Falling scrap from cutoff materials, dropped boards, and carelessly placed objects all have the potential to harm workers operating below. Always keep an eye on what is going on above you.

Avoid walking or working directly below others. Hard hats and eye protection are a must when working where there is a hazard for falling objects. Always be mindful of the possibility of others working below you if you are working from an elevated height. Avoid placing materials, scraps, and tools on elevated work surfaces where they could fall off. Even an unexpected gust of wind can send poorly placed materials flying. Good housekeeping practices are a must.

Other examples of struck-by hazards include nails ejected or deflected inadvertently from a nail gun, vehicles being driven on a jobsite, walls under construction that may fail, and stored materials that are not properly secured. Loads being lifted and moved by cranes and other equipment are a major struck-by hazard on a jobsite. Next, we will look at some of the hazards specifically associated with working around lifting equipment.

Working Safely with Lifting Equipment

On framing jobs that require large trusses, panelized wall or floor sections, or modular construction, cranes are used. Cranes pose safety risks that other equipment does not. However, whenever any equipment is being used to hoist, move, or install materials, extra precautions must be taken to ensure the safety of you and your crew.

Workers using rigging and lifting equipment must be properly trained. ***Rigging*** is the term used to describe the slings, cables, and chains used with a crane to lift heavy objects into place, **Figure 3.6-3**. Chain and wire ropes (sometimes referred to as cable) are used for heavy mechanical equipment, but most of the hoisting performed by framing carpenters is with synthetic slings.

A ***synthetic sling*** is made of nylon or polyester. Synthetic slings are either flat straps or round, like rope, **Figure 3.6-4**. They are soft, flexible, and lightweight, making them easy to handle. Synthetic slings are not affected by water, but they are damaged by exposure to excessive heat or sunlight for long periods of time. They are also easily cut or damaged by abrasion. Slings should be inspected

Goodheart-Willcox Publisher

Figure 3.6-3. A SIP is being hoisted into place using a crane and a web sling.

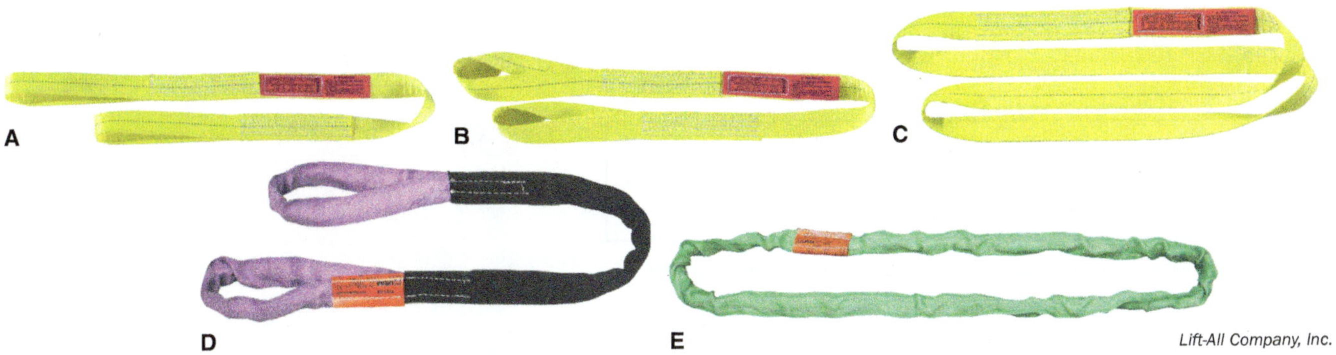

Figure 3.6-4. Types of synthetic slings. A—Standard eye. B—Twisted eye. C—Endless. D—Eye-and-eye round sling. E—Endless round sling.

for damage by a competent person before each use. **Figure 3.6-5** shows some of the defects that might appear. If any of these are seen, the sling should be taken out of service.

Safety is a central concern when using a crane to hoist heavy objects. Even the empty hook on the end of the crane rigging can be deadly if it swings and strikes an unaware worker. All persons not directly involved in the hoisting operation should be clear of the path of the load. One person should be designated as the signal director. That person is responsible for the safety of all workers in the area.

Safety Note

When working near cranes and other lifting equipment, never walk beneath a suspended load. Loads can shift; equipment can fail. The consequences can be deadly.

Do not place your hand inside the sling as it is being tightened around the load. Also, do not attempt to guide the load by pushing or pulling with your hands. A tag line should be attached to the load. A ***tag line*** is a rope attached to the load and held by a person on the ground to prevent the load from twisting and swinging. An object being hoisted by a crane is usually heavy. If you try to handle it, there is a risk of being injured.

Only the person designated to give instructions to the crane operator should do so. Instructions may be given by radio or by approved hand signals. The standard hand signals given in **Figure 3.6-6** are approved by the American National Standards Institute (ANSI).

Personal Protective Equipment (PPE)

Although personal protective equipment (PPE) is the lowest in the hierarchy of controls, it is among the most important and valuable safety control on the jobsite in the event of an accident. PPE is a required safety measure during every phase of construction; however, during the framing stage of a job, risks associated with personal injuries are at a heightened state. Many of the risks already discussed in this chapter, along with the fast pace of framing work, operating outdoors in (sometimes) adverse conditions, and the fact that more people are generally on-site during this phase of construction, are all factors that contribute to a heightened safety risk. Proper use of PPE is critical.

Personal protective equipment (PPE) is any clothing or equipment worn to provide protection against safety hazards. PPE is often the last line of defense against hazards that may otherwise harm a worker. For this reason, the use of PPE should always be taken seriously.

PPE must be maintained, stored, and used properly to be effective. Safety glasses that have been scratched up from bouncing around in the bottom of a toolbox may impede the vision of the person wearing them making them unsafe to use.

INSPECTION CRITERIA FOR WEB SLINGS

The following photos illustrate some of the common damage that occurs to web slings, indicating that the sling should be taken out of service.

THE DAMAGE: **Surface and edge cuts**—It is important to realize that all the fibers in web slings contribute to the strength of that sling. When there have been a significant number of fibers broken in a web sling, as shown here, that sling should be taken out of service.

WHAT TO LOOK FOR: **Broken fibers** of equal length indicate that the sling has been cut by an edge. **Red core warning yarns may or may not be visible with cuts and are not required to show before removing slings from service.**

TO PREVENT: Always protect synthetic slings from becoming cut by corners and edges by using wear pads and other devices.

THE DAMAGE: **Holes/snags/pulls**

WHAT TO LOOK FOR: **Punctures or areas** where fibers stand out from the rest of the sling surface.

TO PREVENT: Avoid sling contact with protrusions, both during lifts and while transporting or storing.

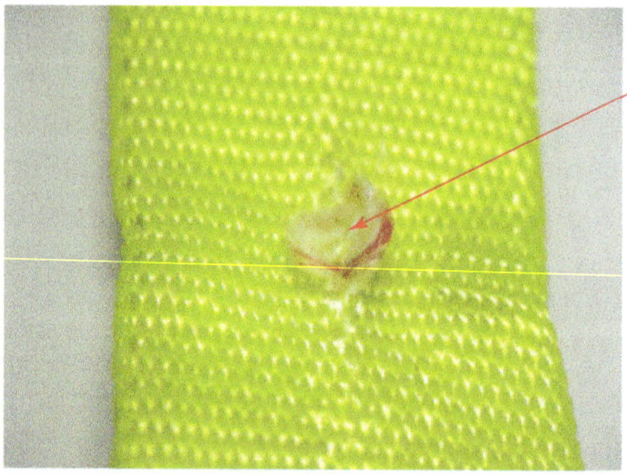

THE DAMAGE: **Abrasion**

WHAT TO LOOK FOR: Areas of the sling that look and feel **fuzzy** indicate that the fibers have been broken by being subject to contact and movement against a rough surface. Affected areas are usually stained.

TO PREVENT: Never drag slings along the ground. Never pull slings from under loads that are resting on the sling. Use wear pads between slings and rough surface loads.

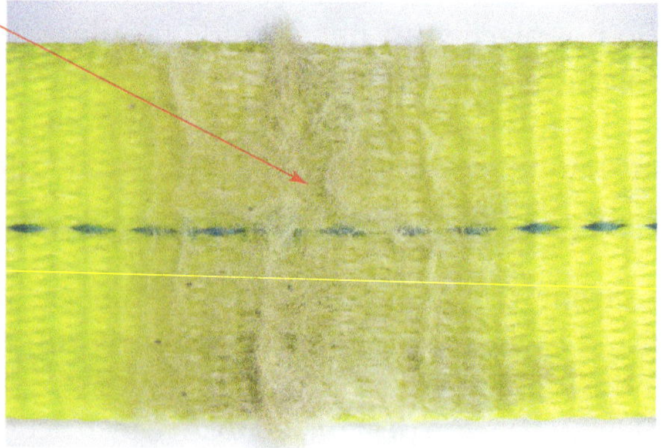

Lift-All Company, Inc.

Figure 3.6-5. Common defects in synthetic slings.

INSPECTION CRITERIA FOR WEB SLINGS

THE DAMAGE: **Heat/chemical**
WHAT TO LOOK FOR: **Melted or charred fibers** anywhere along the sling. Heat and chemical damages look similar, and they both have the effect of damaging sling fibers and compromising the sling's strength. Look for discoloration and/or fibers that have been fused together and often feel hard or crunchy.
TO PREVENT: Never use nylon or polyester slings where they can be exposed to temperatures in excess of 200°F. Never use nylon or polyester slings in or around chemicals without confirming that the sling material is compatible with the chemicals being used.

THE DAMAGE: **Knots** compromise the strength of all slings by not allowing all fibers to contribute to the lift as designed. Knots may reduced sling strength by up to 50%.
WHAT TO LOOK FOR: **Knots** are rather obvious problems as shown below.
TO PREVENT: Never tie knots in slings and never use slings that are knotted.

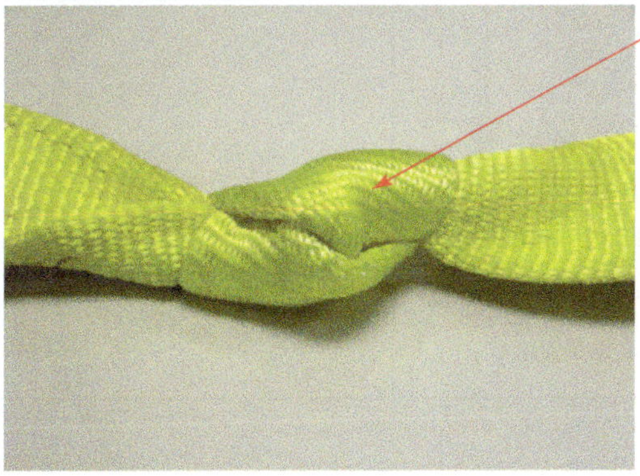

THE DAMAGE: **Broken/worn stitching** in the main stitch patterns of web slings has a direct adverse effect on the strength of a sling. The stitch patterns in web slings have been engineered to produce the most strength out of the webbing. If the stitching in not fully intact, the strength of the sling may be affected.
WHAT TO LOOK FOR: **Loose or broken threads** in the main stitch patterns.
TO PREVENT: Never pull slings from beneath loads where stitch patterns can be hung up or snagged. Never overload the slings or allow the load edge to directly contact the stitch pattern while lifting. Never place a sling eye over a hook or other attachment whose width/diameter exceeds 1/3 the eye length.

THE DAMAGE: **Illegible or missing tags**—The information provided by the sling tag is important for knowing what sling to use and how it will function.
WHAT TO LOOK FOR: If you cannot find or read all of the information on a sling tag, OSHA requires that the sling be taken out of service.
TO PREVENT: Never set loads down on top of slings or pull slings from beneath loads if there is any resistance. Load edges should never contact sling tags during the lift. Avoid paint or chemical contact with tags.

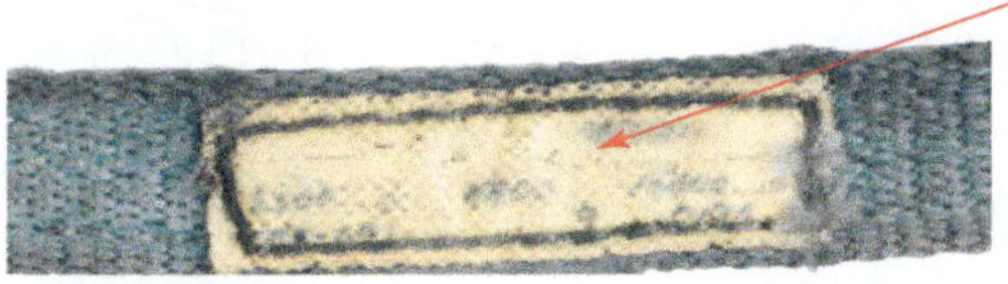

Red core yarns are an **additional** aid to warn of dangerous sling damage. All standard Lift-All web slings have this warning feature. The red core yarns become exposed when the sling surface is cut or worn through the woven face yarns. When red yarns are visible, the sling should be removed from service immediately.

Lift-All Company, Inc.

Figure 3.6-5. *(continued).*

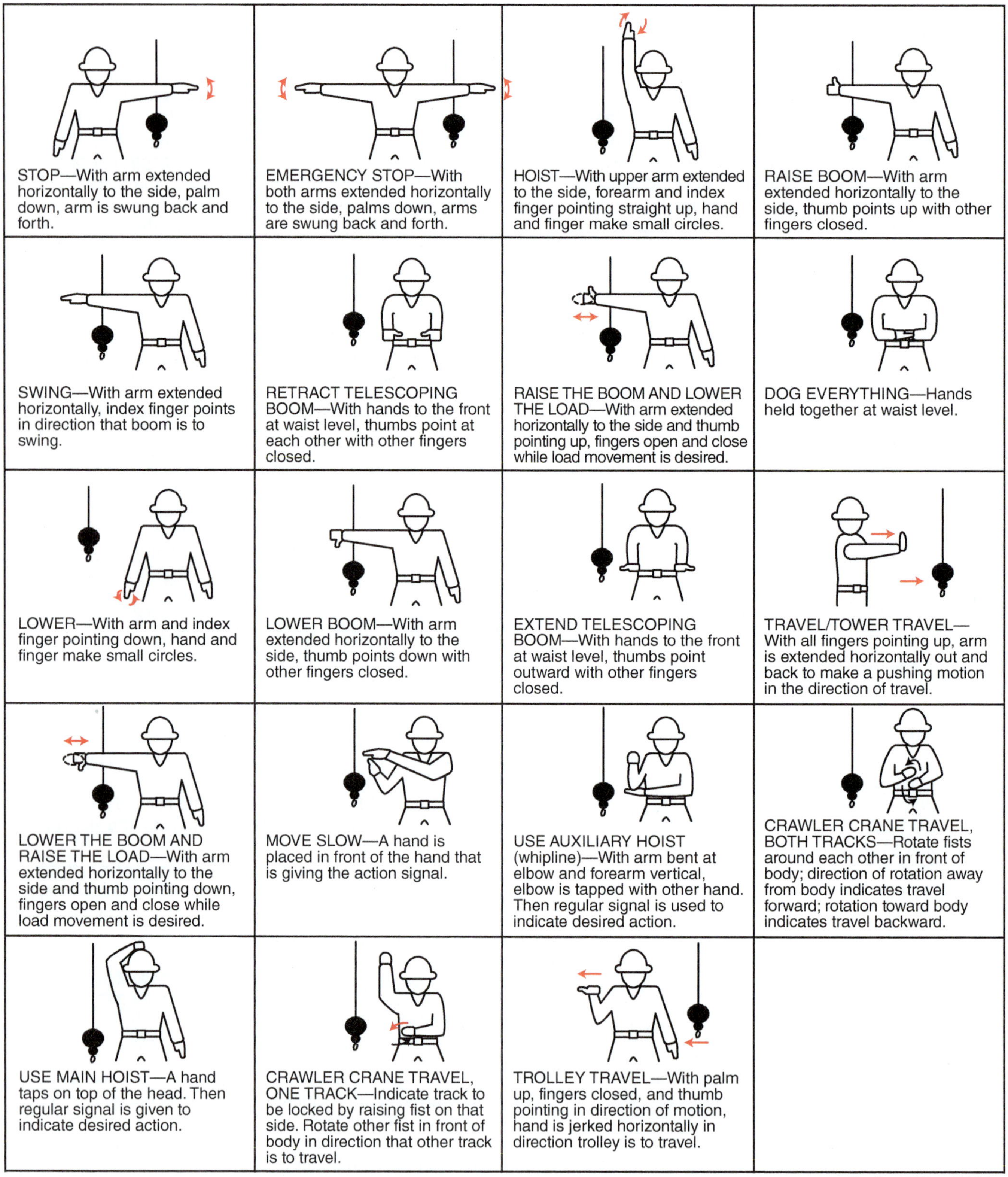

Goodheart-Willcox Publisher

Figure 3.6-6. Approved hand signals to direct a crane operator.

Common items used for PPE for construction include:

- **Safety glasses**—These are to be worn whenever work involves even the slightest potential hazard to your eyes, **Figure 3.6-7**. The American National Standards Institute (ANSI) sets many of the standards that have been adopted by industry. Safety glasses that satisfy OSHA requirements are stamped with ANSI standard number Z87.1. Clean your safety glasses as needed to remove dirt, dust, and other debris that may reduce your vision. Store safety glasses in a bag, box, or case when not in use. If lenses become scratched or pitted or if hinges become loose, replace the safety glasses.
- **Work boots or shoes**—OSHA standards require protective footwear in work areas where foot injuries could occur from falling or rolling objects, sharp objects, or electrical hazards. This footwear consists of special reinforced toes that will withstand a load of 2500 lb. These are required on construction jobs.

MBI/Shutterstock.com

Figure 3.6-7. Using a power tool safely means wearing safety glasses and making sure all guards are in place and working properly. Do not use a corded tool with a damaged cord. When operating a power tool, give your full attention to the work. A moment's distraction can have severe consequences.

- **Hard hat**—Whenever you are exposed to any possibility of falling objects, a hard hat is required. Many construction sites require all persons on the site to wear a hard hat at all times.
- **Dust mask**—A dusk mask, or *particulate mask*, is a disposable filter-type mask worn over the nose and mouth to protect the worker from inhaling dust and other fine particles in the air, **Figure 3.6-8**. A particulate mask is not a respirator and is not adequate protection from contaminants that are immediately hazardous to your health.
- **Respirator**—A respirator is required when removing paint (including non-lead paint), working with asbestos-containing or silica-containing materials, or spraying finishing materials, **Figure 3.6-9**. Many other instances require a respirator. Employers are responsible for developing and implementing a respiratory protection program if respirator use is required.
- **Hearing protection**—Approximately one out of ten people have suffered some hearing loss, and exposure to loud noise is the main cause. The OSHA standards for hearing protection give guidance on when hearing protection is required, starting with noise levels at or above 85 decibels (approximately the noise level of a gas-powered leaf blower or lawn mower). The best way to protect your hearing is to wear hearing protection whenever you are exposed to loud noise. Hearing protection can be foam earplugs or earmuffs for greater protection.
- **Hand protection**—Hand injuries directly affect a carpenter's ability to work, so proper protection is vital. Cuts, nicks, and scrapes are not only painful, but could also lead to infection or reduce the ability of a worker's hands to use tools. Wear gloves of an appropriate type when handling rough materials.

Ground Picture/Shutterstock.com

Figure 3.6-8. A particulate mask filters out dust.

il21/Shutterstock.com

Figure 3.6-9. An approved respirator is required for certain jobs.

LESSON 3.6 Review

Answer the following questions based on information in this lesson.

1. *True or False?* Every person on the jobsite holds a responsibility for managing safety hazards. (3.6-1)
2. What is the most effective control for managing hazards? (3.6-1)
   A. Personal protective equipment
   B. Engineering control
   C. Administrative control
   D. Elimination
3. What is an example of a struck-by hazard? (3.6-2)
   A. Falling objects
   B. Moving vehicles on a jobsite
   C. Scrap materials on an elevated surface
   D. All are correct.
4. *True or False?* A particulate mask is sufficient protection when working with asbestos-containing materials. (3.6-3)
5. Hearing protection is required starting with noise levels at or about _____ decibels. (3.6-3)
   A. 15
   B. 45
   C. 85
   D. 115

LESSON 3.7

Math

LEARNING OUTCOMES

After completing this lesson, you will be able to:

3.7-1 Multiply and divide fractions.

3.7-2 Find the center point on a wall.

TECHNICAL TERMS

total rise

unit rise

Multiplying and Dividing Fractions

Chapter 2, Excavation and Foundations, introduced addition and subtraction of fractions. This lesson discusses multiplication and division of fractions and useful tips for working with fractions on the jobsite. These operations are used regularly when laying out framing assemblies such as wall and sill plates. Often a feature—a window or door—falls in the center of a wall segment. The length or width of a room or other dimension may need to be divided into three or more equal parts. Layout of stair openings and carriages requires multiplication and division of fractions. These are just a few examples of when this math can be of use. Truly, construction workers use the addition, subtraction, multiplication, and division of fractions throughout the home building process.

Multiplying Fractions

To multiply fractions, first multiply the numerators (top numbers) together, and then multiply the denominators (bottom numbers) together.

For example:

$$\frac{3}{4} \times \frac{2}{5} = \frac{3 \times 2}{4 \times 5} = \frac{6}{20}$$

The numerators 3 and 2 are multiplied to result in 6. The denominators 4 and 5 are multiplied to result in 20. To make the result 6/20 easier to work with, reduce the fraction to its lowest terms.

To reduce fractions, write them as an equivalent fraction with the lowest possible denominator. For example, both 6 and 20 can be divided by 2 to make an equivalent fraction with a smaller denominator.

$$\frac{6 \div 2}{20 \div 2} = \frac{3}{10}$$

In this case, 6/20 can be reduced to the lowest terms of 3/10.

To multiply a fraction by a mixed number includes an additional step. Consider the following example:

$$\frac{2}{3} \times 4\frac{1}{2}$$

First, we must change 4 1/2 into an improper fraction to be able to multiple across numerators and denominators as with any common fraction. Recall that an improper fraction is a fraction with a larger numerator than its denominator.

$$4\frac{1}{2} = \frac{9}{2}$$

Then, multiply and reduce the result to the lowest terms.

$$\frac{2}{3} \times \frac{9}{2} = \frac{18}{6}$$

$$\frac{18}{6} = \frac{3}{1} = 3$$

Practice 3-1

To save time, your boss purchased a set of precut deck stair carriages to use for temporary access from the ground to the first story floor platform. They ask you to level the ground to the correct height where the base of the stairs will rest.

You can determine the height of the installed set of stairs by multiplication. The ***total rise*** of the stairs is the overall vertical distance from the ground at the bottom to the floor at the top. The ***unit rise*** of the stairs is the total vertical distance from one step to the next.

The unit rise of the carriages purchased is 7 1/4″. There are 4 risers.

To determine the total rise, multiply the unit rise by the number of risers.

7 1/4″ unit rise × 4 = x

Practice 3-2

It takes 8 3/4 rows of subfloor to cover the width of the home you are building. If each sheet of subfloor is 4 feet wide, what is the width of the building?

$$8\frac{3}{4} \times 4' = x$$

Dividing Fractions

Dividing fractions is not as straightforward as dividing whole numbers. To divide fractions, invert the divisor (swap the numerator and denominator) and then multiply as you would with common fractions. Recall that the divisor is the number that another number will be divided by.

Consider the following example:

$$\frac{3}{4} \div \frac{2}{3}$$

In this example, the divisor is 2/3. To invert, swap the numerator and denominator, and then multiply across.

$$\frac{3}{4} \div \frac{2}{3} = \frac{3}{4} \times \frac{3}{2}$$

$$\frac{3 \times 3}{4 \times 2} = \frac{9}{8}$$

In this example, the quotient is an improper fraction. Recall that a quotient is the result obtained by dividing one number by another. The improper fraction indicates that the value is greater than 1. To express this in its simplest form, convert it to a mixed number.

$$\frac{9}{8} = \frac{8+1}{8} = 1\frac{1}{8}$$

Practice 3-3

How many 14 1/2″ blocks can be cut from a 68 1/2″ board?

68 1/2″ ÷ 14 1/2″ = x

Practice 3-4

How many risers are in a set of stairs with a total rise of 42 3/4″ if the risers are 7 1/8″?

42 3/4″ ÷ 7 1/8″ = x

Converting Measurements

Sometimes it is necessary to combine operations in a single problem. Some such problems also involve more than one type of unit, such as inches and feet. Consider the following example:

In **Figure 3.7-1**, we must find the total length of the highlighted chain dimension, explained in *Lesson 3.8*, found in this bathroom blueprint.

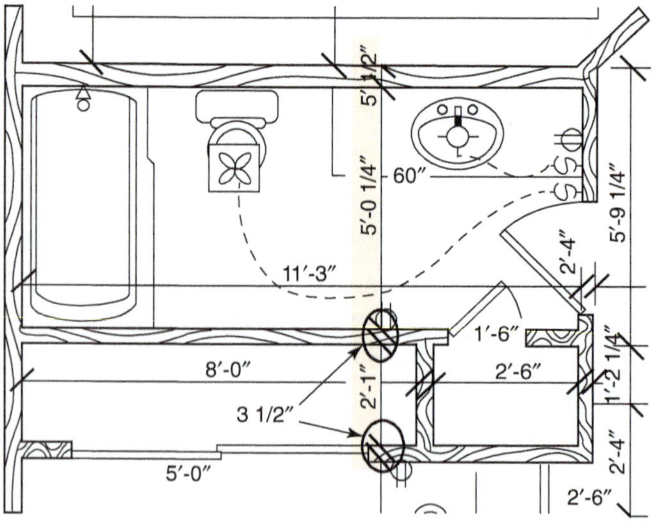

Goodheart-Willcox Publisher

Figure 3.7-1. A section of a blueprint containing a bathroom.

The first step in solving such problems is to decide which operation should be done first. It often helps to write the problem in such a way that it states the order of operations. Do not forget to include the width of each wall. The width for walls framed with 2×4s is 3 1/2″, and the width for walls framed with 2×6s is 5 1/2″.

$$5\ 1/2'' + 5'\text{-}0\ 1/4'' + 3\ 1/2'' + 2'\text{-}1'' + 3\ 1/2'' = x$$

Next, convert everything to the same unit(s). In this case, convert measurements in feet to inches. So, the bathroom dimension measuring 5′-0 1/4″ would be converted to 60 1/4″. The closet dimension measuring 2′-1″ can also be converted to 25″. Now with all measurements in the same unit (inches), we can perform the necessary operations to find the length of the chain dimension.

$$5\ 1/2'' + 60\ 1/4'' + 3\ 1/2'' + 25'' + 3\ 1/2'' = x$$

Where the operations will include adding or subtracting fractions, convert the fractions to their least common denominators.

$$5\ 2/4'' + 60\ 1/4'' + 3\ 2/4'' + 25'' + 3\ 2/4'' = x$$

Now solve the problem, doing all operations in the planned order.

$$96\ 7/4'' = 97\ 3/4'' = x$$

Finally, convert the units to whatever makes sense for the problem. For example, this can be converted to 8′-1 3/4″.

Manipulating Fractions on the Job

When working in construction, it is necessary to manipulate fractions in a short time, and sometimes simply running numbers through your head without writing them down is the quickest, most efficient method. Because construction workers generally work with a small handful of fractions—most commonly halves, quarters, eighths, and sixteenths—equivalent fractions (different forms of the same fraction) can be memorized rather easily.

Finding the Center

Finding the center of a dimension is a common task in construction. Essentially, finding the center involves splitting the dimension into two equal parts. As an example, imagine you are a framing carpenter on the jobsite trying to lay out a wall plate. The wall is 60 1/2″ long. The blueprints call for a door to be centered in the wall. To find the center point using conventional math, you must divide the wall in half.

$$60\frac{1}{2}'' \div 2$$

You could approach this problem using conventional math with the following steps:

1. Convert the mixed number to an improper fraction. Recall that to do this, multiply the whole number by the denominator of the fraction, and then add the numerator.

$$60\frac{1}{2} = (60 \times 2) + 1 = \frac{121}{2}$$

2. Invert the second fraction to get its reciprocal.

$$\text{The reciprocal of } 2 = \frac{1}{2}$$

3. Multiply the first fraction by the reciprocal of the second fraction.

$$\frac{121}{2} \times \frac{1}{2} = \frac{121 \times 1}{2 \times 2} = \frac{121}{4}$$

4. Simplify the resulting fraction, if possible, by dividing both the numerator and the denominator by their greatest common factor. The greatest common factor of 121 and 4 is 1, so this improper fraction cannot be simplified any further.
5. Convert the improper fraction to a mixed number, if desired. To do this, divide the numerator by the denominator, then write the quotient as the whole number, and use the remainder as the numerator of the fraction.

$$121 \div 4 = 30 \text{ with a remainder of } 1$$

$$\text{Therefore, } 60\frac{1}{2} \div 2 = 30\frac{1}{4}''$$

This method yields the correct answer, but it is complex and presents many opportunities for mistakes. Instead, let's break the problem into more manageable steps based on practicality.

A simpler method for determining half of any fraction is to take half of the numerator. Recall that equivalent fractions are: 1/2 = 2/4 = 4/8 = 8/16. Since we cannot use half of the 1 in 60 1/2, convert the 1/2 into an equivalent fraction, in this case 2/4, and then take half of that numerator. So, 2/4 would be halved to 1/4. Armed with this principle, let's take a different approach to our earlier problem, finding the center point on a 60 1/2″ long wall plate.

1. Separate the whole number from the fraction, splitting the number into two columns.

$$60 \quad \Big| \quad \frac{1}{2}$$

2. Take half of each column.

$$60 \div 2 \quad \Big| \quad \frac{1}{2} \div 2$$

$$30 \quad \Big| \quad \frac{1}{4}$$

3. Bring the two results back together by adding the columns.

$$30\,\frac{1}{4}''$$

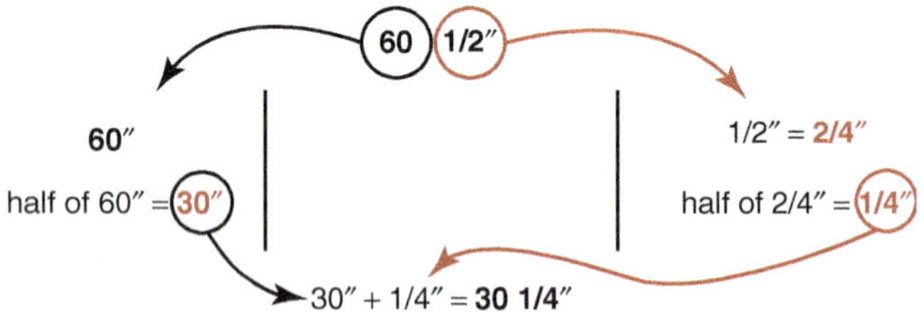

This simplified math yields the same result. This can be a helpful method to use when simple calculations are required. It works the same with other fractions used on the tape measure and can even be used to split dimensions that include feet and inches in half.

For instance, we need to find the center of a wall plate measuring 15′-7 7/8″. Use the same method of splitting the measurement into separate columns to break down this more complex mixed number.

1. Separate feet, inches, and fractions into three columns.

$$15'\text{-}7\ 7/8''$$

$$15' \qquad 7 \qquad \frac{7''}{8}$$

2. Take half of each. Remember to convert fractions with numerators that cannot be evenly halved into equivalent fractions.

$$15' \div 2 \qquad 7'' \div 2 \qquad \frac{7}{8} = \frac{14''}{16} \div 2$$

$$7'\text{-}6'' \qquad 3\frac{1''}{2} \qquad \frac{7''}{16}$$

3. Bring the results back together by adding the three columns. Remember to find a common denominator when adding fractions together.

$$7'\text{-}6'' + 3\frac{1''}{2} + \frac{7''}{16}$$

$$7'\text{-}6'' + 3\frac{8''}{16} + \frac{7''}{16} = 7'\text{-}9\,\frac{15''}{16}$$

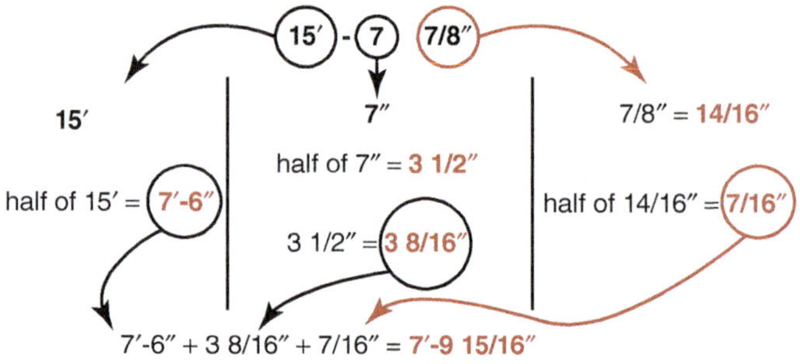

With consistent practice and application, this method can be done quickly in your head, but there is always the option to scribble it down on a scrap piece of lumber.

Doubling Fractions

Occasionally, it may be necessary to double a number with a fraction in it. For instance, if you need to determine the perimeter of a rectangular house when estimating materials, you can simply double the dimension of the long wall, double the dimension of the short wall, and then add the two sums together to get the perimeter. Like dividing fractions in half, there is a simple shortcut when working with most fractions used in construction.

To double a fraction, simply cut the denominator in half. For instance, 5/16″ doubled is 5/8″. When using this method, you may end up with an improper fraction. For instance, doubling 5/8″ yields 5/4″. This number should be converted to 1 1/4″.

LESSON 3.7 Review

Answer the following questions based on information in this lesson.

1. *True or False?* To multiply fractions, there needs to be a common denominator. (3.7-1)
2. Where is the center point on a wall that is 104 5/8″ long? (3.7-2)
   A. 52 5/4″
   B. 52 5/8″
   C. 52 5/16″
   D. None are correct.
3. *True or False?* To double a fraction, cut the denominator in half. (3.7-2)

LESSON 3.8

Print Reading

LEARNING OUTCOMES

After completing this lesson, you will be able to:

3.8-1 Identify chain dimensions and center lines on architectural prints.

3.8-2 Interpret floor plans to accurately identify critical dimensions and details such as window and door locations, partition walls, structural details, and other architectural features.

TECHNICAL TERMS

center line
chain dimensions
floor plan
framing plan

Reading Dimensions Correctly

A solid understanding of print reading is vital to framing a home. Precise locations of windows, doors, walls, fixtures, and other elements are taken from floor plans. If details on the prints are misinterpreted or overlooked, results can be costly. Critical dimensions such as the placement of kitchen or bathroom partition walls can mean the difference between a set of cabinets or shower unit fitting or not. Stairwell openings, fire blocking, structural details, and many other requirements must be met to ensure the home will pass building code inspections. Plumbers, HVAC installers, electricians, and other trades rely on the framing to be built as designed on the prints.

Chain Dimensions and Center Lines

In *Chapter 2, Excavation and Foundations*, you learned that dimensions on prints are expressed along dimension lines. The dimension begins and ends at an endpoint where the dimension line intersects an extension line. These dimension lines are often linked across a drawing linearly to form a continuous chain of dimensions, sometimes intersecting extension lines as they go, **Figure 3.8-1**. *Chain dimensions* are dimension lines linked across a drawing linearly to form a continuous series of dimensions and help keep a drawing organized and neat. A drawing may feature multiple chains of dimensions, each chain communicating dimensions for like-objects along that side. Typically, the overall building dimension is the furthest from the center of the drawing, a second chain often indicates locations of walls and partitions, while a third chain shows the locations of windows and doors along that side of the building.

Just as with individual dimensions, dimensions in the chain begin and end at an endpoint where the dimension line intersects an extension line. Occasionally, a dimension line may intersect an extension line or other dimension line without an endpoint. This indicates that the dimension listed continues beyond this extension line, **Figure 3.8-2**.

When looking at foundation plans, you may notice that windows and doors are located with a dimension that begins and ends at the edges of the masonry opening, **Figure 3.8-3**. Windows and doors in wood-framed walls, such as on a floor plan, are typically located with a dimension to the center of the rough opening, **Figure 3.8-4**. Rather than an extension line, a center line is used. Architects use a *center line* to mark the center of a symmetrical object, a circular feature, or column lines. Center lines are drawn on a plan using a thin line of alternating long and short dashes,

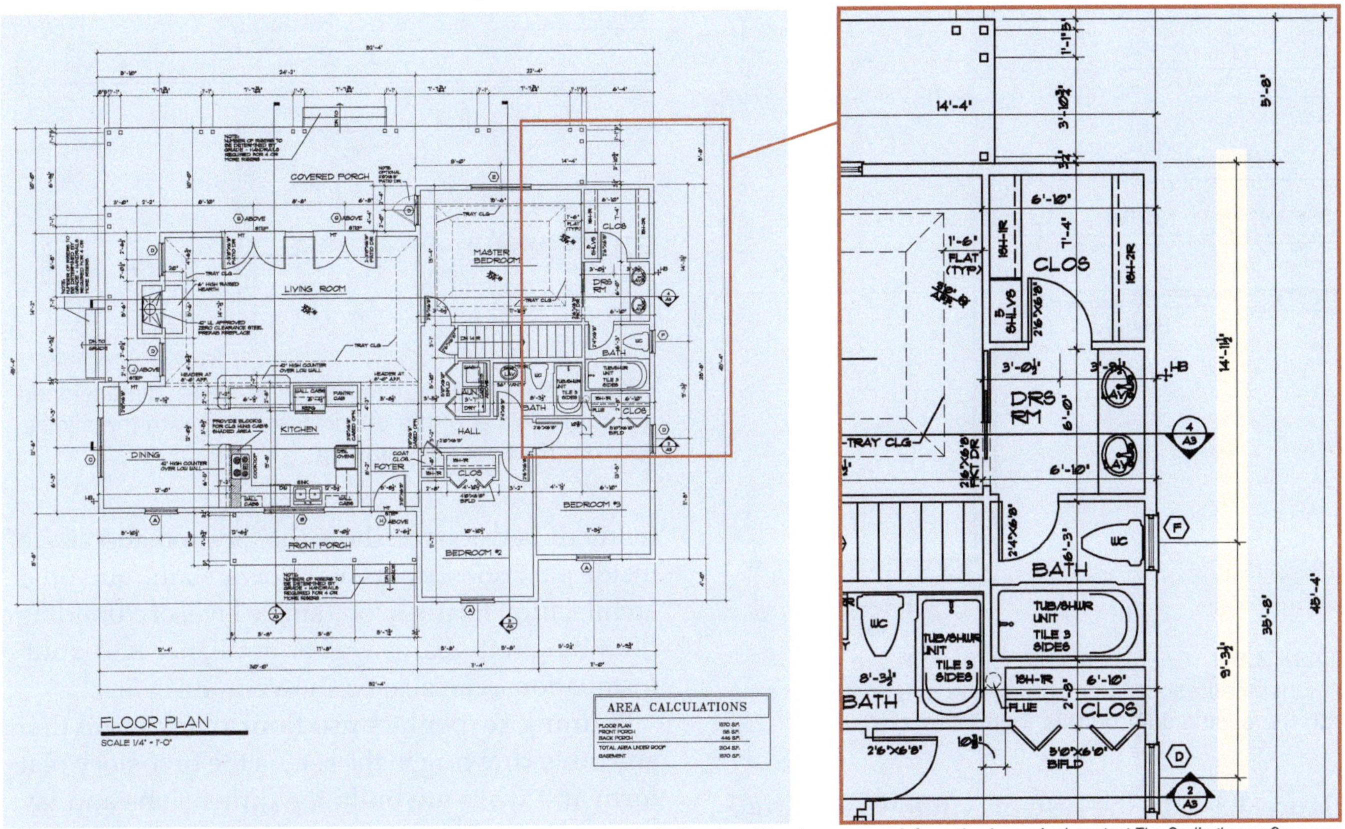

The Garlinghouse Company. The drawings in this chapter cannot be reproduced or built. If copies of the plans or more information is required, contact The Garlinghouse Company.

Figure 3.8-1. Chain dimensions are used extensively in floor plans to keep the drawing organized. The chain dimensions highlighted in this blueprint are linked linearly, expressing the distance from endpoint to endpoint.

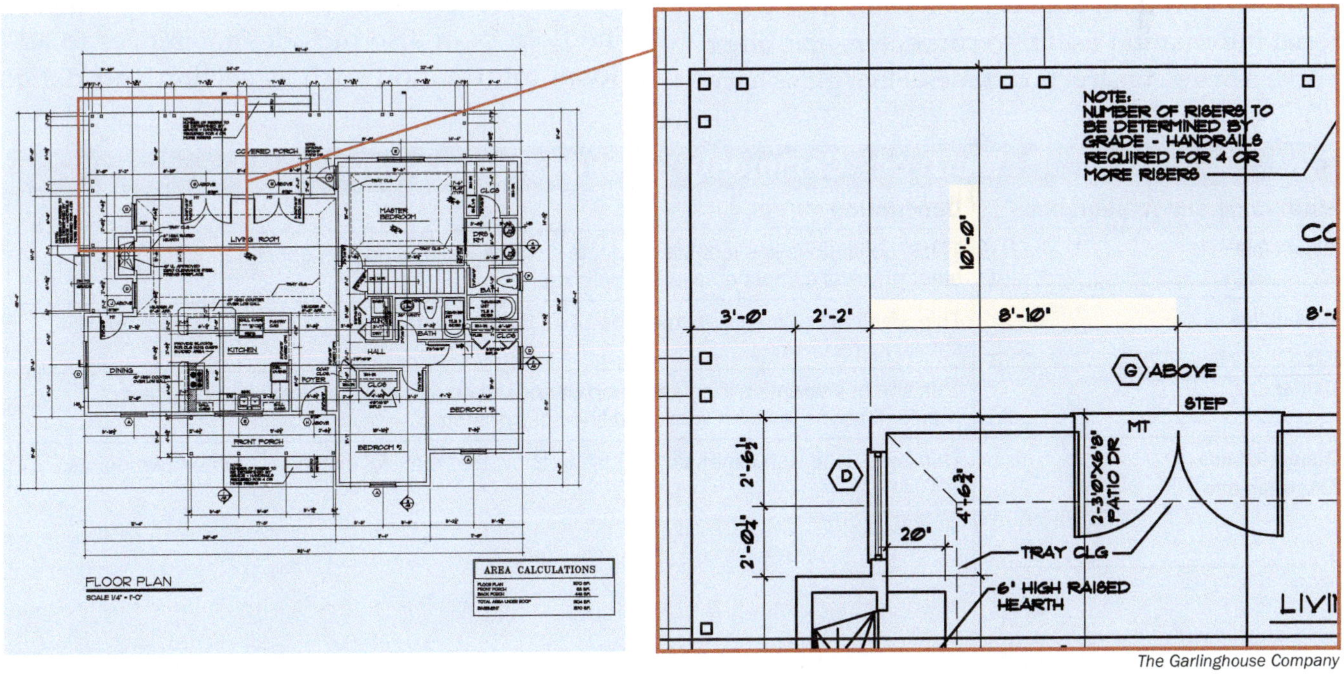

The Garlinghouse Company

Figure 3.8-2. At times, dimension lines will intersect at a point, but that will have no bearing on the adjacent dimension line. The dimension lines highlighted in the blueprint intersect, but because there is no endpoint, the intersection has no bearing on the given dimensions.

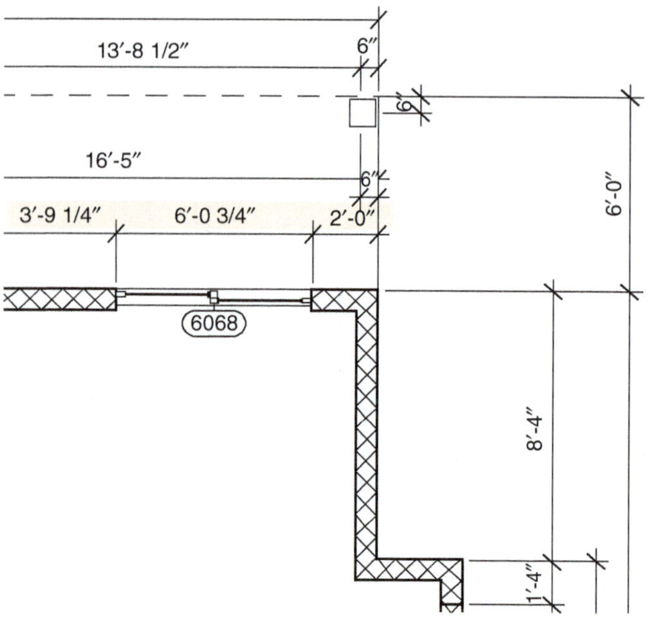

Goodheart-Willcox Publisher

Figure 3.8-3. Openings in masonry walls are dimensioned about the edges of the opening, such as with the sliding door on this foundation drawing.

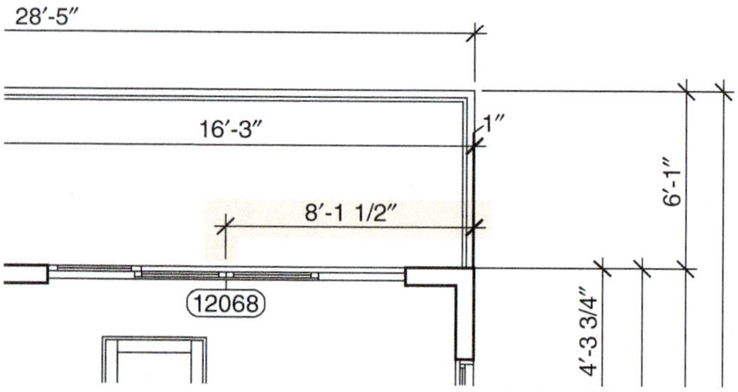

Goodheart-Willcox Publisher

Figure 3.8-4. Openings in framed walls are dimensioned about the center of the opening, such as with the sliding door on this floor plan drawing.

Figure 3.8-5. In the case of windows and doors, a center line is used in place of an extension line.

Interpreting Floor Plans

A *floor plan* is a plan-view drawing that shows the layout, size, and outline of a building and its rooms as seen from above. Floor plans give additional information useful to carpenters and other workers in the construction trades. Floor plans have many dimensions to show the location and size of inside partitions, doors, windows, stairs, and other architectural features for a single level of a building. Plumbing fixtures, as well as appliance and utility installations, can also be shown in this view.

Framing carpenters use floor plans more than any other drawing in the set. As the first-story platform and walls are built, the dimensions and layout of the building are taken from a careful study of the floor plan. The carpenter must interpret the prints correctly to determine details such as how wall intersections should be framed, the type of wall (interior or exterior, loadbearing or non-loadbearing, etc.), sizing and spacing of framing members, and sizing for window and door openings.

The floor plan also includes references to additional information such as section and detail

Line Types Commonly Used on House Plans

Line Name and Appearance	Description
Object line	Thick, continuous line identifies objects such as walls, windows, doors, foundations, and other physical components of the building.
Hidden line	Thin, dashed line identifies the edge of an object located behind another object in the view shown in the drawing.
Center line	Thin line of alternating long and short dashes used to mark the locations of centers of circular features, column lines, and lines of symmetry.
Dimension line **Extension line** Extension line 12'-4" Dimension line	Thin lines used with dimension values to show the "real life" distance between two points.

Goodheart-Willcox Publisher

Figure 3.8-5. Common line types used on construction drawings.

drawings, or the window and door schedule. Referring to these other elements gives the builder key specifications regarding the components of the home. A home will have more than one floor plan if it is a two- or three-story house. **Figure 3.8-6** shows the floor plan and an illustration of the completed residence. Take a moment and match up the orientations of these two drawings.

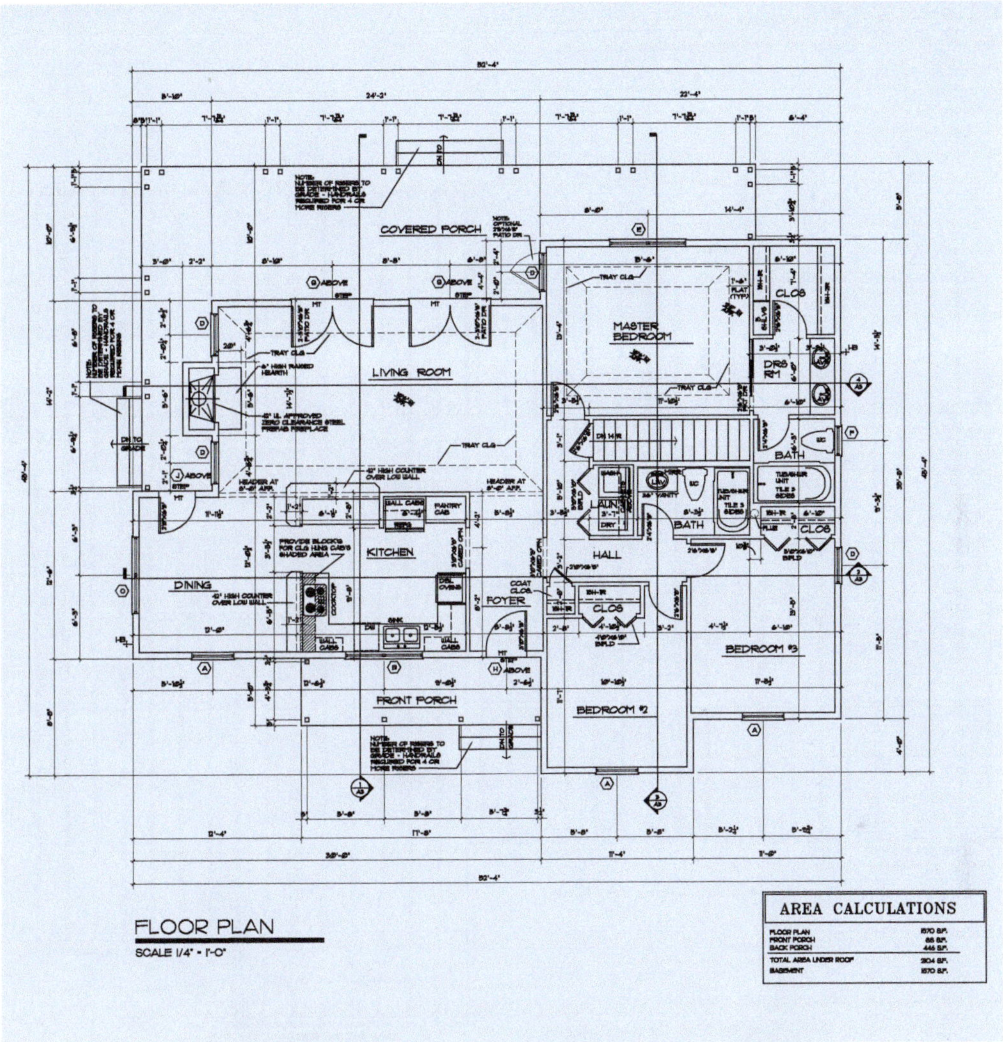

A

B

The Garlinghouse Company

Figure 3.8-6. A—The floor plan for a one-story house. Floor plans do not give details on how assemblies are built; however, they do give overall dimensions, shapes, and sizes of rooms. They also give placement of mechanical systems, cabinetry, and other built-in features of the building. B—A full-color rendering of the completed house.

Framing Plans

Sometimes a house plan will include a *framing plan*, which has drawings showing the size, spacing, number, and location of the structural members of the building's frame, **Figure 3.8-7**. Separate plans may be drawn for the floors,

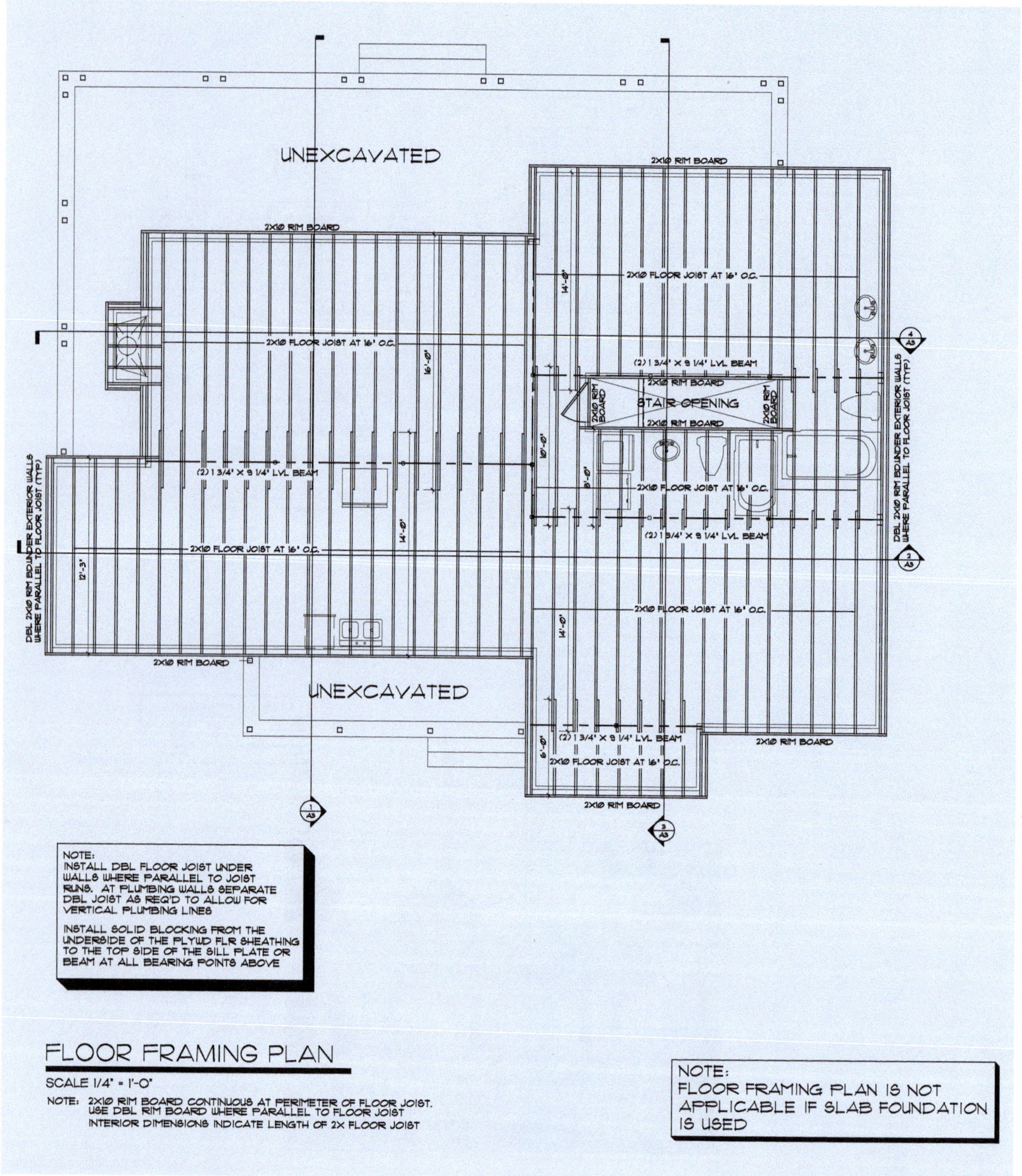

The Garlinghouse Company

Figure 3.8-7. House plans often include framing drawings. The floor framing plan specifies sizes and spacing of joists, girders, and support columns.

ceilings, walls, and roof. When the roof is to be framed with trusses, a truss plan is included, **Figure 3.8-8**. The trusses and other framing members are drawn to show where they will be positioned in the building. Openings for features such as chimneys are shown. Dimensions are not used, but the drawings are made to scale, as with other plan drawings.

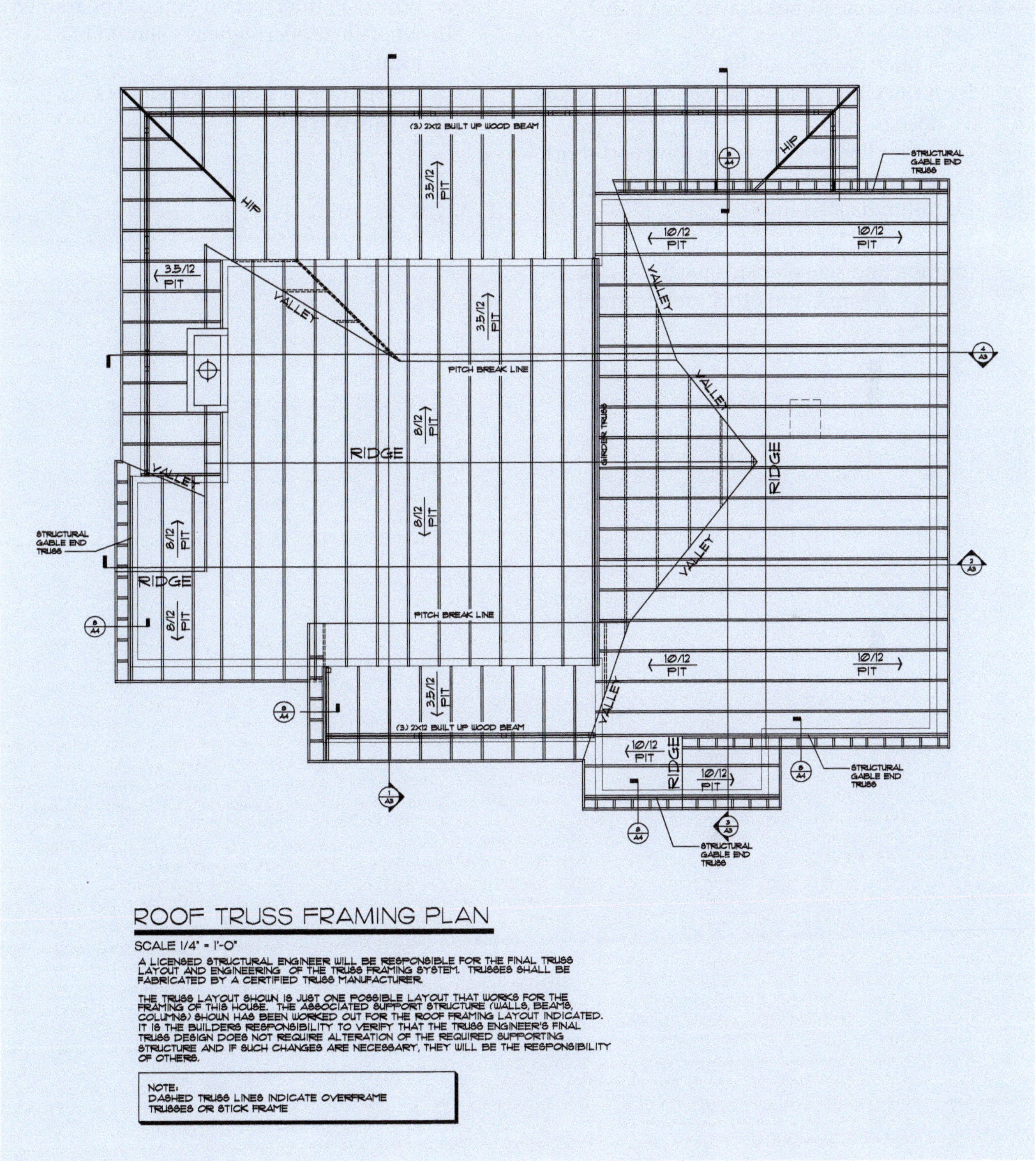

The Garlinghouse Company

Figure 3.8-8. A truss plan shows a carpenter where each truss is to be located and how special situations, such as openings in the roof, are framed.

LESSON 3.8 Review

Answer the following questions based on information in this lesson.

1. *True or False?* Chain dimensions are linked across a drawing in a line. (3.8-1)
2. How are center lines drawn on a plan? (3.8-1)
   A. A thick continuous line
   B. A thick line of alternating long and short dashes
   C. A thin line of alternating long and short dashes
   D. A thin dashed line
3. *True or False?* The framing plan shows the location and size of inside partitions, doors, windows, stairs, and other architectural features. (3.8-2)
4. A carpenter uses a floor plan to determine _____. (3.8-2)
   A. how wall intersections should be framed
   B. where loadbearing walls should be located
   C. the spacing of framing members
   D. All are correct.

LESSON 3.9

Green Building and the Future of Construction

LEARNING OUTCOMES

After completing this lesson, you will be able to:

3.9-1 Identify two alterations from conventional framing used in advanced framing methods.

3.9-2 Summarize the benefits of using an advanced framing system.

TECHNICAL TERMS

advanced framing

structural insulated panel (SIP)

Alternatives to Traditional Framing

Advanced framing is a system of framing that uses a more efficient framing layout to minimize the amount of lumber needed while maintaining the building's structural integrity. This may include eliminating unnecessary studs, headers, and other framing elements.

Using this system brings many benefits. It reduces the amount of material used; board footage of framing lumber can be up to 10% less than that of conventional framing. This method also reduces labor costs. Fewer studs, joists, rafters, and plates equate to less assembly time. Most importantly, an advanced framing system builds more energy-efficient homes to meet energy codes. It has a 10% increase in wall cavity insulation, or approximately 85%, compared to approximately 75% with conventional framing. Some may choose to utilize certain elements of advanced framing, but not all of them.

Green Note

Evaluate waste for other uses on or off the jobsite before disposal. "Wasted" lumber can often be used elsewhere in the project, on another jobsite, or in a craft project offsite.

Conventional framing typically uses 16″ on center (OC) spacing. Advanced framing uses 2×6 studs spaced 24″ OC, **Figure 3.9-1**. If all the framing members are in line vertically, the top plate does not need to be doubled. The studs must be placed directly over the floor joists at each level, and trusses or rafters must be placed directly over the studs. Refer to **Figure 3.9-2** to see how these framing members align.

Advanced framing replaces wood with insulation wherever that can be done without sacrificing structural integrity. Headers are built with a single ply of glue-laminated beam (glulam) or dimensional lumber, **Figure 3.9-3**. Headers are eliminated on non-loadbearing walls and replaced by a single plate and cripple studs, just like the sill framing of a window. This leaves space for insulation where the extra lumber would have been used. Using header hangers instead of trimmer studs can also result in savings. The hanger is nailed to the rough opening stud, and the header slips into the

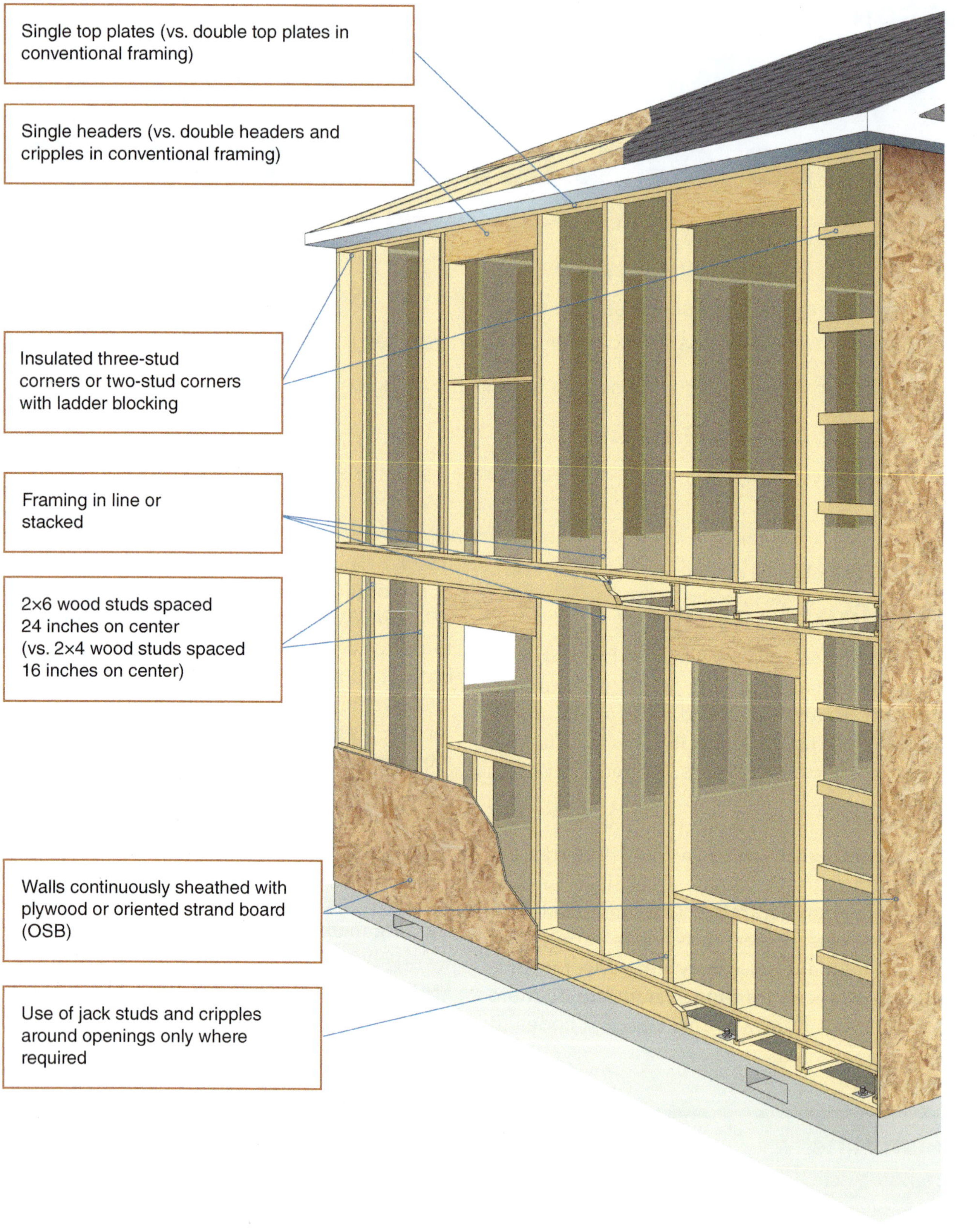

Figure 3.9-1. Advanced framing techniques may include some or all the details shown here.

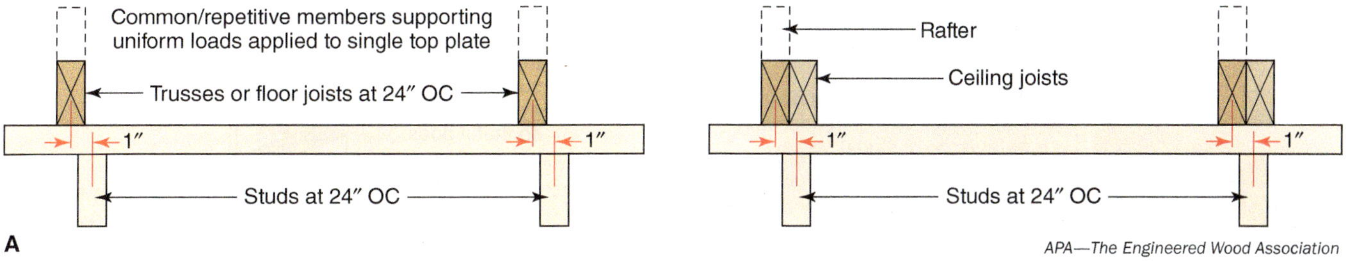

APA—The Engineered Wood Association

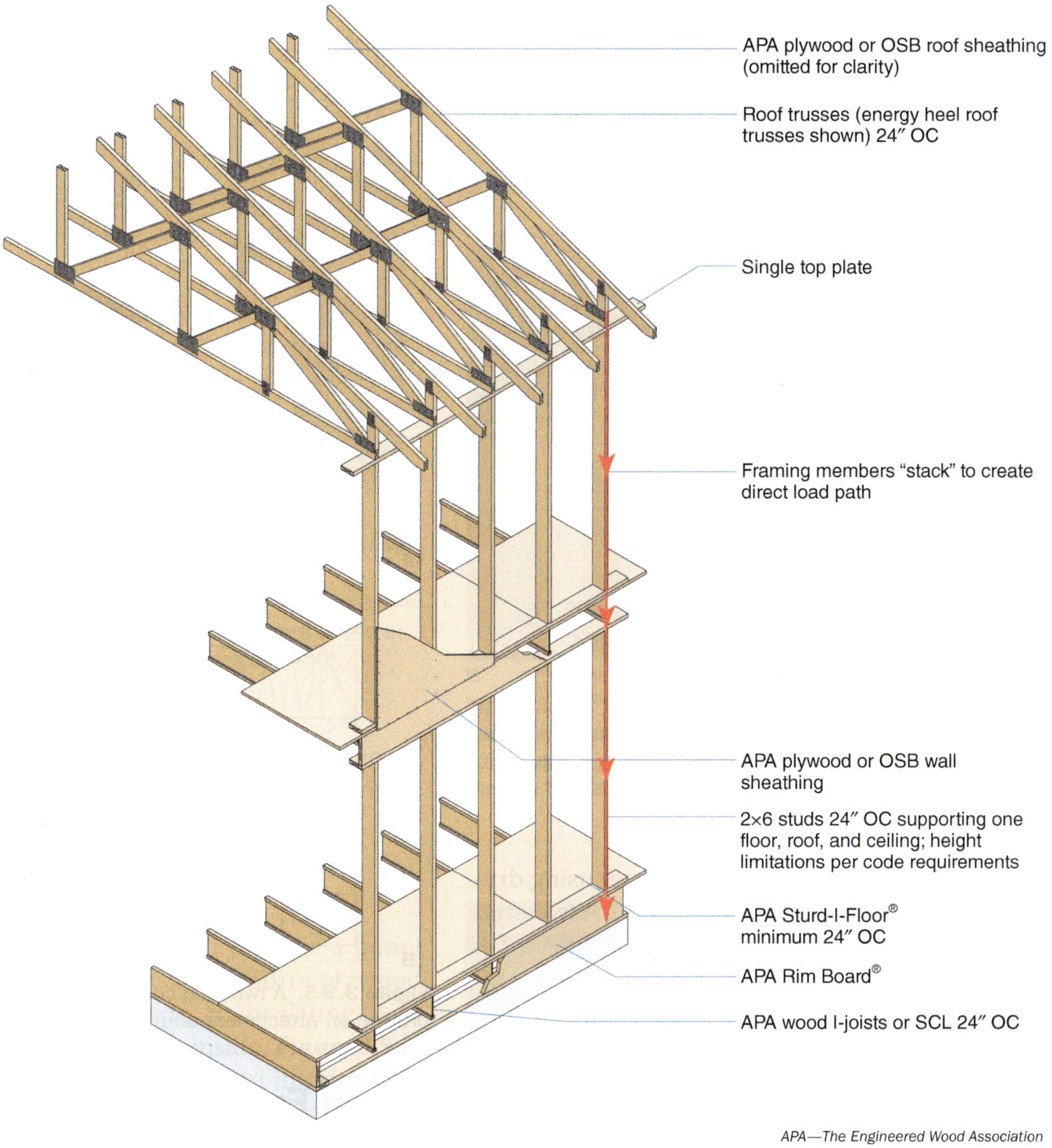

APA—The Engineered Wood Association

Figure 3.9-2. In-line framing 24″ on center (OC). A—Framing members must be centered over the member below within 1″. B—Aligned framing members create a straight load path from the roof to the foundation.

top of the hanger. Steel connecting plates can be used on single top plates to make splices and attach adjoining walls.

To gain maximum insulation, corners are constructed so they can be fully insulated. This can be accomplished with the three-stud corner, shown in framed wall, the cost of building a house with SIPs can be comparable because of the savings in labor. They are especially energy efficient because they have few, if any, openings in the insulated foam and tight-fitting joints in the interior and exterior sheathing. SIPs also save on materials. The National Association of Home Builders has estimated that construction of a conventional 2,000 sq ft house produces about 7,000 lb of waste. With SIPs, there is practically no waste during the framing phase of construction. SIPs can be used for floors, walls, and/or roofs, **Figure 3.9-9**.

A brizmaker/Shutterstock.com

B SV Production/Shutterstock.com

Figure 3.9-9. A—This SIP wall is being lifted into place with a crane. It has all the door and window openings already prepared. B—SIP walls once in place.

LESSON 3.9 Review

Answer the following questions based on information in this lesson.

1. *True or False?* An advanced framing system typically uses 16″ on center (OC) spacing between studs. (3.9-1)
2. *True or False?* A structural insulated panel (SIP) is cheaper than the materials used for a conventionally framed wall. (3.9-1)
3. Using advanced framing methods could result in ____ less framing lumber than that of conventional framing. (3.9-2)
   A. 10%
   B. 20%
   C. 30%
   D. 40%
4. Building a 2,000 sq ft house with SIPs results in ____ waste during the framing phase of construction. (3.9-2)
   A. 1,000 lb of
   B. 2,000 lb of
   C. 7,000 lb of
   D. practically no

Chapter 3 Review

Summary

Lesson 3.1

- Many people who work in the construction industry are business owners. A business can be structured as a sole proprietorship, a partnership, a limited liability company, or a corporation. (3.1-1)

Lesson 3.2

- Building codes give the minimum requirements to which a building must be constructed. Building codes exist to provide for the health, safety, and general welfare of building occupants and other people in the community. (3.2-1)
- Before beginning a construction project, a building permit must be obtained. (3.2-2)
- As work progresses, a building inspector makes inspections to ensure work meets the applicable building codes. (3.2-3)

Lesson 3.3

- After the foundation of a house is built, framing may begin. Three types of framing are platform, balloon, and post-and-beam framing. (3.3-1)
- Common structural framing members include girders, joists, studs, sole plates, top plates, cripple studs, trimmer studs, headers, rafters, and trusses. (3.3-2)

Lesson 3.4

- Engineered lumber products such as plywood, OSB, and LVLs use less timber and can be manufactured from smaller trees than sawn lumber. These products are also stronger, straighter, and more dimensionally stable than sawn lumber. (3.4-1)
- Framing connections must meet building code requirements so the house frame can withstand all forces imposed on it. These connections are made using fasteners, metal framing connectors, and bonding agents. (3.4-2)
- Various nail types are available to accommodate different purposes during the construction of a home. (3.4-3)

Lesson 3.5

- Tools used in the framing process range from basic hand tools to an assortment of power saws and nail guns. (3.5-1)

Lesson 3.6

- The hierarchy of controls should be applied to assess and manage all identified safety hazards on the job. (3.6-1)
- Cranes are sometimes used in residential construction when heavy assemblies need to be lifted. Safety is a central concern when a crane is in use. Workers must remain mindful of struck-by hazards while working and should strive to reduce such hazards. (3.6-2)
- Personal protective equipment (PPE) includes safety glasses, work boots or shoes, hard hats, dust masks, respirators, hearing protection, and hand protection. Appropriate PPE should be selected and used according to the task being performed. (3.6-3)

Lesson 3.7

- Multiplying and dividing fractions are an important part of laying out and estimating material usage in construction. (3.7-1)
- Finding the center of a dimension is a common task in construction. Essentially, finding the center involves splitting the dimension into two equal parts. (3.7-2)

Lesson 3.8

- The dimension begins and ends at an endpoint where the dimension line intersects an extension line. These dimension lines are often linked across a drawing linearly to form a continuous chain of dimensions, sometimes intersecting extension lines as they go. A center line is used to mark the center of a symmetrical object, a circular feature, or column lines. (3.8-1)
- Floor plans have many dimensions to show the location and size of inside partitions, doors, windows, stairs, and other architectural features for a single level of a building. Plumbing fixtures, as well as appliance and utility installations, can also be shown in this view. (3.8-2)

Lesson 3.9

- Advanced framing replaces wood with insulation wherever that can be done without sacrificing structural integrity. (3.9-1)
- Structural insulated panels (SIPs) framing, an alternative to conventional stick framing, uses structural panels with an insulating foam core. (3.9-2)

Apply and Analyze

1. What are the primary advantages to starting a sole proprietorship over an LLC? What are the disadvantages? (3.1-1)
2. Describe how building codes keep residents safe in their homes. (3.2-1)
3. When would a builder choose to use the post-and-beam framing method? (3.3-1)
4. Explain the difference between a cripple stud and a jack stud. (3.3-2)
5. Briefly state the use for each of the following types of engineered lumber: laminated-veneer lumber (LVL), glue-laminated beams (glulams), parallel-strand lumber (PSL), and laminated strand lumber (LSL). (3.4-1)
6. Why would a carpenter not be able to swap framing nails for trim nails if they are short on materials? (3.4-3)
7. A crew of framing carpenters is concerned about the fall hazard while placing rafters on a specific job. Using the hierarchy of controls, starting with the most effective measures, what controls might the competent person on the framing job use to manage this hazard? (3.6-1)
8. Describe three pieces of personal protective equipment (PPE) used when completing the framing phase of construction. (3.6-3)
9. Describe how to find the center point on a wall. (3.7-2)
10. What are three benefits of advanced framing? (3.9-2)

Critical Thinking

1. In small groups, discuss how you would manage a construction business together, or how you would manage a construction business on your own. (3.1-1)
2. Research the strengths and weaknesses of using manufactured wood I-joists as opposed to using traditional framing lumber. When building your own home, which would you choose and why? (3.4-1)
3. Why would it be difficult for some builders to adopt advanced framing methods? (3.9-1)

Communicating about Construction

1. **Speaking and Listening.** Divide into groups of four or five students. Using your textbook as a starting point, research your topic and prepare a report on how to install platform floor framing. As a group, deliver your presentation to the rest of the class. Take notes while other students give their reports. Ask questions about any details that you would like clarified. (3.3-1)
2. **Speaking.** Obtain a drawing of (or draw) a framing assembly. Label the areas on the assembly where these types of components are located: joist, sill, girder, stud, cripple stud, trimmer stud, king stud, sole plate, top plate, and rafter. Be prepared to explain the purpose of each component. (3.3-2)
3. **Speaking.** Obtain literature about OSB, plywood, foamed plastic, and gypsum sheathing. Download this material from the internet or obtain it directly from manufacturers. Also, study books and other reference materials. Prepare a report for the class, based on the information you obtain. Include grades, manufacturing processes, characteristics, and application requirements. Discuss current prices and purchasing information. Be prepared to discuss specifications. Relate these specs to your local code. (3.4-1)

Alysson M/shutterstock.com

CHAPTER 4

Drying In

Chapter 4 Outline

Introduction

Once a home has taken shape through framing and sheathing, it is time to make it weather-tight. The process of installing the first protective layer is referred to as *drying in*. This chapter will discuss the basics of drying in, along with many of the materials and tools used. You will also learn the difference between meeting code requirements and using best practices. You will have the opportunity to dig deeper into construction math, blueprint reading, and safety. With respect to green building, you will learn the importance of a well-constructed building envelope and explore some of the testing methods that are used to improve efficiency. Before getting into these subject areas, we will explore some of the less traditional construction careers that you may want to consider as a career path.

LESSON 4.1

Industry and Careers

LEARNING OUTCOMES

After completing this lesson, you will be able to:

4.1-1 List alternative opportunities for employment in the construction industry.

TECHNICAL TERMS

house flipping

inspector

Niche Opportunities

Many assume that working in the residential construction industry means building and remodeling homes, **Figure 4.1-1**. While that encompasses the big picture, there are also many niche opportunities for individuals who have complimentary skillsets. If you are a person with a knack for building relationships, attention to detail, teaching, writing, being creative, or someone who loves being in front of a camera, you can use these and many other talents in all kinds of construction jobs. Let's explore just a few of the possible niches open to you in your career.

House Flipping

House flipping is nothing new in construction, though it has been popularized as an income stream in recent years by television shows with charismatic actors. ***House flipping*** is the process of buying a home at a bargain price, renovating it, and reselling it to generate a profit. **Figure 4.1-2** shows a home that was in a state of disrepair and brought back to a marketable condition. House flippers tend to target homes to purchase that are not desirable to the average homebuyer. They use construction

FOTOGRIN/shutterstock.com

Figure 4.1-1. Traditional career paths in residential construction include a variety of trades as well as general contracting and remodeling.

Saviz Sarshar/shutterstock.com

Figure 4.1-2. House flippers use construction experience to find economical ways to make a home more marketable.

and estimating knowledge and experience to determine what repairs and alterations will make the home marketable while generating a profit.

People who flip homes enjoy the liberty of being their own boss. They are free to make design choices, determine what changes should be made to a home, and what products to buy. For a house flipper, a penny saved is truly a penny earned. However, their pennies are also investments. When flipping houses, some renovation projects, such as a new kitchen or an additional bathroom, can yield far more return on investment than something like a new water heater or a replacement roof. See **Figure 4.1-3**.

On the downside, house flipping requires a large amount of capital upfront. The home being flipped must first be purchased and all the renovations paid for. Any costs for owning the property, such as paying the utilities and property taxes, must be covered. Sometimes the home does not sell as quickly as expected. It is not until the home is finished and resold that a house flipper sees a profit, that is, if there is a profit. House flippers can also lose money if they did not factor in all expenses or if the home sells for less than anticipated. Flipping homes can be a lucrative career. However, it is highly dependent on a number of factors that can make it a risky venture. A few of the risks associated with house flipping include:

- Failure to accurately project renovation costs.
- Failure to accurately predict property ownership expenses such as utilities, maintenance, or property taxes.
- The tendency for construction project costs to balloon.
- Unforeseen issues or damage such as rot, termites, or structural issues with existing homes.
- Operating in the complexities of the real estate market, which can include sales commissions, transfer taxes, and closing costs.
- Unstable or unfavorable real estate market conditions.
- Fluctuations in interest rates.
- Fluctuations in construction material prices.
- Global, national, and local economic conditions.

hanohiki/shutterstock.com

Figure 4.1-3. Kitchen remodel before and after.

Entertainment or Recreation Industries

While being part of a construction television show is a possibility, such positions may be difficult to secure. However, construction opportunities in the entertainment industry include more than working in front of a camera. Set design and construction for theaters, movies, television, and theme parks requires many of the same skills as building actual homes, **Figure 4.1-4**. Some of these careers require highly skilled craftsmen and experience working with specialized materials or techniques that may require quick assembly or disassembly. If you have ever been to a theme park, consider some of the staged environments that were incorporated into the line que or the ride itself. Those were all built by someone with specialty skills and knowledge in construction. These

Stefano Tammaro/shutterstock.com

Figure 4.1-4. The buildings on this film set have a façade that mimics traditional construction with fine craftsmanship. A closer look reveals their secrets.

jobs can be fun and rewarding, allowing you to incorporate a high level of creativity and style. They can also be demanding and require you to work long or odd hours.

If you are interested in learning more about what kind of specific construction careers might be available in the entertainment and recreation industries, check out websites that specialize in industry job listings. Or use keywords that narrow your results in a regular job search engine. These platforms often feature job opportunities for various roles, including construction workers.

Content Creators

Since the rise of the internet, people have continually found new ways to communicate and connect to consumers. Content creation can be a viable career option if you are an expert in a field, have exceptionally creative talent, are comfortable in front of a camera, or have strong writing skills. Content creation can range from blogging, documenting projects, creating instructional videos, or just putting an entertaining spin on construction work. Social media platforms have opened new opportunities for communicators to create and express their expertise, thoughts, skills, and ideas. The internet offers worldwide exposure but establishing an audience can be very challenging. Much of this type of work requires a tremendous time commitment staging and editing content. There are no guarantees for establishing or maintaining an income stream. This is certainly a nontraditional, niche career option that may require a part-time approach while working in an established career. Success is possible, provided there are favorable circumstances, a good idea, and the optimal combination of experience, skill, creativity, charisma, and perseverance.

Journalism

Although not a large part of the construction industry, journalism is an area that is understaffed. Writing and construction may seem like an unlikely combination; this makes the construction worker who is skilled in writing a real asset to the trades. Quality construction magazines and websites often look for freelance writers to submit articles. Sometimes journalists or freelance writers review tools or products and report their findings. They may write about construction processes or industry changes, such as building code updates or political issues that impact the industry.

Dozens of websites that serve contractors and do-it-yourselfers also publish articles pertaining to residential construction. The National Association of Home Builders (NAHB) is a leading authority on the happenings in the residential construction industry. They regularly publish articles and columns on their website that feature industry research, laws, advice, and calls to action for members.

Other construction career paths for those with writing skills include technical writing, such as construction curricula for technical schools and training organizations, industry reference manuals, or textbooks. Admittedly, this kind of writing is not for the beginner in construction, but if you have an interest, you can make it a long-term career goal.

Teaching and Training

Many larger home developers and organizations, such as the NAHB, undertake construction training for their employers or for industry. In fact, the NAHB has an entire training arm called the Home Builder's Institute (HBI) whose mission includes educating and preparing individuals for construction careers. High school trade programs and technical colleges have building construction, carpentry, masonry, plumbing, and HVAC programs to teach and train students for a career in the building industry. Trade unions also have apprenticeship programs that take less experienced members through a process of training. Construction product manufacturers have educational programs that demonstrate and train contractors on the correct use of their products and showcase new developments, **Figure 4.1-5**. Each of these programs requires a staff of teachers and/or trainers that have a strong grasp on construction skills and principles to communicate and teach their subjects.

Teaching and training jobs can be both fun and rewarding. The ability to communicate your construction expertise is a valuable skill. This is especially true as you develop a deeper and more thorough understanding of your craft. It is not unusual for successful and skilled tradesmen to

A Pennsylvania College of Technology B Simpson Strong-tie

Figure 4.1-5. Teaching as a profession. A—Teaching construction to beginners is a rewarding profession. B—A representative of a company conducts training on the proper use of hangers and connectors.

become teachers. For some, this is a part-time occupation. Others may leave the construction field to teach their skills to others. This may involve returning to school to obtain certification or getting a college degree in education. Some students take construction courses to become proficient in teaching career and technical courses.

Construction Consulting

Construction consulting is another occupation that requires a strong working knowledge in a construction specialty. Again, this is not a career choice for beginners in the field but may be an appealing long-term career goal.

A construction consultant provides expert advice and guidance on construction projects. Their role is to help clients make informed decisions about complex matters that might hinder a project, **Figure 4.1-6**. They must have a good working knowledge of their specialty to ensure a project meets specific goals, regulations, or performance metrics.

Construction consultants may specialize in the following areas:

- Project planning
- Cost estimating/project budgeting
- Construction management
- Permitting and code compliance
- Quality control
- Troubleshooting and problem resolution
- Managing logistics
- Expertise in a trade or specialty

Leszek Glasner/shutterstock.com

Figure 4.1-6. Construction consultants provide expertise to homeowners, builders, architects, and other industry professionals.

A career in construction consulting requires a combination of relevant experience, skills, and knowledge. When a client hires a consultant, they expect to bring in an industry expert who can solve complex issues in a manner that is well-founded, reasonable, and based on a thorough understanding of the issues and the consequences of making a wrong decision.

Sales and Marketing

Product manufacturers, distributors, and retail outlets rely on a sales staff who have a strong working knowledge of construction to sell their products. Custom home builders also employ sales staff to work with clients when choosing finishes and options for their homes. Some spec home builders rely on an internal sales team, rather than real estate agents, to sell their homes.

Working in sales requires effective communication and interpersonal skills. Building relationships is often key to maintaining client accounts. It is also important to have the appropriate technical knowledge of construction techniques and materials to accurately answer questions and solve problems. Sales professionals often collaborate with project managers, architects, and contractors throughout the construction process. They may need to coordinate communication, ensure product specifications are met, and provide support during the construction process.

Sales careers usually require meeting specific targets and goals. In the residential construction industry, this means hitting sales quotas, getting contracts, and generating profits. Many sales jobs offer commissions, meaning you earn money when you make successful sales. Sales can be a fulfilling career path for individuals who like working with other people and are motivated by closing a deal.

Inspector

An ***inspector*** ensures that a building and its systems conform to codes, standards, and local regulations, **Figure 4.1-7**. There are many kinds of inspectors whose work relates to the construction industry. Many of these can be grouped under the general title of building inspector as they all deal with some aspect of the building process, such as electrical, plumbing, or mechanical systems.

Approximately half of all inspectors are employed by local units of government. The inspection process usually begins even before construction starts, when plans are submitted to the building inspector's department for review. The review may result in approval or may require changes to bring the plan into compliance with codes and regulations. Once construction is underway, inspections are made at several stages: foundation, framing, electrical, plumbing, and mechanical system (such as HVAC) installations. Once all work is completed to the satisfaction of the inspector, final approval, often in the form of an occupancy permit, can be issued.

ungvar/shutterstock.com

Figure 4.1-7. A building inspector checks to ensure codes are followed throughout the building process.

Code Note

Inspectors must have a very strong working knowledge of building codes. Building inspectors working in residential construction are required to complete a training program that covers the International Residential Code (IRC) and other relevant building codes and regulations.

Working conditions as an inspector are similar to most building trade jobs, although less physically demanding. Inspectors must be able to climb ladders, work in confined spaces, and carefully navigate through areas cluttered with materials and tools.

In addition to building code inspectors, a significant number of inspectors work for engineering or architectural service firms to verify that a building is constructed to the specifications. Some inspectors provide various energy audits to determine if a building meets green building standards or to find ways it can be made more efficient. Other inspectors provide home inspections for a home buyer or seller to examine a property being offered for sale.

People employed as building inspectors and in the various specialties, such as electrical or plumbing, should have a thorough knowledge of all codes and local regulations that the builder must meet. A background in the building trades is very desirable for those seeking to become inspectors. Some inspectors have academic preparation in such areas as construction technology or construction management. Inspectors need to be certified to provide inspections for the discipline in which they work.

Insurance Adjuster

Insurance adjusters work for insurance companies. When damage occurs to an insured home, the owner often files an insurance claim, **Figure 4.1-8**. If the claim is covered under the homeowner's policy, the insurance company sends an insurance adjuster to survey the damage and determine the extent of the needed repairs. The adjuster estimates the repair costs based on their understanding of building construction. Insurance adjusters document all aspects of the claim process, including their findings, evaluations, and settlement agreements. They prepare and submit detailed reports to the insurance company for review and approval. Adjusters act as the primary point of contact for policyholders throughout the claims process. They provide guidance, answer questions, and address concerns regarding the claim, policy coverage, and settlement options. They maintain regular communication to keep homeowners informed about the progress of their claims.

Mark Skalny/shutterstock.com

Figure 4.1-8. Storms, natural disasters, fires, and other incidents resulting in damage to a home require insurance adjusters to visit the site and assess the damage to estimate the cost of repairs.

Property Manager

A property manager is responsible for overseeing the maintenance and operation of rental properties, which may include single-family homes or apartment complexes. They work on behalf of the property owner or a real estate management company.

Property managers work with tenants to address concerns regarding the building. They perform or oversee all maintenance and repairs to ensure facilities comply with safety and quality standards. Depending on their skill level, they may also perform renovations and regular property inspections.

Along with construction, maintenance, and inspection of properties, property managers must work with tenants, vendors, and owners to manage the daily operation of the rental business. This may include collecting rent, managing finances, record keeping, scheduling deliveries or services, and dealing with emergency situations.

LESSON 4.1 Review

Answer the following questions based on information in this lesson.

1. *True or False?* People who flip homes must consult with the home buyer to decide the types of improvements that will be made to the house before it is put up for sale. (4.1-1)
2. *True or False?* One risk associated with house flipping is fluctuating interest rates.
3. Which personal characteristic is most important to being a successful content creator? (4.1-1)
   A. Charisma
   B. Loyalty
   C. Accountability
   D. Reliability
4. *True or False?* Although not a large part of the construction industry, journalism is an area that is currently overcrowded. (4.1-1)

LESSON 4.2

Construction Fundamentals

LEARNING OUTCOMES

After completing this lesson, you will be able to:

4.2-1 Compare construction best practices with building code requirements.

TECHNICAL TERMS

best practices

flashing

Best Practices

In *Chapter 3, Framing Systems*, we learned that building codes establish the minimum requirements to which a building must be constructed. Although builders are required to satisfy the building code, they are not limited to it. In other words, builders are free to build to a higher standard than the building code. In fact, in many situations, it makes sense to build or install a product or assembly using the best available methods. This means using the most effective technology, products, and techniques that we currently have. We call this using ***best practices***.

Best practices can add significant cost to a project if they require expensive products or are time consuming. However, some best practices require little to no added cost and help ensure a better built home. One example of this is window installation. The IRC requires that windows be installed according to manufacturer's instructions, or if installation is not addressed in the instructions, that pan flashing be installed at the sill and at the sides and top of the window as well. ***Flashing*** is a thin material used to divert water away from vulnerable areas on a building. According to the building code, the window must also be air-sealed on the interior side.

If a window installation meets these requirements, the code has been satisfied. However, the method and quality of flashing installation vary from builder to builder. The building code does specify how these flashings are installed, or what products are best to use. Furthermore, new products are being developed all the time. If the installer is careless or does not understand the importance of getting the details right, they can meet the minimum code requirements and still have a compromised flashing system, allowing water to get into the home or wall system resulting in damage later on, **Figure 4.2-1**.

Using best practices for window installation requires utilizing products and techniques that ensure water is directed to the outside of the building. It is more about understanding the best way to apply products than upgrading to more expensive ones. In this example, few extra costs are involved. It simply requires research and training to make sure the installer understands how to accomplish their work most effectively.

Green Note

Best practices are often required, rather than encouraged, for homes constructed to green building standards.

Sompetch Khanakornpratip/shutterstock.com

Figure 4.2-1. This window was improperly flashed, resulting in a leak to the interior.

Joe Belanger/shutterstock.com

Figure 4.2-2. If grades were given for construction work, using best practices would yield the best grade possible.

Construction best practices encompass far more than window installation. If there is a better way to perform a task than code requires, or more effective products available than code requires, then meeting building code is not the best practice for that task. Best practices cover various aspects of building construction. They can affect how much energy the building uses, how strong and long-lasting it is, aesthetics, how adaptable it is to diverse needs (such as aging in place), how comfortable it feels to live in, and even the health of the occupants.

Another way to look at best practices is to compare them to grades in an educational setting, **Figure 4.2-2**. A certain educational tract might require a grade of C or better to graduate. All students achieving a C or better will meet the minimum requirements. However, these students have room for improvement. Students who achieve an A letter grade have achieved the highest grade possible. This can be compared to code requirements and best practices. Those who have built according to the building code requirements have met the minimum standard. However, there is room for improvement over and above the building code. Using best practices is achieving the highest level possible.

It is worth noting that the building code may require the best method, effectively making it the same as the best practice for a given procedure. However, sometimes using a construction best practice may be cost prohibitive. Few construction budgets allow for achieving the highest level of quality construction for every component and system on a project. In these cases, meeting the minimum standards of the building code is perfectly acceptable. It is up to the builder and architect to determine where to use best practices and

Pro Tip

For many tasks, it does not cost more to use best practices. All workers on a construction project should strive to follow best practices, even if the building is not planned for green certification.

where building to code is best for a project. However, it is always beneficial for builders to know and understand the best practices for any given task. For this reason, ongoing training and education should be part of every builder's game plan, **Figure 4.2-3**.

goodluz/shutterstock.com

Figure 4.2-3. Training programs are one way to teach construction best practices.

LESSON 4.2 Review

Answer the following questions based on information in this lesson.

1. *True or False?* Building codes require windows to be air-sealed on the interior side. (4.2-1)
2. *True or False?* An installer can meet the minimum code requirements and still leave the homeowner with a compromised flashing system. (4.2-1)

LESSON 4.3

The Construction Path

LEARNING OUTCOMES

After completing this lesson, you will be able to:

4.3-1 Explain the requirements to dry in a house.

TECHNICAL TERMS

drying in

water-resistive barrier (WRB)

Drying In

After a home has been framed and sheathed, it must be made weathertight before mechanical systems and materials can be installed. The process of making a home watertight and protected from the elements during construction is called ***drying in***, **Figure 4.3-1**.

Drying in involves installing:

- Roofing surfaces, such as the underlayment and shingles.
- Water-resistive barriers, such as housewrap or coated sheathing on the walls.
- Windows and doors.
- Flashing and sealing on all exterior penetrations and other critical areas where water could enter.

Drying in is a key step in the construction process, as it protects the structure from potential water damage and mold growth and allows wood materials to dry out and stabilize before insulation and finishes are applied. As soon as the building has been dried in, electricians, plumbers, and HVAC technicians can begin installing their systems in the home.

Jon Rehg/shutterstock.com

Figure 4.3-1. Roofing materials await installation on this new home during the dry-in process.

Green Note

One of the most important factors of efficiency is the building envelope. A home can be built from the latest energy-efficient materials and still perform poorly if they are not assembled correctly. The building envelope must be both well-designed and assembled properly to ensure performance. It is pointless for an owner to pay top dollar for a high-performance window if the builder installs it poorly.

Water-Resistive Barrier (WRB)

A ***water-resistive barrier (WRB)*** is a material or system used to provide a primary layer of protection against water intrusion on the exterior walls of a home. A WRB must be applied to prevent water accumulation within the wall and must be built in a manner that provides a way for water to drain to the outside of the wall assembly. The WRB is installed on the outside of the wall sheathing and under the cladding. It is not visible when the job is complete.

The WRB is the final line of defense against water. It must be designed and constructed in a way that ensures a watertight installation. Windows and doors are flashed into the WRB system to keep water out of the framing, **Figure 4.3-2**. Laps or seams in material should be dealt with according to the manufacturer's instructions or the building code.

Materials that can be used for a home's WRB include coated sheathing, housewrap, foam board insulation, peel-and-stick membrane, or liquid-applied coating. The most common of these products are described in *Lesson 4.4*.

Code Note

Building codes require the exterior wall envelope to be designed and built in a manner that prevents accumulation of water in the wall assembly by providing a WRB behind the exterior cladding. There is also a requirement that the WRB has a way to drain water to the outside of the cladding.

Alysson M/shutterstock.com

Figure 4.3-2. Housewrap is being installed on this home as a WRB. Windows are installed so their flashing directs water out to the surface of the WRB.

LESSON 4.3 Review

Answer the following questions based on information in this lesson.

1. *True or False?* Mechanical systems should be installed before the home is made weathertight. (4.3-1)
2. The water-resistive barrier is installed on the _____. (4.3-1)
   A. inside of the wall sheathing and over the cladding
   B. inside of the wall sheathing and under the cladding
   C. outside of the wall sheathing and over the cladding
   D. outside of the wall sheathing and under the cladding

LESSON 4.4

Materials and Fasteners

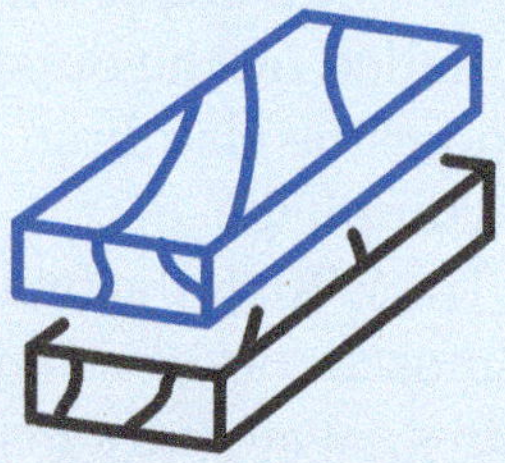

LEARNING OUTCOMES

After completing this lesson, you will be able to:

4.4-1 Differentiate materials used for drying in the walls, roof, and foundation of a house.

4.4-2 Explain why cap nails and staples are used for installing materials.

TECHNICAL TERMS

awning window	hopper window	roofing underlayment
casement window	horizontal sliding window	sash
double-hung window	housewrap	sidelight
felt paper	housewrap tape	synthetic underlayment
fixed window	ice dam	transom
flexible flashing tape	ice-and-water barrier	

Selecting Materials for Drying In

The variety of materials that are available for drying in a home is extensive. Factors to consider when selecting roofing material can include design preferences, climate, and cost. When selecting materials for the WRB, performance factors are important. When selecting windows and doors, quality, cost, and design play a significant role. We will explore common options for these materials. However, note that the extent of options available is beyond the scope of this book. The wise builder is always considering which products best suit their project.

Roofing

Materials used for roofing in residential construction include asphalt shingles; wood shingles or shakes; clay, concrete, or slate tiles; metal roof panels or shingles; and composite shingles made from polymers, rubber, or plastic. Regardless of the material used for the roofing surface, roofing underlayment must be applied over the sheathing unless a coated roof sheathing was used.

Underlayment

Roofing underlayment provides a secondary layer of protection against weather and offers temporary protection before the finished roofing surface is installed.

Traditionally, roofers used felt paper as roofing underlayment, **Figure 4.4-1**. *Felt paper* is a thin, flexible material made from cellulose or fiberglass fibers that have been impregnated with asphalt or bitumen. It is available in rolls approximately 3′ wide. While felt paper is still in use, many builders have opted for a *synthetic underlayment* which is lighter and easier to handle. Synthetic underlayment is made from synthetic materials such as polypropylene or polyethylene. These materials offer enhanced durability, strength, and resistance to tearing, compared to traditional felt, **Figure 4.4-2**.

Christina Richards/shutterstock.com

Figure 4.4-1. Here a roofer installs shingles over felt underlayment.

Oak City Drone/Shutterstock.com

Figure 4.4-2. A variety of synthetic roofing underlayments are available. They have greater tear resistance and are lighter than felt paper.

Owens Corning

Figure 4.4-3. In cold climates, an ice-and-water barrier is applied to the roof edges. It is also advisable to use it on valleys and under other flashing on the roof.

northlight/shutterstock.com

Figure 4.4-4. Ice-and-water barriers protect a home from water backing up behind an ice dam like this one.

Pro Tip

Do not put down underlayment on a damp roof. It may trap moisture and damage the roof.

In climates that receive regular snowfall, building codes require an *ice-and-water barrier*, also called a membrane or shield, to be used along the eves of a roof, **Figure 4.4-3**. Ice-and-water barriers protect against water infiltration caused by layers of ice, called ice dams, that tend to build up along the eves.

An *ice dam* can form after a large snowfall event, **Figure 4.4-4**. Heat escaping from a home can warm the underside of the roof, causing snow to melt. Water from the melting snow makes its way to the eves of the roof where the cool overhang causes it to refreeze. As time goes on, the layer of ice thickens until a dam is formed, trapping water on the roof and allowing it to back up under the shingles. Without the protection of an ice-and-water barrier, this water can make its way into the attic and walls of a home, resulting in damage.

Ice-and-water barriers come in a roll like felt paper or synthetic underlayment. They are typically made of a rubberized asphalt compound with a sticky backing covered by a release paper. During installation, the release paper is removed, and the membrane is adhered to the roof sheathing. This bond produces an area of sheathing that has extra protection against water penetration and seals around roofing fasteners.

Pro Tip

To get the best performance from any roofing material, always study and follow the manufacturer's directions. They are usually included on the packaging material.

Asphalt Shingles

Assorted styles and types of asphalt shingles are the most commonly used roofing material for homes. Asphalt shingles can be classified into four groups:

- 3-tab shingles, also called strip shingles, are the oldest type of asphalt shingle. They have a relatively simple design that vaguely resembles a traditional slate tile roof. 3-tab shingles are being installed on the roof shown in **Figure 4.4-1**.
- Architectural shingles, also known as laminate or dimensional shingles, consist of two layers, offering more weather resistance than 3-tab shingles. They are simpler to install than 3-tab shingles because they have a random pattern. The design of architectural shingles gives more definition and style to a roof surface that vaguely resembles a traditional wood shingle roof, **Figure 4.4-5**. Architectural shingles are among the most popular choice for residential roofing today.
- Luxury shingles, also called designer or premium shingles, are made to mimic the look of slate tile or wood shingles more closely than 3-tab or architectural shingles. This is accomplished by built-up layers and strategically placed shadow lines and layers.
- Performance shingles offer one or more features that improve the performance of the product. This can include colors designed to reflect heat to keep the building cooler; higher impact resistance for climates where hailstorms are common; or shingles with higher wind resistance, fire resistance, or guard against algae growth.

Pro Tip

When laying asphalt shingles, soft-soled shoes that will not damage the surface and edge of the shingles should be worn. Asphalt products are easy to damage when worked on at high temperatures. Try to avoid laying these materials on extremely hot days.

Wood Shingles

At one time, wood shingles and wood shakes were the most popular roofing choice for homes in the United States. Although they are still a viable, environmentally friendly option with a unique and authentic look, **Figure 4.4-6**, wood shingles

Goodheart-Willcox Publisher

Figure 4.4-5. Architectural asphalt shingles do not have tabs but are layered to give them a random textured appearance.

Alfred State College

Figure 4.4-6. Furring is placed for a secure nailing surface and space for ventilation between the wood shingles or shakes and the underlayment.

have fallen out of favor due to their high maintenance and insufficient fire resistance. They are also more expensive and require more labor to install than asphalt shingles.

Pro Tip

Use care when nailing wood shingles. The wood is soft. It can be easily crushed and damaged under the nail heads. Drive the nail just flush with the surface.

Wood shingles are sawn to give them a consistent surface, while shakes are split, which gives them a more rustic appearance. Shingles and shakes are made from cedar, spruce, or pine. They are typically left unfinished, allowing them to weather to a dull gray color over time.

Metal Roofing

Metal roofing is available in panel or shingle form. Although metal roofing panels made of galvanized steel or aluminum are used in residential construction, **Figure 4.4-7**, they are more commonly used in agricultural and commercial applications

Safety Note

Field-cut metal has sharp edges and may have needle-like metal slivers. Wear leather gloves when handling it.

because they lend themselves to large structures with few obstructions like roof intersections, hips, valleys, dormers, chimneys, and other features common on homes.

Metal shingles are available in a variety of shapes and materials. They can be made from aluminum, steel, or copper. Metal shingles are increasing in popularity due to their durability and appearance. They are available in a variety of colors and styles.

Slate, Clay, and Concrete Tiles

Roofing tiles made from slate, clay, and concrete are durable and offer a distinct look, **Figure 4.4-8**. Roofs finished with these materials offer good fire resistance and can last up to 100 years. However, these products are the most expensive roofing options and are very heavy, often requiring reinforced roof framing to withstand the additional weight.

Composite Shingles

Composite shingles can be made of rubber, polymer, or plastic and are made to closely mimic the look and texture of more traditional materials, **Figure 4.4-9**, without the drawbacks of the low fire resistance of wood or the weight of tile.

Composite shingles are the newest of the roofing materials mentioned here. Their use in residential roofing applications has not yet been subjected to long-term field testing.

A M2020/shutterstock.com B Volodymyr_Shtun/shutterstock.com

Figure 4.4-7. A—Metal roofing panels can provide a clean, sleek look. B—More intricately shaped metal panels create a more decorative look.

Monier Group

Figure 4.4-8. Clay tiles are attached to battens spaced according to the size of the tiles.

Safety Note

Use appropriate fall protection and wear shoes or boots with skid-resistant soles when working on pitched roofs.

Materials Used for Water-Resistive Barriers

Examples of commonly used WRBs include housewrap, rigid foam board insulation, coated sheathing systems, and liquid-applied membranes, **Figure 4.4-10**. These materials can vary in function and installation, depending upon the design of the building envelope system. It is critical for the installer to understand the design and intended function of the system and its components to ensure longevity of the building.

Natalia S Wilkinson/Shutterstock.com

Figure 4.4-9. Composite shingles are made with molds to appear to have the texture of natural materials to closely mimic natural roofing materials when installed.

Housewrap

Housewrap is a plastic sheet material designed to protect from bulk water infiltration while allowing water vapor to pass through. This material property promotes drying through evaporation if water manages to get into the wall assembly through a leak in the WRB or if condensation occurs inside the wall.

Housewrap is nailed or stapled to the wall using special fasteners equipped with a plastic washer, **Figure 4.4-11**, for more holding power and to protect the housewrap from tears. All seams of housewrap must be lapped in a shingle fashion to promote drainage to the outside of the wall. *Housewrap tape*, also commonly called builder's tape, is an adhesive tape specifically designed to be used on housewrap or other select materials. Its purpose is to provide a seal for joints and seams to prevent water intrusion and air leakage.

Housewrap is available in a variety of types, each with different material properties and quality levels. It is available in rolls of various widths up to 10′.

Pro Tip

Do not confuse a material designed to stop air movement, such as housewrap, with a vapor barrier. A vapor barrier is designed to stop movement of water vapor, while most housewraps are designed to breathe, to let vapor pass through.

A Typar B Owens Corning C Photo courtesy of Huber Engineered Woods

D Dörken Systems Inc. E fotodrobik/shutterstock.com

Figure 4.4-10. A variety of building materials can be used as a water-resistive barrier. A—Housewrap is commonly used on wall systems. B—Special attention must be given to flashing details when using rigid foam board insulation. C—Coated sheathings provide a surface that is impenetrable to water, though nailing and taping details are critical on these systems. D—Peel and stick membranes adhere directly to the sheathing. E—Liquid-applied membranes are used primarily on commercial projects.

Goodheart-Willcox Publisher

Figure 4.4-11. For housewrap to function as an air barrier, it should have no holes or tears and be nailed or stapled using special fasteners.

Rigid Insulation Board

In colder climates, rigid insulation board is often used on exterior walls for its thermal properties. Rigid insulation comes in a variety of materials with differing characteristics, **Figure 4.4-12**. Which material is used, and how it is used, depends on the design of the wall.

- Extruded polystyrene (XPS) foam—typically blue, pink, or green in color—has an insulation value of R-5 per inch. It is rated for ground contact installation and will not absorb water.
- Expanded polystyrene (EPS) foam is similar to the material used in white Styrofoam® coolers and cups. EPS is a petroleum product that uses recycled polystyrene.

Goodheart-Willcox Publisher

Figure 4.4-12. Rigid insulation board comes in a number of different varieties. A—Extruded polystyrene foam. B—Expanded polystyrene foam. C—Polyisocyanurate.

It is made up of small beads formed during manufacturing.

- Polyisocyanurate (PIR), more commonly known as polyiso, is yellow in color and is often foil-faced. Polyiso will absorb water if kept in contact with moisture continuously; therefore, it cannot be used for ground-contact installations.
- Mineral wool board insulation is made from natural and recycled materials. It is usually brown and has a fibrous appearance, **Figure 4.4-13**. Mineral wool is a material that is spun into fibers from molten rock. It is water-resistant and allows vapor to pass through freely. This makes it a good choice as an insulation over housewrap. Unlike foam, which is toxic when it burns, mineral wool is noncombustible, providing a measure of fire resistance to the wall.

Using rigid exterior insulation board creates challenges for the installer. Of particular importance is how the insulation relates to the WRB and its flashing details around windows and doors.

Green Note

Organize a jobsite recycling program to manage efforts to minimize waste onsite. Designate a place to store materials ready for reuse on the jobsite. Include bins for separating wood, metal, and other scrap material for recycling. Many cities have stores that champion the reuse, restore, and repurpose mentality. These stores will accept salvageable materials as donations and offer them back to the community at a discounted rate. Taking advantage of these programs can reduce the cost of disposal and keep salvageable materials from filling up our landfills.

Serhii Krot/shutterstock.com

Figure 4.4-13. Mineral wool rigid board insulation is gaining popularity in the homebuilding industry as an exterior insulation.

Foam board insulation can be used as a WRB if the installers have paid careful attention to flashing details.

Coated Sheathings

Coated sheathings, **Figure 4.4-10C**, are often used as exterior wall sheathing or roof sheathing in both residential and commercial construction. They are typically made of OSB sheathing coated with a waterproof coating on one side. Seams of the sheathing are taped with a proprietary flashing tape or coated with a liquid flashing after installation, creating a continuous barrier against water infiltration. Builders must follow the manufacturer's guidelines for proper installation and compatibility with other building components, such as flashings, windows, and doors.

As you may have guessed, some of the materials commonly used as WRBs are designed to do more than keep out water. You can learn more about the building envelope and control layers in *Lesson 4.9*.

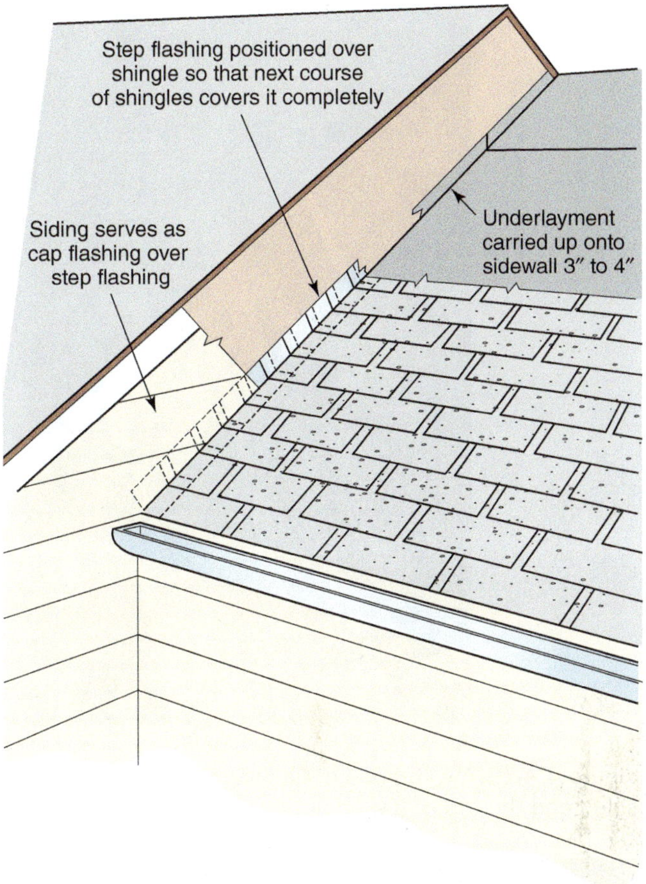

Goodheart-Willcox Publisher

Figure 4.4-14. Metal step flashing is applied to waterproof joints between the sloping roof and the walls. Note that the siding serves as cap flashing.

Flashing

Flashing diverts water away from vulnerable areas on a building, such as where a roof meets a wall or chimney, around windows and doors, or around any penetrations in a home's walls or roof, like around pipes, wires, or vents.

Many flashings are made of sheet metal, such as aluminum, copper, or galvanized steel. Joints in flashing must be lapped in a shingle fashion to keep water directed to the outside as it runs down a slope or vertical plane, **Figure 4.4-14**.

Flexible flashing tape is a specialized adhesive tape used to provide a watertight and airtight seal around openings, joints, and vulnerable areas of a building such as windows, doors, or seams on coated sheathing, **Figure 4.4-15**. Flashing tape does not take the place of rigid metal flashing but is used to supplement or accompany it.

Flashing tape is typically made of a durable and flexible material, such as modified bitumen, butyl rubber, or other synthetic materials. It is designed to withstand exposure to the elements, including UV radiation, temperature changes, and moisture. It uses a strong adhesive, allowing it to adhere to a variety of building materials including wood, metal, plastic, and concrete.

Standard or *straight* flashing tape only flexes in two dimensions, allowing it to be folded around corners, but not stretched in three dimensions. However, special *stretch tapes* are available that allow for three-dimensional installations without cutting, such as at the sill pan for a window or door, **Figure 4.4-16**.

Builders should always check the manufacturers' recommendations for use on flashing tapes as they do not adhere to all products and surfaces.

Windows

In general, windows can be categorized as three basic types:

- Sliding
- Swinging
- Fixed

A B Goodheart-Willcox Publisher

Figure 4.4-15. Two brands of flashing tape. A—Used to create sill pan flashing on a window. B—Used to flash the sides and the top of a window.

Photo courtesy of Huber Engineered Woods

Figure 4.4-16. Standard flashing tapes cannot be stretched in three dimensions like this. Special stretch tapes have been developed to allow cut-free flashing on sensitive areas.

Each of these includes various designs or operation methods. Sliding types include the double-hung window and horizontal sliding window. See **Figure 4.4-17**. Swinging units that are hinged on the side are called casement windows, while those hinged on the top or bottom can be either awning windows or hopper windows.

A B *Marvin Windows*

Figure 4.4-17. A—Double-hung sliding window. B—Horizontal sliding window.

Double Hung

A ***double-hung window*** has two sashes that slide up and down past each other in the window frame. ***Sashes*** are the individual glass panel assemblies of a window, including the frame around the glass panel. With a double-hung window, you can keep the sashes in any open position without them sliding down. This is accomplished by a friction fit between the sash and the frame or by springs and various balancing devices. Double-hung windows are widely used because of their economy, simplicity of operation, and adaptability to many architectural designs. **Figure 4.4-18** shows a window unit consisting of three double-hung windows.

Horizontal Sliding

Horizontal sliding windows have two or more sashes. At least one of them can move horizontally within the window frame, **Figure 4.4-19**. The most common design consists of two sashes that are both movable. In units with three sashes, the center one is usually fixed (not movable).

Noel V. Baebler/shutterstock.com

Figure 4.4-18. Three double-hung units.

Kwangmoozaa /Shutterstock.com

Figure 4.4-19. A horizontal sliding window has at least one movable sash and may have one fixed sash.

Casement

A ***casement window*** has a sash that is hinged on the side and swings outward. Some units consist of two or more windows separated by mullions. Sashes are operated by a cranking mechanism, or a push bar mounted on the frame, **Figure 4.4-20**. Latches are used to close, lock, and tightly hold the sash against the weather stripping.

The swing sash of a casement window permits full opening of the window. This provides good ventilation. Frequently, where a row of windows is desirable, fixed units are combined with operating units. Screens are attached to the inside of standard casement windows.

Awning

Awning windows have one or more sashes that are hinged at the top and swing out at the

Goodheart-Willcox Publisher

Figure 4.4-20. Casement windows are hinged on the side and are operated by a crank or push bar.

bottom, **Figure 4.4-21**. They are often combined with fixed units. Several operating sashes can be vertically stacked in such a way that they close on themselves or on rails that separate the units.

Pro Tip

Consideration of outside clearance must be given to both casement and awning windows. When open, they may interfere with movement on porches, patios, or walkways that are located adjacent to outside walls.

Most awning windows have a *projected action* in which the top rail moves down as the bottom of the sash swings out. Crank and push bar operators are like those on casement windows. Like casement windows, screens are mounted on the inside.

Awning windows are sometimes installed side by side to form a *ribbon* effect (a narrow band of windows, resembling a ribbon) at the top of a wall. Such an installation provides privacy for bedroom areas and also permits greater flexibility in furniture arrangements along outside walls. Often, the ribbon is placed above fixed windows to provide ventilation.

Hopper

The ***hopper window*** has a sash hinged along the bottom and swings inward, like an upside-down awning window. It is operated by a locking handle in the top rail of the sash. Hopper windows are easy to wash and maintain. However, their operation may interfere with drapes, curtains, and the use of inside space near the window. Hopper windows are most often used in basements.

Marvin Windows

Figure 4.4-21. Awning windows have one or more sashes that are hinged at the top.

Fixed

The ***fixed window*** can be used in combination with any of the other units. Its main purpose is to provide daylight and a view of the outdoors. When used in this type of installation, the glass is set in a fixed sash and mounted in a frame that matches the ventilating windows. Large sheets of plate glass are often separated from other windows to form window walls. They are usually set in a special frame formed in the wall opening.

Green Note

Regardless of how energy efficient a window or door unit is, using faulty installation practices can leave a home drafty. Always consult manufacturers' recommendations and follow current Energy Star best practices for window and door installation.

Exterior Doors

Entry doors come in a variety of configurations, **Figure 4.4-22**, offering unique styles, functionality, and customization options. Aside from the traditional single door, examples include:

- **Double doors**—Also known as French doors, this configuration features two doors side by side, typically hinged on the outer sides and opening from the center, **Figure 4.4-22A**.
- **Door with sidelights**—In this configuration, a single door is accompanied by narrow windows called ***sidelights*** on one or both sides. Sidelights provide additional natural light and can enhance the aesthetic appeal of the entryway, **Figure 4.4-22B**.
- **Door with transom**—A door with a ***transom*** has a horizontal window panel positioned above the door. Transoms can

be fixed or operable and come in various shapes, such as rectangular, arched, or elliptical.

- **Sliding door**—Sliding doors are a popular choice for patio or backyard entryways. They consist of one or more panels that slide horizontally along tracks, offering space-saving functionality and a wide opening.

Wood, metal, and fiberglass are the materials of choice for hinged exterior doors. However, sliding doors are often made from vinyl or aluminum.

Exterior doors arrive on the job with the door, frame, and lights (if applicable) assembled as a unit, **Figure 4.4-23**.

Fasteners Used for Drying In

Staples have traditionally been the fastener of choice for installing roofing underlayment and housewraps. They are installed quickly and effectively using a hammer tacker. However, most manufacturers of synthetic underlayment and housewrap now recommend the use of plastic cap staples or nails to improve wind and leak resistance. Plastic cap staples require a special tool for installation. Plastic cap nails are available in a variety of lengths to accommodate various product thicknesses. They consist of a nail with a plastic cap attached to the head of the nail. They can be purchased in strips to use in a nail gun or loose for hand nailing, **Figure 4.4-24**. Cap nails are used

A karamysh/Shutterstock.com

B Goodheart-Willcox Publisher

Figure 4.4-22. A—Main entry doors. B—Frames and doors for main entrances as they appear on elevation drawings.

Pro Tip

After installing large glass units, it is considered good practice to place a large *X* on the glass while the building remains under construction. Use masking tape or washable paint to apply the *X*. This makes the glass more visible and alerts workers not to walk into the glass or damage it with tools and materials.

Jason Finn/shutterstock.com

Figure 4.4-23. This entry door with sidelights arrived on the job as a single unit, ready to install.

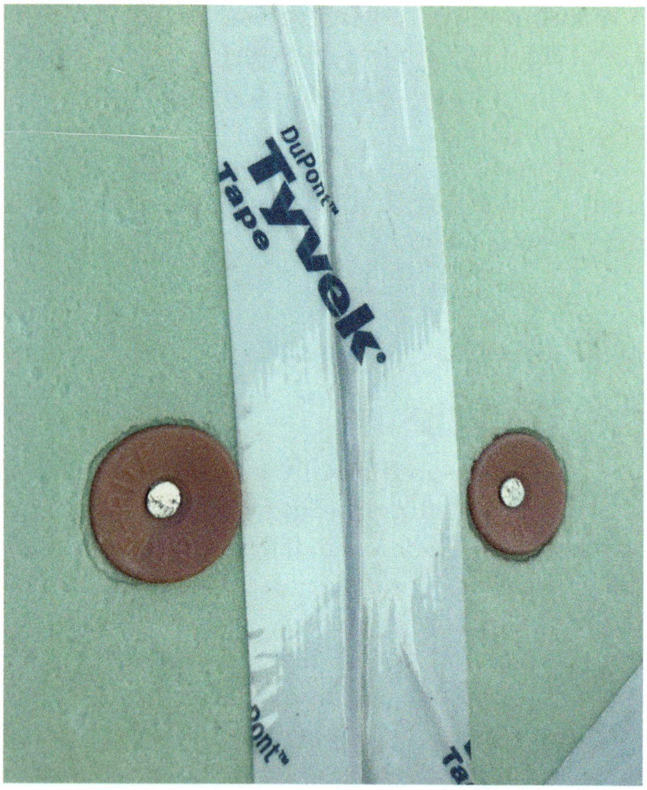

Goodheart-Willcox Publisher

Figure 4.4-24. Cap nails provide better holding strength and weather resistance. Here they are used to fasten foam board insulation.

for sheet goods such as roofing underlayment and housewrap, or for soft, thicker materials such as foam board insulation. Failure to use the recommended fastener and fastener spacing can void product warranties so it is important to follow manufacturers' instructions.

Roofing nails feature a larger head than framing or trim nails, **Figure 4.4-25**. The large head adds holding power when nailing softer materials like asphalt shingles. Roofing nails are typically made of galvanized steel for corrosion resistance. They are available in lengths up to 3″, but most applications require 1 1/4″ to 1 3/4″ lengths.

Adam J/shutterstock.com

Figure 4.4-25. Roofing nails have large heads.

LESSON 4.4 Review

Answer the following questions based on information in this lesson.

1. *True or False?* Architectural shingles are easier to install than 3-tab shingles. (4.4-1)
2. *True or False?* Wood shingles are split rather than sawn, to give them a rustic appearance. (4.4-1)
3. On which of these structures are you most likely to find metal roofing panels? (4.4-1)
   A. Single-family home
   B. Two-family home
   C. Barn
   D. Single-car garage
4. Which statement about mineral wool board insulation is true? (4.4-1)
   A. It is usually blue, pink, or green.
   B. It is water-absorbent.
   C. It prevents vapor from passing through.
   D. It is created from molten rock.
5. *True or False?* OSB coated sheathing has a waterproof coating on both sides. (4.4-1)
6. In horizontal sliding window units with three sashes, _____. (4.4-1)
   A. all three sashes are usually fixed
   B. the center sash is usually fixed
   C. the side sashes are usually fixed
   D. none of the sashes are usually fixed
7. *True or False?* Most manufacturers of synthetic underlayment and housewrap recommend the use of plastic cap staples or nails for installation. (4.4-2)
8. Roofing nails are typically made of _____. (4.4-2)
   A. aluminum
   B. copper
   C. stainless steel
   D. galvanized steel

LESSON 4.5

Tools

LEARNING OUTCOMES

After completing this lesson, you will be able to:

4.5-1 Identify tools used to dry in a home.

TECHNICAL TERMS

aviation snips
cap stapler
hammer tacker
roofing nail gun
staple gun
tape roller

Drying In Tools

Drying in a home requires some specialty tools that are used exclusively for specific materials or applications such as a cap stapler, roofing nailer, or tape roller while others are used extensively throughout the construction process, such as a utility knife or aviation snips. In fact, even if you have no construction experience, you may have used some of these tools.

Cutting Tools

During the process of drying in, a number of materials are cut or trimmed to size using hand tools that can easily be carried in a tool belt. One of the most versatile cutting tools on the construction site is the utility knife. Most construction workers keep a utility knife in their tool belts for quick access when needed.

Another cutting tool that is commonly kept in tool belts for regular use is a pair of aviation snips, **Figure 4.5-1**. ***Aviation snips*** are designed to cut through sheet metal, aluminum, and other thin materials. The color-coded handles on aviation snips indicate the type of cut each snip is designed for. Red-handled snips are designed for cutting curves to the left. Green-handled snips are for cutting curves to the right. Yellow-handled snips are designed for straight cuts.

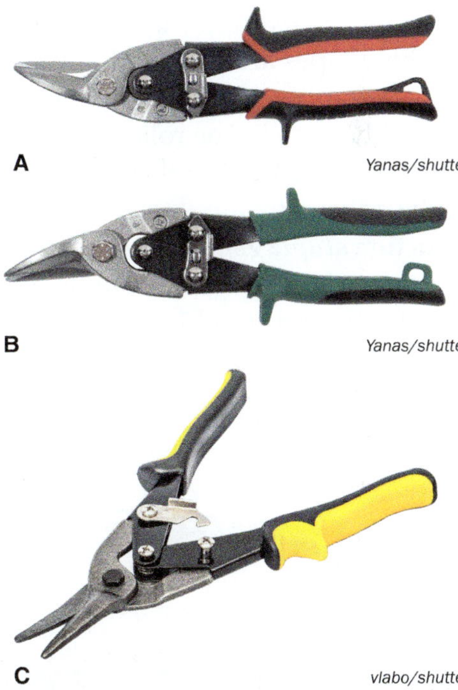

A Yanas/shutterstock.com
B Yanas/shutterstock.com
C vlabo/shutterstock.com

Figure 4.5-1. A—Left-cutting snips. B—Right-cutting snips. C—Straight snips.

Peyker/shutterstock.com

Figure 4.5-2. Tin snips are often used in metalworking tasks for cutting in straight lines.

Some people call aviation snips tin snips. However, tin snips typically have long handles with flat jaws and cutting edges, **Figure 4.5-2**. They are designed for straight cuts in thin-gauge sheet metal and are commonly used in HVAC work, roofing, and general metalworking tasks where straight cuts are predominant.

Various types of electric tin snips or powered metal shears are also available and can greatly reduce fatigue when performing straight cuts and basic metal work. Powered metal shears are discussed in greater detail in *Chapter 9, Finish MEP Systems and Punchlist*.

Attachment Tools

Tools used for attaching various WRB products include the tape roller, staplers, cap staplers, pneumatic cap staple guns, and roofing nail guns. The ***tape roller*** is a hand tool used when applying pressure-activated flashing tape, **Figure 4.5-3**. Many flashing tapes are designed to properly adhere only after applying pressure. The tape roller has a stiff rubber or silicone roller head. The roller is used to apply pressure to the flashing tape, activating the adhesive.

Construction ***staple guns*** are a tool used to shoot a staple into various materials. Staple guns come in various forms and should be selected based on the task. The manual staple gun has a squeeze handle that shoots a staple each time it is fully compressed, **Figure 4.5-4**. This type of stapler is used when precision is more important than speed. An alternative to the manual stapler is the electric staple gun which can shoot the same type of staple at the pull of a trigger. The ***hammer tacker***, **Figure 4.5-5**, uses a compression tip to drive a staple when the tool is swung like a hammer and contacts the surface. These tools are used for tacking down rolled sheet goods such as felt paper.

Many manufacturers of roofing underlayment and housewrap now require their products to be fastened using plastic cap staples or nails. Builders can use a cap stapler for this task. The ***cap stapler*** operates like a hammer tacker but also features a coil of collated plastic caps which automatically feed into the head of the unit before each staple is driven. Another tool that can be used to install plastic cap staples is the pneumatic cap stapler, **Figure 4.5-6**. These staple guns are equipped with a coil or sleeve type magazine of plastic caps.

A

B Photo courtesy of Huber Engineered Woods

Figure 4.5-3. A—A tape roller. B—Flashing tape is sealed to the substrate after it is applied using the tape roller.

Ivanoooooo/shutterstock.com

Figure 4.5-4. The manual staple gun fires a staple each time the handle is squeezed.

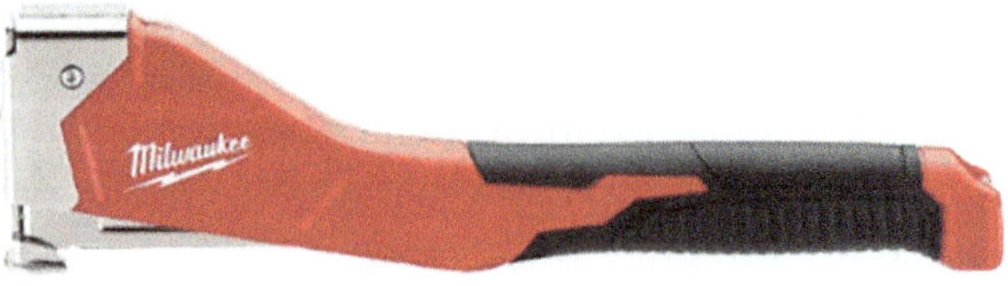

Milwaukee Tool

Figure 4.5-5. Felt paper underlayment is often installed using a hammer tacker.

A B *Goodheart-Willcox Publisher*

Figure 4.5-6. A—Pneumatic cap staplers speed progress and reduce stress on the user. B—A manual cap stapler can be just as effective.

A new cap feeds into the drive chamber each time a staple is driven.

A ***roofing nail gun*** is a pneumatic or battery-operated nail gun that uses a coil of wire collated roofing nails to attach shingles. These nail guns work much like the framing nail guns described in *Chapter 3, Framing Systems*.

When using any nail or staple gun, be sure to follow all recommended safety precautions. These tools release fasteners at a high velocity and must be respected at all times.

LESSON 4.5 Review

Answer the following questions based on information in this lesson.

1. To make a straight cut in a thin material such as aluminum, you should select aviation snips with _____ handles. (4.5-1)
   A. red
   B. blue
   C. yellow
   D. green
2. *True or False?* A manual staple gun is used when precision is more important than speed. (4.5-1)
3. A cap stapler is used when installing _____. (4.5-1)
   A. aluminum flashing
   B. asphalt shingles
   C. wood shingles
   D. roofing underlayment

LESSON 4.6

Safety

LEARNING OUTCOMES

After completing this lesson, you will be able to:

4.6-1 Identify areas where slips, trips, and falls can occur on a construction site.

4.6-2 List methods of fall protection that are approved by OSHA.

TECHNICAL TERMS

anchorage	handrail	rope grab
body harness	lanyard	stair rail
deceleration device	lifeline	stepladder
extension ladder	personal fall arrest system (PFAS)	

Slips, Trips, and Falls

Working from heights is common in construction, **Figure 4.6-1**, but an injury from a fall can occur just walking across the jobsite. Slips, trips, and falls are the leading cause of construction accidents. These injuries can include a sprained ankle or bruised hip from tripping over an extension cord, or a fatal fall while working at an extended height. Falls can occur from scaffolding, roofs, and portable ladders or even into a hole or trench. These are just a few examples of the many fall hazards that occur on construction sites.

Sirisak_baokaew/Shutterstock.com

Figure 4.6-1. Work in residential construction involves performing tasks that are at various heights. Workers must be trained in how to protect themselves from a fall.

Employers have a duty to provide fall protection equipment and training to their employees. However, employees have an obligation to apply their training and to use fall protection when it is required.

Any time a worker has the potential of falling from a height of 6′ or more, fall protection is required by OSHA. An exception to this rule is when employees are working from a scaffold. In that case, fall protection is required when the working surface is at or above 10′.

Training and Prevention

Workers must be trained by their employer regarding the hazards they may experience while working where they may be subjected to falling, or working on a jobsite where others are working from height, so they understand the risks and requirements and know how to take appropriate action.

The responsibility for making sure workers follow safety rules on the job falls on the competent person, designated by the employer. The competent person is trained to know the OSHA safety standards for fall protection. They have the authority to take corrective action when they see a safety violation or hazard. They make decisions regarding how the crew can accomplish the work in a way that protects them and gets the job done in an efficient manner.

Each person working on a construction site should assess the worksite to identify fall hazards. The hierarchy of controls should be used to determine the most effective ways to reduce potential hazards. An example of what this might look like is given in **Figure 4.6-2**.

Often, prevention is the key to protection. Using equipment as it was intended to be used goes a long way toward preventing an injury or safety incident from occurring. Good housekeeping measures help to avoid slips and trips. A messy jobsite can be dangerous, **Figure 4.6-3**. Rain, mud, snow, and ice can all cause hazardous conditions. Even entering and exiting the cab of a piece of equipment can cause a fall resulting in serious consequences. Always use the proper hand grabs and footholds. Maintain at least three points of contact—two hands and one foot, or two feet and one hand—when climbing in or out of machines, or up and down ladders. See **Figure 4.6-4** for correct and incorrect examples.

Hierarchy of Controls: Sample Task Assessment for a Fall Hazard	
Elimination	Is there a way the work can be performed on the ground instead of at height?
Substitution	Is equipment available, such as a boom lift, that would reduce the risk of falling?
Engineering	Can guardrails, safety nets, or scaffolding be installed to provide fall protection in the event of a mishap?
Administrative	Is additional training needed for the task? Does signage or a limited access zone need to be established?
PPE	Do workers need to use a personal fall arrest system for this task?

Goodheart-Willcox Publisher

Figure 4.6-2. Use the hierarchy of controls to assess hazards on the jobsite. Solutions may call for more than one hazard control to be implemented.

MiscMedia/shutterstock.com

Figure 4.6-3. Jobsite clutter is avoidable and can lead to slips, trips, or falls.

Fall Protection Systems

Fall protection is required anytime a worker has the potential to fall from a height of 6′ or more, or if they are working on a scaffold where the platform is 10′ or higher from a lower level. Several fall protection systems can be used to protect workers.

A *Dusan Petkovic/shutterstock.com*

B *RossHelen/shutterstock.com*

Figure 4.6-4. A—The correct way to climb in or out of machinery or on ladders is by maintaining three points of contact. As this worker steps up the ladder, both hands and one foot will remain in contact with the equipment. B—A worker exiting machinery in a hurry is more likely to slip and fall. As he takes the next step down, only one hand and one foot will be contacting the equipment.

Which system to use depends on the circumstances. Conventional fall protection systems include:

- Guardrail systems
- Personal fall arrest systems (harnesses)
- Safety nets

Guardrail Systems

Guardrail systems, **Figure 4.6-5**, are barriers that are put up to stop workers from falling to lower areas. Guardrails must meet the following requirements:

- Top rails must be between 39″ and 45″ from the walking or working surface.
- Top rails must be able to withstand a force of at least 200 lb applied in any outward or downward direction.
- Some type of protection between the top rail and the walking or working surface is required. This can include screens, midrails, mesh, balusters, or equivalent structural members.
  - If midrails are used, they must be installed at a height halfway between the top edge of the guardrail system and the walking or working surface.
  - If screens or mesh are used, they must extend from the top rail to the walking or working surface.
  - Screens, midrails, mesh, balusters, or equivalent structural members must be able to withstand a force of at least 150 lb applied in any outward or downward direction.

Goodheart-Willcox Publisher

Figure 4.6-5. Guardrails on the top level of this scaffolding protect workers from falls.

Personal Fall Arrest Systems

A *personal fall arrest system (PFAS)* is a combination of gear that works together to stop a falling worker. It includes a body harness, a lanyard or other connection device, and an anchorage. In the event of a fall, a ***body harness*** is designed to distribute the forces of a fall across the wearer's entire body, reducing the risk of injury. A full-body harness typically consists of straps that secure around the shoulders, chest, waist, and legs of the wearer, **Figure 4.6-6**. It also includes an attachment point, called a D-ring in the center of the upper back where a lanyard or other connection device can be connected.

Connection devices act as a link between the worker and the anchorage point. Connection devices include lanyards and lifelines. A ***lanyard*** is a length of rope or strap made of synthetic material with locking snaprings on each end. One end of the lanyard attaches to the D-ring on the body harness, while the other end attaches to the anchorage point or to a lifeline. Lanyards often feature a deceleration device, **Figure 4.6-7**. A ***deceleration device*** is

Diana Jo Marmont/shutterstock.com

Figure 4.6-6. A fall protection harness must be adjusted properly to the wearer in order to effectively distribute the forces generated in the event of a fall.

King Ropes Access/Shutterstock.com

Figure 4.6-7. This vertical lifeline features a deceleration device designed to reduce the impact of a fall.

a component that is integrated into a lanyard. Its purpose is to reduce the impact of a fall by absorbing energy through tearing stitches, components, or otherwise deforming when a fall occurs.

Safety Note

Personal fall arrest systems are not typically designed for self-rescue. When using a PFAS on a jobsite, the employer must have a rescue plan in place. The rescue plan outlines how a person will be brought to safety in the event of a fall.

A ***lifeline*** is a cable, rope, or other device that acts as a movable anchorage to improve the mobility of workers using a PFAS. Vertical lifelines connect to a high-anchorage point, such as the peak of a roof, while the other end hangs freely, **Figure 4.6-8**. A worker is attached to a lanyard with a rope grab. A ***rope grab*** is a connecting device that attaches the lanyard to the lifeline. The worker can slide the rope grab up the lifeline when they need to work closer to the top of the lifeline. If they need to descend the lifeline, the worker squeezes the rope grab to unlock it. In the event of a fall, the rope grab grips the lifeline, maintaining the connection between the worker and the anchorage. Only one worker may be attached to a vertical lifeline at any given time.

Goodheart-Willcox Publisher

Figure 4.6-8. A vertical lifeline gives workers more mobility.

Horizontal lifelines allow multiple worker attachments, **Figure 4.6-9**. They can be installed across the length of a roof or wall. OSHA requires that horizontal lifelines be designed, installed, and used only under the supervision of a qualified person.

An ***anchorage***, or anchor point, is a secure point of attachment designed to support the forces generated during a fall. OSHA requires anchorage points to be capable of supporting at least 5,000 pounds per worker attached. The anchorage should be attached to a substantial structural member, such as a beam, girder, roof truss, or rafter. When using a manufactured anchorage, be sure to follow the manufacturer's installation instructions carefully to ensure the design capacity is met, **Figure 4.6-10**.

Safety Note

Follow these safety rules when working on temporary stairs.

- Use handrails when walking up or down stairs whenever possible.
- Do not run up or down stairs.
- Do not carry any items up or down stairs that you cannot see over.
- Make sure stairways are well-lit and uncluttered.
- Stairs and stairways must not have any hazardous projections such as protruding screw or nail heads.
- Do not use stairs under slippery conditions, such as stairs covered with snow or ice.

LESSON 4.6 Review

Answer the following questions based on information in this lesson.

1. *True or False?* A trench is a fall hazard. (4.6-1)
2. Fall protection is required anytime a worker is working on a scaffold where the platform is _____ or higher from a lower level. (4.6-2)
   A. 6′
   B. 8′
   C. 10′
   D. 12′
3. The top rails on a guardrail system must be able to withstand a force of at least _____ applied in any outward or downward direction. (4.6-2)
   A. 100 lb
   B. 200 lb
   C. 300 lb
   D. 350 lb
4. *True or False?* If midrails are used on a guardrail system, they must be installed no more than 10 inches below the top rail. (4.6-2)
5. A(n) _____ for a lifeline is a secure point of attachment designed to support the forces generated during a fall. (4.6-2)
   A. anchorage
   B. body harness
   C. lanyard
   D. transom
6. *True or False?* A disadvantage of fiberglass ladders is that they cannot be used near electrical wires because they conduct electricity. (4.6-2)
7. Which statement about handrails is true? (4.6-2)
   A. They must have at least 2″ clearance between the handrail and the wall or other objects.
   B. They must be mounted at least 24″ above the surface of the stairs at the face of the riser.
   C. They must provide an adequate handhold so that workers can grasp them comfortably.
   D. They are required by OSHA for all stairways in the workplace, regardless of the number of risers.

LESSON 4.7

Math

LEARNING OUTCOMES

After completing this lesson, you will be able to:

4.7-1 Demonstrate the ability to add, subtract, multiply, and divide decimal numbers.

4.7-2 Convert fractional numbers to decimal form and decimal numbers to fractions readable on a tape measure.

TECHNICAL TERMS

decimal fractions

Decimal Fractions

Construction workers continually work with both fractional numbers and decimal numbers. It is common to convert from feet, inches, and fractions of an inch, such as 12′-3 1/16″, to a number that is representative of the same length in decimal form, such as 12.225′. These two dimensions are equivalent. Builders should be comfortable adding, subtracting, multiplying, dividing, and converting these numbers.

Decimal fractions are fractions whose denominators are multiples of ten. If the denominator is 10, the fraction is tenths. If the denominator is 100, the fraction is so many hundredths. Decimal fractions are commonly simply called decimals.

Decimal fractions are often written on a single line with a dot separating digits representing one or more from the decimal fraction. The dot between the whole number and the decimal fraction is the decimal point. Every place to the left of the decimal point increases the value of the digit in that place tenfold. That is why the second place to the left of the decimal point is called the tens place and the third place to the left is the hundreds place, etc. Moving to the right of the decimal point, the place values decrease tenfold. A decimal fraction of 4/10 can be written as 0.4. A decimal fraction of 45/100 can be written as 0.45.

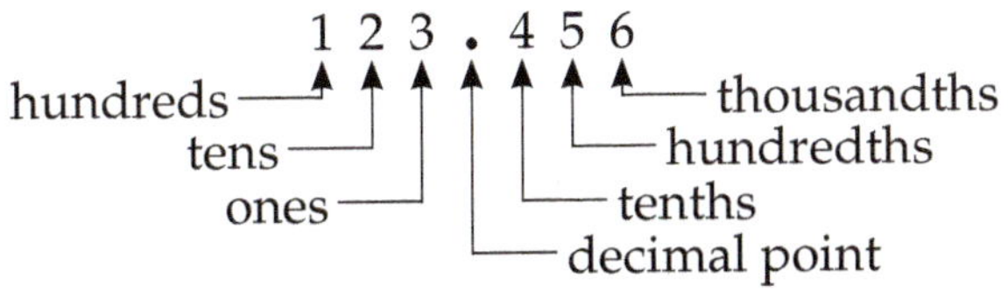

The value of the number in the example above is one hundred twenty-three and four hundred fifty-six thousandths.

Adding and Subtracting Decimals

To add decimals, line up the decimal points in a column, add the numbers, and put the decimal point in the result in the decimal point column.

Example:

$$\begin{array}{r} 1.4 \\ 19.2 \\ +\,31.7 \\ \hline 52.3 \end{array}$$

Subtracting decimals is very similar. Line up the decimal points in the problem and the answer and subtract as usual.

Example:

$$\begin{array}{r} 27.74 \\ -\,2.23 \\ \hline 25.51 \end{array}$$

the need to show many details, section drawings are made to a large scale. **Figure 4.8-2** shows the section view indicated by the cutting-plane line in **Figure 4.8-1**. Most sets of plans include at least one typical wall section if they do not have specific wall sections. A typical wall section shows what would be seen if the section was taken at any point in the wall. A complex structure may need many section drawings to show key details of construction. **Figure 4.8-3** shows the wall section that is indicated by the dashed oval at the left side of **Figure 4.8-2**. The scale used on this drawing is 3/4″ = 1′-0″. This larger scale allows for more detail to be communicated. Note the reference mark 1/A5 in **Figure 4.8-2** corresponds to the drawing number 1 on sheet A5 in **Figure 4.8-3**. These reference marks help the builder navigate the prints and orient them to one another. For instance, further details on the upper wall construction and roof overhang are available as indicated by the reference mark found at the top of **Figure 4.8-3**. Those details can be found on drawing 8, sheet A5 as designated by the dashed oval on this drawing. They are not included in this text.

Detail Drawings

As in sections, ***detail drawings*** show the builder any important and complex construction that cannot be included in plan and elevation drawings. They are also large scale and show how various parts are to be located and connected. Fireplaces, stairs, and built-in cabinets are examples of constructions shown as detail drawings. See **Figure 4.8-4**. Some detail drawings are shown at or near full size. See **Figure 4.8-5**. Detail drawings may be shown in plan view (from above) or elevation view (from the side).

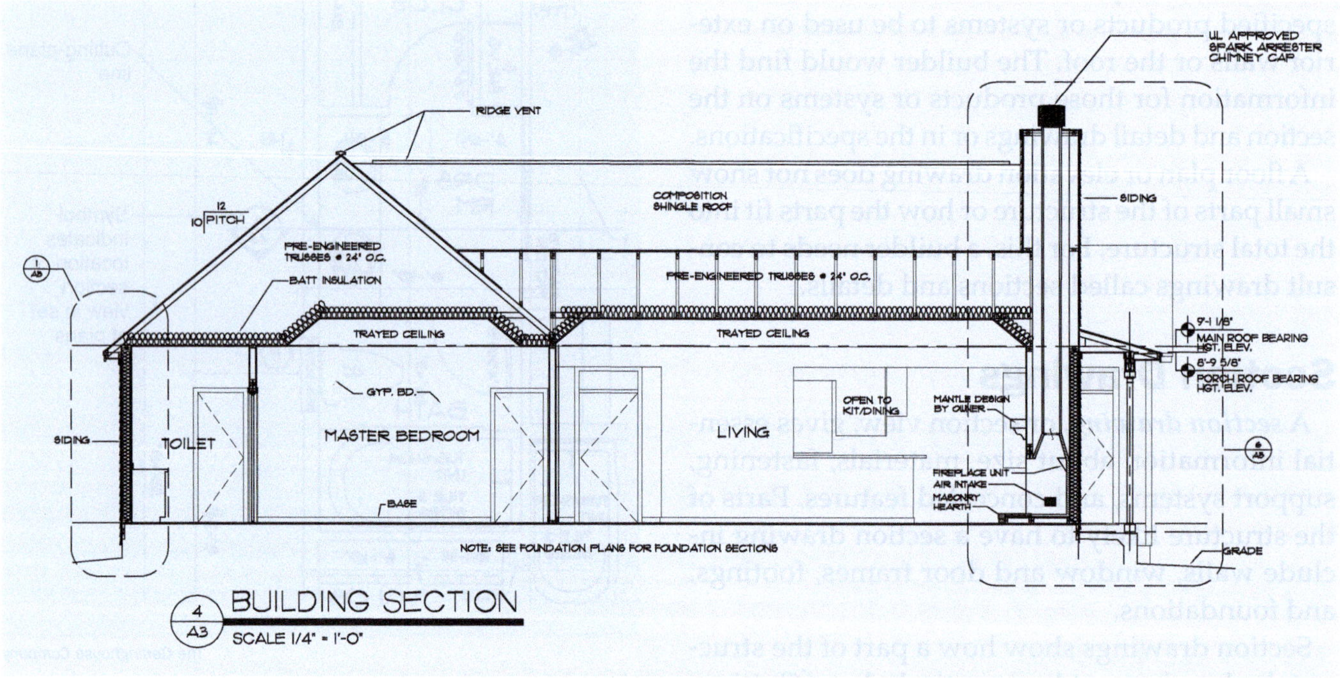

The Garlinghouse Company

Figure 4.8-2. A section view of an entire house helps a carpenter understand the construction.

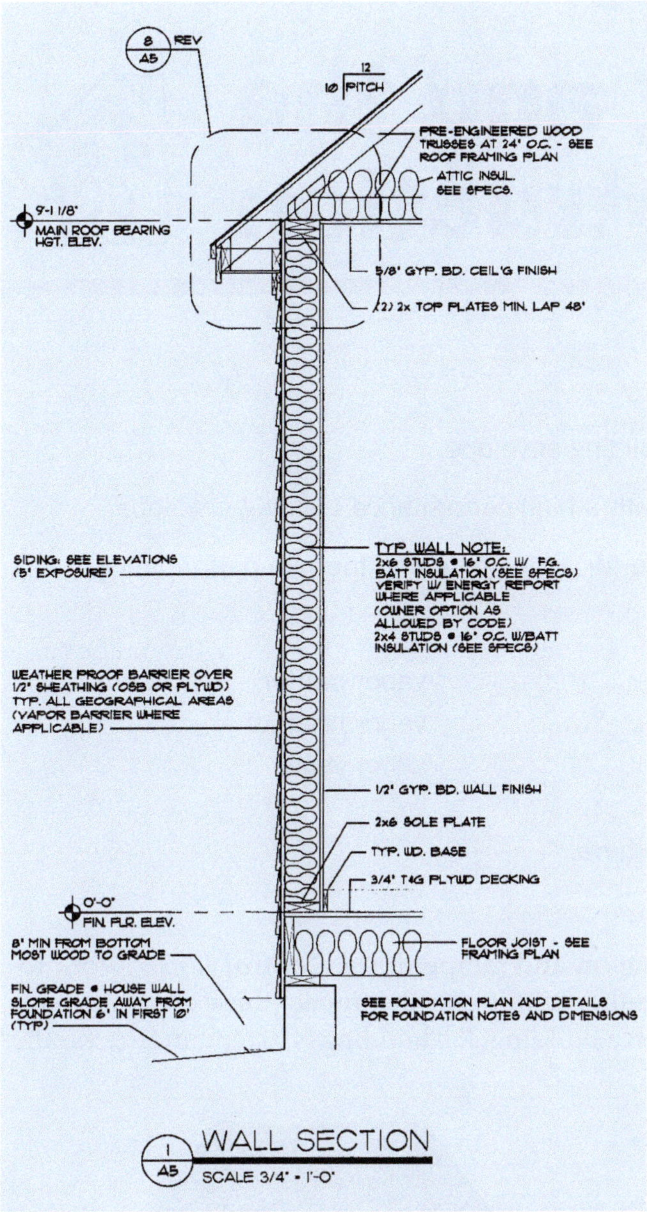

The Garlinghouse Company

Figure 4.8-3. This wall section is indicated by dashed lines and a circle with a notation that this is drawing 1 on sheet A5.

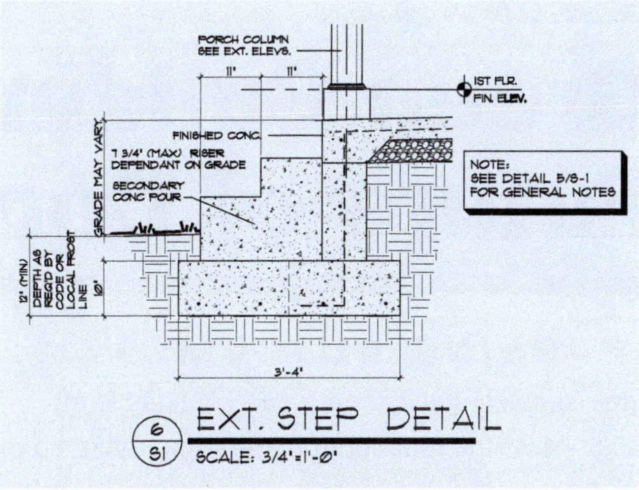

The Garlinghouse Company

Figure 4.8-4. This detail drawing shows how the stairs at the back of the house are to be built.

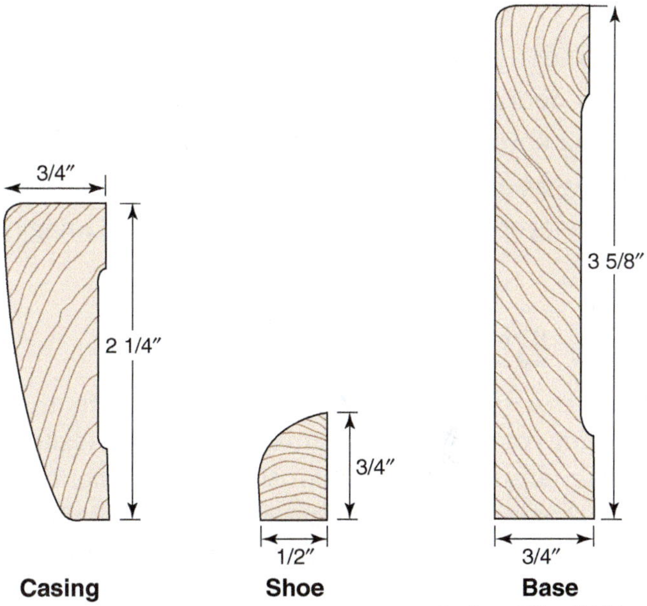

Goodheart-Willcox Publisher

Figure 4.8-5. Some detail drawings are made at or near full size. Shown are cross sections of special millwork needed for finishing a house.

LESSON 4.8 Review

Answer the following questions based on information in this lesson.

1. *True or False?* The ceilings on the first floor of a home are likely to have section drawings. (4.8-1)
2. *True or False?* Section drawings are generally made to a small scale. (4.8-1)

LESSON 4.9

Green Building and the Future of Construction

LEARNING OUTCOMES

After completing this lesson, you will be able to:

4.9-1 List the four control layers that make up the building envelope.

4.9-2 Explain the complications of building a home with a high-performance building envelope.

4.9-3 Describe how a building envelope can be tested for air leakage and thermal performance.

TECHNICAL TERMS

air barrier	diffusion	vapor barrier
blower door test	perm ratings	vapor retarder
building envelope	pressure differential	water vapor
building science	R-value	
control layer	thermal imaging camera	

The Building Envelope

Control layers work together to create the *building envelope*, the separation between the outside environment and the inside environment of an enclosure. Siding, roofing, flashing, insulation, and housewrap are just a few of the components that can make up the layers that keep the inside in and the outside out, **Figure 4.9-1**.

Building science helps us understand the importance of the building envelope. *Building science* is the application of knowledge of the physical world to building processes and materials. Studying building science helps us recognize how environmental changes inside and outside of an enclosure impact the systems that separate them.

The building envelope separates the climate-controlled space inside a building and the unconditioned environment outside, **Figure 4.9-2**. Builders should understand the impact of environmental changes on both sides of the wall system. Major environmental factors that affect a structure include ultraviolet radiation, water, air pressure, water vapor, and temperature. *Control layers* work together to mitigate the impact these elements have on a building. If a building is to remain structurally

studiovin/shutterstock.com

Figure 4.9-1. A building envelope is comprised of all the different components separating the interior of a home from the outside environment.

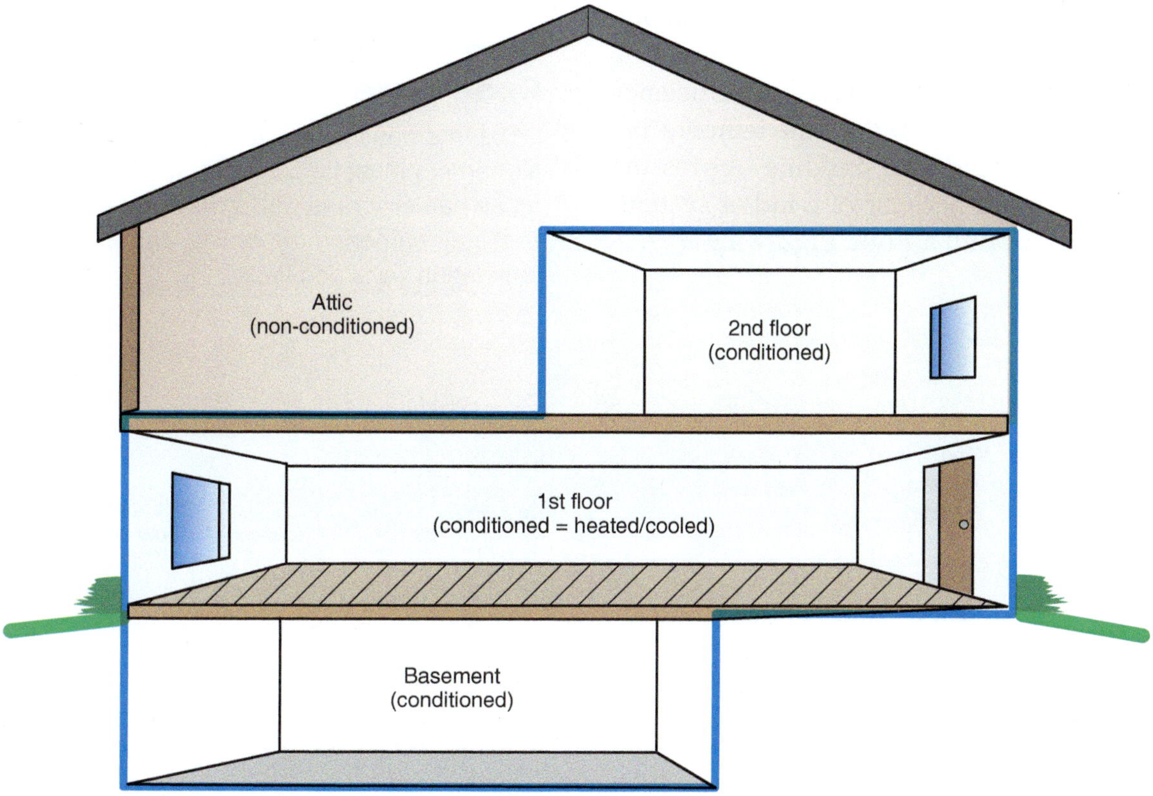

Goodheart-Willcox Publisher

Figure 4.9-2. The building envelope separates conditioned space from unconditioned space.

sound, energy-efficient, and a healthy place of habitation, it must maintain four control layers. In order of relative importance, these layers control water, air, vapor, and temperature.

Water Control

Bulk water can be introduced to the building envelope in a variety of ways. The most obvious method is rain, with wind-driven rain being especially persistent. Homes must be designed and built to protect vulnerable materials from water exposure. Water can also infiltrate a structure through ice. In some cases, water can flow into narrow or porous spaces, often resisting the force of gravity. Consider how a sponge draws water up from a pool of water against the force of gravity. This same principle is at work in construction materials.

Regardless of how water gets into a structure, water leakage into a building is a tremendous threat to a home's structural integrity and the health of its occupants. A building's first line of defense against bulk water is the roofing and siding. It is important to understand that siding should never be considered waterproof. In fact, some systems, such as vinyl siding, allow water to pass through at a variety of junctures. Corners, windows, and doors tend to be weak points for cladding systems. Vinyl siding and some other cladding systems rely heavily on the WRB to keep bulk water from infiltrating. A builder should always assume that some water will leak past the siding. The WRB must be capable of managing water leakage.

Air Control

A *pressure differential* is a difference in air pressure between the inside and outside of a building envelope. Pressure differentials affect a home by forcing air into or out of the structure through any leak in the *air barrier*. The air barrier forms the air control layer of a home and is made up of a combination of materials, depending on the building's design.

The air control layer separates the air inside a building from the air outside the building. The performance of the air barrier also impacts the thermal and vapor control of the house. Leaks in the air control layer may bring with it water (either bulk or vapor) or contaminants, such as dust, pollen, and mold. Air movement through the building envelope also significantly reduces the efficiency of the home's ability to maintain the temperature and comfort level inside the building. A pressure differential can occur because of wind or systems inside a house that pull air out, **Figure 4.9-3**.

Code Note

The International Residential Code (IRC) says that air barriers must be continuous. This means that an air barrier for the wall system must be coupled to the air barrier on the ceiling, as well as the foundation walls and floor.

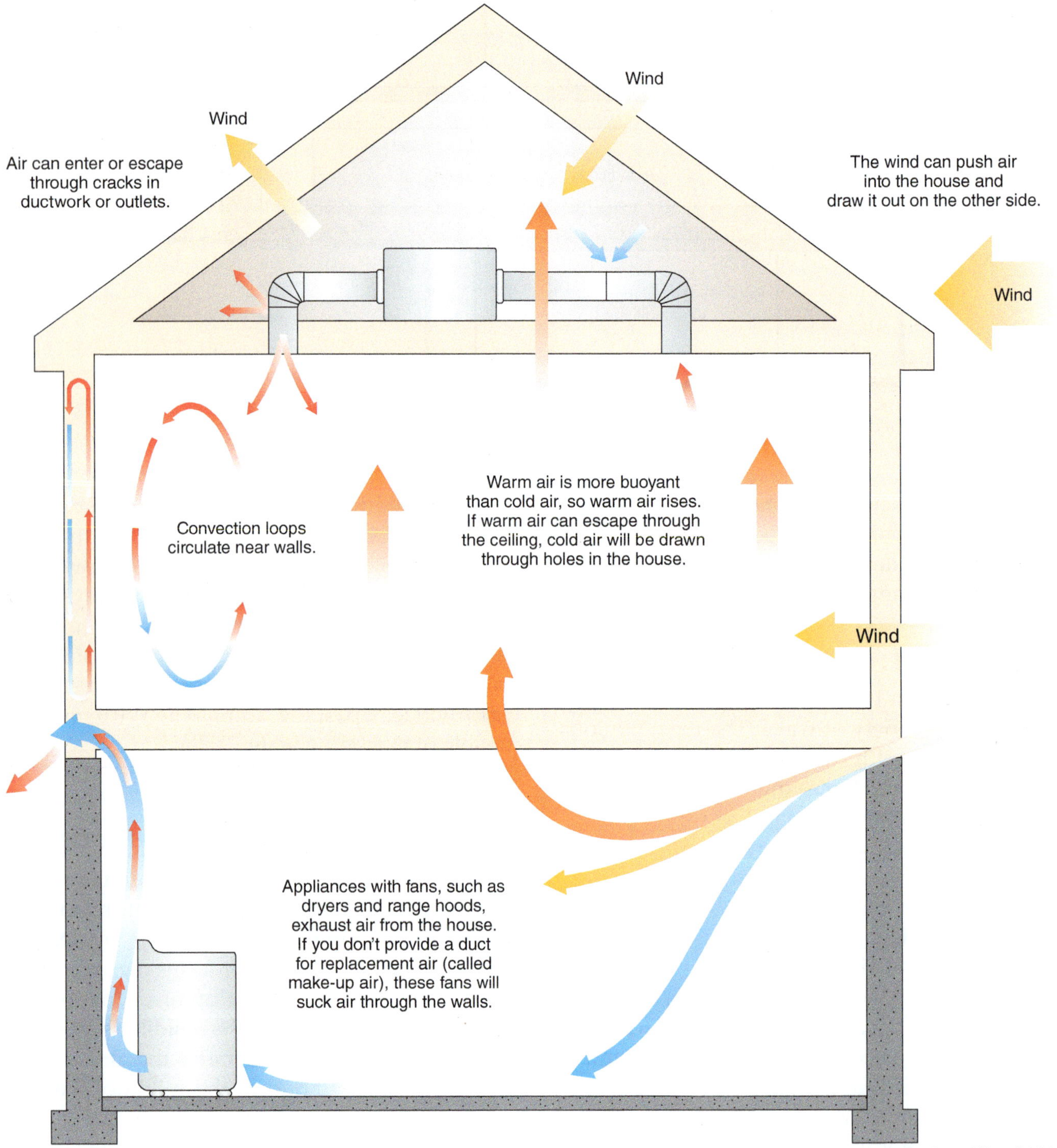

Figure 4.9-3. Pressure differentials create a suction effect that moves air through the building envelope. This image shows forces at work that can create pressure differentials and common air leaks in a home.

Vapor Control

Water vapor is a gaseous form of water always present in the air. It can penetrate wood, stone, concrete, and most other building materials. Water vapor can be a threat to the health of a building and its occupants. High-moisture content in organic materials can lead to mold, mildew, and rot.

Vapor migrates slowly through building materials. It moves from a higher to a lower humidity level; this tendency is called ***diffusion***. Diffusion is accelerated in highly porous materials and is slowed by materials with low porosity. The speed and direction vapor drive occurs in a building can vary depending on environmental conditions. Differences in air pressure, temperature, and especially air leakage through the building envelope can all increase problems caused by water vapor.

Vapor barriers and vapor retarders are membranes through which water vapor cannot readily pass. To be classified as a ***vapor barrier***, a material must have a permeability rating, or perm rating, lower than 1. ***Perm ratings*** measure how much water vapor can pass through a material. A lower perm rating indicates that the material is more resistant to the flow of water vapor. For examples of common building envelope material perm ratings, see **Figure 4.9-4**.

Code Note

Climate zones are an important factor in determining both what classification of vapor barrier or vapor retarder to use and where it is placed in the wall assembly.

Use of vapor barriers in homebuilding is mostly limited to areas where water vapor infiltration through ground sources is a threat, typically under concrete slabs or as ground cover in crawl spaces.

The IRC relies more heavily on ***vapor retarders*** than vapor barriers when listing requirements for wall systems. Vapor barriers have a perm rating of 0.1 or less. Vapor retarders have a perm rating greater than 0.1 but not more than 1. Materials with a perm rating above 1 are considered permeable, allowing vapor to freely pass. Vapor retarders allow some vapor migration from one side to the other. This slows vapor diffusion into the wall but still allows assemblies to dry from the inside out. Building materials, such as kraft-faced fiberglass insulation, some perforated foil-faced products, and some rigid insulations, act as vapor retarders. See **Figure 4.9-4**.

Building Envelope Perm Ratings

Material	Perm Rating	Vapor Barrier	Common Use
Glass	0	Class I	Windows, doors, skylights
Aluminum foil	0.05	Class I	Foil-faced insulation
Polyethylene sheathing	0.06	Class I	Capillary break under concrete slab
Foil-faced polyisocyanurate insulation (PIR)	0.1	Class I	Exterior insulation
Extruded polystyrene (XPS)	1	Class II	Exterior insulation
Kraft-faced insulation	1	Class II	Wall insulation
OSB sheathing	Approx. 2	Class III	Wall sheathing
Expanded polystyrene insulation (EPS)	5	Class III	Exterior insulation
15 lb felt paper	5	Class III	WRB/Shingle underlayment
Standard latex paint	Approx. 7.5	Class III	Interior wall protection
Plywood sheathing	Approx. 10	Class III	Wall sheathing
Wood siding	Approx. 10	Class III	Siding
Housewrap	Range between 5 and 50	Class III or vapor open	WRB/Air barrier
1/2″ drywall	Approx. 50	Vapor open	Interior wall covering

Goodheart-Willcox Publisher

Figure 4.9-4. Vapor barriers have a perm rating of 0.1 or less. Vapor retarders have a perm rating between 0.1 and 1. Materials with a perm rating above 1 are considered permeable.

Thermal Control

Temperature is the primary, though not the only, environmental comfort factor for an enclosure. Regulating the temperature of the home at a comfortable level requires more than just insulation and a heating or air conditioning system. Factors such as relative humidity, solar radiation, type of heat source, and air movement through and in the house all contribute to the comfort of a space.

The thermal control layer is made up of cavity insulation inside ceilings, walls, and floors, as any exterior insulation such as rigid foam board is placed on the outside of walls or under a concrete floor slab. The thermal control layer manages heat loss in a structure (or heat gain in cooling months). A material's resistance to the passage of heat is measured as an ***R-value***. When a building material has a low R-value, it means that the material has little resistance to the passage of heat. If proper water, air, and vapor control is not achieved, the R-value of the insulation may be compromised.

Pro Tip

Although the International Residential Code (IRC) prescribes certain levels of insulation in floors, walls, or ceilings, the codes do not always have the final say. Architects and builders may be able to reduce thermal performance in one area by adding to another. This is accomplished by analyzing the building envelope as a whole using computer software. Values are entered into the program for each variable in the thermal layer. The software generates a report signifying whether the building envelope meets minimum energy compliance. This allows architects and builders to incorporate designs such as cathedral ceilings, which would not otherwise meet minimum code compliance.

High-Performance Building Envelopes

Consider a high-performance building envelope compared to a high-performance automobile. In an old work truck, a poor alignment or an engine misfire might not even be noticeable; however, in a racecar, a poor alignment or misfire could be catastrophic. High-performance vehicles are often more sensitive and less forgiving than the old farm truck.

In a similar fashion, high-performance buildings are also more sensitive than the houses and buildings constructed decades ago. In older buildings, a leak in the WRB might still allow water in, but that water was likely to dry because there may have been little or no insulation in the wall. Air movement through the wall would also be likely, drying out any moisture that was trapped. These buildings were inefficient and often uncomfortable in poor weather.

Today's high-performance building envelopes are both efficient and comfortable. However, because they are built so tightly, air quality can suffer. Air that is trapped inside the building can become stagnant. Some building materials, such as new paint and carpet, can emit toxins called volatile organic compounds (VOCs). Other contaminants inside the house can build up as well. Therefore, building codes require mechanical ventilation systems to control the air exchanges in a building. Systems such as heat-recovery ventilators (HRV) or energy-recovery ventilators (ERV) use heat exchangers to scavenge heat from the exhaust air and transfer it to the intake air and vice versa during the cooler months, **Figure 4.9-5**. They are also equipped with an air filter to remove external contaminants, such as pollen and mold spores. Through mechanical ventilation, the rate of air coming into and out of the building is controlled and conditioned, rather than leaking through the building envelope. See more on HRV and ERV systems in *Chapter 9, Finish MEP Systems and Punchlist*.

Another challenge with high-performance building envelopes is that if there is an air pressure differential between the inside and outside of the house, air will be drawn through any breach in the air barrier. Imagine a balloon that has a pinhole in it. All the air in the balloon will pass through that pinhole to equalize pressure. Research shows that as much as 30 quarts of water can pass through a 1″ × 1″ hole in a drywall air barrier during one heating season, **Figure 4.9-6**. This moisture has the potential to condense and collect in concentrated areas as it cools below the dew point. Because there is

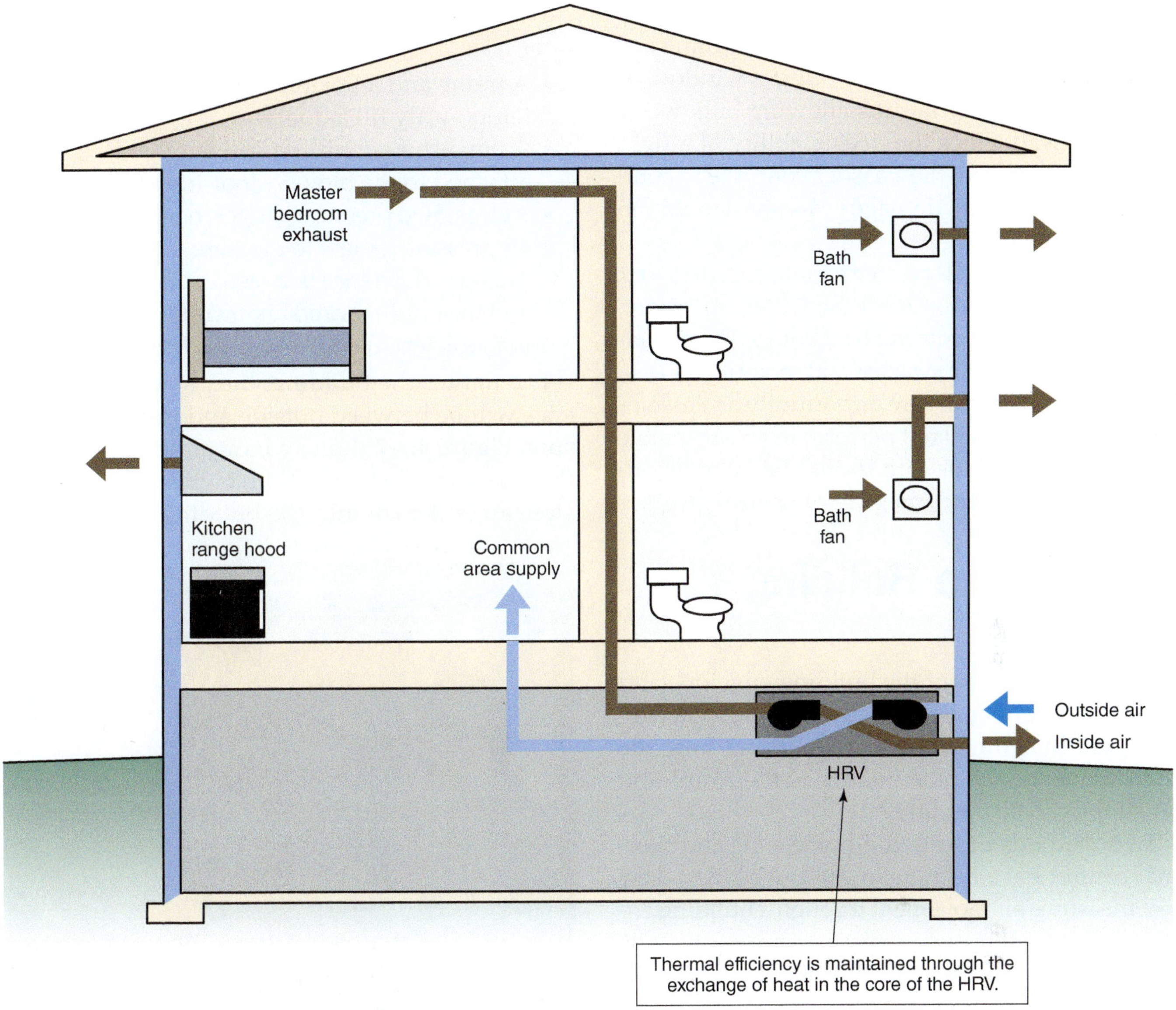

Goodheart-Willcox Publisher

Figure 4.9-5. Heat-recovery ventilators (HRVs) and energy-recovery ventilators (ERVs) scavenge heat from outgoing air, or vice versa in a cooling climate, while bringing fresh, filtered air into the building.

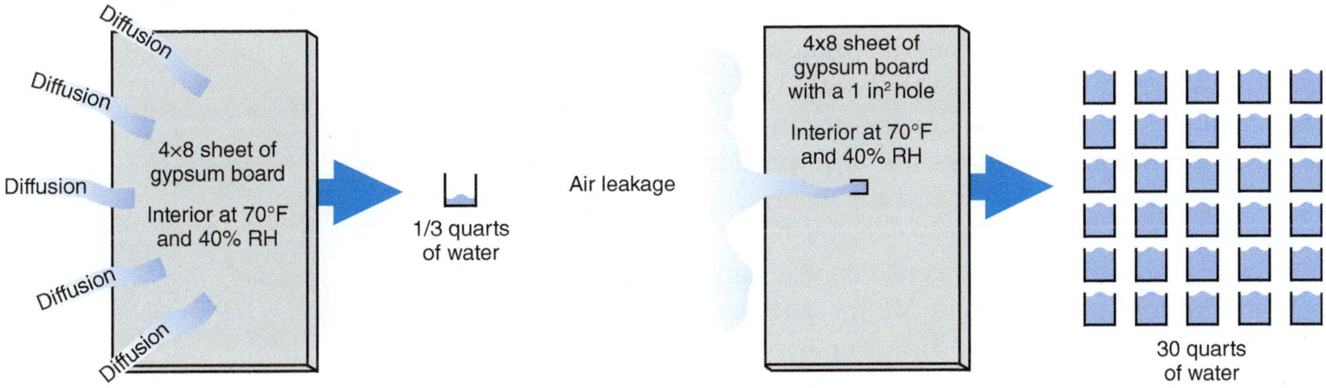

Goodheart-Willcox Publisher

Figure 4.9-6. A leak in the air barrier of a building permits far more water vapor to travel through the building envelope than diffusion.

no air movement inside the wall cavity of a high-performance envelope, drying is limited to diffusion.

A similar scenario can play out if a window or door is not properly flashed allowing bulk water to enter a wall. Since the drying ability of a tight enclosure is limited, these issues continue to compound. Very quickly, damage or harmful mold can result.

The building envelope control layers must work together to properly control the water, air, vapor, and thermal constraints put on a house. The combinations of acceptable systems and materials is vast. However, equally vast are opportunities to assemble a system that will not perform in an acceptable manner. Tremendous care should be taken to ensure the function and integrity of control layers.

Testing the Building Envelope

After learning about the building envelope and its role in keeping a home comfortable, efficient, and healthy, you might be wondering how the builder can be sure the quality of their building envelope can be measured.

Two methods of testing a home's air and thermal control have been developed. A blower door test measures air movement through a building envelope, and a thermal imaging camera can show weak areas in the thermal control layer.

Blower Door Testing

All buildings, no matter how tightly they are constructed, have some air leaks. Excessive leaks can cause one part of the house to feel drafty while the rest of the house may feel comfortable. A ***blower door test*** measures the tightness of a building by mounting a large fan in an otherwise sealed doorway. While pressurizing or depressurizing the building, the system measures the difference in pressure between the interior and the exterior of the building.

Often, energy efficiency incentive programs such as Energy Star, LEED, or NGBS require a blower door test. Many building departments also now require one before issuing a certificate of occupancy.

There are three main components for a blower door test:

- A frame and a flexible panel designed to temporarily fill a doorway
- A powerful variable-speed fan that is attached to the blower-door frame
- At least two pressure gauges: one (the air pressure gauge) to measure the pressure difference between the home's interior and the outdoors and another (the airflow manometer) that measures the fan's airflow

The tighter the building, the greater the difference will be between outside and inside air pressure. **Figure 4.9-7** shows a blower door in place in the doorway of a house. With higher pressure outside, air is drawn into the building through any

The Energy Conservancy

Figure 4.9-7. A blower door consists of an adjustable frame, a flexible covering, and a powerful fan.

cracks or openings. A smoke pencil can be used to detect where the air is flowing into or out of the building.

Thermal Imaging

A ***thermal imaging camera***, **Figure 4.9-8**, is a device that detects a heat signature given off by a building or its components. It allows the user to see variations in temperature of different surfaces.

Thermal imaging cameras help identify areas of heat loss or energy inefficiency. The camera produces a color-coded image, where warmer areas appear as red or yellow, and cooler areas appear as blue or purple. By analyzing these images, potential issues such as insulation gaps, air leaks, or moisture problems can be identified. For example, if there is inadequate insulation in the walls, the camera can show areas of heat loss as cooler spots. This information can be used to prioritize energy-efficient upgrades, seal leaks, improve insulation, and ultimately enhance the energy efficiency and comfort of the home.

Ivan Smuk/shutterstock.com

Figure 4.9-8. A thermal imaging camera can help detect weaknesses in a building's thermal control layer.

Thermal imaging cameras are particularly useful for conducting energy audits, diagnosing HVAC system problems, locating water leaks, and assessing the effectiveness of insulation. They allow users to identify issues with the building envelope and other systems without physically altering the building.

LESSON 4.9 Review

Answer the following questions based on information in this lesson.

1. *True or False?* A building's first line of defense against bulk water is the roofing and siding. (4.9-1)
2. *True or False?* Diffusion of water vapor is accelerated in materials with low porosity. (4.9-1)
3. A low perm rating indicates that the material is _____. (4.9-1)
   A. less resistant to the flow of vapor
   B. more resistant to the flow of vapor
   C. less resistant to thermal loss
   D. more resistant to thermal loss
4. A material's resistance to the passage of heat is measured as a(n) _____. (4.9-1)
   A. H-value
   B. J-value
   C. R-value
   D. T-value
5. *True or False?* Making a newly constructed building completely airtight guarantees high indoor air quality. (4.9-2)
6. How many pressure gauges are required when performing a blower door test? (4.9-3)
   A. At least one
   B. At least two
   C. No more than one
   D. No more than two
7. *True or False?* Thermal imaging cameras can be used to identify areas of energy inefficiency. (4.9-3)

Chapter 4 Review

Summary

Lesson 4.1

- The residential construction industry offers many niche career opportunities that complement or support the core building construction and remodeling sectors. This includes house flipping, entertainment or recreation industries, content creators, journalism, teaching, and many other career opportunities. (4.1-1)

Lesson 4.2

- While constructing a home according to building code meets the minimum requirements, using best practices is preferrable whenever reasonable. (4.2-1)

Lesson 4.3

- The process of making a home weathertight, or drying in, is completed after it has been framed and sheathed. A WRB must be applied to exterior walls to prevent water accumulation within the wall and must be built in a manner that provides a way for water to drain to the outside of the wall assembly. (4.3-1)

Lesson 4.4

- Roofing materials, windows, doors, and the material used for the WRB must be selected in accordance with performance, design preferences, climate, and cost. (4.4-1)
- The use of plastic cap staples or nails is recommended for installing roofing underlayment and housewraps to improve wind and leak resistance. (4.4-2)

Lesson 4.5

- Tools used for drying in include tools used for cutting, such as a utility knife or aviation snips, and attachment tools, such as tape roller, staple guns, hammer tacker, cap stapler, and roofing nail gun. (4.5-1)

Lesson 4.6

- Slips, trips, and falls are the leading cause of construction accidents and can happen anywhere on a jobsite. (4.6-1)
- OSHA-approved fall protection methods include guardrail systems, personal fall arrest systems (PFAS), and safety nets. (4.6-2)

Lesson 4.7

- Construction workers continually work with both fractional numbers and decimal numbers. They must be comfortable adding, subtracting, multiplying, dividing, and converting these numbers. (4.7-1)

Lesson 4.8

- To determine details of how individual parts fit into the total structure, a builder needs to consult section and detail drawings. (4.8-1)

Lesson 4.9

- The building envelope's control layers manage water, air, vapor, and temperature. (4.9-1)
- High-performance building envelopes are both efficient and comfortable; however, they are also more sensitive to moisture and air quality concerns than the houses and buildings constructed decades ago. (4.9-2)
- A building envelope's effectiveness can be tested using blower door testing or thermal imaging. (4.9-3)

Apply and Analyze

1. Would a house flipper be more likely to buy an inexpensive house that needs a moderate amount of renovation and that no other buyers are interested in, or a moderately priced house that requires little work and that several families are interested in purchasing? Why? (4.1-1)
2. Under what circumstances would a quality builder choose to simply meet the minimum standards of the building code rather than use construction best practices? (4.2-1)
3. Explain why it is important for a home to be dried in. (4.3-1)
4. How does an ice-and-water barrier protect the house from damage caused by an ice dam? (4.4-1)
5. How might the use of performance shingles reduce energy costs for the homeowner? (4.4-1)
6. Should you apply pressure when using a tape roller to apply flashing tape? Why or why not? (4.5-1)
7. If you are placing the top of an extension ladder 20′ above the ground, how far should the base of the ladder be from the house. Why? (4.6-2)
8. Convert 12.452″ to the nearest sixteenth of an inch. (4.7-2)
9. If you need to check the plans to see what type of fastener you should use to install a fixture, should you look at the floor plan? Why or why not? (4.8-1)
10. What is the building envelope, and why is it important to green building? (4.9-1)

Critical Thinking

1. Triple-pane windows are expensive. Research costs for comparable double- and triple-pane windows in your area by calling a supplier. Think about local building code requirements and best practices. If you were building a new home, which window would you purchase to install? Consider factors, such as length of home ownership or climate region, that might justify an upgrade from double-pane to triple-pane glass when building a new home. (4.2-1)
2. While working on a remodeling job, you remove an entry door. Your experienced coworker mentions that the door was not flashed properly when it was originally installed. What are some visible signs she may have picked up on? What kind of problems could this issue cause over time? (4.2-1)
3. Explain why it is so important to ensure the function and integrity of control layers. (4.9-1)

Communicating about Construction

1. **Speaking and Listening.** Working in small groups, create a poster that illustrates the different windows described in this chapter. As you work with your group, discuss the meaning of each term. Afterward, display your posters in the classroom as a convenient reference for discussion and assignments. (4.4-1)
2. **Speaking and Listening.** Working with a partner, design a building envelope that meets code requirements for your climate zone. Discuss the various materials that will satisfy requirements for water, air, vapor, and thermal control. Choose materials that you both agree will best suit your design. (4.9-1)

tokar/Shutterstock.com

CHAPTER 5

Exterior Finishes

Chapter 5 Outline

Introduction

As we continue our journey through the building process, a true picture of the home's finished exterior begins to unfold. With the roof and weather-resistive barrier (WRB) complete, the home's cladding can be installed. This process gives the house much of its exterior color and style. As the home's exterior finishes are installed, work on the interior also begins. *Chapter 6, MEP Systems Rough-In*, will highlight the rough installation of the mechanical systems of the house. In this chapter, we take some time to examine common siding, exterior trim, and decking systems used on homes today, along with common materials and tools used during their installation. Like installing roofing and WRB materials, exterior wall finishes and trim require workers to perform their tasks from elevated heights. We will continue to study fall safety hazards by exploring several types of scaffolding and aerial lifts that are used in residential construction.

This chapter details how to lay out a building by using the 6–8–10 method to square corners and highlights the importance of building plumb, level, and square when building a home. This chapter also covers how elevation drawings fit into the construction process, how symbols are used to communicate on all types of construction drawings, and how material selection plays an important role in green building.

LESSON 5.1

Industry and Careers

LEARNING OUTCOMES

After completing this lesson, you will be able to:

5.1-1 Describe the importance of professional development throughout your career.

5.1-2 Identify three strategies to develop your skill set.

5.1-3 Set a goal for your career.

TECHNICAL TERMS

professional development

Long-Term Success in Residential Construction

To this point, you have learned about various opportunities and career paths in the residential construction industry. Landing a construction job is only a step in the process in achieving long-term success in the industry. Although a good employer will train and develop their employees to help them achieve their highest potential, **Figure 5.1-1**, as the employee, your efforts are essential.

Regardless of trade or career path, take the initiative to maximize your professional development. ***Professional development*** is the ongoing process of improving the knowledge, skills, and abilities required for your career to progress. It enables you to pursue the career path that you want. Once you learn all the tasks associated with a particular job, it can be easy to become idle or feel like you are in a position where there is no advancement. Sometimes opportunities for growth come along by chance, but many times you must search them out and prepare for them. What follows are some strategies to build your professional repertoire and maximize your growth potential in a construction career.

Brocreative/Shutterstock.com

Figure 5.1-1. Training for advancement is one sign of a quality employer.

Develop Your Skill Set

Construction workers build their skills starting on day one and continuing throughout their career. Once you have mastered a skill, it is important to remember that there are still opportunities for development in that area. Skills can continuously be improved and strengthened in various areas of construction.

There are many ways to continue developing your skill set throughout your career. For instance, workers can utilize training and educational resources to improve their skills, stay informed of industry advancements and trends, and explore new challenges that include more difficult or technical work. Even after mastering a skill, opportunities exist to improve a technique or deepen knowledge about a particular process, material, tool, or task in general. Workers can also ask their employers about areas for improvement or where training would be beneficial.

Regardless of trade, you can learn more and develop a skill set in numerous ways: through lifelong learning opportunities, by networking and collaboration, or with mindful goal setting. When you reach a level of expertise where you consistently produce high-quality work in an efficient manner, consider how you might contribute to the trade by educating and training others on a particular process, or developing new tools and techniques to advance the trade. Highly experienced builders are the best resource to improve their industry, **Figure 5.1-2**.

Pro Tip

Many construction fields offer or require certifications that prove expertise. These credentials can give you a competitive edge in your field or may be required to provide legal status to operate. Be familiar with the certifications relevant to your field and pursue them if they will help your professional credibility or facilitate new opportunities.

Lifelong Learning

Construction techniques and materials change over time. Keeping up to date with industry developments will help workers stay on top of the trends and technologies that are driving the industry. The

Goodheart-Willcox Publisher

Figure 5.1-2. Many colleges and trade schools offer courses in the building trades. Here students learn framing principles with a lesson from a master framer.

forward-looking builder should be intentional about seeking opportunities for growth. These may include:

- Reading trade journals and other publications in their field
- Subscribing to trusted internet sources, such as blogs, video channels, and websites that feature new products, methods, technologies, and other industry developments
- Attending workshops, seminars, webinars, and conferences
- Enrolling in higher education courses or programs that can complement or improve expertise

Along with discovering new technologies and trends, continuing education and research help to ensure individuals that they are following proper guidelines, building codes, or best practices. It can also provide an avenue for growth into other trade specialties.

Pro Tip

Construction materials and methods are changing faster than ever. To be a valuable member of the industry, construction workers must be prepared to learn as much as possible about the newest materials and latest methods.

Networking and Collaboration

A strong professional network within the construction industry will generate opportunities that would otherwise be inaccessible. Creating this network can be challenging if your exposure to the industry has been limited. However, it is likely that there are established trade associations and organizational networks that are within reach to most areas of the country. Membership in these organizations can be gained through various means. Many of them, such as local chapters of the National Association of Home Builders (NAHB), offer assistance on business management, provide various training sessions, and even connect contractors with jobs to bid or subcontractors to partner with.

Nationally, conferences such as the International Builder's Show, also hosted by the NAHB, are great places to learn and network with others in the industry. Attending industry events is an effective way to connect with people in the industry and expand your network. You are likely to find organized home shows and construction events nearby.

Workers can also benefit by taking opportunities to engage with fellow tradespeople, contractors, and suppliers on the job, **Figure 5.1-3**. These networks often lead to other work opportunities and partnerships for contractors. In fact, any networking within the industry can lead to work referrals, opportunities, and shared knowledge, lending credence to the well-known saying, "It's not what you know, but who you know." Networking benefits can be extensive, though perhaps a more accurate version of this saying would be, "It's what you know *and* who you know!"

michaeljung/Shutterstock.com

Figure 5.1-3. Other contractors and trade workers can be an excellent source for networking.

Setting Goals

When building a career, it is important to have a destination or a general area of exploration in mind. Knowing where you want to end up in your career will help to map out the best way to get there. One goal of this book is to help you understand the many career options available in the construction industry.

It is important to be thoughtful when considering your long-term career goals and creating plans to achieve them. These goals become milestones as you travel your career path. Set milestones that you need to meet along the way, and then be sure to regularly reflect on your progress. Reflection can help keep you focused and motivated as you pursue a rewarding construction career.

As important as career goals can be, you should not overlook the importance of goal setting in your everyday development. Even setting goals for small tasks can be extremely beneficial to your success. However, don't get discouraged when you miss the mark. Even the most experienced trades workers are on a path of continuous improvement. Naturally, these efforts come with high points and low points. Give yourself permission to make mistakes and learn from them. If you persevere, you are sure to improve with time.

LESSON 5.1 Review

Answer the following questions based on information in this lesson.

1. *True or False?* Professional development enables you to pursue the career path that you want. (5.1-1)
2. *True or False?* The construction techniques and materials used today have remained largely unchanged over the past 50 years. (5.1-1)
3. If you are having trouble networking because you have little industry experience, you should _____. (5.1-2)
   A. focus solely on taking courses to build your knowledge
   B. ask for overtime work to hone your skills
   C. seek out an established trade association
   D. hold off on networking until you have more experience
4. *True or False?* Your professional goals must relate to major tasks, such as completing your schooling or earning a license or certification. (5.1-3)

LESSON 5.2

Construction Fundamentals

LEARNING OUTCOMES

After completing this lesson, you will be able to:

5.2-1 List common laws and regulations that builders operating a residential construction business must follow.

TECHNICAL TERMS

environmental permit
erosion control plan
setback
zoning ordinance

Laws and Regulations

Laws and regulations in any industry can be complex, and construction is no exception. Contractors in the residential construction industry are required to follow various laws and regulations that govern different aspects of their work. *Chapter 1, Preparing for Construction*, introduced you to building codes and employer requirements mandated by the Occupational Safety and Health Administration (OSHA) intended to protect workers on the job. *Chapter 3, Framing Systems*, expanded on building code requirements, including the process of permitting and inspection. This lesson introduces zoning laws, environmental regulations, and licensing requirements that builders must also be aware of to varying degrees. Additionally, employers must understand consumer protection laws, insurance requirements, labor laws, and tax laws to satisfy legal requirements when operating a business and employing workers.

Because of the complexity of laws and regulations, anyone starting their own construction business should consult with their local Small Business Administration (SBA) for help understanding legal requirements and regulations that they will need to comply with.

Zoning Laws

A ***zoning ordinance*** is a law that defines and dictates how property can be used. These laws are specific to geographic areas within a municipality and may mandate whether a property can be used for residential, commercial, or industrial use, **Figure 5.2-1**. Zoning ordinances may control the size or height of new construction, or in some cases, preserve a historic district by limiting changes to existing buildings or restricting new construction. Ordinances often regulate the minimum property ***setback***, which is the distance a building must be from the street and the minimum clearances from adjoining properties. The local ordinances must be carefully checked for these requirements during the planning phase of construction.

Consumer Protection Laws

Some states and municipalities have instituted laws that aim to protect consumers from fraud of poor workmanship. These laws may require contractors to register with the authority having jurisdiction (AHJ), provide written contracts, limit the amount of money consumers can pay as a down payment on work, adhere to agreed timelines, and resolve disputes fairly.

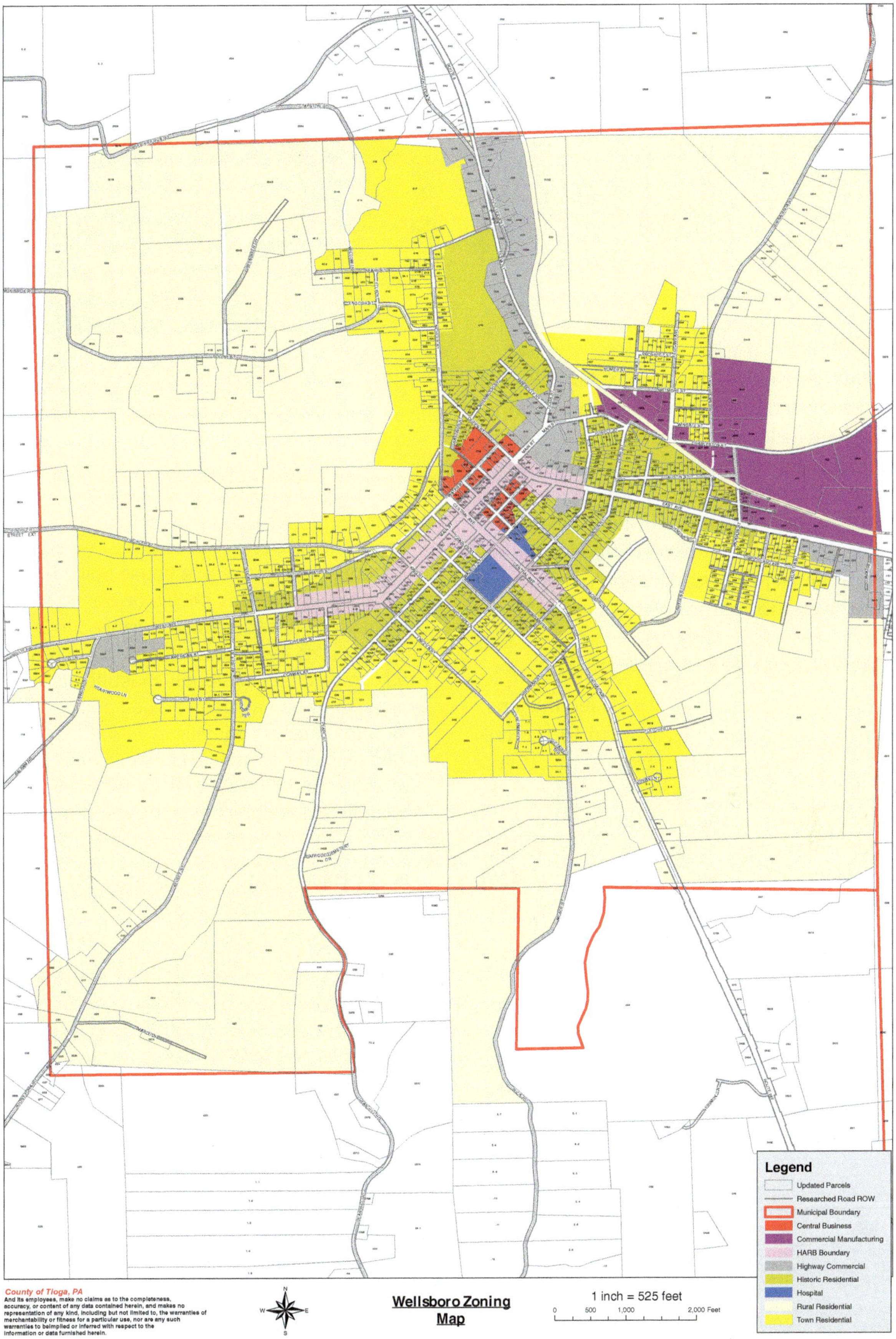

Wellsboro Borough Zoning Map

Figure 5.2-1. This town zoning map illustrates designated land usage and restrictions according to local zoning.

Labor Laws

Throughout history, some employers have treated employees unfairly. To protect workers from unsafe and unfair labor practices, state and federal governments require all employers to follow various labor and employment laws. The requirement for employers to comply with OSHA standards is an example of a labor law. Other labor laws include minimum wage requirements, overtime pay, and antidiscrimination laws.

Tax Laws

All profitable businesses must pay taxes based on income generated by their business operations. Various taxes can be levied against businesses or individuals operating on the local, state, and federal levels. Common taxes include income taxes, sales taxes, payroll taxes, and property taxes. Business owners typically consult with a certified public accountant or other tax professional to help them navigate the tax laws and requirements, **Figure 5.2-2**.

Environmental Regulations

Some trades workers, such as HVAC technicians, handle materials that can be harmful to the environment or the health of themselves or others. Ozone-depleting substances such as refrigerants require special training and certification to work with. Abatement contractors who work with hazardous substances like asbestos or frequently deal with lead must also be certified to perform their jobs, **Figure 5.2-3**.

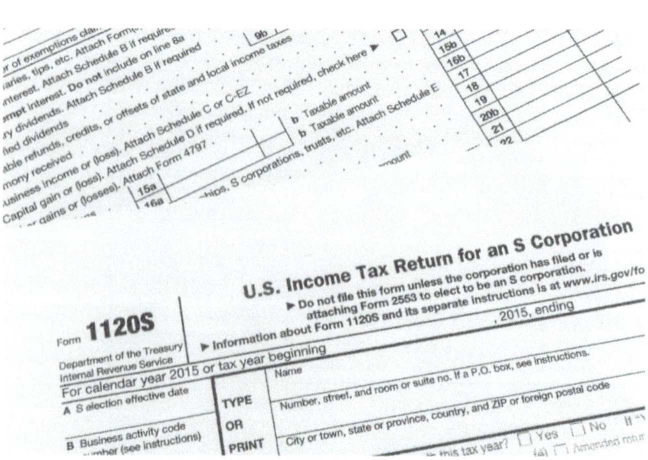

David S Swierczek/shutterstock.com

Figure 5.2-2. Tax laws can be complex and difficult to navigate.

Jamie Hooper/Shutterstock.com

Figure 5.2-3. Abatement contractors are highly regulated and must follow strict protocols to protect themselves and others from harmful substances.

Safety Note

Dealing with hazardous materials is a concern in remodeling projects. In a house built before 1978, painted surfaces should be tested for lead content. Lead-based paint was banned in 1978. Kits for testing lead content in paint are available. Lead-contaminated materials must be disposed of by a person who has been trained and certified to work with materials containing lead. Other possible hazards include chlordane-contaminated lumber; lead solder in old pipes; and asbestos in pipe insulation, old flooring tile, and exterior siding.

Many projects require environmental permits before a project can begin. An ***environmental permit***, such as a storm water pollution prevention plan (SWPPP), ensures that groundwater runoff is properly controlled and flooding of adjacent property will not occur. An ***erosion control plan*** protects the land from soil erosion, **Figure 5.2-4**.

Project size, location, type of construction, and protection laws determine the extent of environmental regulation that must be implemented. Other environmental regulations that can have an impact on a construction site include water conservation, waste management, and protected areas or habitats such as wetlands. These regulations can vary according to state, locality, or even by site. Usually, an engineer or the owner obtains any environmental permits, but anyone changing the site should check to ensure the necessary permits have been secured.

Sallehudin Ahmad/Shutterstock.com

Figure 5.2-4. Erosion control is part of a comprehensive SWPPP required for larger housing projects.

Art Konovalov/shutterstock.com

Figure 5.2-5. Home builders carry the applicable insurance policies to protect against the risks of catastrophic loss.

Licensing Requirements

Many states or local jurisdictions require residential contractors and professionals to be licensed. Obtaining a license typically requires the contractor to demonstrate knowledge, experience, and compliance with regulations regarding their specific trade. Licensing requirements and processes vary according to the authority having jurisdiction (AHJ), which might be the state or the local government. It is common for the AHJ to require a license to practice as a general contractor, electrician, plumber, HVAC technician, roofing contractor, or masonry contractor.

Insurance Requirements

Imagine being the general contractor on a million-dollar home build. As the project was nearing completion, one of your employees left a heater on in the garage over the weekend. The garage caught fire and burned the entire place to the ground, **Figure 5.2-5**. This scenario would be disastrous. However, mistakes happen, and unfortunate outcomes do occur. Most contractors do not have the financial status to absorb such a loss.

Here is where insurance policies come into play. Insurance policies protect the builder and the property owner from major financial loss due to an accident, theft, injury, fire, lawsuit, or other catastrophic event. These policies can be purchased through an insurance broker. Insurance needs vary based on the size of the business and the scope of the project. A client or state may demand that a builder carry a policy that meets their minimum requirements.

LESSON 5.2 Review

Answer the following questions based on information in this lesson.

1. Which type of law may mandate whether a property can be used for residential, commercial, or industrial use? (5.2-1)
   A. Consumer protection law
   B. Labor law
   C. Tax law
   D. Zoning law
2. A law that limits the amount of money a client can be required to pay as a down payment is an example of a _____ law. (5.2-1)
   A. consumer protection
   B. labor
   C. tax
   D. zoning
3. *True or False?* A client can demand that a builder carry an insurance policy to protect against major financial loss due to catastrophic events such as accidents, theft, and fire. (5.2-1)
4. An antidiscrimination law is an example of which type of law? (5.2-1)
   A. Consumer protection law
   B. Labor law
   C. Tax law
   D. Zoning law
5. Licensing requirements may be established by a _____ government. (5.2-1)
   A. local or state
   B. state or federal
   C. local or federal
   D. federal or international

LESSON 5.3

The Construction Path

LEARNING OUTCOMES

After completing this lesson, you will be able to:

5.3-1 List the components that make up the exterior finishes of a home.

5.3-2 Identify the major elements of cornice and cladding systems.

TECHNICAL TERMS

cornice	fascia	porch
deck	frieze board	rainscreen
downspout	gutter	soffit
exterior finish		

Exterior Finishes

The application of the exterior finishes of a home is an exciting and busy time in the construction process. Often, the jobsite is buzzing with activity during this phase of construction. While the outside of the home is being protected and beautified, plumbers, electricians, and HVAC techs are busy at work on the inside of the home, roughing in their mechanical systems. Mechanical rough-ins will be covered in *Chapter 6, MEP Systems Rough-In*.

The term ***exterior finish*** includes all exterior materials of a structure: roofing materials, cornice trim, wall cladding (such as siding), trim around doors and windows, as well as porches and decks. Roofing materials were covered in *Chapter 4, Drying In*, as part of the WRB for the home. This lesson will focus on other key exterior finishes.

Goodheart-Willcox Publisher

Figure 5.3-1. Decorative cornice assemblies, such as the one on this brick Victorian house, were once commonplace in new homes.

Cornices

Cornice is a term for all the parts that enclose or finish the overhang of a roof. It usually includes the fascia board, soffit, and gutter, along with any decorative moldings. Highly decorative cornice systems were once commonplace, **Figure 5.3-1**. However, increased labor and maintenance costs have driven builders and architects to resort to simpler systems.

Cornices provide a finished connection between the wall and roof. The cornice may have multiple functions on a building. Aside from adding architectural appeal, the cornice overhang protects the wall finish from water by directing runoff away from the wall and in many cases capturing it in a gutter and diverting it away from the building. Often, the underside of a cornice is used as an air intake for attic ventilation. In homes built with energy efficiency in mind, the cornice can be a functional shading device, keeping the intense summer sun from coming through the windows. If the building has a gable or gambrel end, then that part of the cornice may be referred to as raked or sloping, **Figure 5.3-2**.

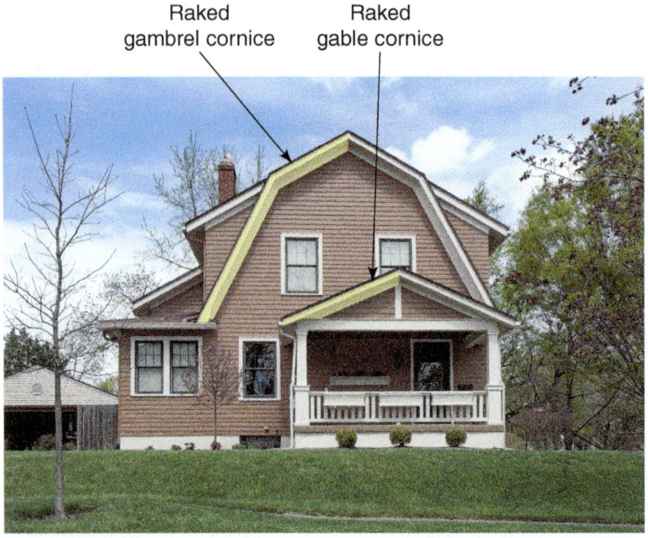

Susan Law Cain/shutterstock.com

Figure 5.3-2. A sloping cornice on a gable or gambrel roof is referred to as a raked cornice.

Wermers Design and Architecture

Figure 5.3-4. A traditionally detailed and proportioned cornice adds architectural appeal to a building.

During the construction process, the cornices of a home are generally finished before the siding is applied. This gives the wall finish a clean termination point. The style of the house determines the cornice design. Modern methods of cornice construction are often simpler and less ornate than those in classical construction. It is important to note that variations can exist with all systems. **Figure 5.3-3** shows structural and trim parts of a box cornice that are constructed of wood or a similar material.

Complex cornice designs may be incorporated into higher-end builds, **Figure 5.3-4**. Notice the architectural details of the cornice return. Incorporating these kinds of details in a project takes a level of time and skill that many builders are not willing or able to invest in, or that clients are not willing to pay for. Although this level of craftsmanship is not seen on many new builds, it is advantageous to mimic the proportions and the look of a well-constructed cornice return when possible.

Cornice systems constructed of a basic lumber structure with a vinyl or aluminum soffit and an aluminum fascia trim, **Figure 5.3-5**, are the most economical and common in homebuilding today. **Figure 5.3-6** shows this system under construction. Vinyl and aluminum systems are common even when vinyl or aluminum siding is not used. These products come prefinished, are easy to work with, and are attractive when installed correctly.

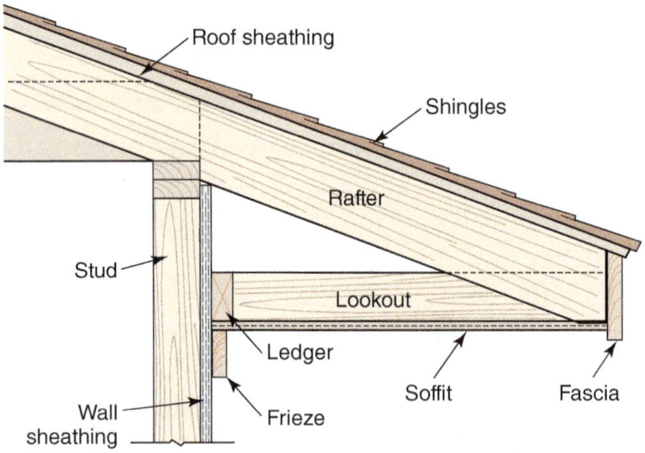

Goodheart-Willcox Publisher

Figure 5.3-3. Parts of a typical box cornice. Soffit materials are often prefabricated, usually from metal, plastic, or plywood.

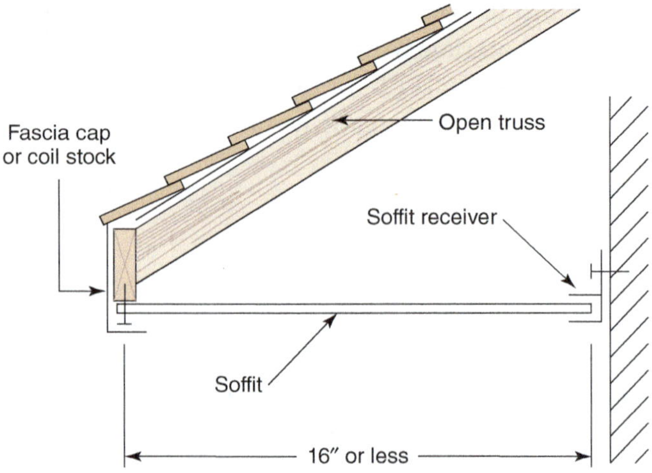

Goodheart-Willcox Publisher

Figure 5.3-5. A closed box soffit is simple and quick to install using vinyl and aluminum systems.

Goodheart-Willcox Publisher

Figure 5.3-6. Typical new construction, closed box soffit and fascia trim installation at the eaves.

Elements of Cornice Systems

The major elements of cornice systems are the fascia, soffit, frieze board, gutter, and downspout, **Figure 5.3-7**. The *fascia* is the main trim member. It is horizontally attached along the ends of the rafters. It is typically made up of a subfascia board or backer and a decorative fascia trim board or cover such as aluminum or vinyl.

The *soffit* closes off the space between the wall and the fascia board. Soffit can be constructed from a range of materials, including vinyl, aluminum, plywood, hardboard, solid lumber, or fiber cement. If the soffit is designed to ventilate the attic space, vented panels are used. Occasionally, a frieze board is incorporated into the design. The *frieze board* is the decorative member that is placed flat against the wall below the soffit.

The cornice at the eve often includes a gutter, or eves-trough to catch and divert water as it runs off the roof. A *gutter* collects the water runoff from the roof and discharges it into a downspout. A *downspout* is a decorative pipe or tube through which water from the roof is directed toward the ground. The downspout dumps the water into a drainpipe or diverter to guide it away from the building.

tokar/Shutterstock.com

Figure 5.3-7. Elements of a typical modern cornice system.

Cladding

Cladding is the finished exterior surface of a home's walls. It gives the building an attractive look and protects the WRB from wind, driving rain, and harmful ultraviolet radiation. Cladding systems used in residential construction include siding, stucco, or exterior insulation finishing systems (EIFS), and masonry veneers, such as brick or stone. The cladding system is the first line of defense against driving rain. However, the WRB is the primary water control layer and is the final line of defense against water infiltration. Some cladding systems are more effective at keeping water off the WRB than others. As stated in *Chapter 4, Drying In*, the WRB must be designed and installed to direct water toward the outside of the wall. This remains true even when the cladding is installed.

Some cladding applications incorporate a *rainscreen*, which is a gap between the cladding and the WRB. A rainscreen allows water to drain freely and provides ventilation behind the cladding, **Figure 5.3-8**. This space is created by applying furring over the WRB. Siding is fastened directly to the furring. The bottom of the rainscreen is covered with an insect screen to keep insects from getting in. Ventilation behind the siding allows residual water to evaporate and escape. The bottom acts as an air inlet for the ventilation air. The top of the rainscreen can be tied into the eave to allow ventilation to continue into the attic space. This air movement keeps the siding dry on the wall side, which reduces cupping in wood siding and ensures longer paint life.

Wall covering and cornice system installations differ based on the materials and type of system that is used. Wood, fiber cement, and composite materials are all installed using similar processes, while vinyl, aluminum, and steel use a different system that incorporates interlocking pieces and receiving channels. Additionally, stucco, EIFS, and masonry veneer systems use altogether different processes and techniques. Each of these cladding systems is described in *Lesson 5.4*.

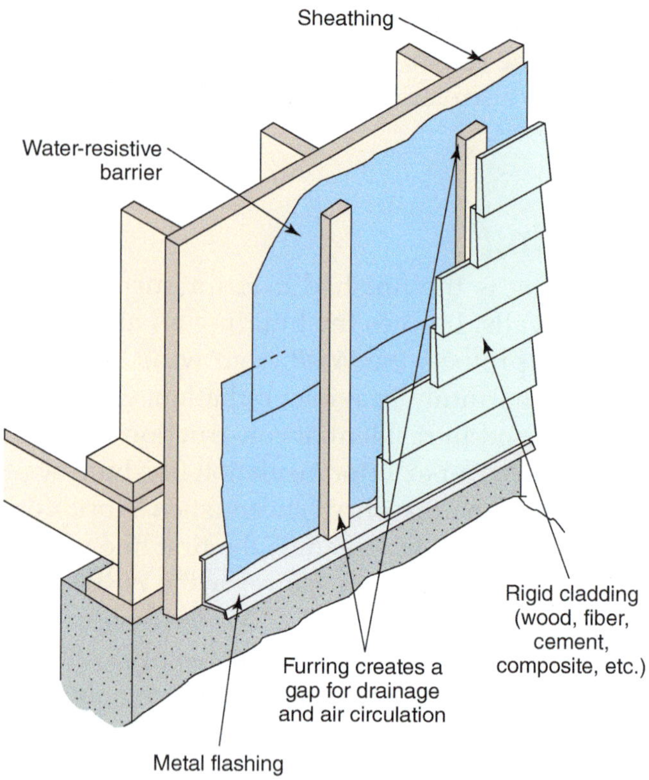

Goodheart-Willcox Publisher

Figure 5.3-8. A rainscreen behind this siding provides a space for water to freely drain and for ventilation, allowing the backside of the siding to dry.

David Papazian/Shutterstock.com

Figure 5.3-9. The porch on this new farmhouse-style home is an important architectural feature of the design.

Porches and Decks

Following the roof, the first part of the home to receive its finish covering is the cornice followed by the walls. Decks and porches are often framed before these finishes are applied; however, if permanent decking or floor surfaces are installed, they must be carefully protected from damage during the rest of construction. It is common for a temporary decking, such as plywood or OSB, to be used for this purpose. The temporary decking is removed after major construction activity is complete, at which time the permanent decking is installed.

Porches and decks add curb appeal, as well as usable outdoor living space to a home. A ***porch*** is typically attached to or built into a building and provides an extension of the main structure, **Figure 5.3-9**. It often has a roof and may be partially enclosed or screened. Porches are often integrated into the design of a home and built using similar materials, having brick or stone foundations and a finished floor, such as painted tongue and groove wood. A ***deck*** can be fully or nearly independent of the design of the home and tends to be built around the landscape of the site, considering privacy and usage needs. Decks can feature multiple levels and are fully, or nearly fully, exposed to the elements, **Figure 5.3-10**.

Trex Company

Figure 5.3-10. This multilevel deck is built on piers, posts, and a framing system composed of girders and joists.

LESSON 5.3 Review

Answer the following questions based on information in this lesson.

1. Which of these features is part of the exterior finish of a house? (5.3-1)
   A. Basement window
   B. Ceiling-mounted light fixture
   C. Wallpaper
   D. Wall cladding
2. The fascia board, soffit, and gutter are considered part of the _____. (5.3-2)
   A. cladding
   B. cornice
   C. deck
   D. porch
3. A _____ collects the water runoff from the roof and discharges it into a downspout. (5.3-2)
   A. gutter
   B. fascia
   C. frieze board
   D. soffit
4. An exterior insulation and finish system would be considered part of a home's _____. (5.3-2)
   A. cladding
   B. cornice
   C. deck
   D. porch

LESSON 5.4

Materials and Fasteners

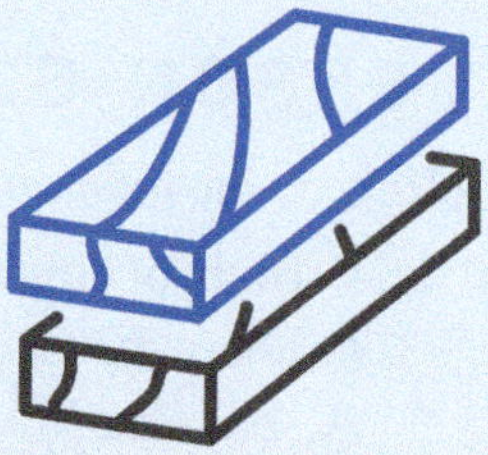

LEARNING OUTCOMES

After completing this lesson, you will be able to:

5.4-1 Identify materials used in exterior finishes.

5.4-2 Classify fasteners commonly used for installing exterior finishes.

TECHNICAL TERMS

buttlock
coil stock
color-match trim nail
composite decking
composite wood siding
cultured stone
exterior insulation and finish system (EIFS)
polyvinyl chloride (PVC) trim
receiving channel
specialty fastener
stucco
top lock
vinyl siding

Exterior Finishing Materials

When choosing the materials that will be used for the exterior finish of a home, architects and builders must consider a variety of factors, including climate, architectural style, maintenance requirements, cost, and personal preference. The decisions regarding materials for exterior finishes are among the earliest design choices to be made.

This lesson examines common materials used for exterior, trim, cladding, and decking.

Green Note

Damaged and lost materials create unnecessary waste on a project. Store materials in a secure location to minimize theft. Protect stored materials from damaging weather. Locate materials away from worker pathways to reduce the chance of accidental damage. Keep materials neatly organized to ensure that the materials can be located when they are needed.

Cornice Materials

Cornices can be comprised of a variety of versatile materials covered in previous chapters such as wood and plywood. They can also be built from materials that are part of cladding systems such as vinyl and fiber cement siding, which are covered later in this lesson. This section focuses on materials primarily used on cornice systems. Some of these materials, such as coil stock, are also used as flashing or trim elsewhere on the exterior of a home.

Aluminum Coil Stock

Bent aluminum ***coil stock*** is a material commonly used as fascia trim to cover and protect the subfascia board. Many contractors prefer to bend their own fascia trim and other siding trim and flashing pieces with a portable field brake. Tools are covered in more detail in *Lesson 5.5*. Aluminum coil stock can be purchased in a variety of colors to customize the finish trim on a project. Coil stock is a roll, or *coil*, of sheet metal that is typically 24″ wide by 50′ long, and it can be cut and bent to form a desired shape.

Pro Tip

Aviation snips can be used to cut aluminum coil stock; however, snips often leave a jagged edge when they cut. A better way is to use a sharp utility knife to score the material several times, and then bend until it snaps for a clean cut. Use a speed square as a straightedge to guide the blade while you score the material for a crosscut. A portable field brake can be used as a straightedge for lengthwise cuts. Be careful not to damage the brake; always start from each end and score toward the center to avoid nicking the brake's hinge.

Polyvinyl Chloride (PVC) Trim Boards

Polyvinyl chloride (PVC) trim boards are a low-maintenance exterior trim material that can be used as an alternative to exterior wood trim boards in applications such as window, door, and fascia trim, **Figure 5.4-1**. These boards are resistant to moisture, rot, and insects. They are paintable and resist UV radiation. PVC boards can be cut and shaped using the same tools used for wood. They can be attached using adhesives, nails, or screws.

It is important to note that PVC boards cannot be used as structural material. They are for decorative uses only.

tokar/Shutterstock.com

Figure 5.4-1. PVC trim boards are a popular alternative to wood.

Gutter Materials

A wide variety of metal gutter styles are available to control roof drainage. Manufacturers have developed gutter and downspout systems that include various component parts. Whole systems can be quickly assembled and installed on the building site.

Materials consist of aluminum or galvanized steel. Gutters are prefinished to match a wide range of colors. Plastic guttering systems are also available.

Gutter systems include inside and outside mitered corners, joint connectors, pipes, brackets, and other items. Parts easily slip together and are generally held with soft pop-rivets or sheet metal screws. **Figure 5.4-2** shows standard parts of a typical gutter and downspout system and how they are assembled.

Cladding Materials

The look of the home is largely dependent on the materials used for cladding. Materials used for exterior cladding can be combined to achieve a particular style or to add texture or character to a home's exterior.

Wood Siding

Wood siding is a traditional material used for the exterior finish of homes in the United States. Depending on the type of wood, wood siding can give a variety of distinct looks that accentuate the architectural style of the house. From the ornate features of the classic Victorian to the rustic appeal of a cabin in the woods, the versatility of wood siding is tremendous. The look of quality wood siding is so desirable that most manufactured siding materials are designed to mimic the look of wood.

Although the appearance of new wood siding may be preferred by property owners, it is not without drawbacks. The cost of wood siding materials, installation, and proper maintenance can be high when compared to other materials. If maintenance is overlooked, wood siding can quickly lose its attractive appearance. Additionally, decay, flammability, and vulnerability to insects can influence the decision to use wood or not. Examples of wood siding types include various profiles of wood board siding, **Figure 5.4-3**, and wood shingle, shake, and plywood siding, **Figure 5.4-4**.

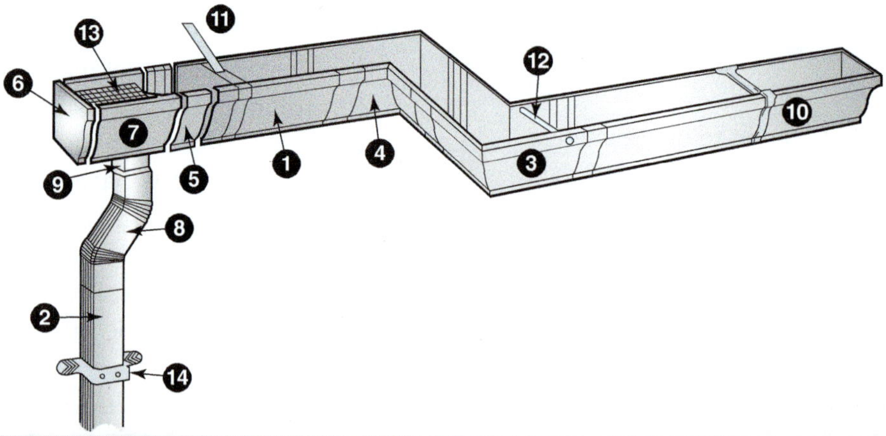

Components of a Metal Gutter System			
Key	**Description**	**Key**	**Description**
1	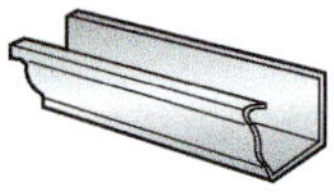 5″ K Gutter	9	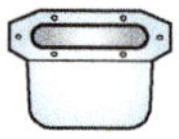 K Outlet tube (with flange)
2	 3″ Square corrugated downspout	10	 5″ K Fascia hanger
3	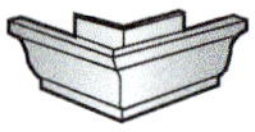 5″ Miter (outside)	11	 5″ K Strap hanger
4	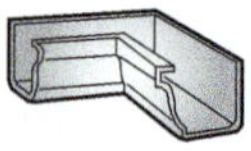 5″ Miter (inside)	12	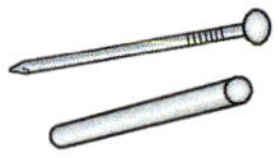 7″ Spike (aluminum) 5″ Ferrule (aluminum)
5	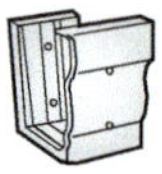 5″ K Slip joint connector	13	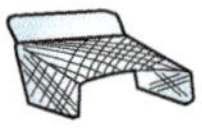 5″ K Strainer
6	 5″ K End cap left or right	14	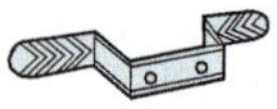 3″ Pipe band (ornamental)
7	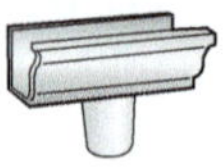 5″×3″ K End section with outlet tube		 Touch-up paint
8	 3″ Square corrugated 75° elbow or 60° elbow style a and b		 Gutter seal (tube or cartridge)

Crown Aluminum Industries

Figure 5.4-2. Metal gutter systems are assembled on-site. A wide variety of parts are available to customize the system to the home.

Edge Views of Various Horizontal Siding

Siding Patterns	Nailing: 6″ and narrower	Nailing: 8″ and wider
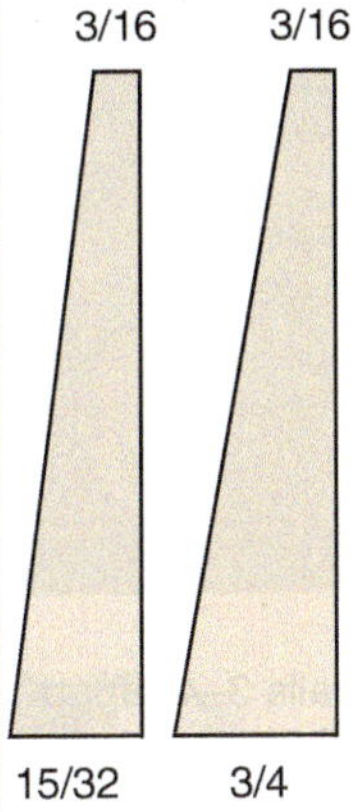 **Bevel or Bungalow** Bungalow ("Colonial") is slightly thicker than Bevel. Either can be used with the smooth or saw-faced surface exposed. Patterns provide a traditional-style appearance. Recommend a 1″ overlap. Do not nail through overlapping pieces. Horizontal applications only. Cedar Bevel is also available in 7/8×10, 12.	Plain  Recommend 1″ overlap. One siding or box nail per bearing, just above the 1″ overlap.	Plain 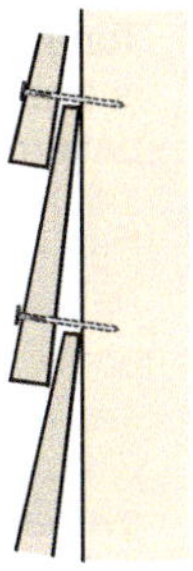 Recommend 1″ overlap. One siding or box nail per bearing, just above the 1″ overlap.
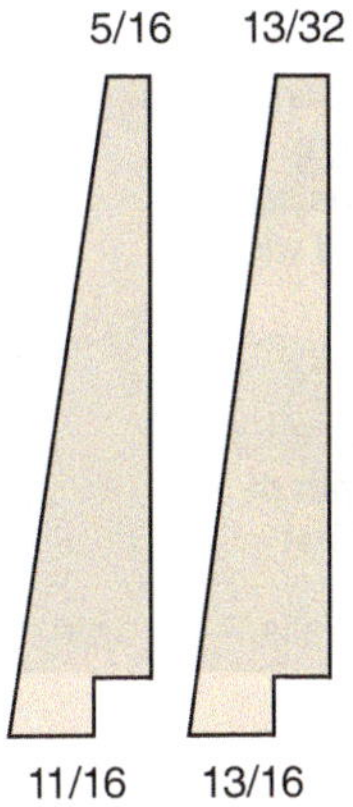 **Dolly Varden** Dolly Varden is thicker than Bevel and has a rabbeted edge. Surfaced smooth or saw textured. Provides traditional-style appearance. Allows for 1/2″ overlap, including an approximate 1/8″ gap. Do not nail through overlapping pieces. Horizontal applications only. Cedar Dolly Varden is also available 7/8×10, 12.	Rabbeted edge 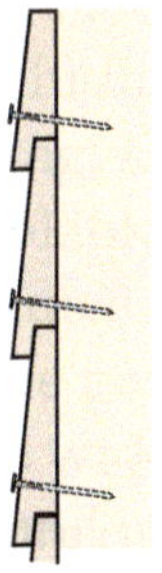 Allows for 1/2″ overlap. One siding or box nail per bearing, 1″ up from bottom edge.	Rabbeted edge 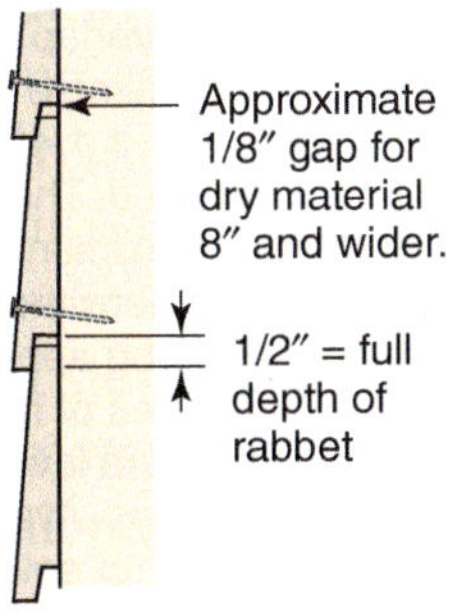 Allows for 1/2″ overlap. One siding or box nail per bearing, 1″ up from bottom edge.
 **Drop** Drop siding is available in 13 patterns, in smooth, rough, and saw-textured surfaces. Some are T&G (as shown), others are shiplapped. Refer to WWPA "Standard Patterns" (G-16) for pattern profiles with dimensions. A variety of looks can be achieved with the different patterns. Do not nail through overlapping pieces. Horizontal or vertical applications.	T&G pattern 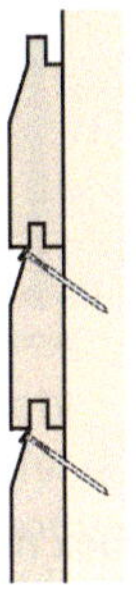 Shiplap patterns  Use casing nails to blind nail T&G patterns, one nail per bearing. Use siding or box nails to face nail shiplap patterns, 1″ up from bottom edge.	T&G pattern 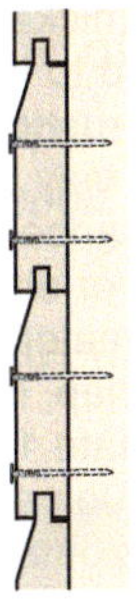 Shiplap patterns 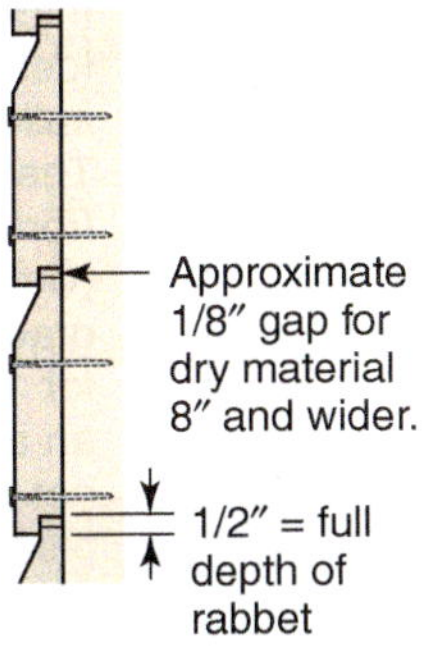 Use two siding or box nails, 3–4″ apart to face nail, 1″ up from bottom edge.

(continued)

Western Wood Products Assn.

Figure 5.4-3. Edge views of six different types of horizontal siding. Nominal sizes are used in figuring footage of siding. Note nailing suggestions.

Lou Oates/shutterstock.com

Figure 5.4-10. The base coat is going on over metal lath. Sufficient pressure is applied to push some of the stucco through the lath.

The three layers include a base or scratch coat 3/8″–1/2″ thick, brown coat 1/4″–3/8″ thick, and 3/8″ color or finish coat, **Figure 5.4-11**. Where applied over a wood-framed wall, two layers of WRB or a WRB with a weep-able drainage plane must be used to allow water to drain from behind the stucco.

Exterior Insulation and Finish Systems (EIFS)

An ***exterior insulation and finish system (EIFS)*** is a type of cladding system used in both residential and commercial construction to provide insulation, weatherproofing, and decorative exterior finish to buildings. EIFS is similar in appearance to stucco and is often referred to as synthetic stucco, **Figure 5.4-12**. EIFS consists of several layers that work together to create a durable and energy-efficient building envelope. These systems offer high R-values, low maintenance, design flexibility, and a variety of colors.

Earlier systems have been redesigned to handle problems with water penetration and retention. The systems currently being installed drain away any water penetration before it can cause structural damage. **Figure 5.4-13** shows two typical EIFS applications in cutaway illustrations.

Installation should be attempted only after carefully studying the manufacturer's instructions. An exterior insulation finish system can be installed over gypsum sheathing, glass mat–reinforced gypsum, plywood, oriented strand board, concrete,

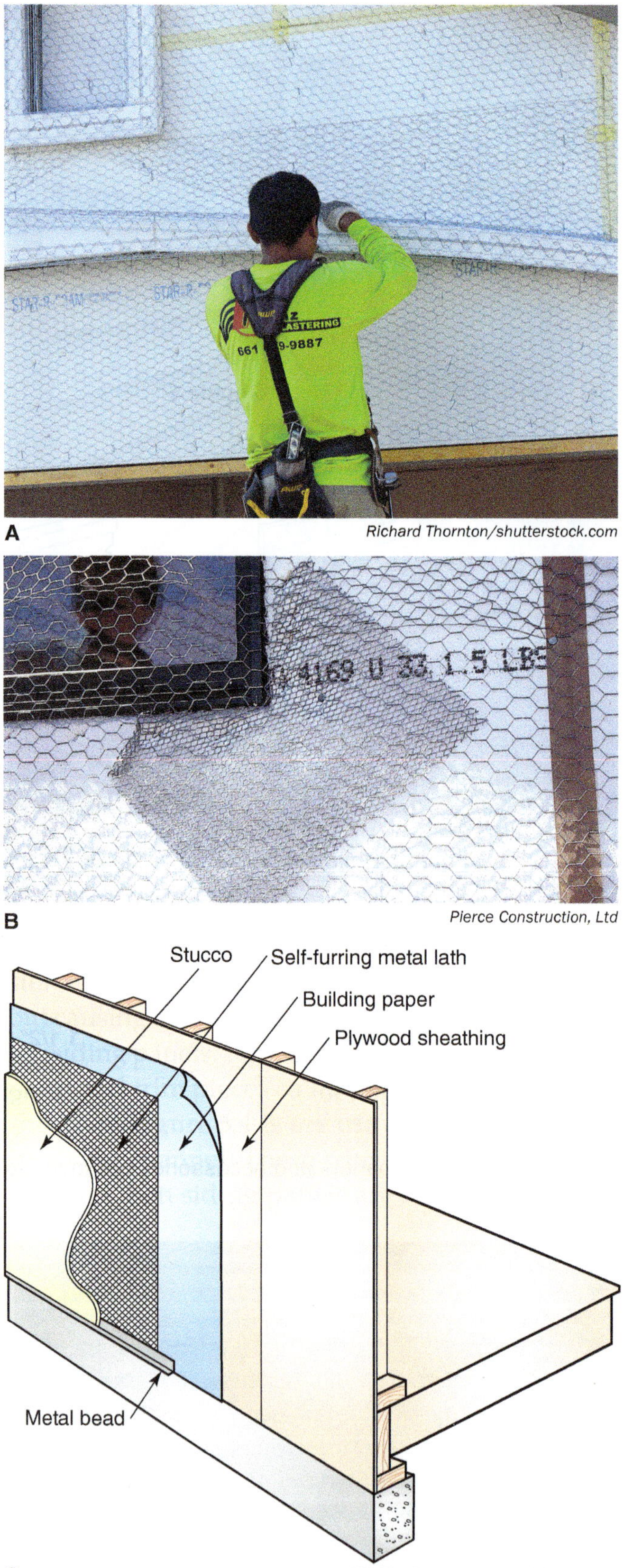

A Richard Thornton/shutterstock.com

B Pierce Construction, Ltd

C APA-The Engineered Wood Association

Figure 5.4-11. Preparation for stucco. A—A worker applies metal lath over insulating board. B—Reinforcement is added at window corners. C—Construction details for stucco finish. Sheathing paper or housewrap is recommended over both interior and exterior plywood sheathing.

brizmaker/shutterstock.com

Figure 5.4-12. EIFS during the installation process. The dark-colored base coat has just been applied below the flashing. The light-colored finish coat has been applied above the flashing.

and concrete block. The surface of the substrate must be flat before beginning installation.

Brick or Stone Veneer

A veneer wall is usually not referred to as a masonry wall, although it is. It is a wall framed in wood or metal to which brick or stone is attached rather than siding. However, the weight of the veneer is supported directly by the foundation, **Figure 5.4-14**. The foundation must be wide enough to provide a base for the masonry units. A gap is left between the masonry and the WRB for water drainage and ventilation. Corrosion-resistant metal ties are used to secure the veneer to the framework. Small openings in the bottom course of the masonry called weep holes allow the escape of any water or moisture that may penetrate the wall.

Cultured Stone

Cultured stone, also called *manufactured stone*, is made of concrete and is designed to resemble natural stone through the use of authentic shapes, textures, and pigments, **Figure 5.4-15**. Cultured stone is available in a variety of product lines that resemble limestone, fieldstone, and other genuine types of stone.

Preparation for a wood-framed wall where cultured stone is applied requires a scratch coat just like a stucco wall. After the proper preparation of the WRB and scratch coat, the cultured stone is adhered directly to the wall using mortar. Joints between the stones are filled and pointed in to create an even and finished appearance, **Figure 5.4-16**.

Deck Materials

Decks are typically framed using pressure-treated framing lumber; however, some deck framing systems incorporate the use of galvanized steel and/or the use of exterior-grade laminated veneer lumber (LVL).

Resistance to decay is a major concern when choosing lumber for a deck. Deck materials are always subject to weather conditions and may not be protected by paint or other coatings. Some woods naturally resist rot, but they are more expensive than commonly used pressure-treated lumber. The most popular choices for decking materials are pressure-treated lumber and composite decking. However, we will also explore some other alternatives.

Pressure-Treated Lumber

A commonly used decking material is 5/4″ pressure-treated Southern Yellow Pine, **Figure 5.4-17**. Pressure-treated deck boards are durable and attractive when maintained. They are one of the most cost-effective materials used for decking.

Pro Tip

Pressure-treated lumber is commonly used for a variety of applications in construction where resistance to rot and decay is required. Aside from decks, common uses include framing connections to masonry surfaces, such as sill plates; fencing; retaining walls; utility structures, such as pergolas and gazebos; as well as piers and docks.

Pressure-treated lumber generally arrives on the jobsite saturated with fluid left from the treatment process. It is important to note that when installed wet, it will shrink as it dries. Wet deck boards can shrink as much as 3/8″. Installers must plan for shrinkage during installation as it can lead to gaps between joints.

After deck boards have had a chance to dry, they are sanded and stained with an appropriate deck stain and sealer to help protect them from moisture

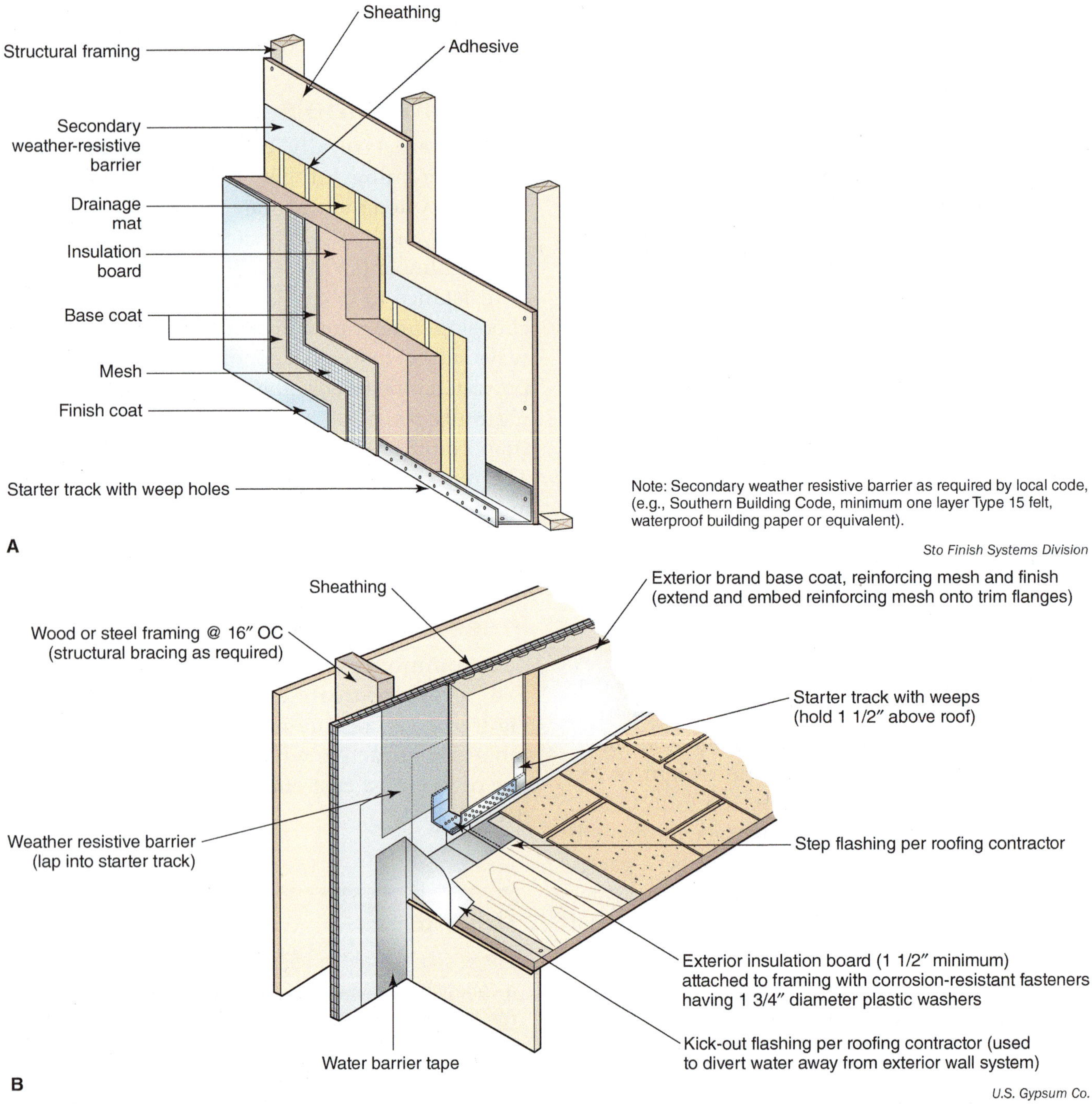

Figure 5.4-13. Typical EIFS installations. A—This cutaway shows the components of an EIFS-sided structure that incorporates a drainage cavity and water-resistive barrier. B—This cutaway illustrates a water-managed EIFS that incorporates grooves in the insulation board to channel any incidental moisture to weep holes.

and UV exposure, **Figure 5.4-18**. A variety of stain types and colors are available.

Composite Decking

Composite decking is a popular alternative to traditional wood decking. It is made from a combination of wood fibers, recycled plastics, and binding agents to produce an attractive, durable, low-maintenance decking material, **Figure 5.4-19**. Its ability to resist fading, warping, and insect damage makes it a popular choice for homeowners and builders.

Additional Decking Materials

A variety of alternatives to pressure-treated and composite decking exist. However, for a variety of reasons such as cost or availability, their use is less widespread than these common decking materials.

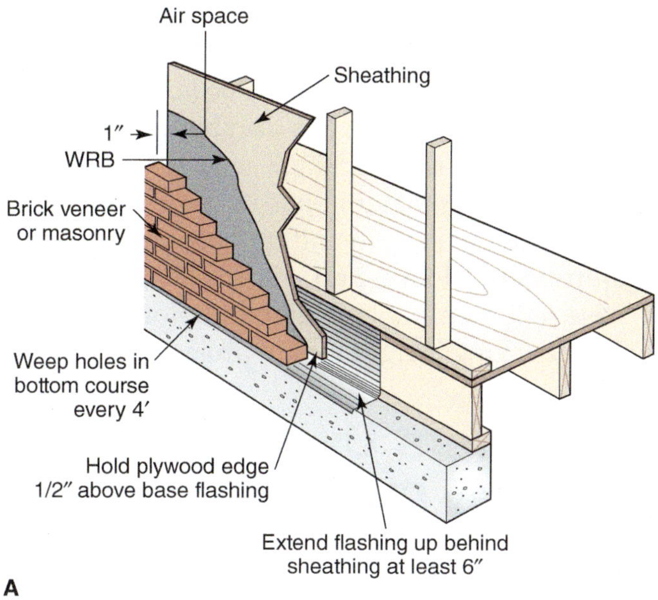

A

B

APA-The Engineered Wood Association

Figure 5.4-14. A—General construction details for brick veneer siding with plywood sheathing. One tie should be anchored to a stud for every 2 sq ft of brick area. B—A brick veneer being installed.

ungvar/shutterstock.com

Figure 5.4-15. Cultured stone being installed on an exterior wall of a home.

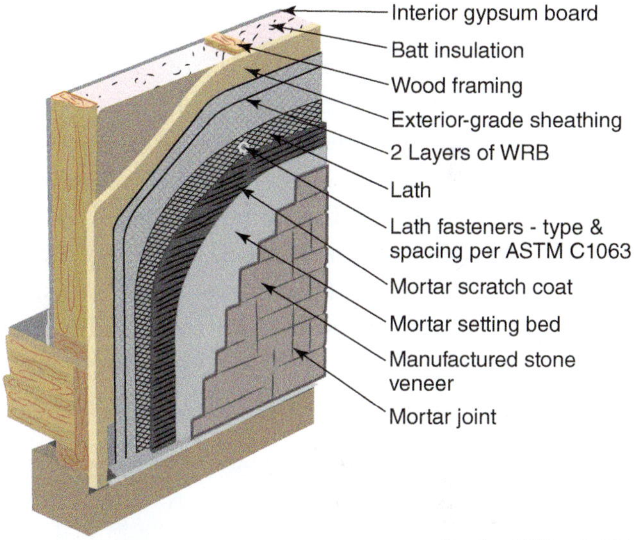

Goodheart-Willcox Publisher

Figure 5.4-16. General construction details for the installation of cultured stone on a wood-framed wall.

Goodheart-Willcox Publisher

Figure 5.4-17. This new deck is being built with pressure-treated lumber. Railing will be installed, and the decking will be sanded and protected with a decking stain/sealer.

Many wood species are naturally resistant to rot and insects. Redwood and cedar are two species native to North America that are sometimes used for decking. Tropical hardwoods from Central and South America are also used occasionally. Species such as ipe (pronounced "E-pay"), cumaru, and tigerwood are imported for use on decks because of their durability and resistance to decay, **Figure 5.4-20**. Using some of these species can showcase unique grain and color variations.

Thermally modified wood is also manufactured as a decking material. The process of thermal modification transforms common species such as yellow

Lauren Blackwell/shutterstock.com

Figure 5.4-18. Pressure-treated decking should be protected from UV rays and water by applying a quality deck stain.

pine and ash into a more stable, rot- and insect-resistant product by baking the wood at a high temperature in a kiln, while using steam to keep the outside surface from drying too quickly. The high temperatures burn off the sugars and resins that bacteria and other organisms feed on. This process modifies the cellular structure of the wood, making it less absorbent to moisture and more stable. Because the wood contains no food source, organisms that would cause rot and decay will not feed on it.

PVC decking is available in a variety of colors similar to vinyl siding. These deck boards can have a solid cellular PVC core or can be hollow with internal reinforcement. It is a durable alternative to wood or composite decking.

Each decking material has its own advantages and disadvantages, including cost, maintenance requirements, looks, and environmental impact. These factors must be weighed before deciding which decking is right for the project.

Fasteners Used for Exterior Finishes

Earlier chapters featured fasteners that are used for a variety of construction applications, specifically screws, nails, and staples. As we continue to consider more construction materials and processes, you will discover many variations of these fasteners that are used for the installation of a specific material. For instance, most wood siding uses a specific type of nail that is not typically used in any other applications.

A ***specialty fastener*** is a fastener designed to be used for a specific purpose. Some specialty fasteners can be used for other purposes as well;

CPG Building Products–TimberTech ReliaBoard

Figure 5.4-19. Composite decking materials offer low maintenance and are easily worked with the same tools as conventional lumber. Surface finishes are available in wood grain or other textures. This installation makes use of a hidden fastener system.

Radoslav Cajkovic/shutterstock.com

Figure 5.4-20. Ipe is a decay-resistant tropical hardwood that can be used as an alternative to pressure-treated or composite decking.

however, they are often defined by their primary intended purpose. Some examples of specialty fasteners used for exterior finishes are described here.

Code Note

Code requires the use of rust-resistant nails for outside finish work. They may be made from aluminum, stainless steel, galvanized steel, or cadmium-plated steel. They can also be sealed with a corrosion-resistant coating.

Color-Match Trim Nails

Installing aluminum trim pieces, such as fascia trim, window trim, or aluminum soffit panels, requires care and attention to detail. Aluminum is a soft metal that expands and contracts with temperature. The installer should be mindful of how fasteners will restrict movement or how they will stand out to the eye. A poor installation job will be evident, **Figure 5.4-21**. Aluminum trim is also easily dented or dimpled with a light strike from a hammer or when it is forced into place. Installers must take care not to damage the material when handling or installing it.

Color-match trim nails are made of aluminum or stainless steel, **Figure 5.4-22**. They are painted to match coil stock colors and are used to secure aluminum trim pieces such as fascia trim, soffit, and other exterior trim pieces. A ***color-match trim nail*** is intended to blend in with the color of the aluminum trim to minimize visibility when installed.

Goodheart-Willcox Publisher

Figure 5.4-21. Oil canning, or buckling, is a common problem when working with aluminum or vinyl materials. The sheet material buckles and creates waves, rather than lying flat. Proper framing and installation can eliminate substandard work such as this.

Goodheart-Willcox Publisher

Figure 5.4-22. Color-match trim nails are used to attach sheet metal trim covers, such as fascia trim.

When installing these trim nails, predrilling through the aluminum trim is recommended to avoid dimpling the face of the aluminum sheet metal. The hole should be larger than the shank of the nail, but smaller than the head. Use a flat-faced trim hammer and drive nails in with a light touch to avoid damaging the delicate surface of the aluminum trim.

Siding Nails

A quality wood siding installation requires fasteners that provide a clean, corrosion-free appearance for many years. Siding nails are made of stainless steel or galvanized steel to prevent corrosion, **Figure 5.4-23A**. They have a thin shank to avoid splitting thin siding pieces and a small head to help them blend in when paint or stain is applied, **Figure 5.4-23B**. Siding nails typically have a ring or spiral shank to prevent the nail from backing out over time.

A Goodheart-Willcox Publisher

B Goodheart-Willcox Publisher

Figure 5.4-23. A—The small head of the siding nails helps them to blend in, especially after paint is applied. B—In this example, metal corner caps are being used instead of corner trim boards.

Deck Fasteners

Deck fasteners must withstand all types of weather and seasonal changes while keeping the decking stable and secure throughout its years of service. One thing all deck fasteners have in common is that they are made from corrosion-resistant materials such as stainless steel, or they have a corrosion-resistant coating.

Methods for securing decking include:

- **Stainless steel or hot-dipped galvanized nails**—Though not used as often as screws, nails can be used for deck fastening. Because they tend to back out over time, any nails used for decking should be ring or spiral shanked.
- **Deck screws**—Deck screws are available in a variety of lengths and gauges. Most available options are coated with a ceramic, polymer, or epoxy coating, but can also be made from stainless steel. They feature a flat head that is typically Torx drive or combination drive. Deck screws are installed with a screw gun and do not require a pilot hole.
- **Hidden fastener systems**—The consumer desire for a smooth, unblemished deck surface and the increased popularity of composite decking have given rise to a variety of options for fastening deck material that conceals the head of the fastener. These systems use specialized tools or hardware that allows the decking to be attached from the side or bottom of the deck board, **Figure 5.4-24**.

Brad Ingram/shutterstock.com

Figure 5.4-24. This composite decking is being installed using a hidden fastener system.

LESSON 5.4 Review

Answer the following questions based on information in this lesson.

1. Gutters are commonly made of which materials? (5.4-1)
   A. Aluminum or galvanized steel
   B. Aluminum or stainless steel
   C. Galvanized steel or stainless steel
   D. Stainless steel or copper
2. Which statement about new wood siding is true? (5.4-1)
   A. It is inexpensive to maintain.
   B. It is highly resistant to decay.
   C. It is highly versatile.
   D. Its installation costs are minimal.
3. *True or False?* Compared to standard wood siding, composite wood siding expands and contracts more with temperature and humidity changes. (5.4-1)
4. *True or False?* An exterior insulation and finish system is similar in appearance to wood siding. (5.4-1)
5. Which statement is *true* about siding nails? (5.4-2)
   A. They are commonly made of aluminum.
   B. They often have a spiral shank.
   C. They generally have a thick shank.
   D. They typically have a large head.

LESSON 5.5

Tools

LEARNING OUTCOMES

After completing this lesson, you will be able to:

5.5-1 List common hand and power tools used to apply exterior finishes to a home.

TECHNICAL TERMS

portable field brake
router
sheet metal hand bender
snap lock punch
vinyl siding removal tool

Tools Used to Apply Exterior Finishes

Much like the tools used for drying in, builders use some specialty tools that only serve one function while installing the exterior finishes to a home. This lesson features some of these specialty tools, along with a few that are more versatile.

Hand Benders

A ***sheet metal hand bender***, also called a *seamer*, is a tool used to bend and shape thin metal flashing and trim, **Figure 5.5-1**. For longer, more precise bends, a portable field brake is used.

Portable Field Brake

A ***portable field brake***, also called a *sheet metal brake*, is a piece of equipment that is commonly used by siding contractors to put sharp, precise bends in sheet metal, forming it into flashing or finish trim profiles for the exterior of a home, **Figure 5.5-2**. Brakes can be outfitted with coil rollers, cutters, and measuring devices to make them efficient and easy to use. Using a brake gives contractors the ability to customize aluminum trim pieces into a variety of shapes and profiles. Bending trim, rather than buying it, can save money if an experienced tradesperson is doing the work.

Goodheart-Willcox Publisher

Figure 5.5-1. Hand bending tools can help put crisp edges on short bends when installing aluminum trim.

Vinyl Siding Removal Tool

The ***vinyl siding removal tool***, also referred to as a *siding unlocking tool*, is a small hand tool with a hooked blade that allows the user to unhook an installed piece of vinyl siding for removal without removing the course above it, **Figure 5.5-3A**. The hooked blade of the tool is slipped between two courses of vinyl siding, allowing the user to gently stretch the panel and separate the connection, **Figure 5.5-3B**. Once disconnected, the user can access the fasteners holding the siding panel that needs to be removed.

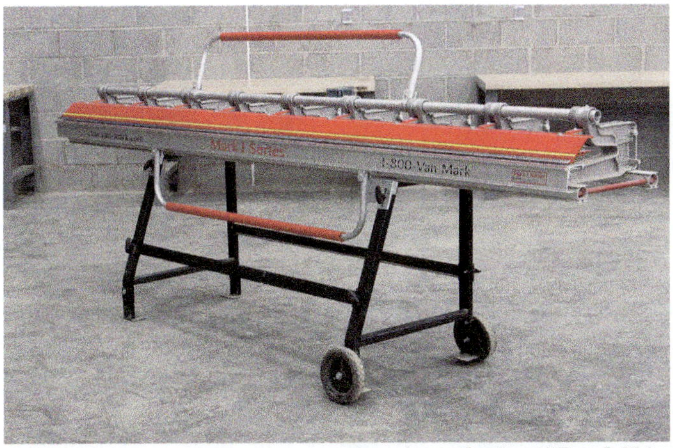

A Goodheart-Willcox Publisher

B Goodheart-Willcox Publisher

C Goodheart-Willcox Publisher

D Goodheart-Willcox Publisher

Figure 5.5-2. A—A sheet metal brake is used to bend profiles from aluminum coil stock. B—Material is clamped in the jaw of the brake and the hinge is lifted by the user to create the desired bend. C—Custom trim, such as unique window or door casing covers, cannot be purchased and must be bent on-site. D—Accessories like this coil roller and cutter can speed the process and improve quality considerably.

A Malco Tools, Inc

B Malco Tools, Inc

Figure 5.5-3. A—The vinyl siding removal tool. B—This tool is used to separate the lock between two panels of installed vinyl siding.

Snap Lock Punch

It is necessary for vinyl siding installers to cut the nailing fin from the top of a siding panel when they reach the top of a wall, or around the bottom of an obstacle such as a window. When this is the case, a ***snap lock punch*** is used to form a locking tab on the vinyl siding panel, **Figure 5.5-4**. This locking tab mates into a corresponding trim piece, **Figure 5.5-5**.

Router

A ***router*** is a power tool used to cut irregular shapes and to form various contours on edges,

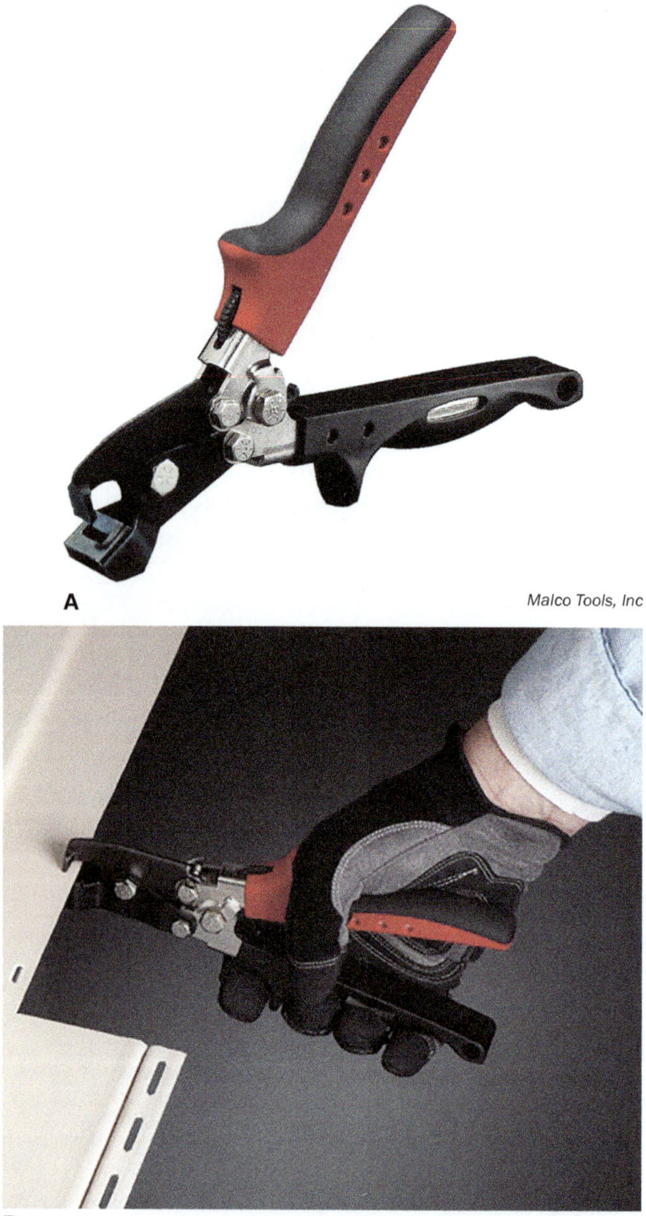

A Malco Tools, Inc

B Malco Tools, Inc

Figure 5.5-4. A—The snap lock punch. B—This tool creates a locking tab that locks into a vinyl utility trim.

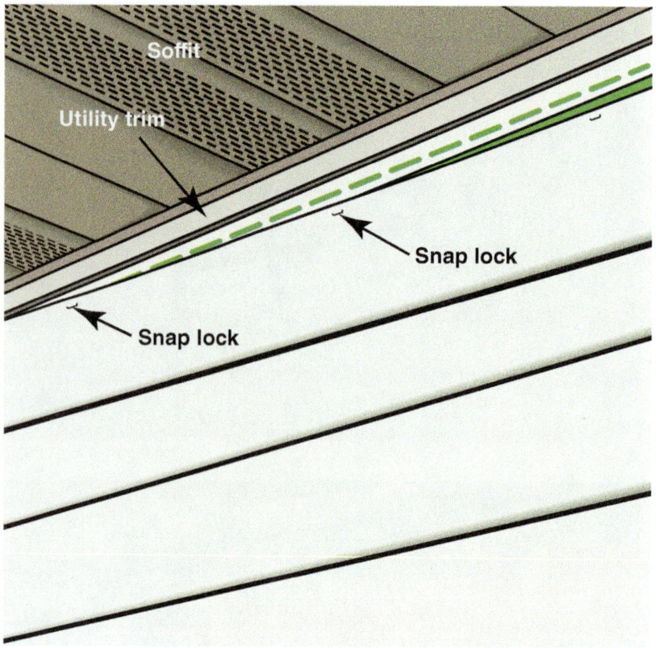

Goodheart-Willcox Publisher

Figure 5.5-5. The locking tabs created by the snap lock punch mate into the utility trim to fasten the top panel of siding.

such as when shaping the edge of a cut deck board, **Figure 5.5-6**. When equipped with special guides, routers can be used to cut dados, grooves, mortises, and dovetail joints. On a router, a collet-type chuck

Safety Note

Follow these safety rules when working with routers:

- Wear safety glasses and hearing protection.
- Securely mount the bit in the chuck and make sure the base is tight.
- Be sure the motor is properly grounded.
- Be certain the work is securely clamped so it will remain stationary during the routing operation.
- Place the router base on the work, template, or guide—with the bit clear of the wood—before turning on the power. Firmly hold the router when turning on the motor. Starting torque could wrench the tool from your grasp.
- Hold the router with both hands and smoothly feed it through the cut in the correct direction.
- When the cut is complete, turn off the motor. Do not lift the machine from the work until the motor has stopped.
- Always unplug the router when mounting bits or making adjustments.

is used to hold a bit. The chuck is connected to an electric motor that spins the bit at a high revolutions per minute (RPM). The bit is equipped with blades that produce the desired cut.

Hundreds of different types of router bits are available, **Figure 5.5-7**. Straight bits are used when cutting dados and grooves. Some bits for shaping and forming edges have a pilot tip that guides the router. The router motor revolves in a clockwise direction (when viewed from above) and should be fed from left to right when making a cut along an edge.

A Elena_Nik/Shutterstock.com

B Goodheart-Willcox Publisher

Figure 5.5-6. A—A router is used to cut irregular shapes and edges. B—Putting a rounded edge on a deck board is just one of many uses for a router.

OlekStock/shutterstock.com

Figure 5.5-7. Router bit shapes are available for everything from square dados and rabbets to decorative edges and special joinery tasks. Bits with carbide cutting edges give better results and last longer than steel cutting edges.

LESSON 5.5 Review

Answer the following questions based on information in this lesson.

1. *True or False?* A portable field brake is used for making large, precise bends in sheet metal. (5.5-1)
2. A vinyl siding removal tool has _____. (5.5-1)
   A. a straight blade
   B. a hooked blade
   C. two prongs
   D. three prongs
3. *True or False?* When viewed from above, a router revolves in a counterclockwise direction. (5.5-1)

Goodheart-Willcox Publisher

Figure 5.6-6. Scaffolding formed with metal brackets and an aluminum plank. Since there is no railing, a personal fall arrest system should be worn when working on this type of scaffolding.

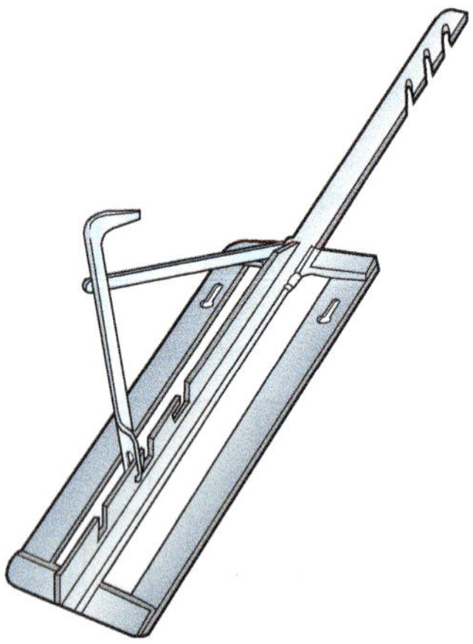

A Goodheart-Willcox Publisher

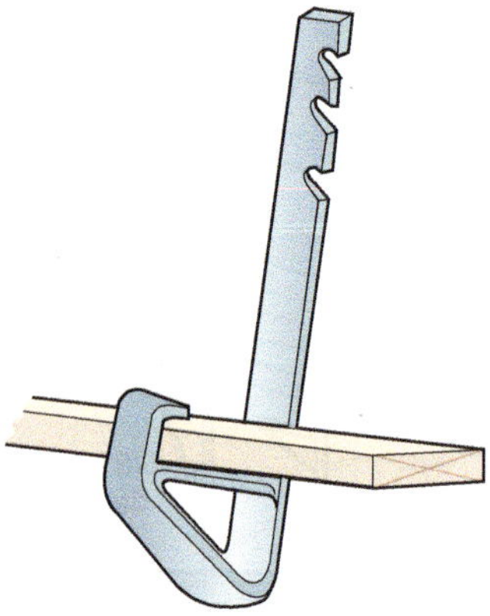

B Goodheart-Willcox Publisher

C Goodheart-Willcox Publisher

Figure 5.6-7. Roofing brackets. A—Bracket with slots to slip over nails driven into the roof frame. The adjustable arm is used to level planking on roofs of different pitches. B—Bracket with fixed angle for steep roofs. C—Brackets and planks installed for a re-roofing job.

Goodheart-Willcox Publisher

Figure 5.6-8. A ladder jack used to provide a low-level scaffold. Two identical ladders must be used. The plank is located inside of the slope of the ladders.

Aerial Lifts

An ***aerial lift*** is a type of mobile scaffolding that provides a convenient way to access elevated locations. They use a hydraulic system to adjust their height and position. **Figure 5.6-11** shows an aerial lift raised by a scissor jack that is activated by a hydraulic cylinder. **Figure 5.6-12** shows another type of aerial lift, usually called a boom lift, or basket lift. Some of these mobile scaffolds are set up to be towed behind a truck; others are self-propelled by a gas, propane, or electric motor.

Many aerial lifts incorporate anchor points into their railing system. These anchor points are intended to be used with a positioning device as a restraint system and do not meet the impact requirements for a personal fall arrest system. Falling from the scaffold could produce enough force to overturn the unit. The railing is intended to act as the fall arrest system. Never work outside the railing unless you have proper fall protection.

A Trudeau Construction Co B Trudeau Construction Co

Figure 5.6-9. Ladder jacks projecting from the front side of ladders. A—A personal fall arrest system should be worn at all times when using a scaffold without a guardrail. B—Ladder jacks used with a metal platform.

Safety Note

When working on or around scaffolding, observe these rules:

- All scaffolds should be built under the direction of a competent person.
- Follow the design specifications listed in local and state codes.
- Inspect scaffolds each day before use.
- Provide adequate pads or sills under scaffold posts.
- Plumb and level scaffold members as each is set.
- Equip all planked areas with proper guardrails, toeboards, and screens when required.
- Power lines near scaffolds are dangerous. Consult the local power company for advice and recommendations.
- Never use ladders or makeshift devices on top of scaffold platforms to increase the height.
- Be certain the planking is heavy enough to carry the load with a safe span length.
- Planking should be lapped at least 12″ and extend at least 6″ but less than 12″ beyond all supports.
- Before moving a rolling scaffold, remove all persons, materials, and equipment from the platform.
- The height of the rolling scaffold platform should not exceed four times the smallest base dimension.

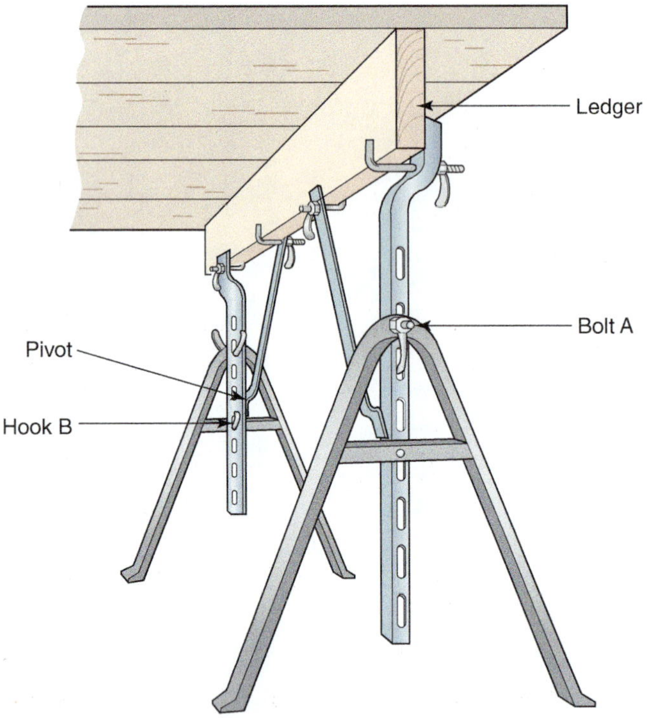

Patent Scaffolding Co

Figure 5.6-10. Trestle jacks used for low platforms. To raise the jacks, loosen bolt A and move hook B to another slot. Note that the ledger is firmly clamped to the jacks.

lmfoto/shutterstock.com

Figure 5.6-11. Mobile scaffolding can be quickly positioned, raised, and lowered. This scissor lift is self-propelled.

Safety Note

When working on or around aerial lifts, observe these rules:

- Only operate aerial lifts if you have been trained and authorized to do so.
- Inspect the lift before each use for any signs of damage or malfunction.
- Ensure the lift is set up on stable ground, and use outriggers or stabilizers as required.
- Wear appropriate personal protective equipment (PPE), including a hard hat and a safety harness.
- Keep a safe distance from power lines and other overhead hazards.
- Never exceed the lift's maximum load capacity or extend the platform beyond its designated limit.
- Use caution when operating in adverse weather conditions.

A Dmitry Kalinovsky/shutterstock.com

B Bannafarsai_Stock/shutterstock.com

Figure 5.6-12. Aerial lifts increase productivity and give safe access to difficult-to-reach places. A—Telescoping boom lifts have an extendable boom that can reach great distances. The largest of these machines can reach up to 185′. B—Articulating boom lifts have a knuckle that allows for versatility in tight spaces.

LESSON 5.6 Review

Answer the following questions based on information in this lesson.

1. A toprail installed _____ above the working surface of a scaffold would meet OSHA requirements. (5.6-1)
   A. 20″
   B. 28″
   C. 36″
   D. 42″
2. *True or False?* The footings of a scaffold must be able to support the weight of the scaffold when it is loaded. (5.6-1)
3. *True or False?* The most common type of scaffolding used in construction today is made of PVC uprights with stainless steel braces. (5.6-1)
4. Wall brackets that support a scaffolding system with a wood platform should be spaced no more than _____ apart. (5.6-1)
   A. 3′
   B. 6′
   C. 8′
   D. 10′
5. Which type of scaffolding utilizes a hydraulic system to adjust its height and position? (5.6-1)
   A. Aerial lift
   B. Ladder jack
   C. Sectional steel
   D. Trestle jack

LESSON 5.7

Math

LEARNING OUTCOMES

After completing this lesson, you will be able to:

5.7-1 Describe the consequences of faulty construction due to a lack of accuracy with regard to plumb, level, or square.

5.7-2 Use the Pythagorean theorem to calculate the length of a side of a right triangle, given lengths of the other two sides.

TECHNICAL TERMS

hypotenuse	right angle
Pythagorean theorem	right triangle

Plumb, Level, and Square

When building a home, achieving plumb, level, and square construction is critical for ensuring structural integrity, functionality, and a quality appearance. These concepts were briefly discussed in *Chapter 1, Preparing for Construction*. However, their importance cannot be overstated. This lesson revisits these concepts and continues to develop methods to enhance accuracy for building plumb, level, and square structures.

> **Pro Tip**
>
> When using a spirit level to check for plumb or level, getting the bubble between the lines is not always enough. Like many building materials, the fluid inside a spirit level expands and contracts with changing temperatures. The fluid expansion and contraction can change the size of the air bubble in the vial, making it smaller on hot days and larger on cold days. For best accuracy, the bubble should be centered on or between the lines.

Plumb

When something is plumb, it is perfectly vertical. Ensuring plumb construction is important for aesthetic reasons because in comparison with other parts of the building, misalignment may be easily noticed.

Consider a row of porch columns. From the front of the building, you would likely not be able to notice one that leaned toward or away from you. However, if you were approaching the building from the side, you might easily spot the leaning column because of its misalignment with the others.

Plumb is also important for the stability of a structure, **Figure 5.7-1**. The forces generated on loadbearing walls are transferred perpendicular to the earth. If a wall is out of plumb, it is not best equipped to transmit that load from the top of the wall to its base. It is not necessary to achieve perfection with regard to plumbing walls and columns; however, if a wall or column is too far out of plumb, the structure may fail a building code inspection.

Level

As with plumb, it is important for builders to ensure building components are accurately leveled. Practically speaking, a floor that is built out of level may be detectable just by walking across the room. Details, such as properly crowning floor joists or accurately placing and floating concrete, can play a part in how level a floor is.

Some builders prefer to install gutters level so they align with the other cornice trim, others prefer to slope them slightly toward the downspout

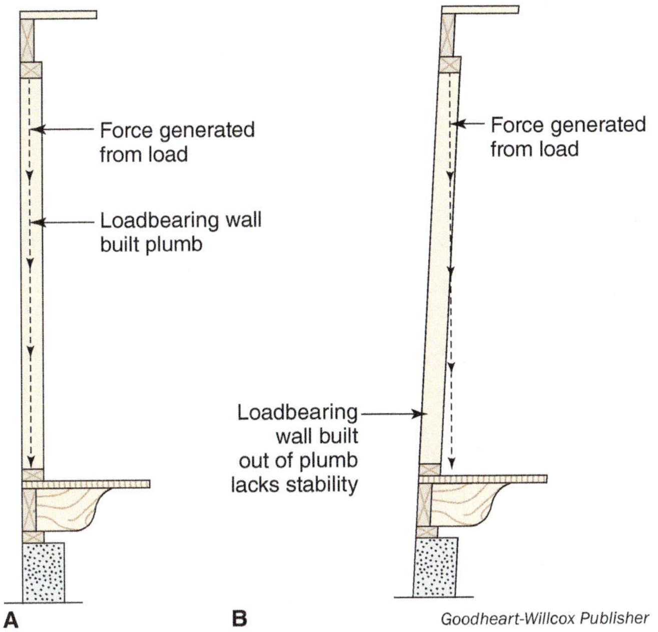

Goodheart-Willcox Publisher

Figure 5.7-1. A wall built plumb (A) has more stability compared to a wall built out of plumb (B).

Susan Law Cain/shutterstock.com

Figure 5.7-3. Note the horizontal reference lines created by the courses of siding on this home. Using these level reference lines, imagine how easy it would be to pick out if one of the windows were installed an inch or two higher than the ones beside it.

drain. If the gutter sloped away from the downspout drainage, dripping, clogging, and overflow issues are likely to occur, **Figure 5.7-2.**

Siding can be a challenge to install if walls were built out of level or if windows have not been installed level in relation to one another. Horizontal siding creates a straight reference line that extends all the way around the house with each course of siding, **Figure 5.7-3.** These horizontal reference lines should indicate quality and consistency when a critical eye scans the home's exterior. The relative distance from the bottom or top of a window or door should be the same for all like-sized windows and doors. The size of the top course should maintain consistency where it meets the soffit from one end of the building to the other. These small details are important tells with regard to the overall quality of the construction.

tokar/shutterstock.com

Figure 5.7-2. Cornices feature many horizontal lines. If something is built out of level, it can be easy to spot.

These examples convey the importance of leveling precision. Today's technology can help builders maintain a high degree of accuracy when it comes to building plumb and level. Laser levels have overcome some of the challenges associated with using spirit levels and plumb bobs, such as leveling over long spans or settling a plumb bob in the wind. However, all tools used for these operations need to be carefully maintained and protected from damage on the job. Calibrated instruments can become inaccurate when dropped or abused.

Square: Working with Right Angles

A ***right angle*** is a 90° angle. Right angles are common in construction. For example, builders use right angles to lay out foundations and square up walls when framing, when laying out floor or ceiling tile, or squaring up a deck with the house, **Figure 5.7-4.**

A triangle having a right angle is called a ***right triangle***. Right triangles are useful in laying out square (90°) corners. Two terms associated with

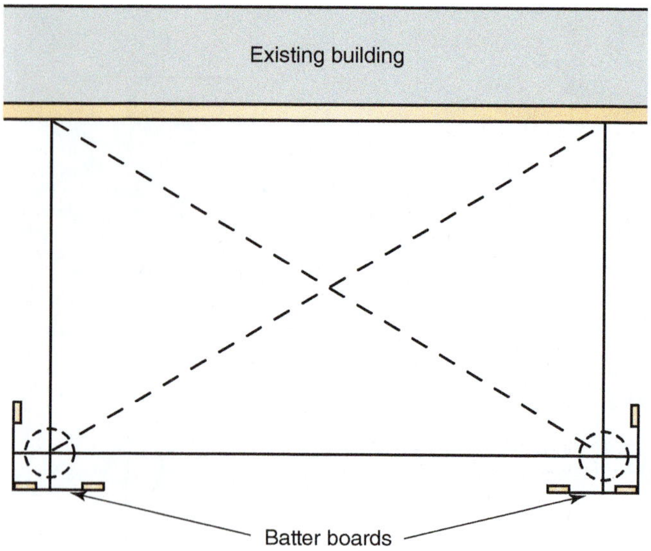

Goodheart-Willcox Publisher

Figure 5.7-4. When building a deck, it is important that it is properly squared to the house.

right triangles that are important to know are side and hypotenuse. The ***hypotenuse*** is the diagonal line in a right triangle that connects both sides.

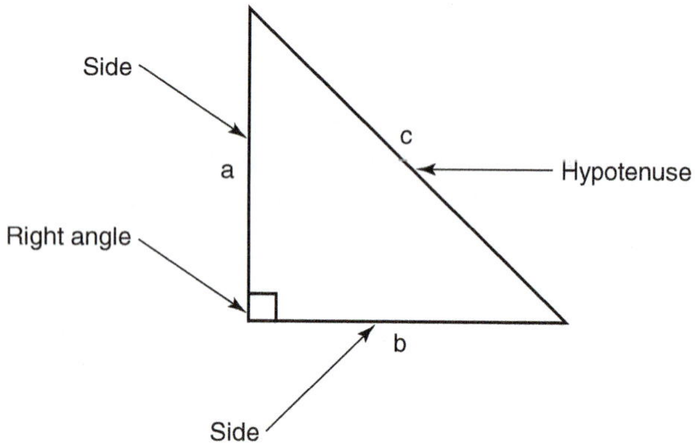

Goodheart-Willcox Publisher

A right triangle can only have one 90° corner. The total of all three corners is always 180°, so if one is 90°, the other two must add up to 90° together.

The ***Pythagorean theorem*** is a geometric principle that relates to right triangles. Named after the Greek mathematician Pythagoras, the Pythagorean theorem states that the sum of the squares of the sides of a right triangle is equal to the square of the hypotenuse. To help keep track of the Pythagorean theorem, it is common to label the two sides *a* and *b* and the hypotenuse *c*. Then the theorem can be stated as the equation $a^2 + b^2 = c^2$. If the lengths of two sides of a right triangle are known, the Pythagorean theorem can be used to find the length of the third.

For example:

On a right triangle, side *a* equals 3′ and side *b* equals 6′. Use the Pythagorean theorem to find the length of the hypotenuse (side *c*).

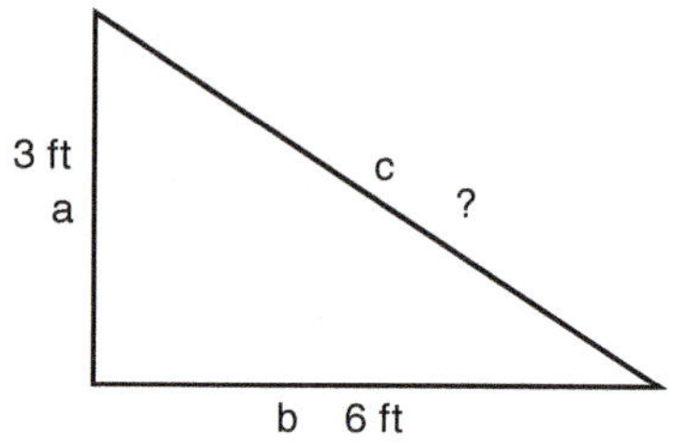

Goodheart-Willcox Publisher

$$a^2 + b^2 = c^2$$
$$(3 \times 3) + (6 \times 6) = c^2$$
$$9 + 36 = c^2$$
$$45 = c^2$$
$$\sqrt{45} = c = 6.7082$$

Using a calculator to find the answer, round to one decimal place. The hypotenuse is 6.7′.

The same calculations can be done to find side *b* when side *a* and the hypotenuse *c* are known. If two sides of a right triangle are known, the Pythagorean theorem can be used to find the length of the remaining side. If side *b* and the hypotenuse are known, $a^2 + b^2 = c^2$ can be rearranged to $a^2 = c^2 - b^2$.

For example:

On a right triangle, side *b* equals 3″ and hypotenuse *c* equals 5″. Use the Pythagorean theorem to find the length of side *a*.

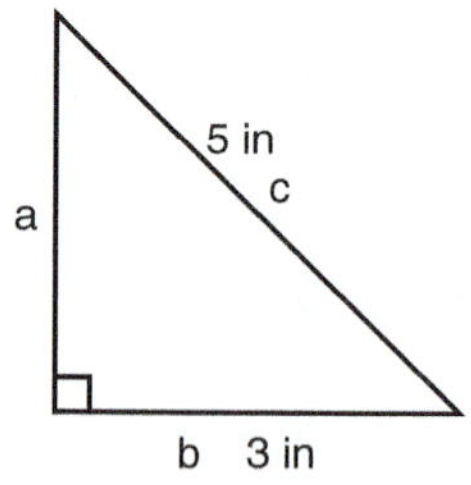

$$a^2 = c^2 - b^2$$
$$a^2 = 25 - 9$$
$$a^2 = 16$$
$$a = 4$$

Goodheart-Willcox Publisher

Practice 5-1

One side of a right triangle measures 24′, and the hypotenuse measures 30′. What is the length of the other side of the triangle?

Using Pythagorean Theorem to Lay Out Square Corners

The Pythagorean theorem can be used to verify that a corner is 90° by measuring along the two sides, then checking the length of the hypotenuse between them. This is usually simplified by using either 3 and 4 units or 6 and 8 units as the sides. If 3 and 4 units are used, the hypotenuse of a square corner is 5 units. If 6 and 8 units are used, the hypotenuse is 10 units.

For example:

Lay out the square corner using the 6–8–10 method, shown in **Figure 5.7-5**. First, measure and mark 6′ along one building line already established. Next, measure 8′ on the intersecting building line and mark that location. Then, measure the diagonal distance from one mark to the other. Move the building line as needed until the diagonal line measures 10′. At that point, the corner formed by the intersecting lines is square. Use a mason's line stretched between the first two screws as a reference if needed. Finally, establish the two remaining building lines by measuring the lengths and squaring the corners.

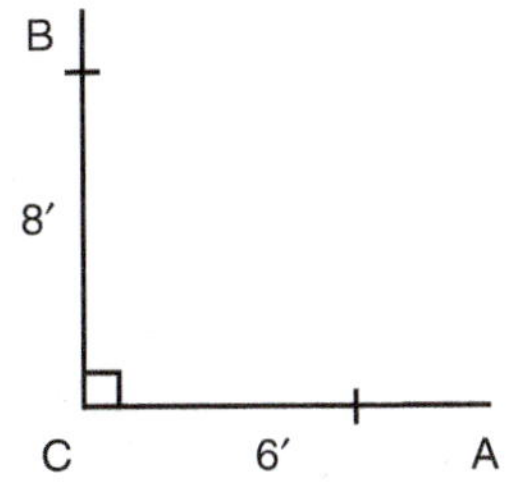

Step 1. Drive a stake at point C (the corner).
Step 2. Draw a line (tight string) toward A.
Step 3. Measure 6′ and mark the line.
Step 4. Draw a line toward B.
Step 5. Measure 8′ and mark the line.
Step 6. Measure the hypotenuse between the two marks and adjust the lines so the hypotenuse is 10′.

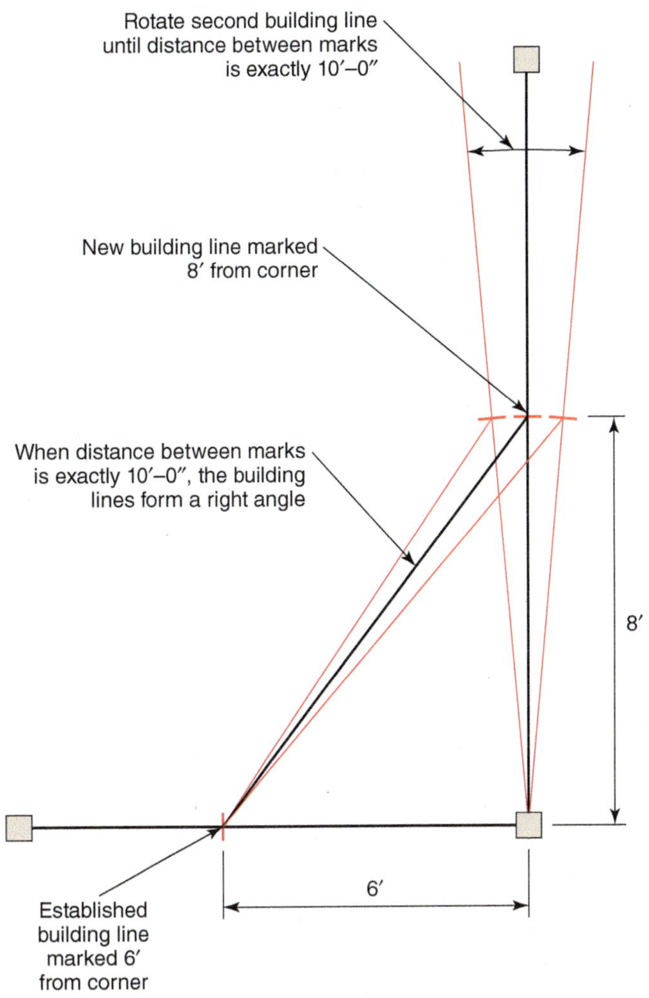

Figure 5.7-5. To check whether a corner is square, mark 6′ along one building line and 8′ on the intersecting building line. If the distance between the marks is 10′, the corner is square.

Pro Tip

As an alternative to the 6–8–10 method, you can use a construction calculator or app to calculate the overall diagonal measurement between opposing corners of a building. To get the diagonal dimension, use the "Rise," "Run," and "Diag" functions.

Practice 5-2

A carpenter preparing to build a deck wants to ensure that the frame of the structure is square to the house. The length of the deck is 26′, represented at points A and B in **Figure 5.7-6**. The deck's frame will project 12′ from the side of the home, which will be represented by a stake driven at point C. To determine the location of point C, determine the length of the hypotenuse from point B to point C. Convert your answer to feet and inches.

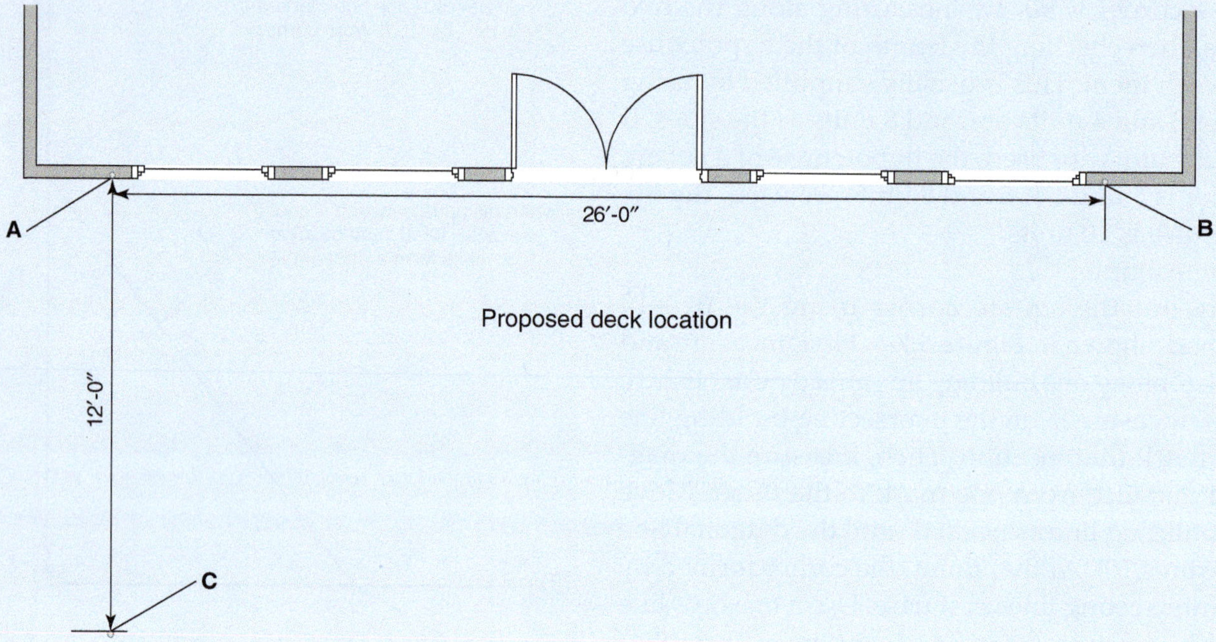

Goodheart-Willcox Publisher

Figure 5.7-6. Determine the length of the hypotenuse.

LESSON 5.7 Review

Answer the following questions based on information in this lesson.

1. *True or False?* A loadbearing wall must be plumb to properly transfer the load from the top of the wall to its base. (5.7-1)
2. A right angle is a _____ angle. (5.7-2)
   A. 45°
   B. 90°
   C. 120°
   D. 180°
3. *True or False?* A right triangle can have two 90° corners. (5.7-2)
4. If side *a* of a right triangle equals 9″ and the hypotenuse equals 15″, side *b* must be _____. (5.7-2)
   A. 8″
   B. 12″
   C. 17″
   D. 18″
5. When using the Pythagorean theorem to verify that a corner is 90°, if you draw a 6′ line from the corner to point A and draw an 8′ line from the corner to point B, the corner will be square when the distance between the two points is _____. (5.7-2)
   A. 6′
   B. 8′
   C. 10′
   D. 12′

LESSON 5.8

Print Reading

LEARNING OUTCOMES

After completing this lesson, you will be able to:

5.8-1 Explain what a builder can determine by studying elevation drawings.

5.8-2 Identify common symbols and abbreviations found on construction drawings.

TECHNICAL TERMS

cut line

elevation drawing

reference mark

window and door schedule

Elevation Drawings

Applying exterior finishes to a home requires careful preparation by the builder. Accommodations for specific cornice and trim details, types of siding, and locations for decks and other features need to be made as construction progresses. Many of these details, and more, can be found on the elevation drawings of a home.

A building ***elevation drawing*** is a drawing showing the outside walls of the structure, giving a visual representation of its height, shape, and architectural details. These drawings are scaled so that all elements appear in their true relationship. Generally, the various elevations are oriented to the site by labeling them according to the direction they face. However, when plans are not designed for a specific location, the terms "front," "rear," "left side," and "right side" are used. An example of an elevation drawing is shown in **Figure 5.8-1**.

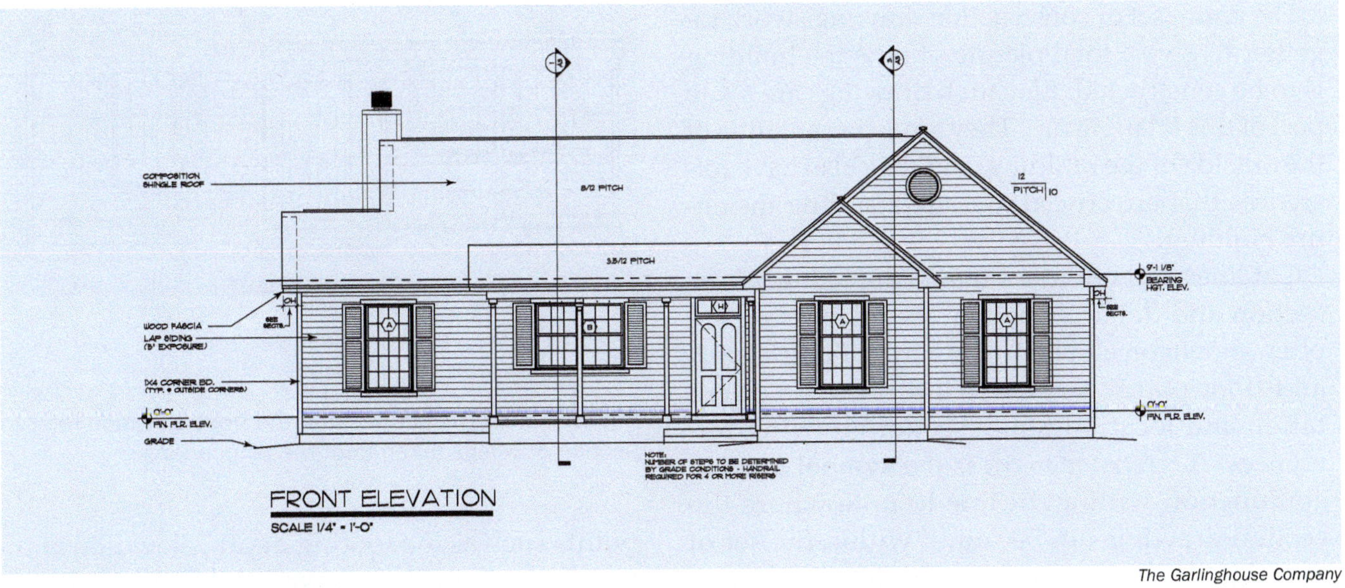

The Garlinghouse Company

Figure 5.8-1. In stock plans, elevations are usually marked "front," "rear," "left side," and "right side" because the architect does not know which direction the house will face. Elevations may carry notes indicating special features, code requirements, or reminders. *(Continued)*

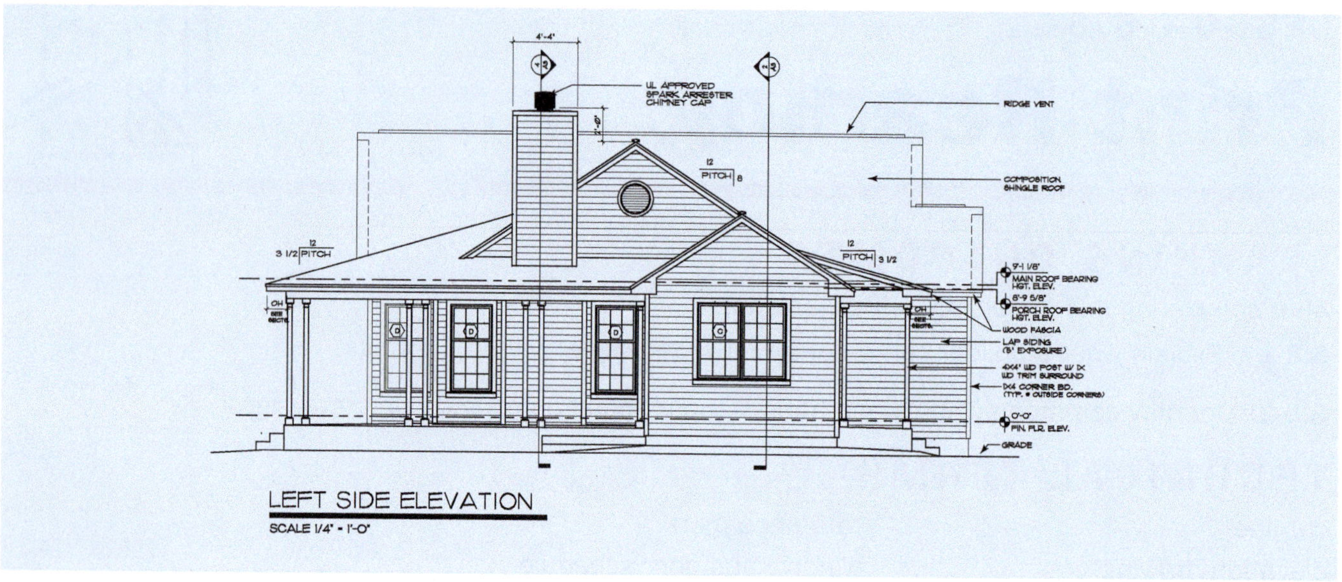

The Garlinghouse Company

Figure 5.8-1. *(Continued)* In stock plans, elevations are usually marked "front," "rear," "left side," and "right side" because the architect does not know which direction the house will face. Elevations may carry notes indicating special features, code requirements, or reminders.

By studying the elevation views, a builder can determine the following information:

- Floor levels
- Grade lines
- Window and door heights
- Roof slopes
- Exterior features, including types of materials used on wall and roof surfaces

The entire set of construction drawings work together to give a total picture of how the building is to be constructed. Elevation drawings are a key part of this relationship. They are representative of the outside of the building, and they also have references that are crucial to understanding the entire building.

Cut lines and reference marks depicting where section and detail drawings are taken from are often shown on elevation drawings, **Figure 5.8-2**, and floor plans. A ***cut line*** indicates the orientation and location that a section drawing references. A ***reference mark*** is the symbol used in conjunction with a cut line to note where the section drawing can be found within the set of working drawings. Site-specific elevation drawings will give elevation measurements for critical points such as the footing depth, elevation at the top of the foundation wall, and finished floor elevations.

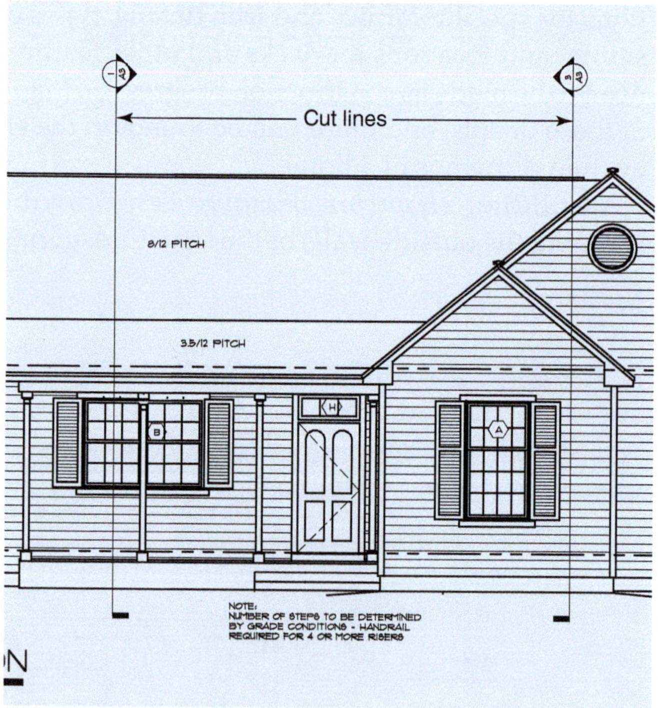

The Garlinghouse Company

Figure 5.8-2. The cut lines on the front elevation refer to section drawings taken from these locations.

References to the window and door schedule may also be found on elevation drawings as well as floor plans and foundation plans. A ***window and door schedule*** is a reference table that includes, among other things, the manufacturer's numbers and rough opening sizes for each unit or combination, **Figure 5.8-3**. An identifying letter is located at each opening on the plan. The same letter is used in the schedule to identify the window or door.

Symbols and Abbreviations

Chapter 1, Preparing for Construction, introduced how symbols and abbreviations represent features and materials to be used in construction. This lesson explains the symbols used in construction drawings that show how materials are represented in different views and details.

As previously stated, since architectural plans are drawn to a small scale, materials and construction particulars are seldom shown as they actually appear. Also, it would require too much time to produce drawings of this nature. The architect, therefore, uses symbols to represent materials and other items and certain approved shortcuts, called *conventional representations*. These simplify the illustration of assemblies and other elements of the structure.

In **Figure 5.8-4**, notice that the same material may be represented differently in plan, elevation, or section view drawings. **Figure 5.8-5** demonstrates the use of architectural symbols as they appear in plan view drawings. Symbols used for plumbing

DOOR SCHEDULE

IND	SIZE	TYPE	ROUGH OPENING	REMARKS
D1	(2) 1-6x6-8	PASSAGE	3'-2" x 6'-1Ø 7/8"	-
D2	2-Øx6-8	PASSAGE	2'-2" x 6'-1Ø 7/8"	-
D3	(2) 2-Øx6-8	PASSAGE	4'-2" x 6'-1Ø 7/8"	-
D4	2-4x6-8	PASSAGE	2'-6" x 6'-1Ø 7/8"	-
D5	2-6x6-8	PASSAGE	5'-2" x 6'-1Ø 7/8"	-
D6	2-8x6-8	PASSAGE	2'-1Ø" x 6'-1Ø 7/8"	-
D7	2-8x6-8	POCKET	2'-1Ø" x 6'-1Ø 7/8"	-
D8	5-Øx6-8	BI-PASS	5'-2" x 6'-1Ø 7/8"	-
ED1	2-8x6-8	EXTERIOR FIRE DOOR	2'-1Ø" x 6'-1Ø 7/8"	SELF CLOSING HINGES
ED2	3-Øx6-8	EXTERIOR SERVICE DOOR	3'-2" x 6'-1Ø 7/8"	SELF CLOSING HINGES
ED3	3-Øx6-8	EXTERIOR	3'-2" x 6'-1Ø 7/8"	W/(2) 15" SL & ARCH TRANSFORM
ED4	(2) 2-8x8-Ø	FRENCH	5'-4" x 6'-1Ø 7/8"	TEMPERED GLASS
ED5	9-Øx8-O	GARAGE-OVHGD	9'-Ø" x 8'-Ø"	OVERHEAD GARAGE DOOR
ED6	16-Øx8-O	GARAGE-OVHGD	16'-Ø" x 8'-Ø"	OVERHEAD GARAGE DOOR
ED7	6-Øx6-8	SLIDING GLASS DOOR	6'-2" x 6'-1Ø 7/8"	-

WINDOW SCHEDULE

IND	MANUFACTURER	SIZE	TYPE	ROUGH OPENING	REMARKS
W1	ANDERSEN	TW2Ø32	TILT WASH DOUBLE HUNG	2'-2 1/8" x 3'-5 1/4"	-
W2	ANDERSEN	TW2846	TILT WASH DOUBLE HUNG	2'-1Ø 1/8" x 4'-9 1/4"	-
W3	ANDERSEN	TW2846-3	TILT WASH DOUBLE HUNG	8'-5 3/4" x 4'-9 1/4"	-
W4	ANDERSEN	TW2852	TILT WASH DOUBLE HUNG	2'-1Ø 1/8" x 5'-5 1/4"	-
W5	ANDERSEN	TW2852-2	TILT WASH DOUBLE HUNG	5'-7" 15/16 x 5'-5 1/4"	-
W6	ANDERSEN	TW3Ø52	TILT WASH DOUBLE HUNG	3'-2 1/8" x 5'-5 1/4"	-
W7	ANDERSEN	DHP41Ø42	PICTURE	4'-11 7/8" x 4'-5 1/4"	-
W8	ANDERSEN	DHT2810	TRANSFORM	2'-1Ø 1/8" x 1'-1Ø 1/2"	-
W9	ANDERSEN	3-Ø x 1-8	CUSTOM ARCH TRANSFORM	-	-
W10	ANDERSEN	5-1Ø x 1-8	CUSTOM ARCH TRANSFORM	-	TEMPERED

NOTE: ALL WINDOWS ARE SERIES 400

Goodheart-Willcox Publisher

Figure 5.8-3. Window and door schedules give detailed information about the size and type of window or door to be used. The location is denoted on the print by an alphanumeric code, which can be cross-referenced to the schedule.

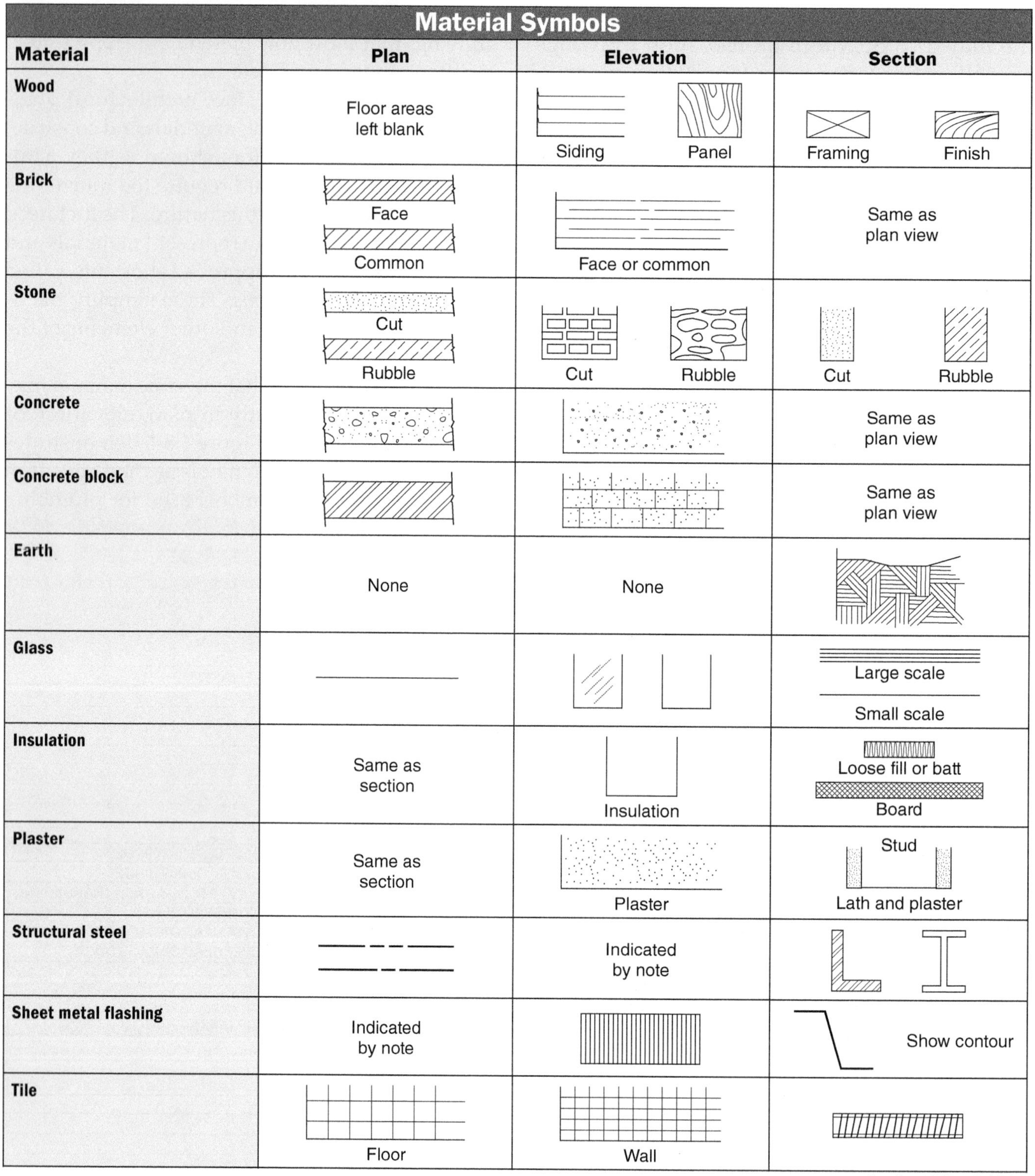

Material Symbols			
Material	**Plan**	**Elevation**	**Section**
Wood	Floor areas left blank	Siding; Panel	Framing; Finish
Brick	Face; Common	Face or common	Same as plan view
Stone	Cut; Rubble	Cut; Rubble	Cut; Rubble
Concrete			Same as plan view
Concrete block			Same as plan view
Earth	None	None	
Glass			Large scale; Small scale
Insulation	Same as section	Insulation	Loose fill or batt; Board
Plaster	Same as section	Plaster	Stud; Lath and plaster
Structural steel		Indicated by note	
Sheet metal flashing	Indicated by note		Show contour
Tile	Floor	Wall	

Goodheart-Willcox Publisher

Figure 5.8-4. Symbols are used to represent things that are impractical to draw.

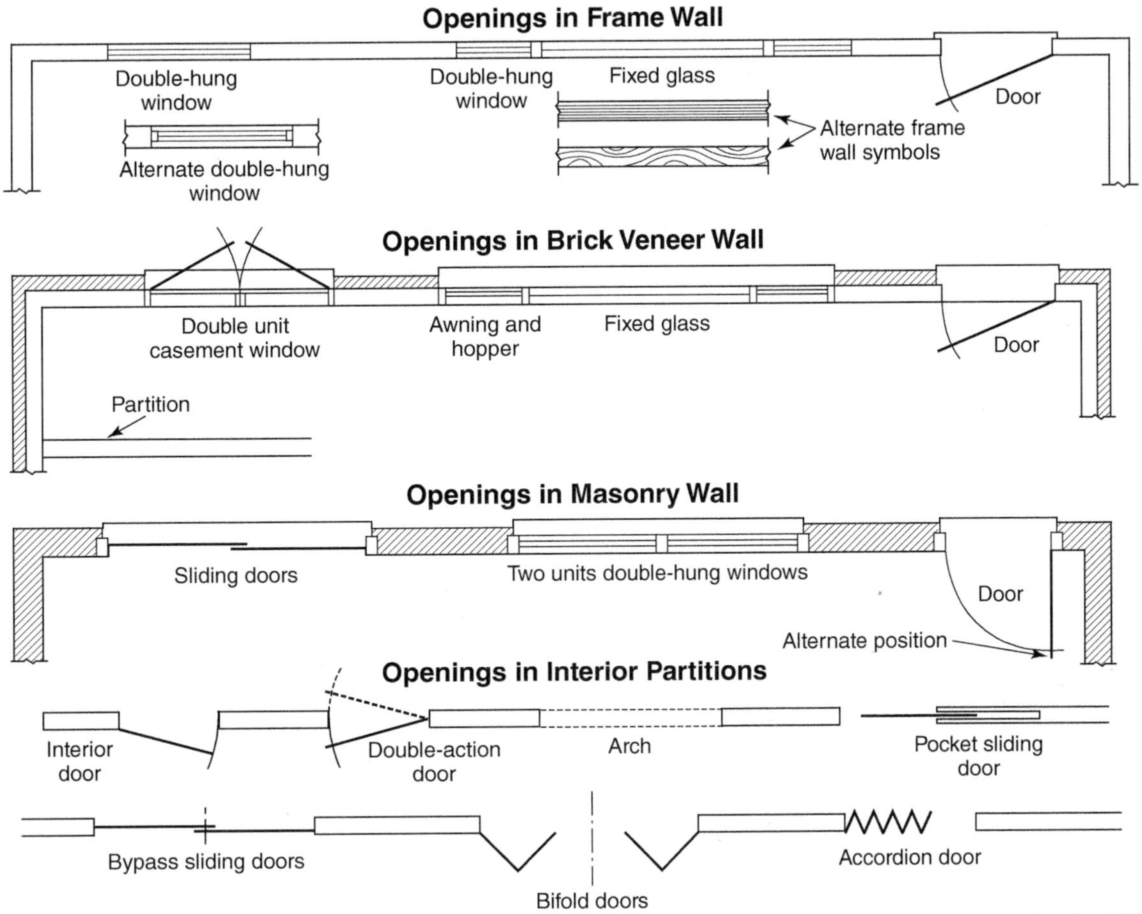

Goodheart-Willcox Publisher

Figure 5.8-5. How windows and doors are presented in plan views. The swing of doors (the direction they swing) is always shown.

and HVAC fixtures and systems are shown in **Figure 5.8-6**. Electrical symbols are used to communicate switch types and other fixtures, **Figure 5.8-7**. Although the symbols shown in the figures are generally accepted as standard, some drafters use variations of these symbols. Most drawing sets include a key for all symbols used.

Abbreviations help the architect to keep the drawings neat and organized. Typically, a list of abbreviations used on the set of construction drawings is found on the title page, though it is helpful to be familiar with commonly used abbreviations. These abbreviations are listed in **Figure 5.8-8**.

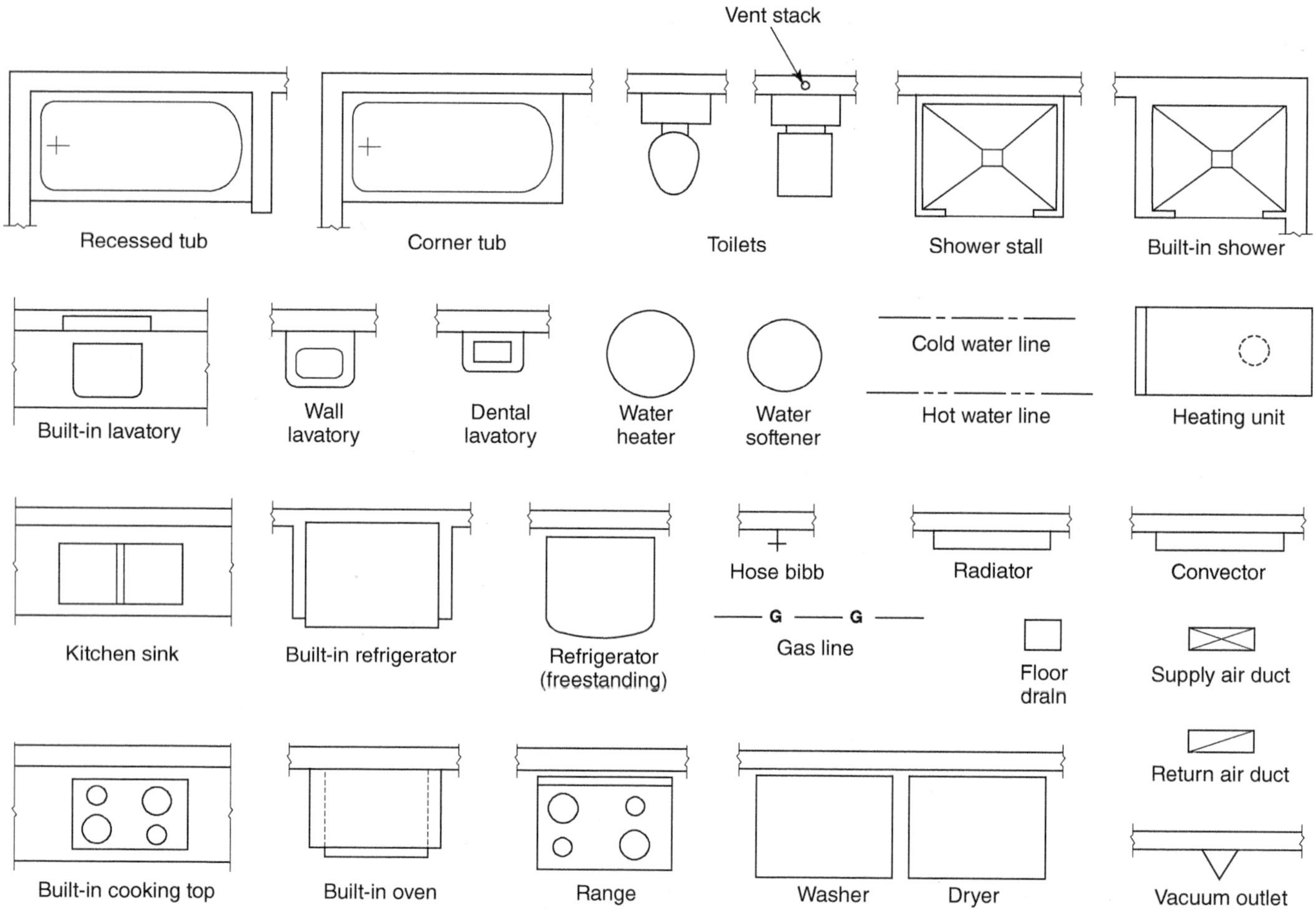

Goodheart-Willcox Publisher

Figure 5.8-6. These symbols are used to indicate plumbing fixtures, appliances, and mechanical equipment.

Electrical Symbols

	Ceiling outlets for fixtures		Lighting panel
			Power panel
	Wall fixture outlet	S	Single-pole switch
PS	Ceiling outlet with pull switch	S_2	Double-pole switch
PS	Wall outlet with pull switch	S_3	Three-way switch
	Duplex convenience outlet	S_4	Four-way switch
WP	Waterproof convenience outlet	S_P	Switch with pilot light
1, 3,	Convenience outlet 1 = single 3 = triple		Push button
R	Range outlet		Bell

Goodheart-Willcox Publisher

Figure 5.8-7. A sample of common electrical symbols.

Common Abbreviations					
Abbreviation	**Meaning**	**Abbreviation**	**Meaning**	**Abbreviation**	**Meaning**
ALT	Alternate	DR	Door	NOM	Nominal
BSMT	Basement	EL, ELEV	Elevation	REF	Reference
BM	Beam	EXT	Exterior	REV	Revision
BLDG	Building	GR	Grade	STD	Standard
CKT	Circuit	HDR	Header	THK	Thick
COL	Column	LAV	Lavatory	WC	Water closet
DIA, ∅	Diameter	LVL	Level	WWF	Welded wire fabric

Goodheart-Willcox Publisher

Figure 5.8-8. Many abbreviations are used on construction drawings.

LESSON 5.8 Review

Answer the following questions based on information in this lesson.

1. By studying the elevation drawings, the builder can determine the _____. (5.8-1)
   A. height of the kitchen ceiling
   B. number of bathrooms
   C. size of the bedrooms
   D. slope of the roof
2. What does the electrical symbol S_3 indicate? (5.8-2)
   A. Three outlets in a room
   B. Three-way switch
   C. Three switches that operate one light
   D. Three lights operated by one switch
3. *True or False?* Cut lines indicate areas where the frame will need to be cut to allow the electrician to install wires. (5.8-2)
4. What does the abbreviation "COL" stand for? (5.8-2)
   A. Column
   B. Closet
   C. Circuit or load
   D. Crane or lift

LESSON 5.9

Green Building and the Future of Construction

LEARNING OUTCOMES

After completing this lesson, you will be able to:

5.9-1 Explain three methods used in green building to improve energy efficiency.

5.9-2 Describe the environmental impact of various materials throughout their life cycle.

TECHNICAL TERMS

low-impact material
renewable material
thermal bridge
volatile organic compound (VOC)

Material Selection and Sourcing

Material selection and sourcing plays a major role in the green building process. The goal of green building is to generate the lightest possible impact on natural resources. This is true, not only while the building is in operation, but also as it is built, and even when it reaches the end of its usability as a home. The materials a builder uses to construct a home will make an impact on the natural resources used before, during, and after a home is lived in. Therefore, which materials are used, how they are made, where they come from, how they perform when installed, how long they will last, and how they are disposed of are all factors to consider when building green.

Energy Efficiency

A home consumes energy month after month and year after year, so the impacts of energy use build up over time. For this reason, green home programs place great emphasis on energy efficiency. Choosing materials that perform efficiently is perhaps the most obvious form of green building. Many methods can be used in green building to improve energy efficiency. Some of the most effective are listed here:

- **Improve the air barrier**—A home that does not leak air is far more energy-efficient than one that does.
- **Reduce thermal bridging**—A *thermal bridge* is an area in the building envelope where there is a higher level of heat transfer compared to surrounding materials. Thermal bridges can occur where insulation starts and stops, such as at a stud or wall plate, **Figure 5.9-1**. Thermal bridging can be reduced through better insulating practices, alternative product choices, such as structural insulated panels (SIPs) or insulating concrete forms (ICFs), and the use of exterior board insulation.
- **Ceilings, walls, and floor assemblies with high R-values**—The use of more or higher-quality insulation is a way to reduce thermal losses in a building.
- **Energy-efficient windows and doors**—Triple-pane glass, **Figure 5.9-2**, low E-coatings, and argon-filled thermal panes are a few examples of products that improve window and door performance.

Arturs Budkevics/shutterstock.com

Figure 5.9-1. Each rafter in this room creates a break in the insulation. These thermal bridges reduce the performance of the insulation.

Gulpa/shutterstock.com

Figure 5.9-3. Smart thermostats can monitor usage and make adjustments based on efficiency.

Ambiento/shutterstock.com

Figure 5.9-2. Triple-pane glass is available as an option for high-efficiency window assemblies.

- **Reflective roofing**—Some roofing products are designed to reflect thermal radiation from the sun better than others.
- **High-efficiency mechanical systems and appliances**—As technology improves, heating and air conditioning systems, as well as appliances, are becoming more energy-efficient.
- **Smart thermostats and home controls**—Integrating smart user interfaces into homes is becoming more affordable and mainstream in homebuilding. Smart home components can measure usage and make adjustments that can reduce usage, **Figure 5.9-3**.

Sustainable Material Sourcing

Sustainable material sourcing involves considering the origin, harvesting (where applicable), production methods, and resources used when selecting materials.

Renewable Materials

Responsible material sourcing involves selecting materials that are renewable whenever possible. ***Renewable material*** is made from raw materials that are managed and harvested responsibly and can be replenished in a relatively short timeframe. Lumber, for instance, is considered renewable when responsibly managed and harvested, **Figure 5.9-4**. Bamboo, cork, and natural fibers like cotton and wool are also examples of renewable products.

Reclaimed or Recycled Materials

Materials such as reclaimed wood or reused bricks reduce the energy and resources required to manufacture and process new materials, **Figure 5.9-5**. It also diverts construction waste from landfills. According to the US Environmental Protection Agency (EPA), construction and demolition (C&D) debris accounted for about 600 million tons

Aastels/shutterstock.com

Figure 5.9-4. Responsibly managed timber is a renewable resource.

Valdas Jarutis/shutterstock.com

Figure 5.9-5. Reclaimed wood flooring can offer unique character to a new home.

of waste annually, representing 23% of the total waste generated.

Examples of recycled materials that can be used include recycled steel, composite decking, products containing recycled glass, carpets, roofing, and insulation.

Green Note

In selecting building materials for remodeling, choose those made from recycled material wherever possible. By doing so, you save natural resources and help to create markets for recycled products. This is a key requirement to every recycling program.

Low-Impact Materials

Low-impact material is material that has a low environmental impact throughout its entire life cycle, **Figure 5.9-6**. Low-impact materials are produced near the jobsite and reduce fuel consumption and emissions compared to a product that must be shipped across the country or globe. How a product is harvested, mined, or extracted plays a role in its environmental impact.

Durability is a big factor in a material's life cycle impact. If a product must be replaced prematurely, both its life cycle and its replacement product's life cycle factor into the home's resource consumption.

It is also important to consider the disposal and recyclability of products to measure their overall effect on the environment.

Irina Desimple/shutterstock.com

Figure 5.9-6. Cork is considered a low-impact material. This cork flooring is both functional and attractive.

Green Note

Associations and organizations that offer certifications and ratings for product sustainability are becoming more common. It is not unusual to see labels with various green certifications. Be sure to research the validity of the certification for the intended purpose.

The US Environmental Protection Agency has produced a website that allows users to verify federally recognized environmentally preferrable product labels.

Nontoxic Materials

Many construction materials such as paint, carpet, and glues emit volatile organic compounds. A ***volatile organic compound (VOC)*** is a carbon-containing chemical that readily evaporates into the air. When VOCs enter the air, they react with other elements to produce ozone, which causes air pollution and a host of health issues including breathing problems, headache, nausea, and burning, watery eyes. Some VOCs have been linked to not only cancer but also to kidney and liver damage.

Materials that emit the lowest possible VOC off-gassing must be selected. Green building materials that have very low or no VOCs help keep the pollutant load low in the house, **Figure 5.9-7**. It is important to achieve a healthy and comfortable environment inside the building.

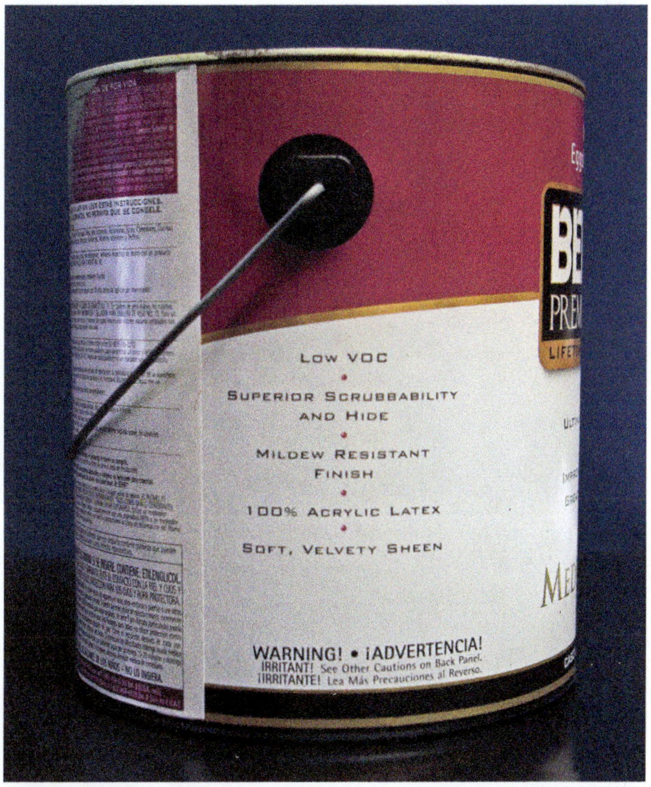

Goodheart-Willcox Publisher

Figure 5.9-7. In green construction, products with low or zero VOCs are used to improve indoor air quality. Water-based paint with less than 250 grams of VOCs per liter of paint is considered low-VOC paint.

Waste Reduction

Some products and materials generate less jobsite waste than others. SIPs, panelized wall systems, and precast concrete foundation walls, for instance, are examples of products that generate little to no waste on-site compared to their counterparts, **Figure 5.9-8**.

Generating less waste on-site has several advantages:

- Often, scrap material that would be thrown away on the job is able to be reduced or reused in a factory setting.
- Transportation emissions to truck excess waste away from the job are reduced.
- Less waste enters landfills.
- Emissions to process the waste at the landfill are eased.

Green Note

ICF foundations and even ICF above-grade walls are popular for green homes. Compared with conventional walls, ICF walls provide greater resistance to heat transfer. Thus, the use of ICFs reduces the amount of energy required for heating and cooling throughout the life of the structure. In addition, ICF construction produces less waste than that produced by conventional form construction.

Water Conservation

Fresh water is a limited resource in many parts of the world. Excessive water use can put stress on local water supplies, especially in places that are susceptible to drought. Green builders and architects should choose fixtures and appliances that are designed to reduce water consumption while meeting the needs of homeowners. These may include low-flow toilets, faucets, and showerheads, **Figure 5.9-9**.

A Enercept, Inc.

B BuildSMART, LLC

Figure 5.9-8. A—This building is being constructed with SIPs for the floors and walls. B— Workers install a factory-assembled panelized wall.

FotoDuets/shutterstock.com

Figure 5.9-9. Dual-flush toilets are one example of water-saving fixtures.

Other strategies that are sometimes employed to reduce water consumption are:

- Rainwater harvesting to use for flushing toilets and watering plants.
- Graywater systems, which recycle shower, sink, and laundry water to irrigate lawns and gardens.
- Thoughtful landscape design, which includes the use of native and drought-resistant plants, **Figure 5.9-10**.

When builders focus on energy efficiency and sustainable material sourcing for home construction, they can help protect the environment,

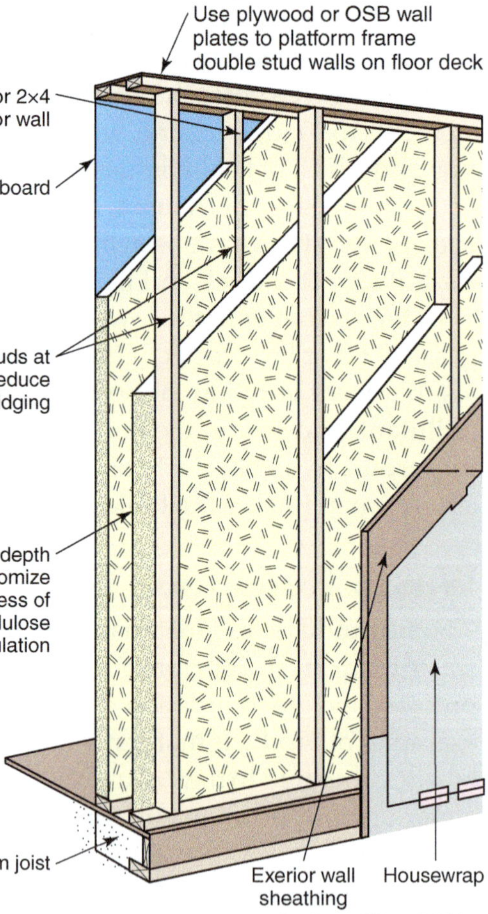

A
Goodheart-Willcox Publisher

Figure 5.9-10. This illustration shows one of a variety of methods that can be used for their construction. A—Double stud walls allow for additional insulation in exterior walls. *(Continued)*

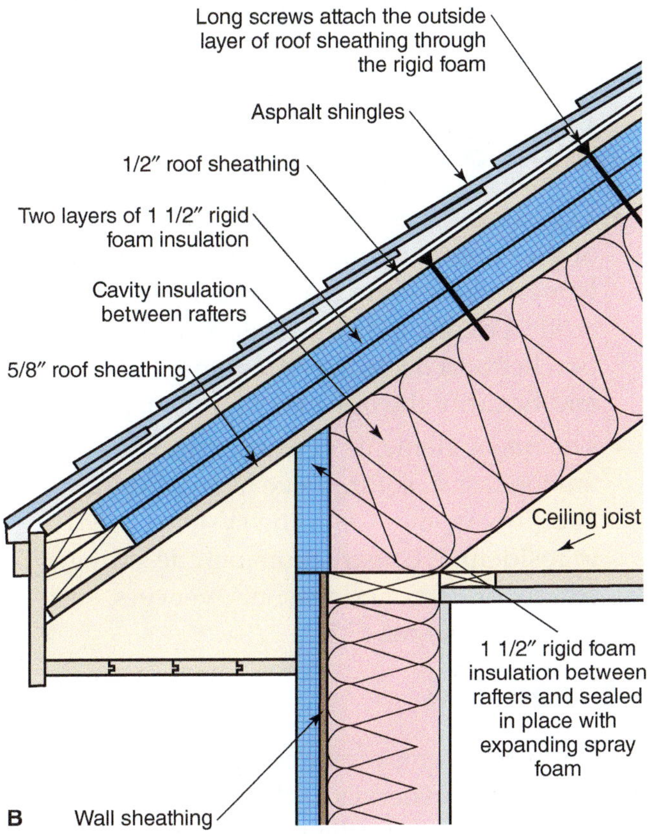

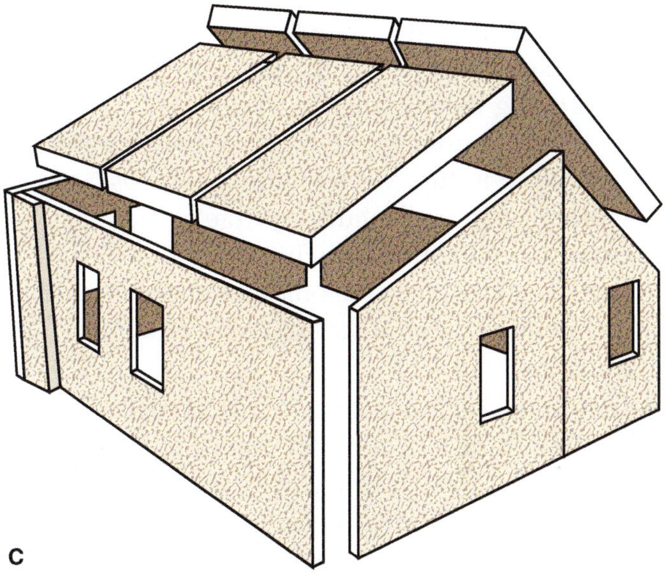

Goodheart-Willcox Publisher

Figure 5.9-10. *(Continued)* B—An insulated roof deck can improve efficiency and reduce condensation issues if built properly. C—SIPs are available in different thicknesses and can be used for walls, ceilings and floors.

conserve resources, and make homes healthier for people to live in. These principles can be implemented on any construction project, not just green building projects, to improve the building process and build in a more sustainable and responsible manner. As we look toward the future of home construction, we will continue to see many of these practices become requirements through building codes and other regulations.

LESSON 5.9 Review

Answer the following questions based on information in this lesson.

1. *True or False?* The lower the R-value of the insulation used in a house, the longer the building will retain heat. (5.9-1)
2. One way to make windows more energy-efficient is to use _____. (5.9-1)
   A. single-pane glass
   B. a high E-coating
   C. argon-filled thermal panes
   D. copper frames
3. *True or False?* Bamboo is an example of a nonrenewable material. (5.9-2)
4. *True or False?* Building materials that have low VOC off-gassing help keep the pollutant load low in the house. (5.9-2)
5. One recommended strategy to help reduce water consumption is to _____. (5.9-2)
   A. include large areas of grass in the landscaping
   B. install high-volume faucets and showerheads
   C. use graywater systems for drinking
   D. harvest rainwater for watering plants

Chapter 5 Review

Summary

Lesson 5.1

- Professional development enables you to pursue the career path you want by improving the knowledge, skills, and abilities required for your career to progress. Following industry developments helps workers stay on the cutting edge of the trends and technologies that drive the industry. The forward-looking builder should be intentional about seeking opportunities for growth. (5.1-1)
- Individuals can develop their skill set by reading trade journals and other publications in their field; attending workshops, seminars, webinars, and conferences; and enrolling in higher education courses or programs that complement or improve expertise. (5.1-2)
- It is important to be thoughtful when considering your long-term career goals and creating plans to achieve them. Set milestones that you need to meet along the way and then regularly reflect on your progress. (5.1-3)

Lesson 5.2

- Builders must be aware of zoning laws, environmental regulations, and licensing requirements. Furthermore, employers in the construction industry must understand consumer protection laws, insurance requirements, labor laws, and tax laws to satisfy legal requirements to operate a business and employ workers. (5.2-1)

Lesson 5.3

- Exterior finishes include all exterior materials of a structure. Roofing materials, cornice trim, wall cladding (such as siding), trim around doors and windows, as well as porches and decks work together to protect and beautify the outside of a home. (5.3-1)
- The major elements of cornice systems are the fascia, soffit, frieze board, gutter, and downspout. Cladding systems used in residential construction include siding, stucco, or EIFS, and masonry veneers, such as brick or stone. (5.3-2)

Lesson 5.4

- Cornice systems can be comprised of a variety of versatile materials, including plywood, fiber and cement siding, coil stock, and PVC trim. Gutters can be aluminum, galvanized steel, or plastic. Cladding systems can consist of wood, composite wood, fiber cement, stucco, veneer, stone, and more. (5.4-1)
- Specialty fasteners are used for installing many exterior finishes. Examples include color-matched trim nails, siding nails, and various deck fastening systems. (5.4-2)

Lesson 5.5

- Tools used to apply exterior finishes include sheet metal bending tools and the snap lock punch and siding removal tool for installing vinyl siding. When building a deck, it is common to use a router for shaping edges of boards. (5.5-1)

Lesson 5.6

- Various forms of scaffolding and aerial lifts are frequently used when applying exterior finishes. It is vital to receive proper training before using this type of equipment. (5.6-1)

Lesson 5.7

- When building a home, achieving plumb, level, and square construction is critical for ensuring structural integrity, functionality, and a quality appearance. (5.7-1)
- The Pythagorean theorem can be used to verify that a corner is 90° by measuring along the two sides and then checking the length of the hypotenuse between them. (5.7-2)

Lesson 5.8

- By studying the elevation drawings, a builder can determine floor levels, grade lines, window and door heights, roof slopes, and exterior features, including types of materials used on wall and roof surfaces. (5.8-1)
- Most drawing sets include a key to all the symbols used. A list of abbreviations used on the set of construction drawings is found on the title page. (5.8-2)

Lesson 5.9

- Material selection and sourcing play a major role in the green building process. Which materials are used, how they are made, where they come from, how they perform when installed, how long they will last, and how they are disposed of are all factors to consider when building green. (5.9-1)
- Renewable material that is managed and harvested responsibly can be replenished in a relatively short time. Materials such as reclaimed wood or reused bricks reduce the energy and resources required to manufacture and process new materials, as well as diverts construction waste from landfills. (5.9-2)

Apply and Analyze

1. What are three strategies you can use to keep current with industry developments? (5.1-2)
2. Can an employer require an untrained employee to remove old asbestos insulation from pipes in a building? Why or why not? (5.2-1)
3. The architect is recommending that cornices be installed on a new home, but the client is concerned that it might be a waste of money. Is the client correct? Why or why not? (5.3-2)
4. Explain how a porch differs from a deck. (5.3-2)
5. Are bacteria and other organisms likely to feed on thermally modified wood? Why or why not? (5.4-1)
6. The foreman reminds the worker installing a gutter to give it a slight slope. In which direction should the gutter slope, and why? (5.4-1)
7. How can using a portable field brake help save money on the aluminum trim applied to a new home? (5.5-1)
8. What is the employer's responsibility regarding employees who work on a scaffold? (5.6-1)
9. Explain how to use elevation drawings to identify which window or door should be installed in an opening in the building. (5.8-2)
10. Describe what sustainable material sourcing involves. (5.9-2)

Critical Thinking

1. Reflect on your own career aspirations and goals. Create a visual representation of a career path in the construction industry. What challenges do you foresee in reaching your career goals? (5.1-3)
2. Why might some contractors choose to use a boom lift on a specific job rather than a ladder or scaffolding? (5.6-1)
3. Considering the importance of plumb construction for both aesthetic reasons and structural stability, how might builders strike a balance between achieving perfect alignment and acceptable tolerances for construction projects? What factors should be taken into consideration when determining acceptable tolerances for plumbness in walls and columns? (5.7-1)

Communicating about Construction

1. **Reading and Writing.** Create an informational pamphlet on the different opportunities and resources available to help you develop your skill set throughout your career in the construction industry. Include images in your pamphlet. Present your pamphlet to the class. (5.1-2)
2. **Speaking and Writing.** Compare and contrast the available types of exterior wall finishes. Include the pros and cons of each type of material. Prepare an oral or written presentation. (5.4-1)
3. **Reading and Speaking.** As a class, create a list of sustainable materials used in green construction. Divide into groups of two to four students and choose one or two of the materials to research. Gather relevant data and prepare a debate with one or two group members on the pro side and one or two on the con side of the chosen material. Include topics such as cost, sustainability, accessibility, and dependability as part of your reasoning. Conduct a debate for the class. Be prepared to answer questions from your peers. (5.9-2)

ungvar/Shutterstock.com

CHAPTER 6

MEP Systems Rough-In

Chapter 6 Outline

Introduction

As you learned in *Chapter 5, Exterior Finishes,* while the exterior finishes of a home are being applied, work continues inside the building as well. This chapter features the installation of mechanical rough-in components of a home. In this chapter, we will explore what makes up the mechanical systems of a home, as well as some of the tools, fasteners, and materials used during this phase of construction. We will also describe some of the print reading skills that are helpful to mechanical contractors. An important consideration for these contractors is to be mindful of electrical hazards when working with or around electrical systems and components. You will learn about the hazard of electrocution on the job and learn methods of safeguarding against electrical shock. Additionally, you will begin to use some of the math skills you have developed for estimating material quantities on the job. In the green building section of this chapter, we will consider some of the technologies that are being used to harness natural resources and improve a home's efficiency. First, we begin by looking at trade organizations that support the residential construction industry, along with some credentials that are available to you as you develop your skill set.

LESSON 6.1

Industry and Careers

LEARNING OUTCOMES

After completing this lesson, you will be able to:

6.1-1 Explain the role of trade organizations in the residential construction industry.

TECHNICAL TERMS

standards organization

trade organization

trade union

Trade Organizations

In the residential construction industry, organizations and associations play an important role. Some of these organizations and associations, such as the Occupational Safety and Health Administration (OSHA), have already been mentioned. This chapter explores some of the more influential organizations that play an important role in the industry.

Trade organizations provide several benefits to the residential construction industry. A ***trade organization***, or trade association, is a member-based collective of professionals and businesses who promote common goals and address challenges faced by their industry. Trade organizations exist at local, regional, national, and even international levels. They provide a platform for collaboration, advocacy, education, and networking for their members. Doing so allows them to strengthen the collective influence of their members and promote the development, growth, and sustainability of the industry. The following are some of the most influential trade organizations in the residential construction industry.

Pro Tip

Many contractors benefit from professional networking through their local trade organizations. These organizations act as a place for general contractors and subcontractors to connect, share training and ideas, and collaborate on job opportunities.

National Association of Home Builders

The National Association of Home Builders (NAHB), **Figure 6.1-1**, is the largest trade association in the residential construction industry. It represents builders, remodelers, developers, and associated professionals. The NAHB operates at local, state, and national levels, providing educational resources, networking opportunities, expertise, and advocacy.

The NAHB was founded in the early 1940s and represents more than 140,000 members. Each year, the organization sponsors the International Builders' Show, the largest event of its kind, offering educational, networking, and product showcase opportunities for all segments of the industry.

The NAHB holds notable influence in the residential construction industry. The organization is

JHVEPhoto/shutterstock.com

Figure 6.1-1. The NAHB is the leading trade organization in the residential construction industry.

not affiliated with a specific political party but does advocate for the industry through local, state, and national governments, addressing legal and policy issues. They are also affiliated with several other organizations with complementary goals:

- Home Innovation Research Labs specializes in enhancing the technology and cost-effectiveness of housing in the United States through development, testing, and evaluation of new materials, methods, standards, and equipment.
- The Home Builders Institute (HBI) is the nation's leading nonprofit building industry educator dedicated to workforce development. HBI is responsible for creating and managing a wide range of educational and vocational training initiatives.
- The National Housing Endowment is a philanthropic organization that assists the housing industry's development of more effective approaches to home construction. Its primary focus is on advancing education and training for upcoming leaders in residential construction and expanding the collective understanding of housing-related matters.

The NAHB has also founded the International Housing Association, which expands its vision for overcoming challenges in the residential construction industry to a global scale. In this way, it partners with other like-minded organizations and builds membership internationally.

Specialty Trade Organizations

Although the NAHB is the largest and most comprehensive trade organization, other organizations and associations focus on a segment or specialty of the industry. To learn more about any of these organizations, visit their website:

- National Association of the Remodeling Industry (NARI)
- National Kitchen and Bath Association (NKBA)
- Independent Electrical Contractors (IEC)
- National Electrical Contractors Association (NECA)
- Plumbing-Heating-Cooling Contractors Association (PHCC)
- Mechanical Contractors Association of America (MCAA)
- Sheet Metal and Air Conditioning Contractors National Association (SMACNA)
- Air Conditioning Contractors of America (ACCA)
- American Society of Heating, Refrigerating, and Air Conditioning Engineers (ASHRAE)
- American Institute of Architects (AIA)

Standards Organizations

In construction, it is important that the products used and the installation of those products meet certain standards. In the simplest terms, ***standards organizations*** exist to create uniformity across products and procedures by establishing criteria for industry participants to abide by. Some of the most influential standards organizations are listed here:

- **International Code Council (ICC)**—In *Chapter 3, Framing Systems*, you learned how building codes help ensure that construction meets specific minimum standards. The ICC is the largest source of model codes and standards for the construction industry. Aside from developing and publishing model codes, their work includes product evaluation, accreditation, technology, training, and certification.
- **Lumber associations**—American Lumber Standard Committee Incorporated (ALSC), Engineered Wood Association (APA), Western Wood Products Association, Southern Forest Products, American Institute of Timber Construction (AITC) Association, California Redwood Association, National Hardwood Lumber Association, and others produce standards for sawn and engineered lumber.
- **Portland Cement Association (PCA)**—Supports sustainability, innovation, and safety while working to improve cement manufacturing, distribution, infrastructure, and economic growth.
- **American National Standards Institute (ANSI)**—A nonprofit organization that administers and coordinates voluntary standards. They work in collaboration with participants from various industries and

the government to identify and develop standards.

- **ASTM International**—A product standards organization that has more than 12,000 standards operating globally.
- **National Fire Protection Association (NFPA)**—Produces codes, standards, information, and knowledge resources on fire, electrical, and related hazards.

Some specialty trade organizations, such as NAHB and ASHRAE, also create standards and influence codes for the building construction industry. One example is the National Green Building Standard.

Trade Unions

A ***trade union*** is an organization that represents and advocates for the interests of workers in a particular trade. It acts as a collective voice for its members and works to protect their rights, negotiate better working conditions, and advocate for fair wages, benefits, and workplace policies. Even though trade unions are more significantly represented in commercial construction, they do operate in residential construction as well. Trade unions often help to balance workforce needs as workers are often connected with employers (and vice versa) through a union. Some of the trade unions that may operate in residential construction include:

- United Brotherhood of Carpenters and Joiners of America (UBC)
- International Brotherhood of Electrical Workers (IBEW)
- Laborers' International Union of North America (LIUNA)
- International Union of Painters and Allied Trades (IUPAT)
- United Association of Journeymen and Apprentices of the Plumbing and Pipe Fitting Industry (UA)
- International Association of Sheet Metal, Air, Rail and Transportation Workers (SMART)

LESSON 6.1 Review

Answer the following questions based on information in this lesson.

1. *True or False?* The National Association of Home Builders advocates for the residential construction industry through various levels of government. (6.1-1)
2. Which organization specializes in enhancing the technology and cost-effectiveness of housing in the United States by development, testing, and evaluation of new materials, methods, standards, and equipment? (6.1-1)
   A. National Housing Endowment
   B. Home Innovation Research Labs
   C. Independent Electrical Contractors, Inc.
   D. Home Builders Institute
3. Which organization is the largest source of model codes and standards for the construction industry? (6.1-1)
   A. National Fire Protection Association
   B. American National Standards Institute
   C. Portland Cement Association
   D. International Code Council
4. The ____ is a nonprofit organization that administers and coordinates voluntary standards. (6.1-1)
   A. Portland Cement Association
   B. Engineered Wood Association
   C. American National Standards Institute
   D. International Code Council
5. *True or False?* The primary focus of ASTM International is advancing education and training for upcoming leaders in residential construction. (6.1-1)
6. *True or False?* The United Brotherhood of Carpenters and Joiners of America is a trade union that operates in residential construction. (6.1-1)

LESSON 6.2

Construction Fundamentals

LEARNING OUTCOMES

After completing this lesson, you will be able to:

6.2-1 Discuss the value of credentials and certifications in the building trades.

TECHNICAL TERMS

credential

trade certification

Trade Credentials and Certifications

Imagine you are a general contractor who is considering hiring an electrician as a new employee or subcontractor. How can you tell if the person you are considering is proficient? You might take their experience claims at face value, though this could be risky to judge the extent of their understanding and skills based on their own word. You might ask for professional references, but these can take valuable time and effort. These strategies are helpful, but if the applicant has applicable trade certifications or other credentials, much of the guesswork is taken away.

Credentials are a type of documentation that recognizes the completion of, qualification for, or experience in a specific area of expertise. Credentials can include both professional and academic qualifications and may take the form of a college degree, a certificate of completion, an award, a license, or an endorsement. They may be awarded by educational institutions, professional organizations, regulatory bodies, or government agencies.

A ***trade certification*** is a credential that is more specific to an area of expertise or an accrediting body. Trade certifications are professional accreditations awarded to individuals who have demonstrated a specific level of knowledge, skills, and competency in a particular trade or profession. These certifications are typically issued by professional associations, trade organizations, or certifying bodies. Trade certifications can validate your professionalism, expertise, and experience.

To become certified, a candidate is typically required to demonstrate technical skill and experience in the specified area of expertise. Some certifications require a minimum level of education that can take the form of online training, webinars, or in-person course work. Trade certifications demonstrate that an individual has met predefined standards and criteria set by the certifying body. Examples of trade certifications include Certified Electrician, Licensed Plumber, Vinyl Siding Institute Certified Installer, or Certified Safety Manager. For instance, NARI offers the certifications shown in **Figure 6.2-1**.

Aside from securing employment and demonstrating competence, trade credentials and certifications can help advance your career and keep you abreast of industry changes. Many specific tasks within your field of expertise may even require a credential or certification to perform a task. For instance, HVAC technicians who maintain, service, repair, or dispose of equipment that could release refrigerants into the atmosphere must be certified by the Environmental Protection Agency (EPA), as must remodelers who work on projects that have the potential to disturb lead-based paint. A safety manager is required to earn a credential issued by OSHA if they wish to provide workers with OSHA 10 or OSHA 30 cards, **Figure 6.2-2**, which are themselves credentials that are required to work on some jobs.

National Association of the Remodeling Industry Certifications	
Certification Type	**Description**
Master Certified Remodeler (MCR)	The highest level of certification NARI offers. Designed for professionals who have maintained their Certified Remodeler (CR) credential for a minimum of 10 years, hold another NARI certification, and have demonstrated leadership within NARI or their local community.
Certified Remodeler (CR)	Designed for highly experienced industry professionals who provide a full range of remodeling services. Often pursued by owners or upper management of a remodeling organization.
Certified Kitchen and Bath Remodeler (CKBR)	Designed for highly experienced industry professionals who provide remodeling services specific to kitchens and bathrooms. Often pursued by owners or upper management of a remodeling organization.
Certified Remodeling Project Manager (CRPM)	Designed for professionals overseeing the administrative aspects of a project: customer satisfaction, scheduling subcontractors, recordkeeping, risk management, financial controls, etc.
Certified Lead Carpenter (CLC)	Designed for experienced field professionals who oversee the technical aspects of the project, including personnel management and material flow.
Universal Design Certified Professional (UDCP)	Designed for professionals who create renovations to improve ease of use, safety of users and caregivers, and accessibility.

National Association of the Remodeling Industry

Figure 6.2-1. Certifications available through the National Association of the Remodeling Industry.

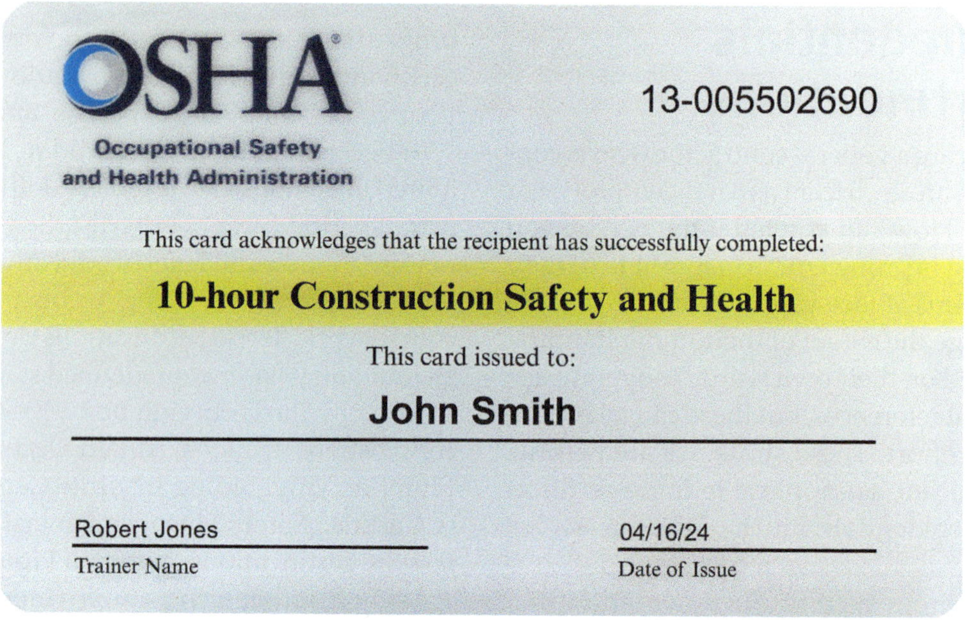

Goodheart-Willcox Publisher

Figure 6.2-2. OSHA issues credentials that acknowledge coursework.

LESSON 6.2 Review

Answer the following questions based on information in this lesson.

1. Which of these is an example of a credential? (6.2-1)
   A. College degree
   B. Pay stub
   C. Work permit
   D. Union membership
2. *True or False?* To be valid, a credential must be awarded by a government agency. (6.2-1)
3. Certified Electrician is an example of a _____. (6.2-1)
   A. trade certification
   B. trade association award
   C. union appointment
   D. professional reference

LESSON 6.3

The Construction Path

LEARNING OUTCOMES

After completing this lesson, you will be able to:

6.3-1 Describe the MEP systems installed during the rough-in phase of the construction of a new home.

6.3-2 Explain the differences in components and operation of a forced-air heating system and a hydronic heating system.

TECHNICAL TERMS

central heating
circuit
condenser
conduction
convection
direct heating system
drainage, waste, and venting (DWV) system
electric radiant heat
electric resistance baseboard heating
electrical current
evaporator
forced-air perimeter system
heat exchanger
hydronic perimeter system
hydronic radiant system
manifold
MEP system
National Electrical Code (NEC)
potable
radiation
register
service entrance
stub-in
vent piping
vent stack
water supply system

MEP Rough-In

Mechanical systems, also known as *MEP systems* (short for mechanical, electrical, and plumbing systems), are the components of a home that work together to provide comfort and convenience to a building's occupants.

Before work can begin on interior finishes, systems must be installed to manage heating, cooling, electricity, and plumbing, **Figure 6.3-1**, along with any wiring required for technology upgrades. The building industry refers to these as the mechanical systems of a building. Fixtures, finishes, and controls for these systems are not installed until after the wall, floor, and ceiling finishes have been applied. See *Chapter 9, Finish MEP Systems and Punchlist*.

Goodheart-Willcox Publisher

Figure 6.3-1. Before insulation materials are installed, MEP trades must perform their rough-in installations. In this structure, the electrical wiring, outlet boxes, and rough plumbing have been installed.

Electrical Wiring

The electrical wiring system consists of conductors (wires), boxes, and various devices that control the distribution of electricity throughout the building. The system provides the ***electrical current*** (flow of electrons) that powers lights, receptacles, heating and air-conditioning units, and appliances.

When doing any type of electrical work, it is necessary to be familiar with all local building code requirements. The ***National Electrical Code (NEC)*** is the model electrical code used by most municipalities. The *NEC* is a collection of rules and standards developed by experts to ensure the proper installation of electrical conductors and related devices.

Electricity generated at a power plant enters the home through conductors called the ***service entrance***. Service conductors may be strung from poles or buried in the ground. An electrical service entrance consists of the conductors that bring power from a transformer to the home. **Figure 6.3-2** is an illustration of all the components of the service entrance at the dwelling. Installation of these components is the responsibility of the electrician. The service extends only from the transformer, through a metering device, to a distribution or entrance panel. **Figure 6.3-3** shows a meter enclosure, where the service first enters the building. The entrance panel houses a main breaker and a circuit breaker for each of a home's circuits, **Figure 6.3-4**.

From the entrance panel, power is distributed to the home through individual circuits. **Figure 6.3-5** shows a sketch of a simple circuit. A ***circuit*** is a loop or continuous path from the source of electric power (the panel), through various loads or current-using devices (such as lights or electric motors), and then back to the source. The source in this case is the service panel containing the circuit breakers. Branch circuits include all the loads on one run of conductors. The circuit begins at the distribution panel, where it is connected to a circuit breaker. An electrician first installs the junction, switch, and receptacle boxes. Then, conductors

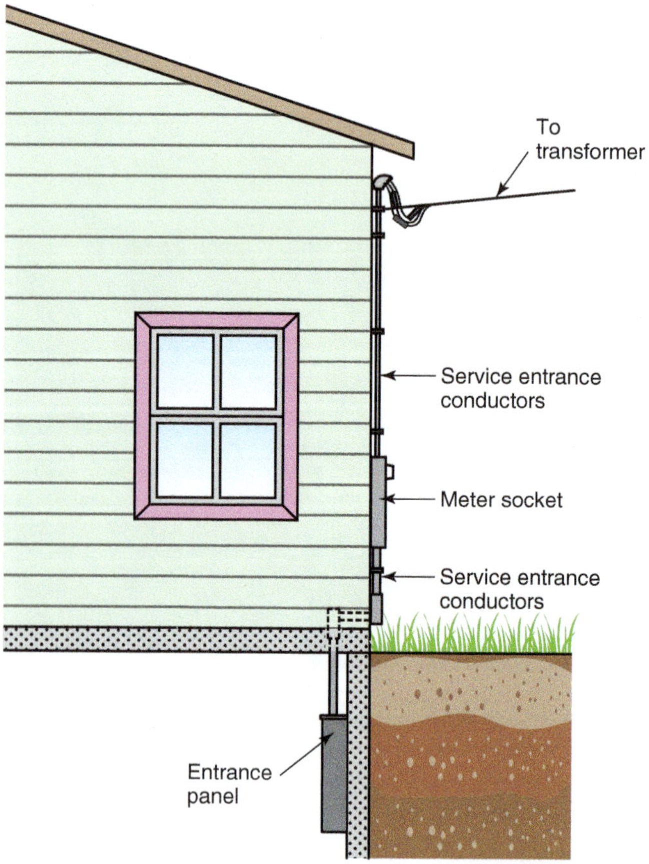

Goodheart-Willcox Publisher

Figure 6.3-2. Components of an above-ground service installation.

Code Note

Electrical outlets must be placed according to local building codes, typically specifying distances from corners, doors, and other features. For example, outlets are usually required to be spaced no more than 12 feet apart along walls in living spaces.

Christian Delbert/Shutterstock.com

Figure 6.3-3. The exterior electrical connection to this home is made at the meter enclosure that houses the electrical meter. Conductors protected by an underground conduit enter the enclosure and exit to the distribution panel located inside the home. The meter measures and displays the amount of electricity used.

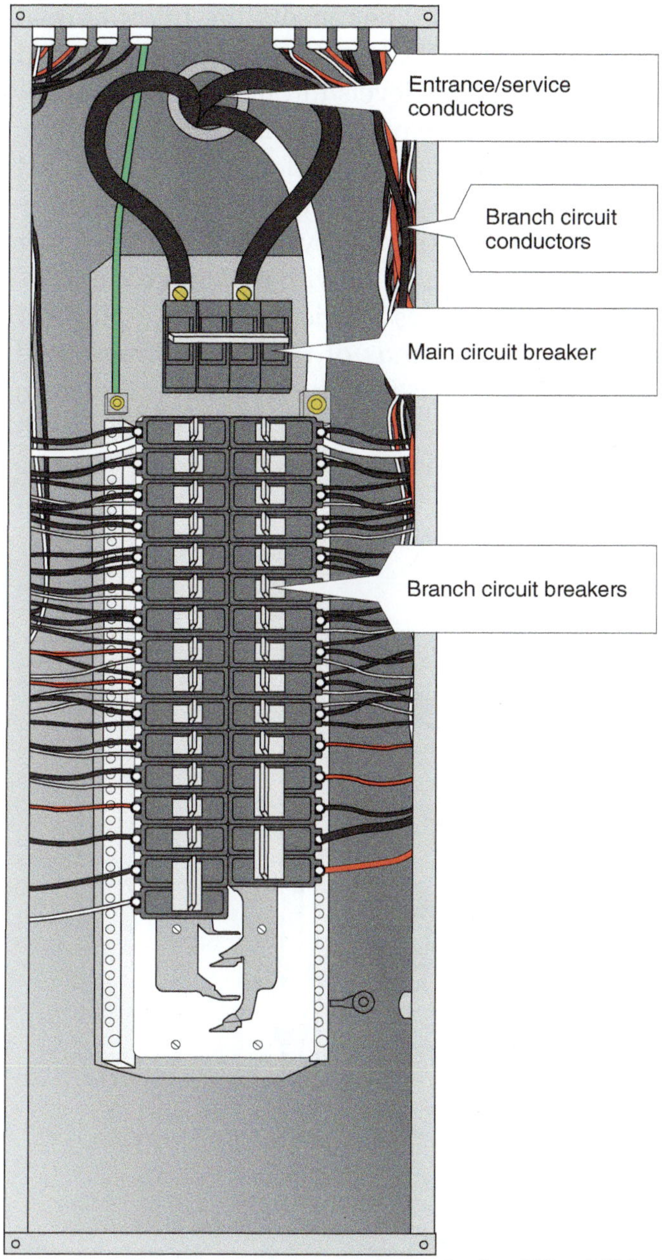

Goodheart-Willcox Publisher

Figure 6.3-4. The service entrance or distribution panel is located inside the building. Each branch circuit is connected to a circuit breaker (black objects in the box). The heavy cables inside the box near the top are the conductors that bring power from the meter. The large black block just below the cables is the main breaker. When connections have been completed, a protective panel cover is placed over the installation.

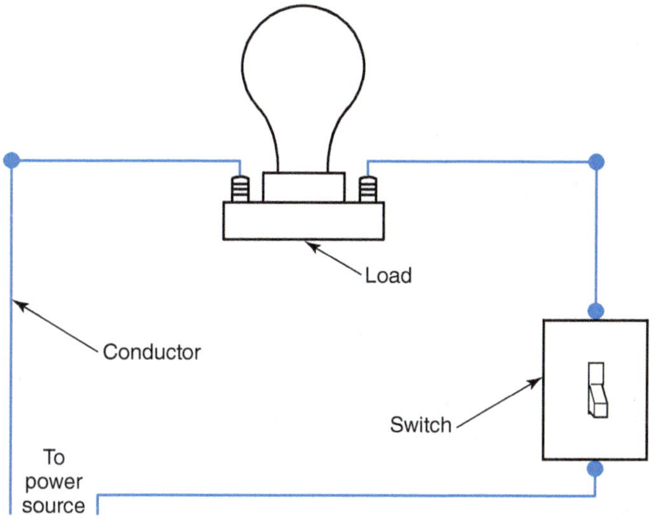

Goodheart-Willcox Publisher

Figure 6.3-5. A circuit is simply a path from a source of power, through energy-using devices, back to the power source.

A Photo courtesy of Robert Bosch Tool Corporation®

B Goodheart-Willcox Publisher

Figure 6.3-6. A—An electrician often must drill holes through the building frame to run electrical cable or conduit. B—Nonmetallic cable must be properly fastened to framing members.

are run (installed) through the building frame into these boxes. An electrician drills holes through the wood framing members so cables can pass through, **Figure 6.3-6A**. The conductors are run to each box and securely connected to them, **Figure 6.3-6B**. Electrical cables must also be securely fastened to framing members.

Plumbing

Plumbing systems include all the piping and fixtures that supply water for drinking, cooking, bathing, and laundry. Also included are drainage pipes that provide a means of disposing of wastewater, either through a municipal sewer or a local septic system. See **Figure 6.3-7**. To install rough plumbing, plumbers must cut holes and notches in the building frame. It is important for a plumber to work closely with a carpenter to be sure that the holes will not weaken joists and studs.

Regulations for installing plumbing systems are covered by several agencies. The International Code Council (ICC) sponsors the International Plumbing Code (IPC), which has been adopted at the state and local levels throughout the country. Some states still have plumbing codes based on model codes that existed before the IPC was first published. Since codes may vary in their requirements from one community to another, a plumber must always be familiar with the local code.

Plumbing requires a variety of skills. A plumber must use many kinds of tools, equipment, and materials. Further, plumbers must be skilled in woodworking, metalworking, pipe threading, soldering, and brazing. Plumbers must make accurate

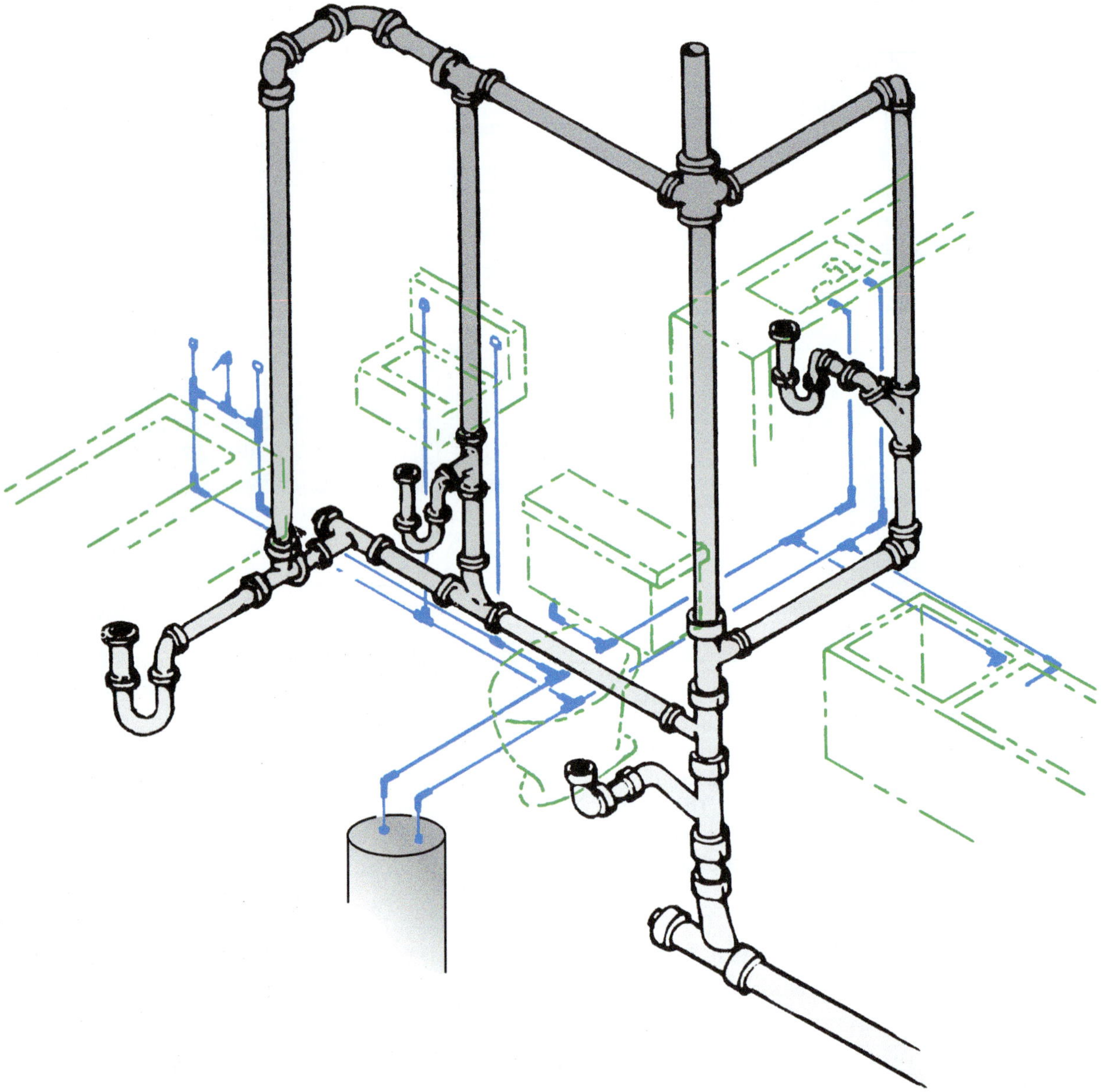

Goodheart-Willcox Publisher

Figure 6.3-7. A pictorial view of a plumbing system. The large pipes are parts of the drainage, waste, and venting (DWV) system. The upper portions of the DWV comprise the venting. The small pipes are the water supply system.

measurements and calculations. They must add and subtract dimensions, figure pipe offsets, and determine the volumes of tanks.

Plumbing includes two subsystems that have important differences:

- Water supply system
- Drainage, waste, and venting (DWV) system

The ***water supply system*** distributes water under pressure throughout the structure for drinking, cooking, bathing, and laundry. Supply pipes, fittings, and joints must be able to withstand the pressure. In a two-pipe system, one pipe carries cold water and the other hot water. Supply systems are watertight to prevent leakage.

Drainage piping, commonly referred to as the ***drainage, waste, and venting (DWV) system***, carries away wastewater and solid waste from bathrooms, kitchens, and laundries. This subsystem is not under pressure; it relies on gravity. DWV systems, like supply systems, are watertight. ***Vent piping*** is important for DWV systems, permitting air to circulate in the pipes. This prevents back pressure and siphoning of water from traps, **Figure 6.3-8**. Under certain conditions, it also prevents the introduction of wastewater into the ***potable*** (fit to drink) water supply. Plumbing codes are specific about the sizes and locations of vent piping.

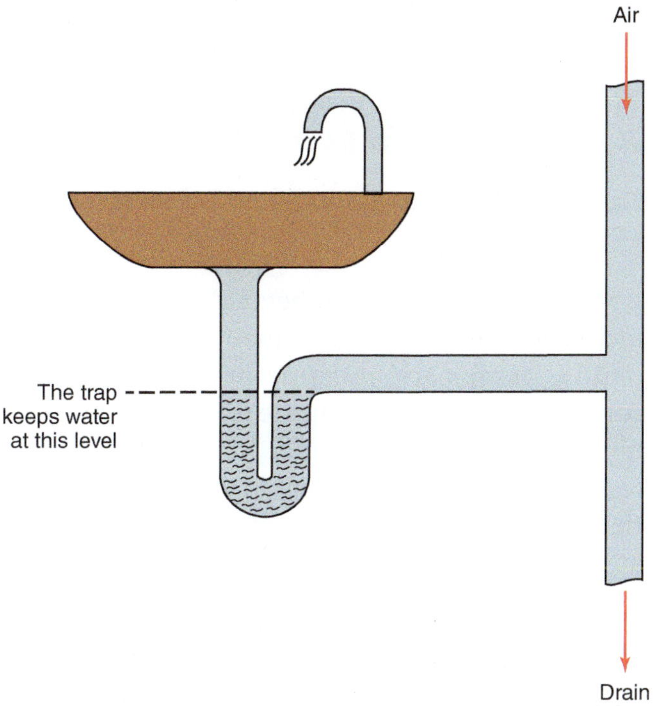

Goodheart-Willcox Publisher

Figure 6.3-8. The vent pipe allows atmospheric air pressure to enter the system, preventing the weight of the water flowing out of the drain from sucking (siphoning) all of the water out of the trap.

Code Note

Drainage pipes must be sloped correctly to ensure proper drainage and prevent backups or clogs. Vent pipes must also be installed to allow the free flow of air and equalize pressure within the plumbing system, preventing suction and odors.

Plumbing installation is usually done by a plumbing contractor. It can begin as soon as the building is closed in. Measurements are made to determine the exact location of pipes and fixtures.

Plumbing is concealed within the building frame. The single exception to this rule is plumbing below first-floor joists. Here, runs can be made below, rather than through, joists, unless the basement will be finished as living space. However, use of floor trusses allows mechanical systems to be run through the open spaces of the truss braces, **Figure 6.3-9**.

Some bathroom fixtures, such as bathtubs and showers, are too large to fit through doorways of framed-up partitions. They must be placed prior to installation of studs for partitions, **Figure 6.3-10**.

The plumbing rough-in includes installing all pipes to the point where connections are made with the plumbing fixtures. All runs are placed and tested with pressurized air or water to make sure no leaks exist. The vent pipes are connected to a large riser, or ***vent stack***, that extends through

Goodheart-Willcox Publisher

Figure 6.3-9. The open spaces in truss joists simplify plumbing runs in a basement or crawl space.

Goodheart-Willcox Publisher

Figure 6.3-10. This tub/shower unit was installed before the steel studs for the interior partition were erected.

ungvar/Shutterstock.com

Figure 6.3-11. Water supply and DWV stub-ins are ready for installation of a bathroom lavatory. Note the protective metal plates on the studs. These protect the pipes from wallboard fasteners.

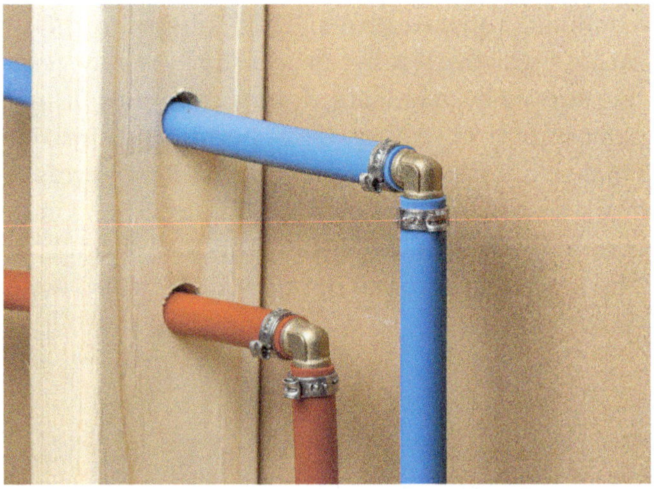

J.J. Gouin/Shutterstock.com

Figure 6.3-12. PEX pipe is a versatile choice for supply-side plumbing.

the roof. It serves as an exhaust for any odors from the waste system. The pipe ends that extend out from the walls and above the floors are known as ***stub-ins***. **Figure 6.3-11** shows stub-ins where fixtures will be connected. Typically, DWV pipes are PVC. Supply-side piping has traditionally been plumbed using copper tubing or chlorinated polyvinyl chloride (CPVC) pipe; however, most supply-side plumbing is now installed using cross-linked polyethylene (PEX) pipe, **Figure 6.3-12**.

Pro Tip

PEX pipe can be installed using a variety of connection types. Always be sure to check the manufacturer's instructions to ensure compatibility between tubing, fittings, and connection types.

Many buildings located beyond the reach of municipal water systems have their own wells. Wells require pumps and pressure tanks to maintain a constant water supply, **Figure 6.3-13**. The pressure tank keeps the system under constant pressure so it can be delivered anywhere in the building at sufficient force to operate fixtures and water-using appliances.

Heating, Ventilation, and Air Conditioning

Installation of heating, ventilation, and air conditioning (HVAC) is an important part of the building process. Without it, buildings would

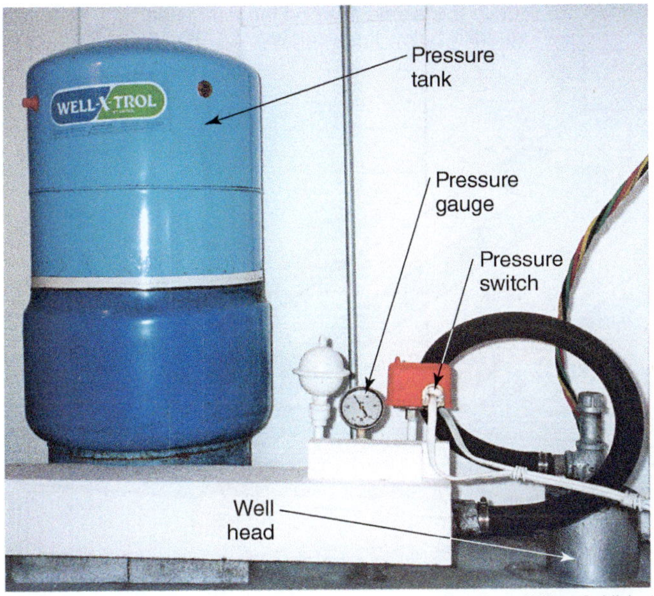

Goodheart-Willcox Publisher

Figure 6.3-13. A typical setup for a well water system. A submersible pump is located in the well.

not be comfortable places to work or live. HVAC systems control temperature, humidity, and indoor air quality. Technology allows for automatic control of systems so that buildings are kept comfortable year-round, regardless of how hot or cold the weather might be.

Heating and cooling of buildings depends on the fact that heated matter—whether air, liquid, or solid—passes its heat on to cooler matter. Heat travels by three methods: radiation, conduction, and convection. In review:

- *Radiation* moves heat by wave motion, similar to light or radio waves. A source produces heat and sends it through the air. As in a microwave, the air is not heated, but solid objects absorb the heat.
- *Conduction* is the movement of heat through substances from one end to the other. Heat moving through a metal spoon placed in a bowl of hot soup is an example of conduction.
- *Convection* is the movement of heat by way of a carrier, such as air or water. This carrier may move naturally or, in the case of furnaces and air conditioners, by fans or blowers.

Most buildings have central heating. Space heating is the use of heating devices, such as fireplaces, stoves, or portable heaters in individual rooms. In *central heating*, a single, large heating unit produces the heat. The heat is then circulated throughout the structure by perimeter heating or radiant heating. Several types of circulating systems are used with central heating:

- Forced-air perimeter heating
- Hydronic perimeter heating
- Hydronic radiant heating

Forced-Air Perimeter System

In a *forced-air perimeter system*, air is circulated through large ducts that direct warmed air to every room in the building, **Figure 6.3-14**. The burner delivers high-temperature combustion gases to the furnace heat exchanger. The *heat exchanger* is a system used to transfer heat from the source to the air. Heated air from the exchanger collects in a large sheet metal chamber called a plenum. A blower forces the heated air in the plenum through tubes or passages, called ducts, into the living spaces of the building. At the same time, the blower draws cooled air back to the furnace through a series of ducts called the air return. An air filter located in the air return filters out dust particles picked up from the heated spaces. **Figure 6.3-15** shows a cutaway view of a gas furnace.

Furnaces may be oil-fired, gas-fired, or electric. Oil-fired units require a storage tank for the fuel. The tank is usually located either underground outside the building or in the basement near the heating unit. For basement tanks, a filler pipe is usually installed that extends through the wall to the outside for easy access during fuel delivery. Gas-fired units may receive their fuel supply directly from a gas main supplied by a gas utility or from a storage tank located on the property. Electric furnaces use the resistance of conductors to produce heat, much like an electric stove or oven.

Registers used with forced-air furnaces consist of a grille installed over duct openings in the rooms. They usually have shutoffs for controlling airflow. Some have adjustable vanes that control the direction of the airflow as it exits the duct.

As air is heated, it becomes less dense. That means a cubic foot of warm air is not as heavy as a cubic foot of cooler air. This causes warm air to rise and cooler air to settle toward the floor.

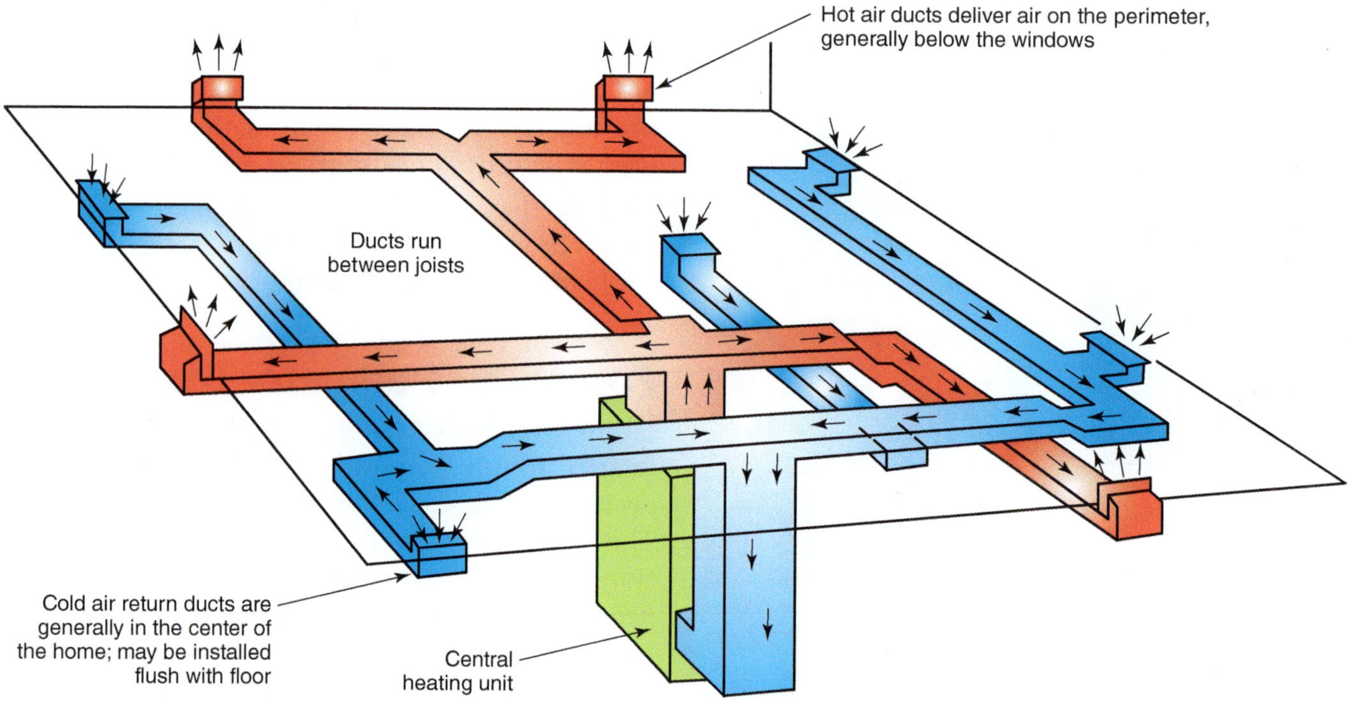

Goodheart-Willcox Publisher

Figure 6.3-14. A forced-air perimeter heating system. Warm-air ducts deliver heated air to rooms under a slight pressure. The cold air return duct brings air into the furnace for heating each time the circulating blower runs.

Since warm air rises, heating registers are more efficient if located at floor level, **Figure 6.3-16**. Since cold air sinks, cold air returns should also be located at floor level. Registers should be located where air is not discharged directly on room occupants.

Installation of a forced-air furnace usually follows the installation of the ductwork. An HVAC technician installs the unit. The installer must be able to make mechanical connections and electrical hookups, **Figure 6.3-17**. The same person may also install the ductwork. This requires a knowledge of certain metal trades as well as a thorough grounding in electrical principles and wiring practice.

Pro Tip

During the rough-in phase of construction, coordination between contractors and subcontractors is crucial for ensuring the smooth progress of the project. Coordination of schedules and clear communication is critical.

Hydronic Perimeter Heating System

A ***hydronic perimeter system*** uses water as a medium to move the heat from the heating unit, **Figure 6.3-18**. The heating unit is called a boiler for the simple reason that it heats water much like a teakettle sitting on a burner. The heated water is piped to radiators that transfer heat to the rooms.

Many hydronic systems include a ***manifold***, **Figure 6.3-18C**. This device balances the delivery of heated water to different zones (areas) of the building. The manifold consists of an inlet pipe fitting with several outlets for connecting one pipe with others. Pipes from each outlet are directed to different zones of the building. Each of these zones has its own circulating pump controlled by a thermostat for that zone.

Hydronic systems require an expansion tank to serve as a reservoir for the heated water's increase in volume. A pressure-relief valve is also part of the system. It opens to release excess water pressure that might otherwise damage the system.

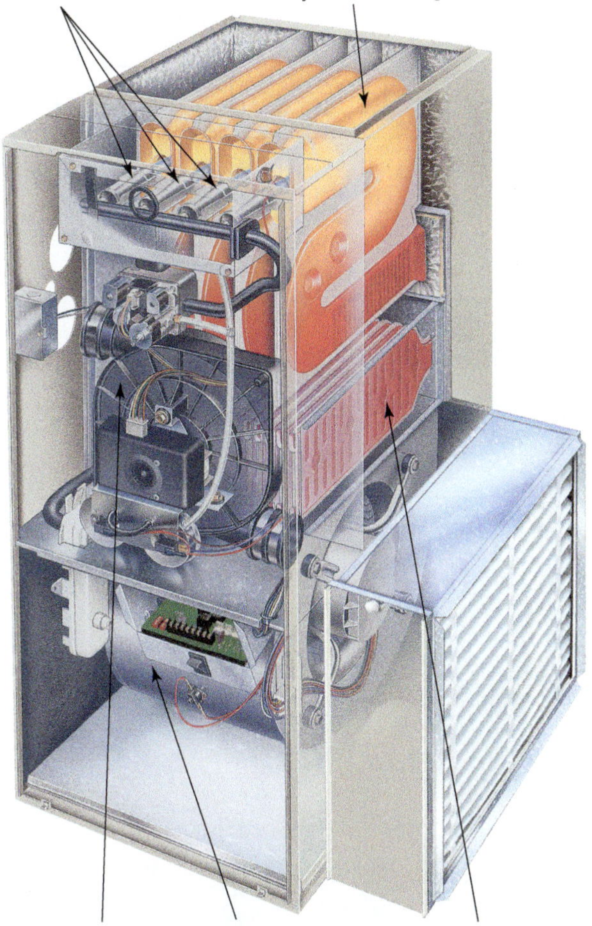

Carrier Corporation, subsidiary of United Technologies Group

Figure 6.3-15. Components of a high-efficiency gas furnace.

Claude Huot/shutterstock.com

Figure 6.3-16. Warm air registers usually have vanes to direct the flow of heated air and a door that can be positioned to reduce or stop airflow.

Goodheart-Willcox Publisher

Figure 6.3-17. An installer making electrical connections during furnace installation.

A pressure-reducing valve limits the pressure of the incoming freshwater supply. Air purge valves are used to remove any air that has been introduced into the system. These valves are placed at high points in the piping where air is likely to collect, **Figure 6.3-19**.

Hydronic Radiant Heating System

A *hydronic radiant system* is similar to a hydronic perimeter system. However, the piping is installed in a structural part of the building. Hydronic radiant systems can be installed several different ways:

- On-grade over a thick underlayment of rigid insulation. Concrete is then poured over the tubing.
- Under subflooring with insulation beneath the heating tubes. Wood or other

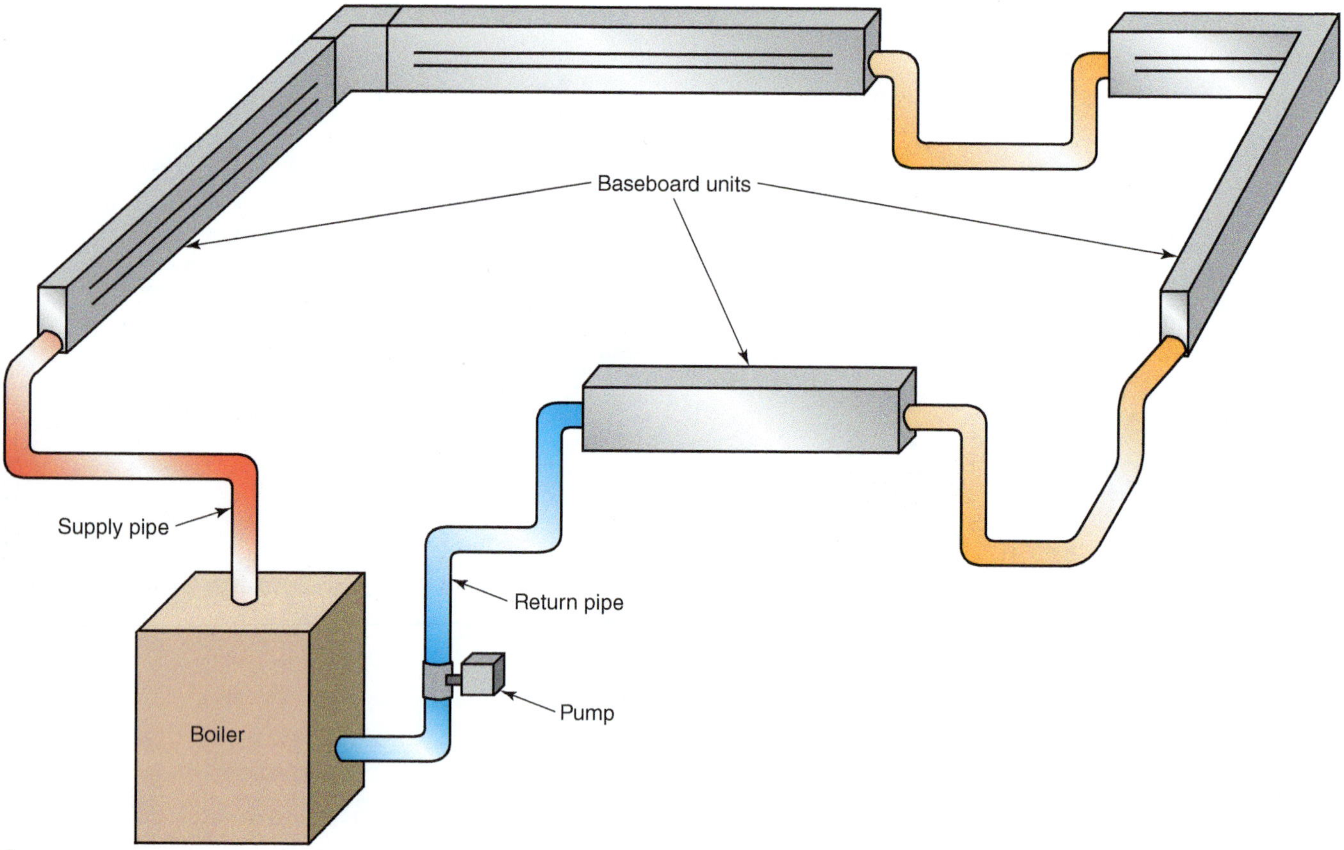

A

Goodheart-Willcox Publisher

B United Technologies Carrier

C United Technologies Carrier

Figure 6.3-18. A—A basic hydronic system. B—This boiler is teamed with a hot water storage unit. The boiler delivers heated water to the tank until it is used by the hydronic heating system. A heat exchanger coil in the tank supplies domestic hot water, eliminating the need for a separate water heater. C—This manifold delivers heated water to five different pipes, creating five different heating zones for the house.

types of flooring can be laid over the piping system after provisions are made to prevent damage or crushing of the pipe or tubing. When installing wood or other types of flooring, installers must be careful not to drive fasteners through the heating pipes.

Such systems are usually installed as part of new construction. However, they can be part of a remodeling project. Piping is usually PEX, which is laid out in a serpentine pattern that covers the area of installation from side to side, **Figure 6.3-20**.

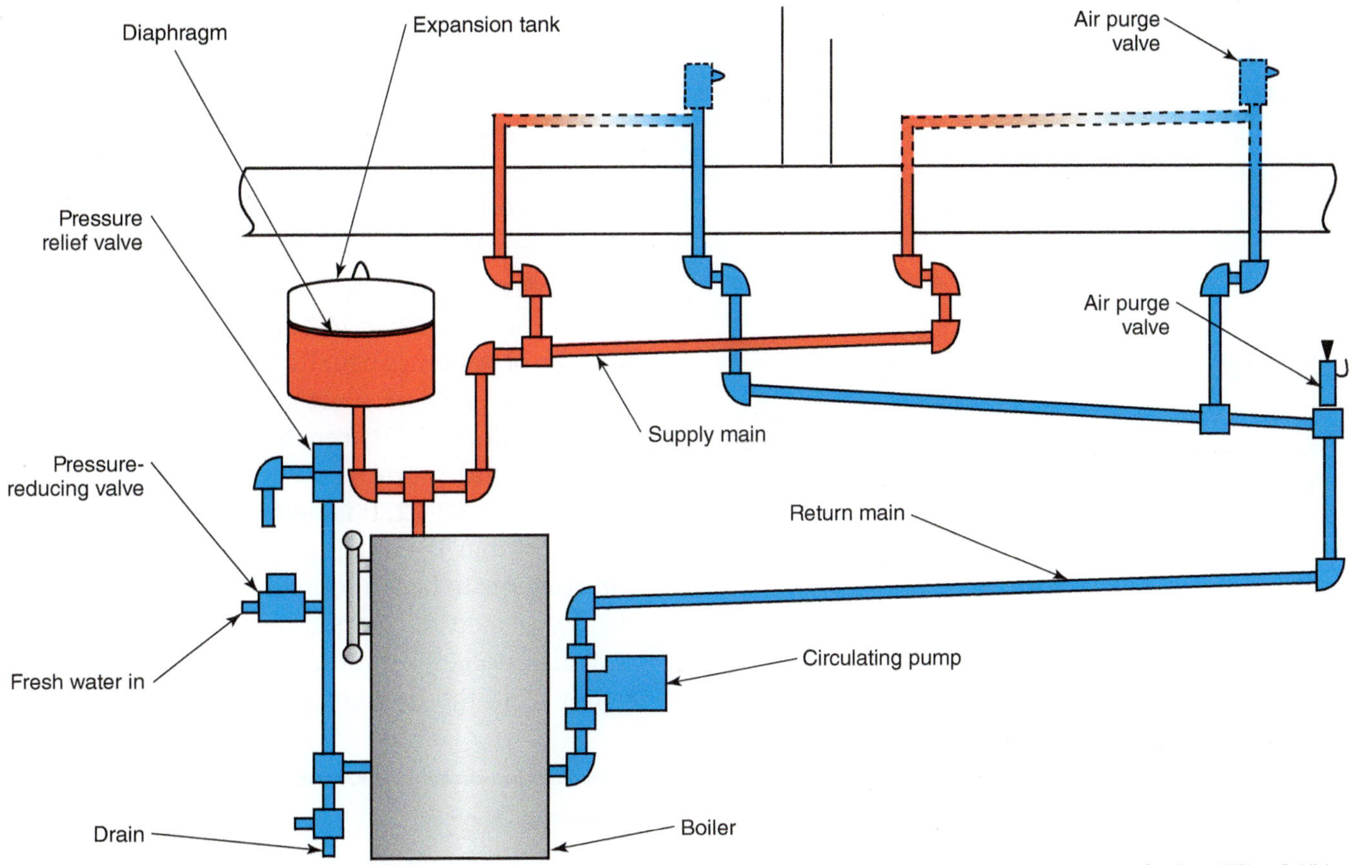

Goodheart-Willcox Publisher

Figure 6.3-19. The major components of a hydronic heating system.

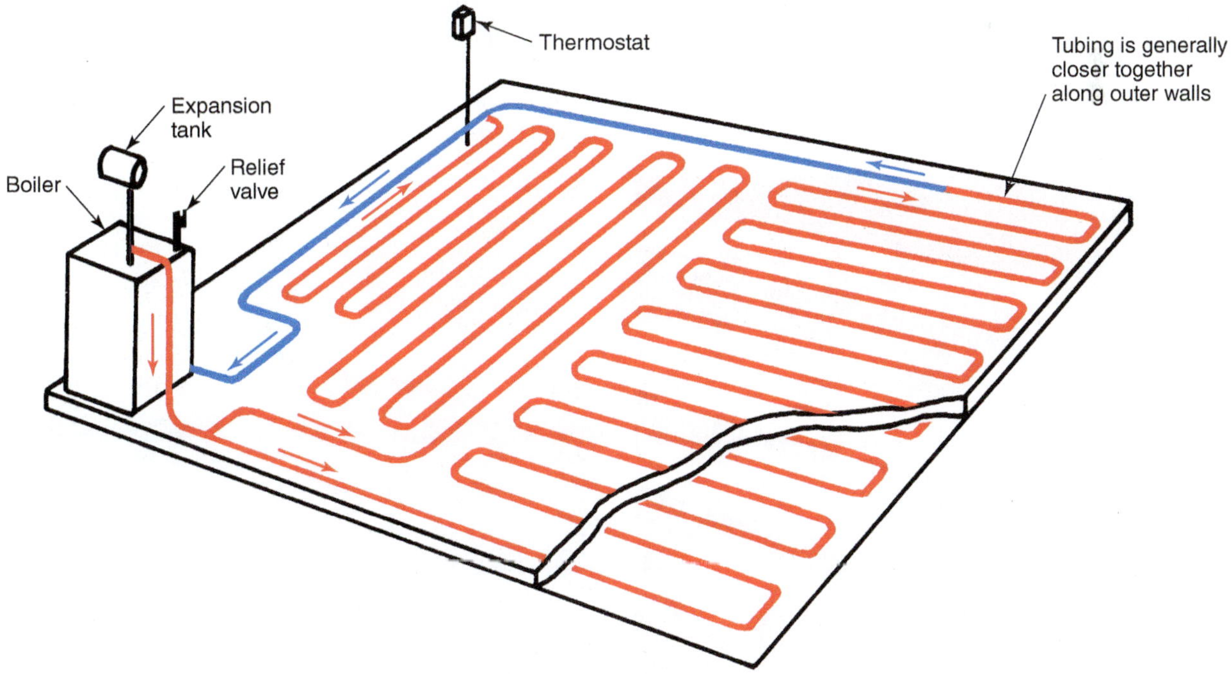

Goodheart-Willcox Publisher

Figure 6.3-20. This drawing shows a hydronic radiant heating system as it would be installed in a slab floor. Note the location of the thermostat.

Pro Tip

To reduce heat loss in radiant installations, insulation is installed on the cold side of the heating elements. In applications where radiant elements are embedded in concrete, the insulation is installed before the slab is poured.

Direct Heating Systems

Direct heating systems distribute heat directly to objects without using ductwork or a piping system. This category includes many basic types of heating, such as stoves and fireplaces. It also includes electric resistance baseboard heating and radiant electric heat.

Electric resistance baseboard heating consists of a heating element wrapped in a tubular casing. Metal fins spaced along the length of the casing radiate the heat to the room air. Electric baseboard radiators come in various lengths—from 1′ to 12′—and are rated from 100 W/ft to 400 W/ft, **Figure 6.3-21**. Like hydronic systems, radiators are located around the room's outside walls, especially under windows to counteract drafts. This system has several advantages:

- Affordable to buy.
- Easy to install.
- Easy to zone heat. Some models have thermostatic controls so that rooms can be individually heated.
- Self-contained. Chimneys and piping or fuel storage are not needed.
- Quiet operation.

The main disadvantage is the operating cost. In some areas, electric rates make this type of heating system too expensive to consider. Consequently, this type of system is usually used only as a backup.

Another type of electric heating known as ***electric radiant heat*** transfers heat by moving it through solid matter, such as a concrete or tile floor, **Figure 6.3-22**. It is designed to directly heat objects, rather than using radiators and convection currents to heat a room. Heating elements are usually placed in floors but can also be installed in walls and ceilings similar to hydronic radiant systems. As the resistance elements heat up, the

Green Note

The local climate, design of a home, and type of heating system can drastically impact a home's energy use. Homes built in climates that are colder for more of the year will have an increased benefit through the use of energy-focused design principles and higher efficiency systems.

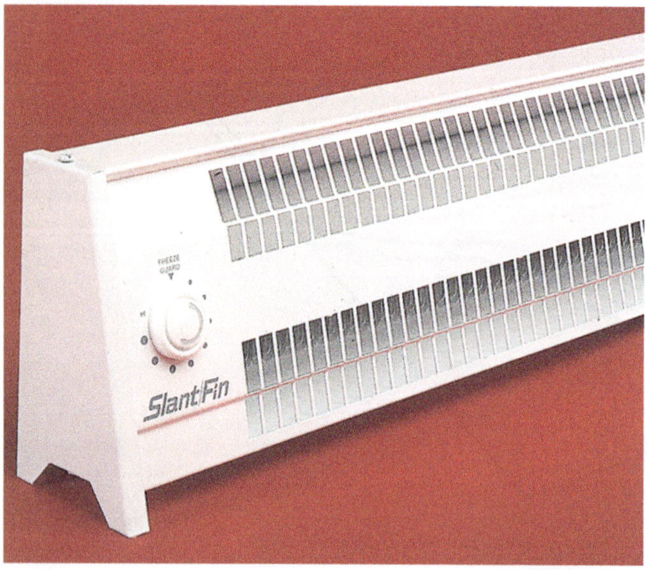

Slant/Fin Corp.

Figure 6.3-21. An electric baseboard unit is sometimes used as auxiliary heat.

brizmaker/Shutterstock.com

Figure 6.3-22. Electric radiant heat has various applications. Here it is covered by a self-leveling grout to prepare the floor for its finish material.

heat radiates to the concrete or wood floor, wall material, or ceiling material. Ceiling or wall radiant electric heating elements may be encased in plaster. This provides a medium through which the heat can radiate.

Air Cooling Systems

Cooling systems move heated air from inside a building to the atmosphere outside. **Figure 6.3-23** shows an air-conditioning unit located outside a home. It contains a compressor, a condenser, and related valves. The compressor is operated by an electric motor. In many cases, the cooling system uses the same air distribution system (ductwork) as the heating system, **Figure 6.3-24**. A system of tubes called an *evaporator* is located in

Goodheart-Willcox Publisher

Figure 6.3-23. A typical central air-conditioner unit located outside of the house. The condenser completely surrounds the compressor and its related components. This arrangement increases the efficiency of the unit.

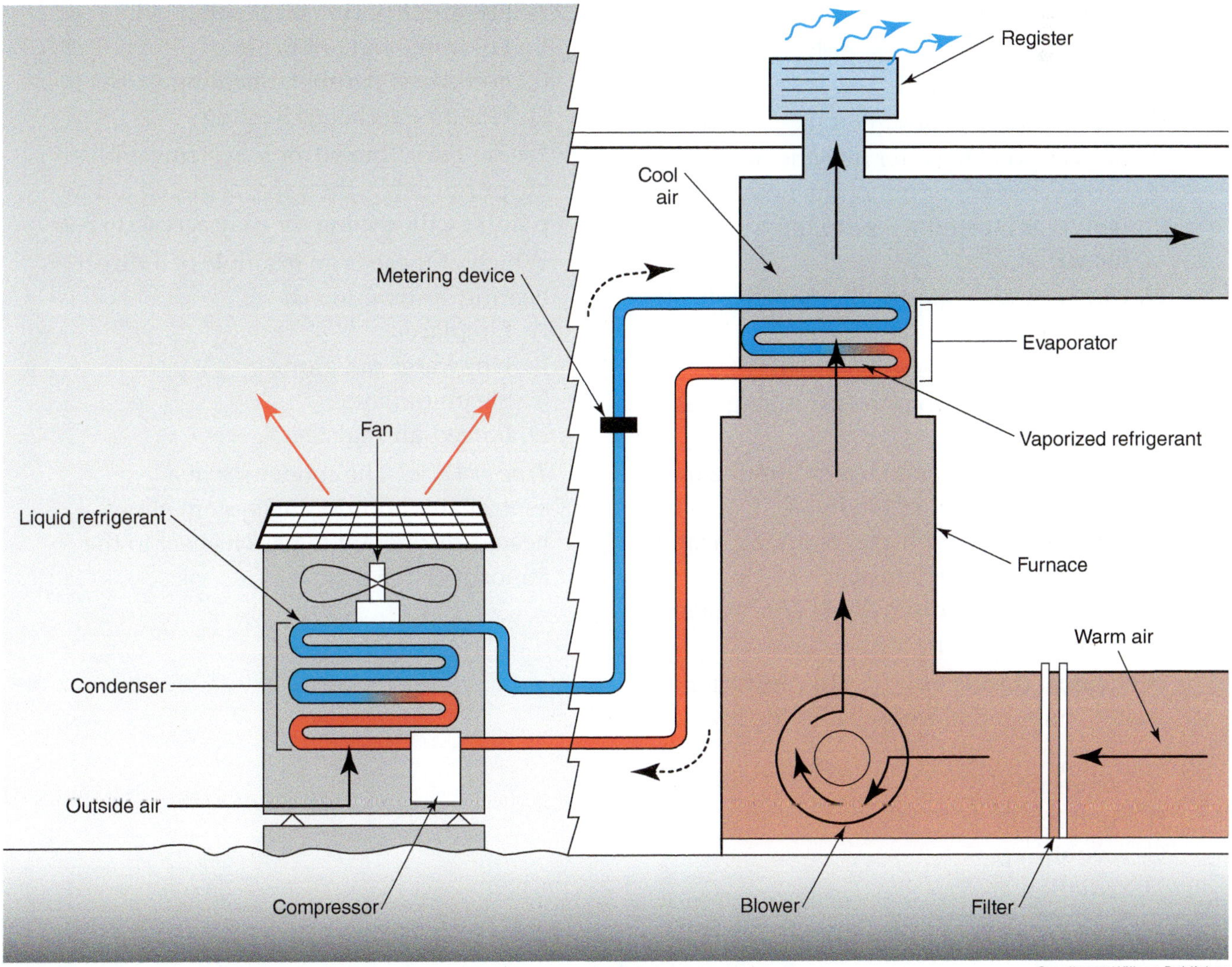

Goodheart-Willcox Publisher

Figure 6.3-24. A central air-conditioning unit is combined with a forced-air distribution system. A coil located in the furnace receives cold refrigerant. Air forced through the coils is cooled and delivered to rooms by way of the duct system.

the air-conditioning unit where return air from the rooms can pass over it.

When the central unit is operating, cooled, liquid refrigerant is pumped through the evaporator. Since it is colder than the room air passing over it, the refrigerant absorbs heat from the air and evaporates. This can be compared to water evaporating as it absorbs heat from a stove and turns to steam. Then, the hot refrigerant gas is drawn back to the outside unit. Here, it passes through another series of coils and fins called a ***condenser***. The condenser acts like the radiator of a car, allowing the heated refrigerant to pass its heat to the atmosphere. In doing so, the refrigerant cools and returns to a liquid state.

From the condenser, the refrigerant is pumped at a high pressure to the evaporator in the furnace. This completes one cycle of the air conditioner. The cycle is continuous as the building's air is passed over the cooling coil.

LESSON 6.3 Review

Answer the following questions based on information in this lesson.

1. *True or False?* In a circuit, after the current passes through a device, such as an electric motor, it travels to the ground, where it dissipates. (6.3-1)
2. *True or False?* The water supply system is a two-pipe system. (6.3-1)
3. Drainage piping is designed to _____. (6.3-1)
   A. deliver potable water into the house
   B. carry wastewater out of the house
   C. deliver unheated water to the water heater
   D. carry heated water from the water heater to a fixture
4. What does a vent stack do? (6.3-1)
   A. Exhaust odors from the plumbing waste system
   B. Transfer smoke and steam from the oven and stovetop out of the house
   C. Carry smoke and fumes from the heating system out of the house
   D. Deliver fresh air into the HVAC system to aid circulation
5. When heat travels by _____, the air is not heated, but solid objects absorb the heat. (6.3-2)
   A. convection
   B. conduction
   C. revolution
   D. radiation
6. Which method of heating uses large ducts to deliver heat to every room in the building? (6.3-2)
   A. Hydronic perimeter heating
   B. Hydronic radiant heating
   C. Forced-air perimeter heating
   D. Hot-air baseboard heating
7. *True or False?* In hydronic systems with an air-purge valve, the valve is placed at low points in the piping for easy access. (6.3-2)
8. Which of these is an example of a direct heating system? (6.3-2)
   A. Fireplace
   B. Hot water radiator
   C. Steam radiator
   D. Forced-air register
9. *True or False?* The condenser in a central air-conditioning system allows heated refrigerant to pass its heat to the atmosphere. (6.3-2)

LESSON 6.4

Materials and Fasteners

LEARNING OUTCOMES

After completing this lesson, you will be able to:

6.4-1 Describe common components and materials required for the MEP rough-in.

6.4-2 List common fasteners and adhesives used during the MEP rough-in.

TECHNICAL TERMS

- boiler
- circuit breaker
- compression fitting
- condensing unit
- conductor
- conduit
- duct seal compound
- ductless mini-split
- furnace
- heat pump
- pipe dope
- valve
- ventilation system

MEP Rough-In Components and Materials

Many components make up the systems in a home. This chapter features the common components and materials used for the rough-ins of the electrical, plumbing, and HVAC systems of a typical home. It should be noted that variations exist.

Electrical System Components and Materials

A circuit is the path through which current flows from the source back to the source. It includes the conductors, switches, receptacles, and any appliances or lights. During electrical rough-in, all provisions must be made to house and provide power for switches, lights, receptacles, and appliances. This includes the entrance panel and circuit breakers, conductors throughout the house, and switch/receptacle boxes. All other electrical fixtures, including switches, lights, receptacles, and appliances, are installed during the finish electrical phase of the project.

Safety Note

Electrical current can be extremely dangerous if a grounded person accidentally touches bare conductors or other energized parts of a circuit. Always follow these safety rules when you work around electricity:

- Make certain that a circuit is de-energized before working on it; turn off the circuit breaker or remove the fuse.
- Remove all jewelry, watches, rings, and any other metal objects before beginning work.
- Keep work areas dry and clean.
- Wear boots with insulated soles.
- Use only wood or fiberglass ladders, never metal.

Switches and receptacles are installed in boxes. Various types of boxes and box covers are available, **Figure 6.4-1**. Their purpose is to enclose devices that control electric current and protect connections between two or more conductors. Covers and trim

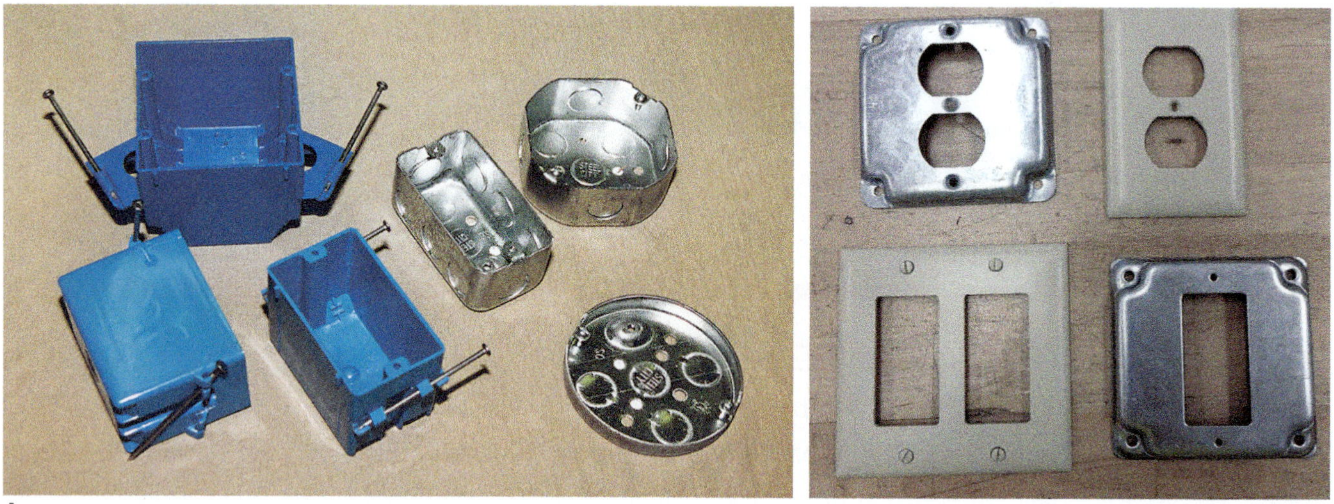

A Goodheart-Willcox Publisher B Stephen P. Kielar

Figure 6.4-1. A—Several types of plastic and metal boxes used for either new construction or remodeling work. B—Covers and trim plates protect devices and conductor connections in the boxes.

plates conceal and protect devices that are placed in boxes. Electric current is transferred between boxes by conductors. ***Conductors*** are materials that carry current. In a residence, conductors are typically metal wires enclosed in conduit, a metal sheath, or a plastic sheath, **Figure 6.4-2**.

Circuit breakers are devices that protect conductors from overload by shutting off power to the circuit if an excessive current draw occurs, **Figure 6.4-3**. A short circuit can occur when a connection comes apart or perhaps a rodent chews through the insulation, causing an increase in the amount of current flowing through the circuit. The circuit breaker or fuse will open the circuit, stopping all current to protect the wires from heating up and causing a fire.

Conduits or raceways are pipes or enclosed channels through which conductors are run, **Figure 6.4-4**. Conduit is made of steel or polyvinyl chloride (PVC) rigid tubing. Lengths of conduit can be cut and bent as needed to fit the structure and can be buried underground for subsurface applications. Conduit lengths are connected with special couplings and fittings.

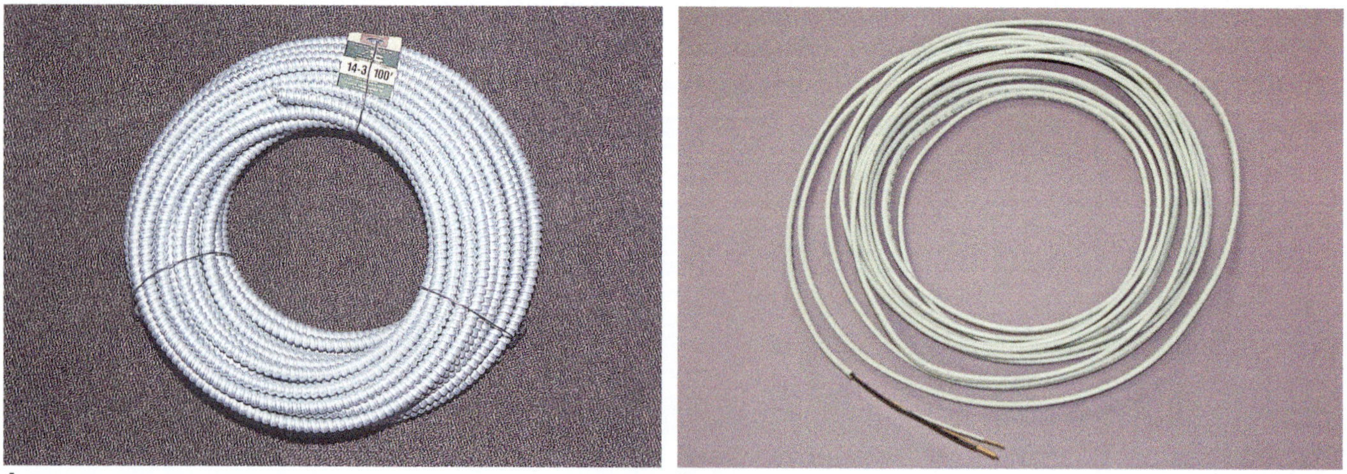

A B Goodheart-Willcox Publisher

Figure 6.4-2. Electrical conductors are the copper or aluminum wires that provide a path for current in the building's circuits. Conductors within the various types of cables are available in different sizes and quantities. Codes vary depending on which types can be used. A—Armored cable, also called BX, has a flexible metal cladding protecting several insulated conductors. B—Nonmetallic (NM) cable is similar to armored cable but has a plastic protective covering.

nattapan72/Shutterstock.com

Figure 6.4-3. Circuit breakers protect circuits and equipment from overloads.

Goodheart-Willcox Publisher

Figure 6.4-4. Conduit is used to protect electric wires. Here, several boxes are connected with polyvinyl chloride (PVC) conduit.

Connectors are fastening devices that ensure tight connections between two or more conductors or lengths of steel conduit, **Figure 6.4-5**. Wire nuts form a sound mechanical and electrical connection between conductors. Color coding is used to identify the proper wire nut for appropriate conductor sizes. Couplings are used to connect steel conduit, while other fittings are used to securely clamp armored or nonmetallic cable to electrical boxes.

Plumbing System Components and Materials

Several types of plumbing materials are used during the rough-in phase:

- Pipe and tubing for supply lines carrying pressurized water to faucets and fixtures.
- Large-diameter pipe to drain away wastewater.
- Valves for controlling water supply.
- Washers and O-rings for both supply and drainage systems.
- Tape, compounds, and putties to seal connections against leaks.

The type of material used for pipes and pipe fittings vary, depending on codes. Supply system pipe may be made of copper, chlorinated polyvinyl chloride (CPVC), or cross-linked polyethylene (PEX). In residential construction, DWV pipe is typically made of PVC.

- Copper tubing is available as rigid, straight lengths or in flexible coils, **Figure 6.4-6**. It is sold in a variety of diameters and seven different weights. Only three weights—types K, L, and M—are typically used for water supply lines. Copper tubing can be cut with a pipe cutter or hacksaw. Flexible tubing is often joined using ***compression fittings***, connectors that seal joints by compressing a ring of soft material. The fitting slides over the tubing. Another method is to sweat solder the fittings. Only lead-free solder is permitted, to avoid the hazard of lead contaminating water supplies.
- Chlorinated polyvinyl chloride (CPVC) is a buff-colored thermoplastic that is rigid, light, easy to handle, and resistant to cracking in freezing conditions. Connections may be threaded or solvent welded. While PVC is

A Goodheart-Willcox Publisher B Goodheart-Willcox Publisher

Figure 6.4-5. Fittings connect and hold conductors, cables, and conduit. A—Wire nuts are used to connect two or more conductors. B—Couplings are used to connect lengths of conduit, and connectors are needed to attach conduits and cables to boxes.

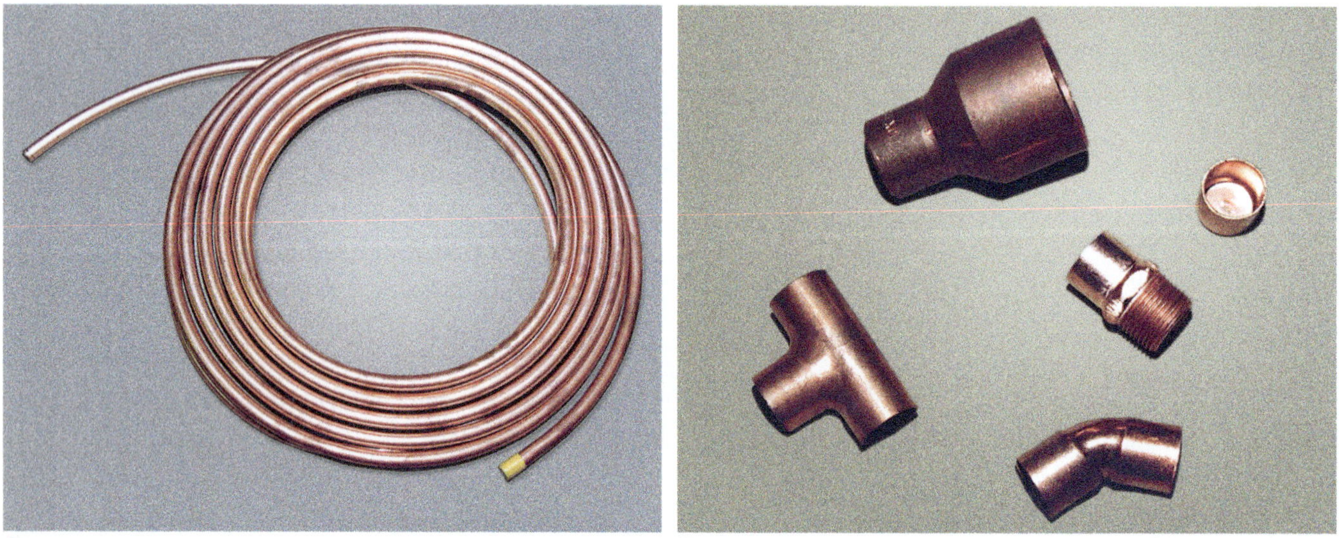

A Goodheart-Willcox Publisher B Goodheart-Willcox Publisher

Figure 6.4-6. Typical copper materials. A—Copper tubing. B—Forged copper fittings. These must be sweat soldered to connect to the pipe.

only rated for cold water applications (and can be used as supply-side piping), CPVC is suitable for hot and cold water piping and is rated for temperatures of up to 180°F and a pressure of 100 psi. CPVC and PVC supply-side piping has largely been displaced by the use of PEX on new construction projects.

- Cross-linked polyethylene (PEX) tubing is popular for water supply piping because it is flexible, lightweight, and connects easily. It can be used for hot and cold water supply lines. Connections are typically made with crimped connectors, **Figure 6.4-7**. A centralized manifold is usually installed to distribute the water.
- Polyvinyl chloride (PVC) has lower thermal expansion that makes long runs easier to control. It is joined with a two-step primer/solvent and suitable fittings. See **Figure 6.4-8**.

Valves are devices that control the flow of water in the water supply system, **Figure 6.4-9**. They are installed where the water main enters a building as well as at certain places in the lines so that water can be shut off to faucets, drainpipes, and appliances.

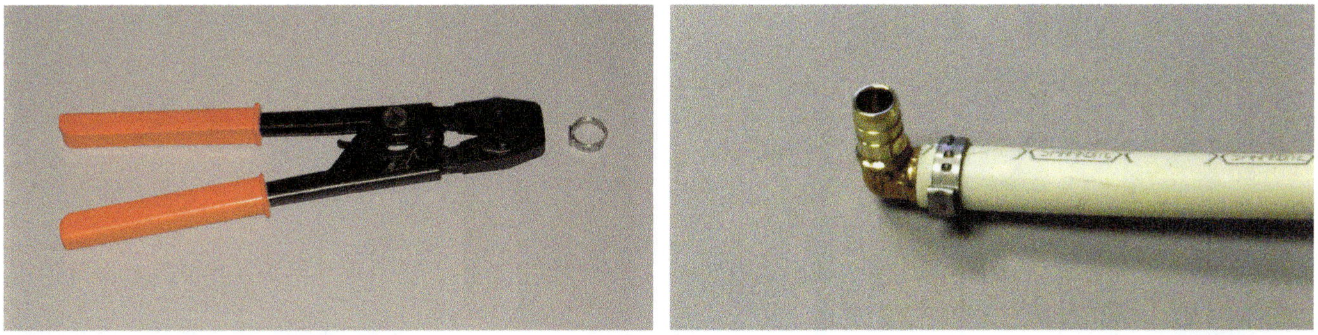

A Goodheart-Willcox Publisher B Goodheart-Willcox Publisher

Figure 6.4-7. PEX tubing is frequently used for residential plumbing. A—A special crimping tool is used to tighten the stainless-steel clamps that secure the PEX to fittings. B—PEX tubing with fitting.

A Goodheart-Willcox Publisher B Goodheart-Willcox Publisher C Goodheart-Willcox Publisher

Figure 6.4-8. Plastic fittings are used in drainage systems. A—Wye. B—Closet flange. C—P-trap.

Alex Kolokythas Photography/Shutterstock.com

Figure 6.4-9. Ball valves are designed to turn on or off a water supply quickly and easily.

The home's domestic water heater is also installed during rough-in. Domestic water heaters are appliances designed to heat water for use at the tap. Water heaters come in different types, sizes, and fuel sources. The most common types are storage tank water heaters and tankless (on-demand) water heaters, **Figure 6.4-10**.

- Storage tank water heaters have an insulated tank that holds and heats the water. The tank is equipped with a heating element (electric) or a gas burner (natural gas or propane) at the bottom of the tank. The heated water rises to the top and is drawn from a hot water outlet when needed. Cold water enters the tank through a tube near the bottom, ensuring that there is always a ready supply of hot water. These water heaters are available in various sizes, typically ranging from 20 to 80 gallons, and are generally installed in basements, utility closets, or garages.
- Tankless water heaters, as the name implies, do not have a tank to store hot water. Instead, they heat the water directly as it flows through the unit. When a hot water tap is turned on, cold water enters the

Basic Tool List

Striking tools	Lineman's or electrician's hammer
Drilling tools	Right-angle drill Cordless drill Drill bits, various sizes ▪ Wood twist ▪ Metal ▪ Masonry ▪ Spade ▪ Bit extenders
Soldering and wire-joining tools	Soldering iron Soldering gun Propane torch Solder, rosin core Soldering paste Crimping tool Wire stripper
Fastening tools	Standard screwdriver Phillips screwdriver Offset screwdriver Torque head screwdriver Adjustable wrench Allen wrenches Socket/ratchet wrenches Box end wrenches
Measuring tools	Folding ruler Carpenter's extension ruler Steel tape 12″ ruler or meter stick Wire gage Multimeter
Cutting and sawing tools	Files Crosscut saw Jab saw Hacksaw Circular saw, 7″ Compound miter saw Reciprocating saw Pocketknife or electrician's knife Cable cutters Wood chisel Wire strippers Cable strippers
Pliers	Slip-joint pliers Lineman's pliers Side cutting pliers Diagonal pliers Long nose pliers End cutting pliers Locking jaw pliers
Special and miscellaneous tools	Fish tape wire puller Wire pulling lubricant Conduit or pipe cutter Reamer Conduit bender (hickey) Fuse puller Tap and die set Flashlight Plumb bob Test light, continuity tester Level Conduit threader Trouble light Gas generator, about 1500 W Portable space heater Assorted wood or fiberglass ladders Wire grips Chalk line Tool pouch

Goodheart-Willcox Publisher

Figure 6.5-1. An electrician's toolbox should be stocked with tools from this basic list.

Necessary cutting tools include saws, files, chisels, snips, and pipe cutters, **Figure 6.5-6**. Boring tools are used for making large holes in wood framing members. Some cutting and boring tools should be suitable for metal-cutting since steel framing is often found in residential construction. A portable band saw is often used to cut steel pipe. The ***portable band saw*** is a power tool with a toothed blade made from a continuous band of metal stretched between two wheels. As the saw operates,

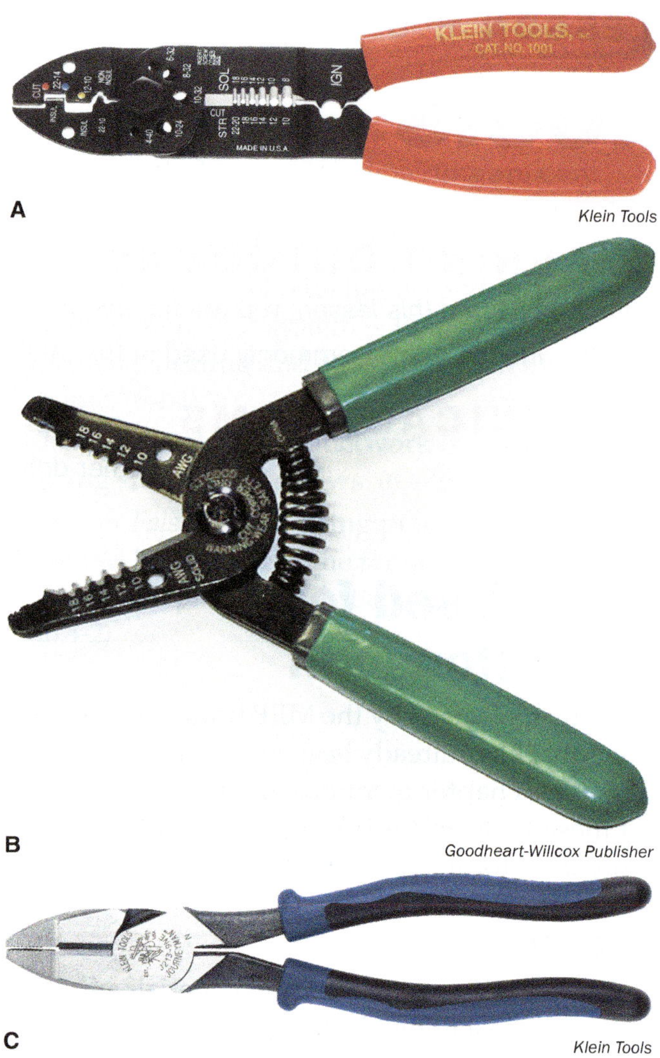

A *Klein Tools*

B *Goodheart-Willcox Publisher*

C *Klein Tools*

Figure 6.5-2. A—Multipurpose tools are used to cut electrical wire, strip insulation, and crimp connectors onto conductors. B—This wire stripper has openings for wire from 10 gauge down to 18 gauge. C—A pair of side cutting, or lineman's, pliers is an essential tool for everyone doing electrical installations.

Milwaukee Tool

Figure 6.5-3. Right-angle drills fit between studs for precise drilling.

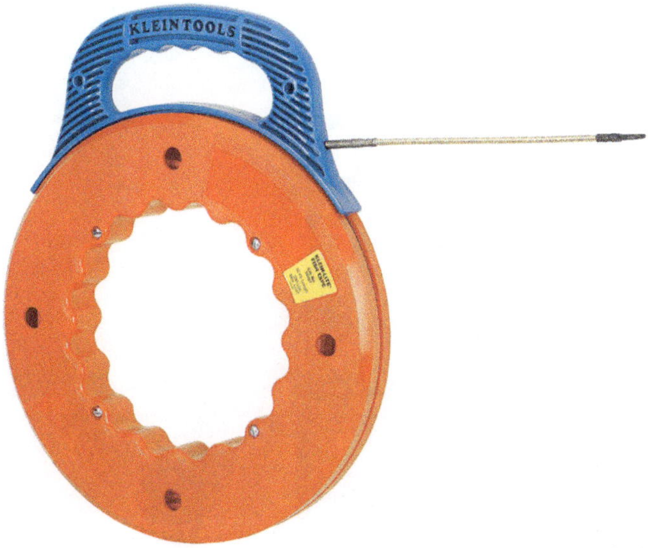

Klein Tools

Figure 6.5-4. A fish tape is used to pull wires through conduit.

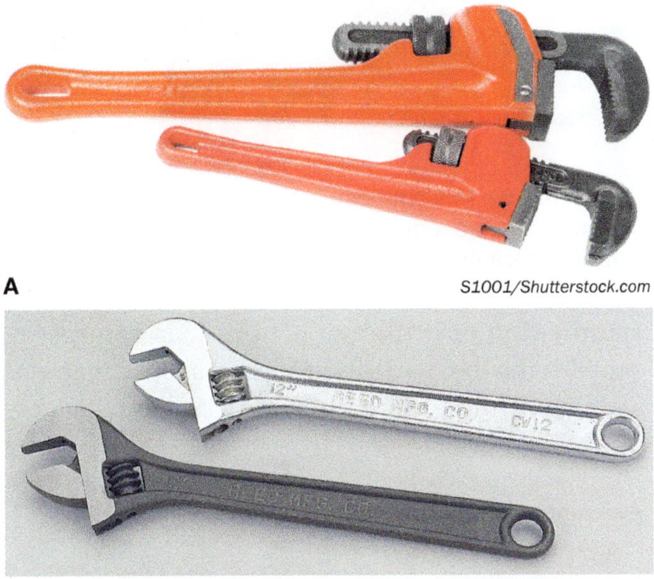

A S1001/Shutterstock.com

B Reed Manufacturing Co.

Figure 6.5-5. Wrenches are a plumber's most-used tools. A—Pipe wrenches are used to tighten threaded fittings on metal pipe. B—Adjustable wrenches work on fittings with flats.

the wheels turn the band, enabling the operator to saw through material, **Figure 6.5-7**.

Reaming and threading tools are used on metal pipe. Reaming the end of a pipe removes the burrs left inside by cutting. Dies are used for cutting threads on steel pipe.

Various types of wrenches are designed for holding and turning pipe and pipe fittings. Pipe wrenches, pliers, locking pliers, chain wrenches,

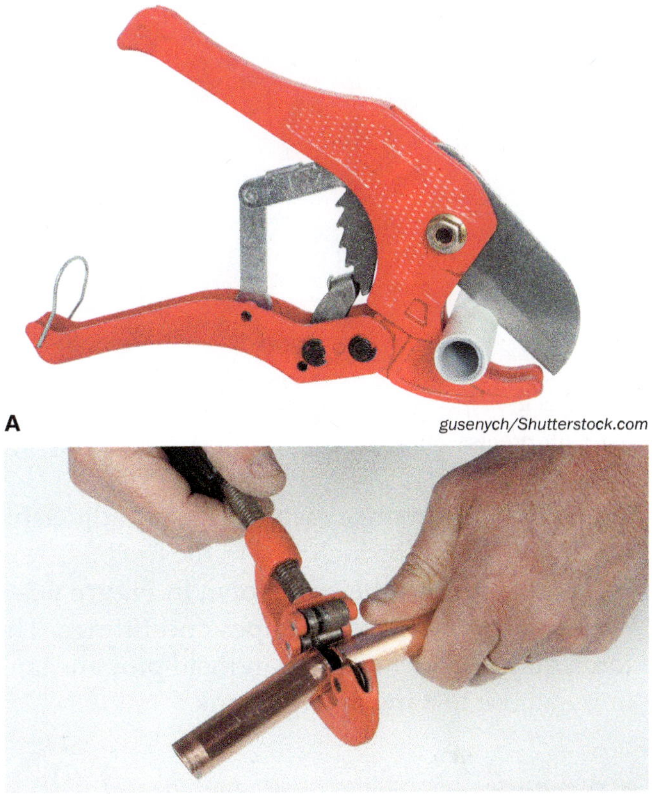

A gusenych/Shutterstock.com

B Hans Geel/Shutterstock.com

Figure 6.5-6. Pipe cutters. A—Plastic pipe cutter. B—Tubing cutter.

Safety Note

Follow these safety rules when working with portable band saws:

- Before use, inspect the portable band saw for any damage, defects, or loose parts. Ensure that the blade is properly installed and tightened according to the manufacturer's instructions. Do not use the saw if it is damaged or in poor condition.
- Securely clamp or hold the workpiece in place using a vise or clamps to prevent it from moving during cutting.
- Maintain a firm grip on the saw with both hands and use a stable stance to support your body weight.
- Operate the saw at an appropriate speed for the material being cut. Avoid using excessive force or pushing the saw too quickly, as this can cause the blade to bind or kick back.
- Be aware of the location of the blade at all times and keep hands and fingers clear of the cutting path. Never reach under or over the blade while it is in motion.

Photo courtesy of Robert Bosch Tool Corporation®

Figure 6.5-7. The portable band saw is a versatile tool used for precise cuts.

strap wrenches, monkey wrenches, and adjustable wrenches should be included in the toolbox.

Torches, such as the one shown in **Figure 6.5-8**, are used to solder copper pipes and fittings. The torch may be attached to a handheld propane tank or by a hose to a much larger tank.

Safety Note

When working with a torch, always be aware of what is behind the workpiece. The heat of a torch flame will quickly ignite many building materials. Make sure the work area is safe and clear of any flammable items. Locate a nearby fire extinguisher before lighting the torch.

Occasionally, components will need to be mounted to a masonry wall or a pipe will need to go through one. For this job, a ***hammer drill*** is used, **Figure 6.5-9**. Hammer drills combine the rotational motion of a standard drill with a rapid hammering or pulsating action. This hammering enables the tool to drill through masonry materials using a carbide-tipped masonry drill bit.

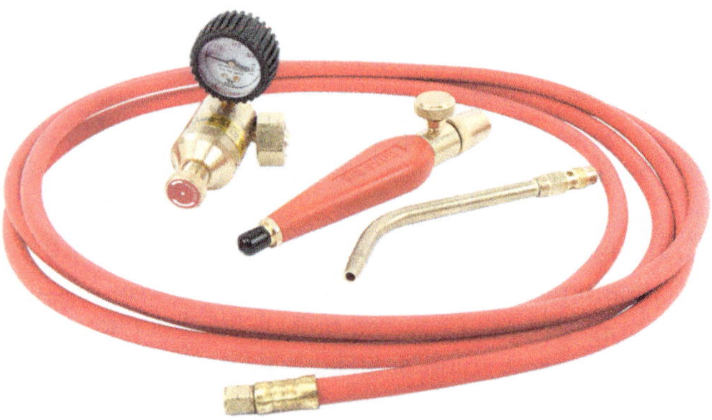

Forney Industries, Inc.

Figure 6.5-8. A plumber's torch kit may consist of either a torch and a handheld tank or a torch, hose, and regulator attached to a larger tank.

Photo courtesy of Robert Bosch Tool Corporation®

Figure 6.5-9. This cordless rotary hammer drill is fitted with a dust extraction system, which reduces exposure to harmful silica dust.

Safety Note

Follow these safety rules when working with rotary hammer drills:

- Wear safety glasses and hearing protection.
- Maintain good balance and a tight grip on the handles. Bits can bind in the hole, tearing the tool from your grasp and possibly knocking you off balance.
- Steadily push into the workpiece, pulling back frequently to clear away chips.
- Keep bits sharp.
- Disconnect the drill from power while installing or removing bits.
- Never lock the trigger in the *on* position.
- Check that the switch is in the *off* position before plugging it in.

LESSON 6.5 Review

Answer the following questions based on information in this lesson.

1. *True or False?* If you need to cut a wire and then strip the sheathing from the end so that you can make an electrical connection, you must use two different tools. (6.5-1)
2. *True or False?* A right-angle drill is designed to be used in tight quarters. (6.5-1)
3. Which power tool is often used to cut steel pipe when installing plumbing? (6.5-1)
   A. Compound miter saw
   B. Portable band saw
   C. Standard table saw
   D. Portable circular saw
4. A _____ is used to cut threads on steel pipe. (6.5-1)
   A. die
   B. chisel
   C. chain wrench
   D. reamer

LESSON 6.6

Safety

LEARNING OUTCOMES

After completing this lesson, you will be able to:

6.6-1 Describe the primary safety hazards associated with electricity and how to avoid them.

TECHNICAL TERMS

arc flash
electric shock
electrocution
ground fault
ground-fault circuit interrupter (GFCI)
lockout (LO)
lockout/tagout (LOTO)
tagout (TO)

Electrical Hazards

MEP trades install and work with electrical components regularly. However, all construction workers must be aware of electrical hazards. Workers who are not aware of the various electrical risks that can be present in their workplace are more susceptible to electrical hazards.

Electric Shock

Electricity moves in closed circuits, and its normal route is through a conductor. ***Electric shock*** occurs when the body becomes a part of the circuit. The degree of injury depends on the path the current takes through the body, the amount of current that flows through the body, and the amount of time the body is in contact with the circuit. A person can experience different levels of electric shock depending on the amount of current flowing throw them, **Figure 6.6-1**.

Electric shock normally occurs in one of three ways; when an individual comes into contact with:

- Both wires of an electric circuit.
- One wire of an energized circuit and the ground.
- A metallic part that has become energized by contact with an energized conductor.

Secondary injuries often occur as a result of electric shock. These include:

- Burns where the electricity enters or exits the body. Electrical burns are serious and require immediate medical attention.
- Falls resulting from a shock that occurs while the worker is in an elevated position, such as a ladder or scaffold.

Electrocution occurs when an electric shock becomes so severe that it results in death. Approximately 8% of all construction worker deaths on the job are attributable to electrocution. It is important to be aware of the electrical hazards that exist on the jobsite.

Fire

Electrical fires can be a real threat to worker safety. Electrical fire hazards can occur because of

Effects of Electric Shock

Level of Current	Effects on Human Body
1 mA	Threshold of feeling. Slight tingling.
5 mA	Shock felt, but not painful yet. Involuntary muscle movements.
10–20 mA	Painful shock. Sustained muscle contraction. Inability to release grip.
100–300 mA	Paralysis of respiratory muscles. Can be fatal. Severe internal and external burns.
2 A	Cardiac arrest (heartbeat stops). Internal organ damage. Death is probable.

Goodheart-Willcox Publisher

Figure 6.6-1. The effects of an electrical shock can range from a slight tingling to death.

conductors, such as extension cords, being undersized to safely carry the current being used, improperly sized circuit breakers, too many devices being plugged into a circuit, faulty wiring, or a faulty electrical device.

Arc Flash

An *arc flash* is a quick discharge of electrical energy through the air caused by a high-voltage gap and a breakdown between conductors. An arc flash emits thermal radiation (heat) as well as bright, intense light that can cause burns. High-voltage arcs can also generate significant pressure waves.

Explosion

When electricity ignites an explosive mixture of materials in the air, an explosion can occur. Hazardous conditions can exist when vapors from fuel, solvents, or other materials have been allowed to collect.

Avoiding Electrical Hazards

Now that you have learned the potential dangers of an electrical incident, consider the conditions that are often responsible for common electrical incidents such as:

- Contact with power lines.
- Insufficient ground-fault protection.
- The path to ground being absent or broken.
- Improper use of extension and flexible cords.
- Failure to operate equipment as directed.
- Failure to follow lockout/tagout procedures.

Contact with Power Lines

Power lines, both buried and overhead, carry high-voltage power. Contact with these lines can be fatal. Operating or moving equipment, such as cranes, aerial lifts, and excavators, or even tools such as ladders and aluminum paint rollers, within a range that can contact power lines requires extreme caution. Warning signs raise awareness and help protect workers, **Figure 6.6-2**.

gobigoo/Shutterstock.com

Figure 6.6-2. Be mindful of the location of power lines, both overhead and underground.

Safety Note

Follow these safety rules to avoid electrical hazards:

- Call 811 to dispatch utility companies to locate buried power lines before any jobsite excavation.
- Watch for power lines that are overhead and mark any that are buried.
- Maintain a distance of 10′ from overhead power lines.
- Assume lines are energized unless verified otherwise.
- Never use a metal ladder when doing electrical work. If the ladder accidentally contacts a current-carrying conductor, serious injury or even death could result. See **Figure 6.6-3**.
- Ensure lines are de-energized when working near them.

WP-7824/Shutterstock.com

Figure 6.6-3. Always carry ladders horizontally, not in an upright position. Aluminum ladders like this one should not be used near power lines.

Ground-Fault Protection

A *ground fault* is an unintentional electrical discharge to ground; the path can occur through a grounding conductor, or through any conducting pathway, including a person, to ground. This connection creates an alternative path for electric current, bypassing the circuit. Ground faults can occur because of damaged insulation, faulty equipment, or improper wiring. They can also occur if tools or electrical connections get wet. Ground faults can lead to electric shock, fires, or damage to electrical systems. *Ground-fault circuit interrupters (GFCIs)*, **Figure 6.6-4**, are designed to detect and interrupt the flow of electric current when a ground fault is detected, thereby preventing potential harm to people or property. Construction work can cause damage to extension cords and power tools through normal wear and tear. Without ground-fault protection, electrical hazards are more likely to occur.

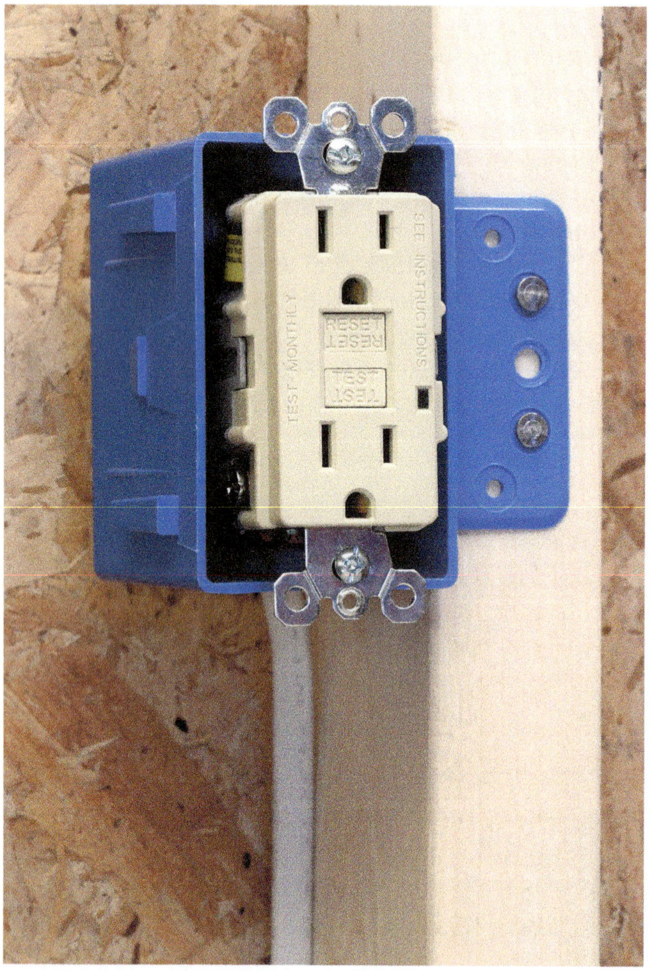

Brandon Blinkenberg/Shutterstock.com

Figure 6.6-4. Ground-fault circuit interrupters (GFCIs) may be built into circuit breakers or receptacles.

Safety Note

Follow these safety rules when working with ground-fault circuit interrupters:

- Use GFCIs on all 110-volt power sources.
- Test GFCIs regularly to ensure proper operation.
- Use tools and equipment that have a double-insulated marking on the label or body of the tool, **Figure 6.6-5**.
- Use tools and equipment according to manufacturer specifications.
- Perform a visual inspection of all electrical equipment, checking for damaged cords, broken grounding prongs, or other damage.

Path to Ground

The ground connection for electrical systems, tools, and equipment is a safety mechanism that helps to prevent electric shock. In the event of a fault or malfunction, the ground connection provides a path of least resistance for the electrical current to return to ground. In the absence of proper grounding, fault current can pass through a worker's body.

Use of Extension and Flexible Cords

Regular use of extension and flexible cords on the job can lead to deterioration, causing wires to become loose or exposed. This deterioration introduces hazardous conditions that can pose a

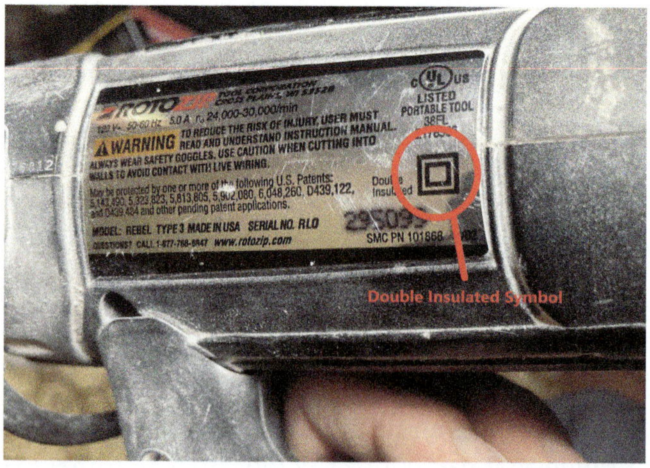

Goodheart-Willcox Publisher

Figure 6.6-5. Look for the symbol indicating tools are double insulated.

Safety Note

Follow these safety rules to ensure proper grounding:

- Ensure all power supply systems, electrical circuits, and electrical equipment are properly grounded.
- Inspect electrical systems regularly to make sure that the path to ground is continuous.
- Do *not* remove ground prongs from extension cords or power tool cords.
- Ensure all exposed metal parts of electrical equipment are properly grounded.

safety risk. All extension cords used in construction should have three wires (two conductors and a ground wire), be designed for hard usage, and not be modified.

Safety Note

Follow these safety rules when working with extension cords:

- Use factory-made cords.
- Only use extension cords rated for hard or extra-hard use.
- Only use 3-wire type extension cords.
- Do *not* use cords to lower tools or any other items from elevated surfaces.
- Use the plug end to remove cords from receptacles. Do not yank on the cord to unplug it.
- Check cords regularly to ensure they meet these requirements, **Figure 6.6-6**. Remove any cords from service that do not comply.

Operate Equipment as Directed

Electrical equipment should always be used for its intended purposes. Safety features provided by the manufacturer are only reliable when the tools or equipment are used according to their instructions. Improper usage includes fabricating extension cords from household wire or receptacle boxes, using ungrounded adapter plugs, improperly sizing circuit breakers, modifying cords or tools, and using damaged tools or equipment.

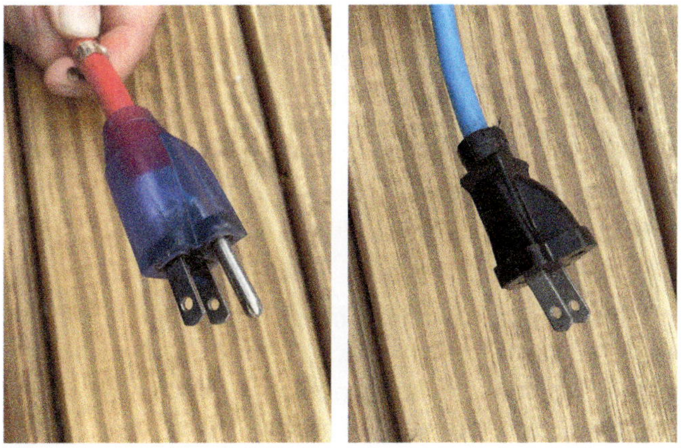

A Goodheart-Willcox Publisher B Goodheart-Willcox Publisher

Figure 6.6-6. A—A serviceable extension cord with the ground prong intact. B—This extension cord is missing the ground prong and should be removed from service immediately.

Safety Note

Follow these safety rules when working with electrical equipment:

- Do *not* use tools and equipment that fail to meet OSHA safety standards.
- Use equipment according to the manufacturer instructions.
- Do *not* use equipment outdoors that is labeled for indoor use only.
- Do *not* modify cords or equipment.
- Inspect tools and equipment to ensure they have not been altered.

Lockout/Tagout

Lockout (LO) is the practice of locking a mechanism or an electrical switch in the open (or off) position so that maintenance or service work can be performed safely. ***Tagout (TO)*** is the practice of placing a tag on a mechanism or an electrical switch to inform others that service work is in process and that this mechanism or switch's position should not be changed.

Often, lockout and tagout are combined into ***lockout/tagout (LOTO)***, in which an electrical switch is locked open and tagged. LOTO is a safety procedure designed to protect workers from the unexpected release of energy by using locks and tags to isolate energy sources such as electrical panels, cords, switches, circuit breakers, or mechanical components to prevent accidental start-up of equipment or energizing a circuit when it is not safe to do so, **Figure 6.6-7**. This helps prevent incidents such as electrocution, crushing, or other types of injuries caused by the sudden release of energy.

Safety Note

Follow these standard lockout/tagout safety rules:

- Never ask someone else to shut off a power source for a system you are working on.
- Assume a power source is live until you directly shut it off.
- Ensure you are up to date on your organization's LOTO training.
- Make sure you are identified on any lockout/tagout devices that are used.
- Follow your organization's approved LOTO procedures.

Rob Byron/Shutterstock.com

Figure 6.6-7. LOTO procedures must be followed in situations with the potential for accidental release of harmful energy.

LESSON 6.6 Review

Answer the following questions based on information in this lesson.

1. *True or False?* Any direct exposure to electricity is considered electrocution. (6.6-1)
2. An arc flash emits _____. (6.6-1)
   A. bright light without thermal radiation
   B. thermal radiation without light
   C. both bright light and thermal radiation
   D. neither bright light nor thermal radiation
3. *True or False?* A person can act as a conducting pathway for electricity to the ground. (6.6-1)
4. *True or False?* Tagout is the practice of locking a mechanism or an electrical switch in the open position so that maintenance or service work can be performed safely. (6.6-1)

LESSON 6.7

Math

LEARNING OUTCOMES

After completing this lesson, you will be able to:

6.7-1 Calculate the area of various shapes as they apply to estimating construction materials.

TECHNICAL TERMS

area	diameter	square foot
circumference	radius	

Estimating Materials Using Area

One of the most common calculations in construction is that of *area*. Area is a measure of total surface. Many materials used in construction are sold in quantities based on how many square feet they cover. Products that cover a surface, such as wall sheathing, roofing, siding, paint, troweled-on adhesives, and floor coverings are just a few examples of products whose quantities are estimated and purchased according to how many square feet they cover.

Pro Tip

Area calculations are among the most common math functions in construction. Any time a surface is covered with a material, an area calculation was likely done to estimate how much material was required.

One *square foot* is a unit of measurement that represents an area equal to 1′ × 1′. The same area (one square foot) is contained in something measuring 6″ × 2′, or 1″ × 12′. The calculation of area depends on the shape in question. For simple geometric figures like rectangles, triangles, circles, or parallelograms, there are specific formulas to calculate their areas.

The area of a surface is always measured in units of square inches, square meters, square feet, and so on. When a number is squared, that means it is multiplied by itself. For example, 3 squared (3^2) is 9. Square units are written with a superscript 2, indicating units × units.

Finding the Area of Squares and Rectangles

To find the area of a square or a rectangle, multiply the number of units it is wide by the number of units it is long.

Area of square or rectangle = Length × Width

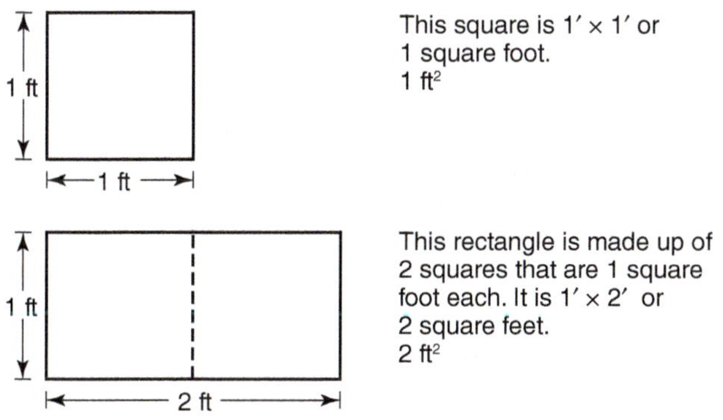

Goodheart-Willcox Publisher

Pipe and Fitting Symbols

Fitting or Valve		Fitting or Valve	
Elbow-90°		T-outlet down	
Elbow-45°		Cross	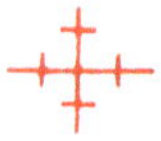
Elbow-turned up		Reducer-concentric	
Elbow-turned down		Reducer-offset	
Elbow-long radius		Connector	
Elbow with side inlet-outlet down		Y or wye	
Elbow with side inlet-outlet up		Valve-gate	
Reducing elbow		Valve-globe	
Sanitary T		Union	
T		Bushing	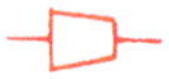
T-outlet up		Increaser	

Goodheart-Willcox Publisher

Figure 6.8-5. Pipe and fitting symbols that appear in drawings.

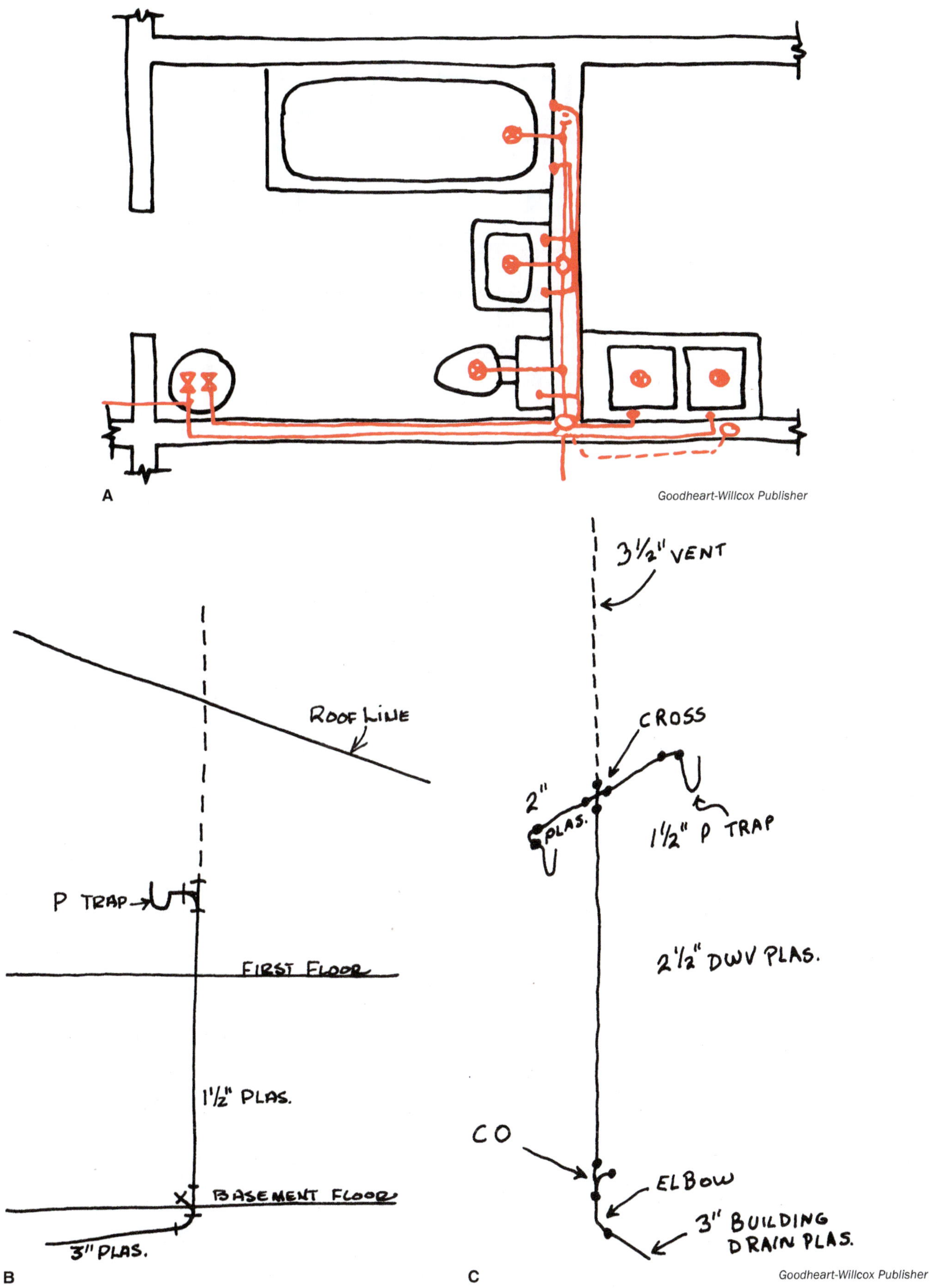

Figure 6.8-6. Three types of sketches. A—Plan view. B—Riser diagram. C—Isometric sketch.

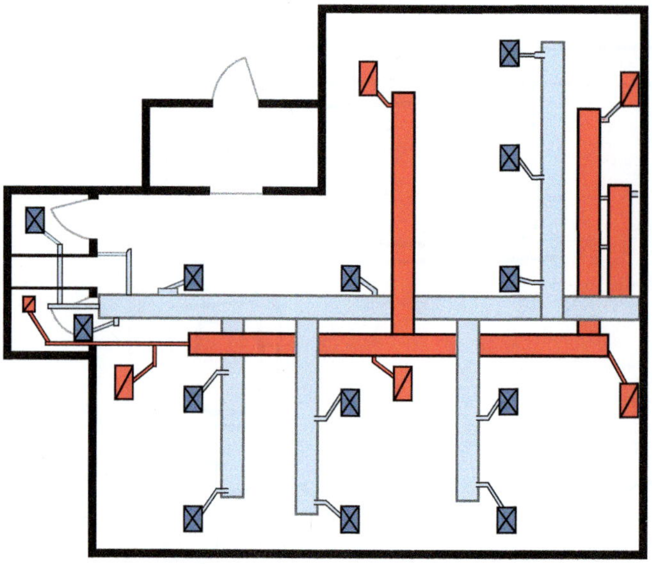

Goodheart-Willcox Publisher

Figure 6.8-7. A shop drawing of HVAC supply and return ducting.

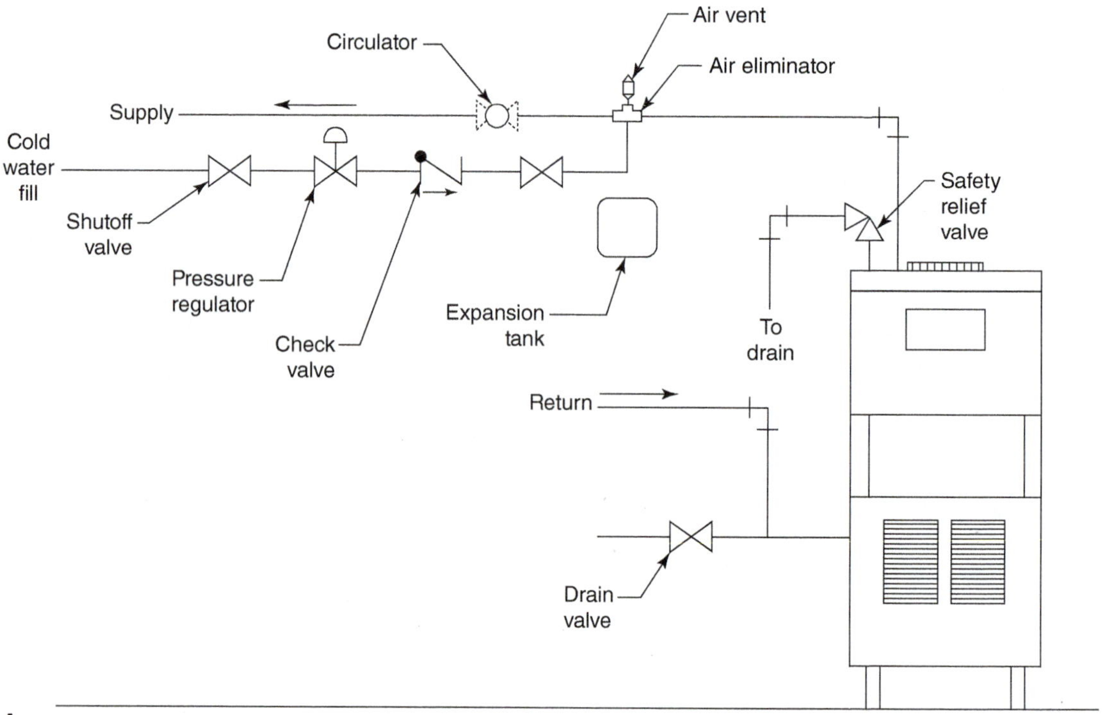

A

Goodheart-Willcox Publisher

Figure 6.8-8. Shop drawings. A—A hydronic boiler system. *(Continued)*

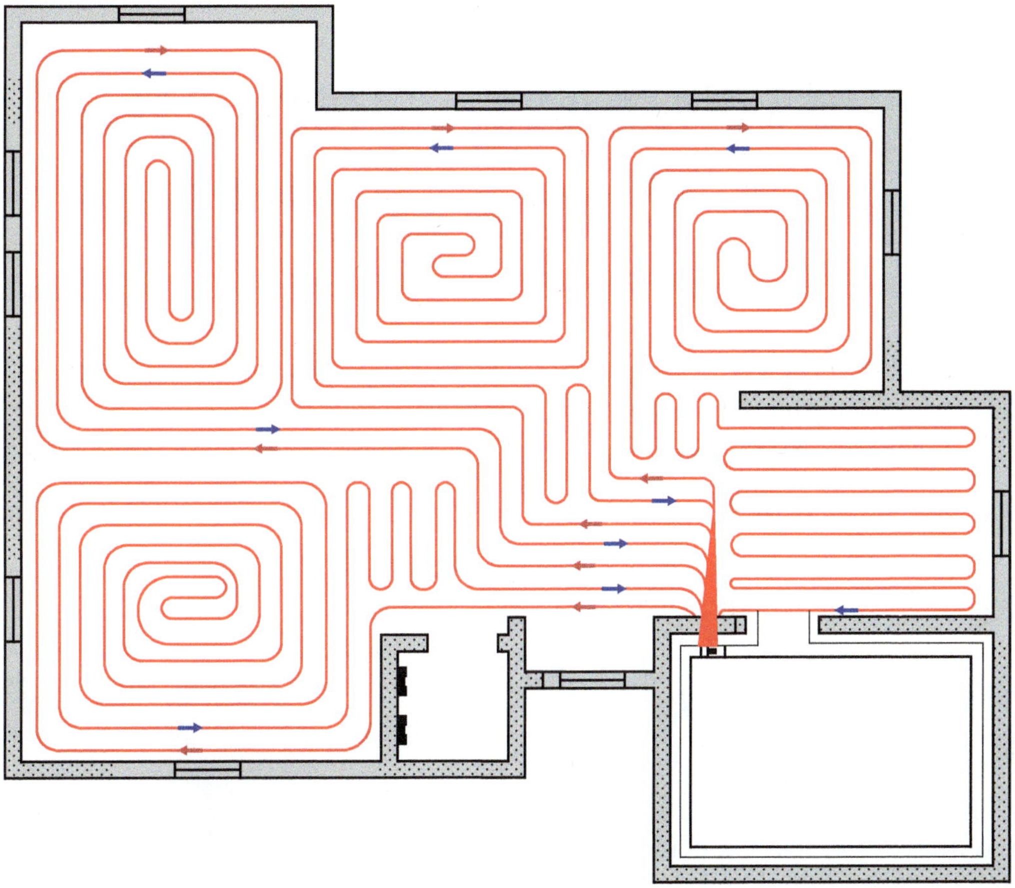

B

Figure 6.8-8. *(Continued)* B—A radiant floor tubing layout.

Goodheart-Willcox Publisher

LESSON 6.8 Review

Answer the following questions based on information in this lesson.

1. *True or False?* If the electrical print of a ceiling shows a circle with the letter V inside, it represents a vent. (6.8-1)
2. Details about a chandelier that will be installed in a dining room can be found on the _____ schedule. (6.8-1)
   A. lighting
   B. panel
   C. equipment
   D. circuit
3. When looking at plumbing plans, you see an empty square drawn on the floor of a utility room. What does it represent? (6.8-1)
   A. Sump pit
   B. Drain
   C. Air duct
   D. Pedestal sink
4. *True or False?* A set of residential working drawings will typically include a set of HVAC plans. (6.8-1)

LESSON 6.9

Green Building and the Future of Construction

LEARNING OUTCOMES

After completing this lesson, you will be able to:

6.9-1 Describe technology that is being used for green building applications in the MEP trades.

TECHNICAL TERMS

modulation

phantom power consumption

Improving Efficiency of MEP Systems

The HVAC field is continually developing technologies to improve efficiency, comfort, and automation for residential heating and cooling systems. Over the past several decades, significant advancements have been made in these areas. The following technologies have been implemented on a broad scale in green construction.

High-Efficiency Condensing Boilers and Furnaces

A condensing boiler or furnace is a type of heating appliance that is highly efficient and designed to extract maximum heat from the fuel it burns, **Figure 6.9-1**. On a traditional boiler or furnace, heat is lost through the exhaust created from the combustion processes.

Traditional units typically have efficiency ratings ranging from 78% to 84%. This means that they convert 78% to 84% of the fuel's energy into heat, while the remaining fuel is lost through the flue gases.

Condensing units are designed to maximize heat recovery from the flue gases using a secondary heat exchanger. This results in significantly higher efficiency. Condensing units achieve efficiency ratings of 90% or higher, with some models even exceeding 95% efficiency.

caifas/Shutterstock.com

Figure 6.9-1. The compact design of this condensing boiler allows it to be mounted to the wall.

Another factor that has improved efficiency is modulation. Traditional furnaces and boilers operate at full capacity when they are on and shut down when the heating demand stops. ***Modulation*** refers to the ability of the heating system to adjust its heat output based on the desired indoor temperature, effectively matching the required heating load. Modulation is achieved using variable gas valves and variable speed blowers. These units compare outdoor air temperature with indoor air temperature and adjust their output using modulating control systems. This type of operation significantly reduces lost efficiency due to system components repeatedly heating up to operating temperature and then cooling down when the system shuts off. It also creates a more comfortable living environment by more precisely meeting the heating requirements of the home. Modulation has been incorporated into cooling systems through the use of variable refrigerant flow systems and multistage compressors.

Ductless Mini-Split Systems

Ductless mini-split systems have gained popularity due to their flexibility and energy efficiency, **Figure 6.9-2**. They allow for individual room control, making them ideal for retrofitting older homes, room additions, or areas with specific heating and cooling requirements. Mini-splits offer options to supplement existing systems or challenging layouts.

Geothermal Heat Pumps

A heat pump transfers heat from one place to another using refrigeration principles. It operates based on the concept that heat can be moved against the natural flow of temperature, allowing it to provide both heating and cooling functions. Heat pumps are known for their energy efficiency because they move heat rather than generate it by burning fuel. Electrical energy is required to power the compressor and other components; however, the heat pump can produce more heating or cooling energy than the electrical energy it consumes.

Air-source heat pumps move the heat in the air outside the home as a source for operation. Because there is little heat energy available to move in subfreezing conditions, air-source heat pumps have reduced efficiency for heating in colder climates.

Geothermal heat pumps utilize the relatively stable temperature of the ground or a nearby water source to heat or cool homes, **Figure 6.9-3**. They are highly energy-efficient, even in colder climates. Geothermal systems circulate a fluid through underground pipes to exchange heat with the ground, providing both heating and cooling capabilities.

Though geothermal heating systems are efficient, they are not suitable for every building site. Piping

Pro Tip

Conducting a life cycle cost analysis to evaluate the long-term costs and benefits of different MEP system options, including installation, operation, maintenance, and replacement costs, can help identify the most cost-effective and sustainable solutions over the building's lifespan.

A   Andrey_Popov/Shutterstock.com

B   BNMK 0819/Shutterstock.com

Figure 6.9-2. A ductless mini-split system. A—A small condensing unit on the exterior. B—A wall-mounted head unit inside.

Studio Harmony/Shutterstock.com

Figure 6.9-3. Geothermal fields or wells are used to maintain a consistent source of heat for efficient operation of the heat pump.

used to exchange ground-source heat is extensive and must be installed by drilling wells or excavating large areas of ground. Installation costs can also be extensive for these systems.

Smart and Connected Systems

The rise of smart technology has revolutionized HVAC systems. Smart thermostats, for instance, enable homeowners to create a schedule and control temperature settings remotely using smartphones or other devices, **Figure 6.9-4**. They often feature advanced scheduling capabilities and can learn users' preferences to optimize energy usage. Additionally, these systems can integrate with home automation platforms, allowing for seamless control and coordination of HVAC systems with other smart devices.

Smart energy monitoring systems enable homeowners to track their energy usage, giving them insight into how energy is used in their homes and suggesting slight changes to curb excessive use and waste.

Advanced home controls can open or close window shades at the ideal time in coordination with HVAC systems to maximize or minimize solar heat gain. Automated lighting systems can adjust lighting levels based on the availability of natural light or turn off lights in rooms that are not being used.

Phantom power consumption—power use that occurs when electrical devices are in standby mode—is significant in many modern homes. Smart plugs and outlets can reduce energy wasted by cutting power to televisions and computers when they are not in use, for example. These devices can also be used to schedule the operation of appliances, reducing energy bills by using power during times when rates are lowest.

By using smart technology, homeowners can gain insight and control over their energy consumption, maximizing the efficiency of the systems and features of their specific home.

Goodheart-Willcox Publisher

Figure 6.9-4. Smart thermostats are an affordable method of reducing energy consumption.

LESSON 6.9 Review

Answer the following questions based on information in this lesson.

1. *True or False?* When a traditional boiler is used, heat is lost through the exhaust created from the combustion process. (6.9-1)
2. Condensing units achieve efficiency ratings of ____. (6.9-1)
   A. 48% or lower
   B. 62% to 73%
   C. 78% to 84%
   D. 90% or higher
3. *True or False?* Ductless mini-split systems allow for individual room control. (6.9-1)
4. Which statement about smart thermostats is false? (6.9-1)
   A. They can integrate with home automation platforms.
   B. They enable homeowners to track their energy usage.
   C. They are a significant source of phantom power consumption.
   D. They can learn users' preferences to optimize energy usage.

Chapter 6 Review

Summary

Lesson 6.1

- Trade organizations play an important role in residential construction by providing a platform for collaboration, advocacy, education, and networking, strengthening the collective influence of their members and promoting the development, growth, and sustainability of the industry. (6.1-1)

Lesson 6.2

- Trade credentials and certifications can be a big benefit to career progression, as well as staying current with industry changes and fulfilling specific task requirements within your field of expertise. (6.2-1)

Lesson 6.3

- Mechanical, electrical, and plumbing systems, known as MEP systems, all have components that must be installed before a home is insulated and interior surfaces are applied. This is referred to as the rough-in phase of construction. (6.3-1)
- Major HVAC rough-in components include ducting for forced-air and make-up air systems, primary piping for hydronic systems, and heating appliances with support equipment. (6.3-2)

Lesson 6.4

- MEP systems rough-in requires a wide variety of materials and components: service entrance components, switch and receptacle boxes, conductors, and conduit for the electrical system; pipes, valves, fasteners, and some fixtures for the plumbing system; and ductwork and tubing for HVAC systems. (6.4-1)
- The MEP rough-in uses a variety of fasteners from electrical cable staples to sheet metal screws and concrete and masonry anchors. Different sealants are used to maintain proper connections between pipes and fittings. (6.4-2)

Lesson 6.5

- Along with common tools, MEP trades use hand and power tools specific to their trade, such as a fish tape and hammer drill. (6.5-1)

Lesson 6.6

- Working with or near electricity can be dangerous. Hazards include electrical shock, electrocution, fires, arc flashes, and explosions. (6.6-1)

Lesson 6.7

- Many materials used in construction are sold in quantities based on how many square feet they cover. One square foot represents an area equal to $1' \times 1'$. (6.7-1)

Lesson 6.8

- MEP plans are not always included in a set of residential drawings. These trades rely heavily on experienced tradespersons and shop drawings to render plans for the layout and execution of their work. (6.8-1)

Lesson 6.9

- HVAC is one of the advanced trades with regard to development in green technology. (6.9-1)

Apply and Analyze

1. Should you join a trade organization when you start working in the residential construction industry? Why or why not? (6.1-1)
2. Is it a good idea for people just getting started in the residential construction industry to pursue credentials? Why or why not? (6.2-1)

3. While reviewing the plans for a new home, you notice that the DWV system does not seem to include vent piping. Should you ask your supervisor about this? Why or why not? (6.3-1)
4. A plumber is installing a hydronic perimeter system in a new home, and the homeowner, who is trying to save money, asks whether it is really important to include an expansion tank. What should the plumber tell them? (6.3-2)
5. When instructed to connect two sections of PVC pipe, a plumber's new apprentice applies PVC glue to both pieces and then puts them together firmly and allows them to set for an appropriate length of time. Is the connection likely to leak? Why or why not? (6.4-2)
6. Today, you will be helping a plumber bore holes in the framing members in a new home that is being constructed. When selecting your cutting tools for the day, what types of materials should they be able to cut through, and why? (6.5-1)
7. The electrician you are working with often quizzes you on safety matters. Today, she asks you how electric shocks normally occur and wants you to give her an example of each. What should you tell her? (6.6-1)
8. A carpenter's apprentice is asked to determine the perimeter of a semicircular window that has a diameter of 20". The apprentice works it out and tells the carpenter that it has a perimeter of 62.8". Is the apprentice right or wrong, and how do you know? (6.7-1)
9. When a plumber looks at the architectural plans for the new home that is being built, they notice that the plans do not include pipe drawings. What should the plumber do at this point, and why? (6.8-1)
10. An energy audit of an existing home indicates a significant degree of phantom power consumption is occurring and affecting the homeowner's energy bills. Explain what energy consumption is and how the homeowner could use smart technology to help reduce their energy bill. (6.9-1)

Critical Thinking

1. Repairing plumbing can be a messy business. What precautionary measures would you take if you had to repair a water supply system? What about a drainage, waste, and venting (DWV) system? (6.3-1)
2. What are the implications of owning and living in a home with an HVAC system that performs poorly? (6.3-1)
3. List three safety and health issues that an electrician might face. What precautionary measures should they take to protect themselves from each? (6.6-1)

Communicating about Construction

1. **Reading.** With a partner, create flash cards for the key terms in the chapter. On the front of the card, write the term. On the back of the card, write the phonetic spelling text. (You may use a dictionary for this.) Practice reading aloud the terms, clarifying pronunciations where needed. (6.3-1)
2. **Speaking.** Make a series of safety posters that provide information about the dangers of electric shock. Explain your posters to the class. (6.6-1)
3. **Listening, Writing, and Speaking.** Search online for local unions that may have workshops, guest speakers, or presentations on HVAC installation and new materials. Look online for instructional videos. Also, your local library may have multiple media covering various techniques. Watch videos, attend a presentation, and take notes. Listen for terms that you have learned from your textbook. Write a short summary of how the HVAC field continues to increase efficiency and be prepared to present your findings to the class. (6.9-1)

ungvar/Shutterstock.com

CHAPTER 7 Interior Insulation and Surfaces

Chapter 7 Outline

Introduction

Upon completion of the MEP systems rough-in phase of construction, the home is ready to be insulated and have the interior surfaces such as drywall, tile, and other wall and floor finishes applied. Hanging drywall and installing floors are physically challenging tasks, as are many within construction. This chapter highlights the physical demands placed on workers who build homes as a career and then examines commonly used materials, tools, and fasteners for interior insulation and surfaces and their installation processes. As with other phases of construction, materials must be stored and handled properly to ensure worker safety. This chapter will discuss those practices as well as fire prevention and protection strategies.

When installing interior surfaces, as well as during all phases of construction, the ability to accurately estimate materials is a desirable skill that will benefit workers throughout their careers. This chapter covers how to build on those skills by calculating volume.

Finally, this chapter covers green building methods to add extra insulation to a home, effectively superinsulating the structure for minimum heat loss.

LESSON 7.1

Industry and Careers

LEARNING OUTCOMES

After completing this lesson, you will be able to:

7.1-1 Outline the steps to apply for a job.

TECHNICAL TERMS

cover letter
portfolio
reference
résumé

Applying for a Position

Preparing to enter the workforce involves a number of steps. Having a well-written résumé and cover letter are important items to get started. Before applying for a job, you will also need to organize and update your portfolio. A ***portfolio*** could include photos, documentation, or samples of projects you have participated in or completed to showcase your experience and achievements to potential employers. A portfolio should demonstrate your skills and capabilities; it is not meant to only be a list of achievements. Each piece in a portfolio should tell a story about your experience and interest in construction. Finally, preparing for and practicing interviewing techniques will ensure you are well prepared.

Creating a Résumé

A ***résumé*** is a brief outline of your education, work experience, and other qualifications for work. A well-written résumé can help you stand out among applicants and move you on to the interview process. A résumé contains several sections, **Figure 7.1-1**. Make sure to proofread or ask a friend to review your résumé to ensure it is precise, is without errors, and conveys how you are the best applicant for the position.

Most employers request an electronic résumé submitted via email, through third-party job-search platforms such as LinkedIn or Indeed, or through the employer's website.

Pro Tip

When posting a résumé to an online job-search site, be sure to save it as a PDF or as "text only" without any formatting. Employers may search for key terms that match their descriptions of an ideal job candidate, so you want to ensure the file is accessible.

Along with the résumé, you need to develop a list of references. A ***reference*** is an individual who will provide credible information about you to a prospective employer. A reference can be a teacher, school official, previous employer or coworker, or any other adult outside your family who can speak to your work ethic or character.

Employers typically ask for at least three references. Before submitting someone's contact information, always get permission from each person to use their name as a reference. Be mindful of people's privacy and only share the sensitive identifying and contact information of your references with an employer who has requested them. References may also write letters of recommendation for you. These give an employer a more in-depth look at your skills.

Writing a Cover Letter

The ***cover letter*** is a letter sent to an employer along with a résumé to provide more contextual details about why you are the most effective candidate for an open position. It is often the first contact you have with a potential employer and can make a lasting impression, **Figure 7.1-2**.

Michael J. Garcia

134 Lincoln Street
Wilton, CA 93232

(212) 555-1234
mjgarcia22@e-mail.com

Career Objective

To obtain an apprenticeship in the carpentry field.

Work Experience

Builder's Supply, Holloton, CA — September 2024–present
Retail Sales Associate
- Provide assistance to customers in hardware, plumbing, and electrical departments.
- Review department inventory.
- Maintain neat and orderly displays.

Builder's Supply, Holloton, CA — March 2023–September 2024
Receiver/Stocker
- Unloaded, stored, and stocked merchandise.
- Operated fork trucks and hand trucks.

Tony's Grocery Store, Wilton, CA — May 2022–March 2023
Cashier
- Operated checkout scanner and bagged groceries.
- Answered customer questions.
- General stocking and cleaning throughout store.

Education

Associate Degree in Construction Technology — May 2022
Oceanside Community College
- GPA: 3.22/4.0
- Coursework included residential carpentry, wall and roof systems, and green building.
- Participated in SkillsUSA chapter.

Community Service

Habitat for Humanity, volunteer, summers of 2020, 2021, 2022
Wilton Food Bank, volunteer, 2021–present

References

Available upon request.

Goodheart-Willcox Publisher

Figure 7.1-1. A well-prepared résumé succinctly summarizes your work experience, education, and related information.

LESSON 7.2

Construction Fundamentals

LEARNING OUTCOMES

After completing this lesson, you will be able to:

7.2-1 Describe the physical demands of working in the residential construction industry.

TECHNICAL TERMS

dehydration
exoskeleton
frostbite
heat exhaustion
heat stroke
hypothermia

Physical Demands

In *Chapter 6, MEP Systems Rough-In*, you learned about how trade credentials can make an employee more valuable to employers by preparing them to better perform their work. Another way employees can improve their performance is by taking care of their physical body. Construction work can be physically demanding. Workers often face physical challenges due to the nature of their tasks. Lifting, climbing, stooping, kneeling, and prolonged periods of standing are all part of the job. Construction workers are expected to handle heavy and cumbersome materials and tools, sometimes while working at great heights. On many sites, outdoor work continues in the heat of summer and through the cold of winter. The following are some suggestions to prepare for the physical demands of construction work.

Heavy Lifting and Ergonomics

Construction workers often need to lift and carry heavy or awkward materials, tools, and equipment. This can put strain on their muscles and joints, increasing the risk of injuries. A musculoskeletal disorder (MSD) is an injury or illness that affects the muscles, bones, tendons, ligaments, and other components of the musculoskeletal system. These injuries can occur from repetitive movements or actions, poor posture, overexertion, and vibrations, among other things. MSDs can include repetitive strain injuries and conditions such as:

- Lower back pain/injury
- Carpal tunnel syndrome
- Tendinitis
- Muscle strains
- Tennis elbow
- Rotator cuff injuries

To reduce the strain on workers' bodies, companies have been developing specialty equipment. An ***exoskeleton***, for example, is a framework fitted and worn by a worker to reduce the fatigue of lifting or working overhead. These devices mimic the joints of the human body and boost muscle strength using springs, air pressure, or electric motors and actuators. This technology is intended to assist workers when performing tasks that can strain the body, such as lifting and holding heavy sheets of gypsum in place or performing overhead work for extended periods. With arm-strength exoskeleton technology, moving a stack of CMU block would be less physically demanding for laborers. The residential construction industry has yet to broadly adopt this technology, but these tools may eventually become standard on every physically demanding job, **Figure 7.2-1**.

Safety Note

To reduce the likelihood of developing a serious injury while working in a construction career, follow these guidelines:

- Be mindful about the limitations of your body. Some jobs will be uncomfortable and require physical exertion, which is different from injury. However, if you suspect a task is likely to cause an injury, perhaps you can be productive on another assignment while your body recovers.
- Do not strain to lift more than you are comfortable with. Ask for help if an object is too heavy to move alone. Consider using equipment to lift or move materials if possible.
- You've probably heard someone say, "Lift with your legs, not with your back!" When lifting a heavy object stand close to the load, bend your knees, and grasp the object firmly. Then, lift by straightening your legs and keeping your body as close to vertical as possible, **Figure 7.2-2**. To lower the object, reverse the procedure.
- Avoid bending at the waist when lifting or lowering objects. When carrying a heavy load, do not turn or twist your body. Make adjustments in position by shifting your feet.
- Be proactive about health and fitness. Flexibility and strength training, as well as drinking plenty of water, can reduce the risk of injuries on the job. As Benjamin Franklin said, "An ounce of prevention is worth a pound of cure."

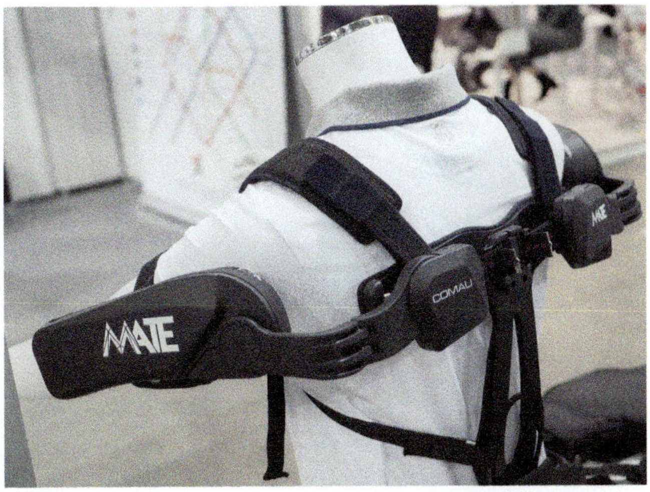

Antonello Marangi/Shutterstock.com

Figure 7.2-1. Exoskeletons are one example of how construction workers can benefit from mechanical assist devices.

Weather Conditions

Adverse weather conditions can create a dangerous jobsite. Snow, ice, or rain can cause slips or other injuries. Lightning, wind, and hail also pose risks to workers; these dangerous conditions require a change in work environment to manage. While snow can be removed, **Figure 7.2-3**, lightning cannot be coaxed to strike elsewhere. If there is a lightning storm or dangerous winds, the job is often shut down or interior work is prioritized.

Temperature extremes, however, create conditions that are often worked through by construction crews. Working outside through the heat of summer and cold of winter can be challenging for

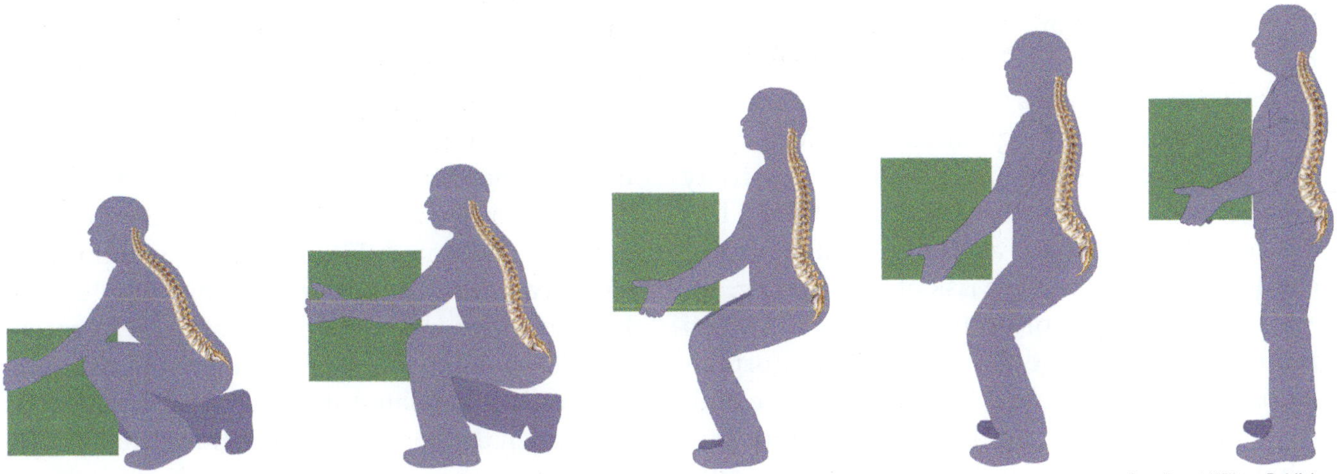

Goodheart-Willcox Publisher

Figure 7.2-2. Proper lifting and carrying techniques are important to avoid injury.

Goodheart-Willcox Publisher

Figure 7.2-3. Snow has been cleared from this scaffolding before use to avoid slips and falls while this crew installs trusses.

many construction workers. Appropriate clothing and thoughtful planning are essential for workers in temperature extremes.

Cold Weather

Construction work in cold conditions can pose several dangers. ***Hypothermia*** is a condition that occurs when the body loses heat faster than it can produce it, resulting in a dangerously low body temperature. Symptoms include shivering, confusion, slurred speech, and drowsiness. In severe cases, it can be life-threatening. ***Frostbite*** occurs when skin and underlying tissues freeze, leading to numbness, discoloration, and tissue damage. Exposed skin and extremities are at risk of frostbite in cold weather.

For jobs during winter and in colder climates, high-quality layers are critical to managing body heat. A heavy parka may initially seem like the most logical choice, but layers can help you regulate your body temperature throughout the day. Even if outdoor temperatures are low, the physical activity during a job will likely make you sweat. With layers, removing one layer as needed will maximize comfort and minimize the risk of overheating.

Keeping the core of your body (torso) warm tends to help manage your overall comfort, allowing for free movement of arms and the ability to manage body heat through the removal of a hat or gloves. Bib overalls and vests are popular with construction workers who work in cold temperatures. Tight-fitting clothing should be avoided because it reduces circulation and can make you colder.

Staying dry in cold weather is important to maintaining your safety and comfort. Wear clothing appropriate for the work and weather conditions. Rain gear can keep you dry and make a wet, rainy day much easier to manage. Keep extra clothing handy. If you get wet, put on dry clothes. Proper footwear is crucial. Insulated boots are available with various concentrations of insulation. Waterproof boots are popular but can restrict the ability for feet to dry from condensation and sweat.

Pro Tip

Take care of footwear to ensure the best protection. Leather conditioners and treatments can help to keep leather boots from cracking and absorbing water from walking through the snow or working in the rain. Ensure your work boots get dried out thoroughly each day after work.

When it comes to gloves, there are different types that are ideal for certain trades. Some jobs, like operating equipment, may not require the same kind of dexterity as pulling nails or screws out of a tool belt. The equipment operator may need warmer gloves because they are moving around less than the roofer and producing less body heat in their extremities.

Drink lots of warm fluids while working to stay hydrated. This is often overlooked in the winter, but a hydrated body will maintain its temperature much better than a dehydrated one.

Hot Weather

During summer and in hotter climates, keeping cool can be difficult. The most important thing to do when working in a high-heat environment is to stay hydrated, **Figure 7.2-4**. Sometimes this can mean drinking water or drinks with electrolytes even when you aren't thirsty. OSHA recommends drinking one cup of water every 20 minutes to avoid heat-related illnesses. ***Dehydration*** occurs when the body loses more fluids than it takes in, leading to an imbalance in the body's electrolytes and impairing its ability to function properly. Dehydration can lead to fatigue, dizziness, muscle

Nong2/Shutterstock.com

Figure 7.2-4. Hydration is one of the most important safety measures you can take when working in high-heat environments.

Safety Note

Heat illness is dangerous, and precautions should be taken in extreme temperatures. Signs of heat illness include:

- Headaches
- Nausea
- Weakness or dizziness
- Heavy sweating or hot, dry skin
- Elevated body temperature
- Thirst
- Decreased urine output

If you or a coworker begin to show these signs, take a break, hydrate, and find a cool spot. Seek medical attention if needed.

cramps, and impaired cognitive function, impacting workers' ability to perform safely.

While on a job in hot weather, wear light-colored, loose-fitting clothes. A hat is also helpful to maintain your temperature by shading your face and neck. Be mindful of shady spots when setting up workspaces on a jobsite, if possible. For example, if you are the person cutting for a crew of carpenters, avoid setting up in the sun if you can. Working in a shaded area can save you a lot of discomfort.

Heat exhaustion is a heat-related illness that can occur with prolonged exposure to high temperatures, combined with physical exertion. Symptoms include heavy sweating, weakness, dizziness, nausea, headache, and fainting. If left untreated, heat exhaustion can progress to heat stroke, a life-threatening condition. ***Heat stroke*** is a severe and life-threatening heat-related illness that occurs when the body's temperature regulation system fails, leading to a dangerously high body temperature. Symptoms include confusion, rapid heartbeat, hot and dry skin (in contrast to heat exhaustion, where the skin is usually moist from sweating), and unconsciousness. Heat stroke requires immediate medical attention.

Working in temperature extremes gets easier as your body acclimates. However, it is important to recognize when conditions are becoming dangerous. Pay close attention to weather forecasts and know how to prepare yourself for those conditions.

LESSON 7.2 Review

Answer the following questions based on information in this lesson.

1. *True or False?* Musculoskeletal disorders can be caused by long-term exposure to vibrations on a jobsite. (7.2-1)
2. What is a common sign of a repetitive stress injury? (7.2-1)
   A. Difficulty breathing
   B. Persistent fatigue
   C. Blurred vision
   D. Lower back pain
3. Numbness and discoloration are signs of _____. (7.2-1)
   A. hypothermia
   B. frostbite
   C. heat exhaustion
   D. heat stroke
4. *True or False?* Tight-fitting clothing is recommended in cold weather to help keep you warm. (7.2-1)
5. When working in hot weather, OSHA recommends drinking one cup of water every _____ minutes to avoid heat-related illnesses. (7.2-1)
   A. 10
   B. 20
   C. 30
   D. 45
6. *True or False?* A person experiencing heat stroke will exhibit heavy sweating. (7.2-1)

Department of Energy R-Value Recommendations

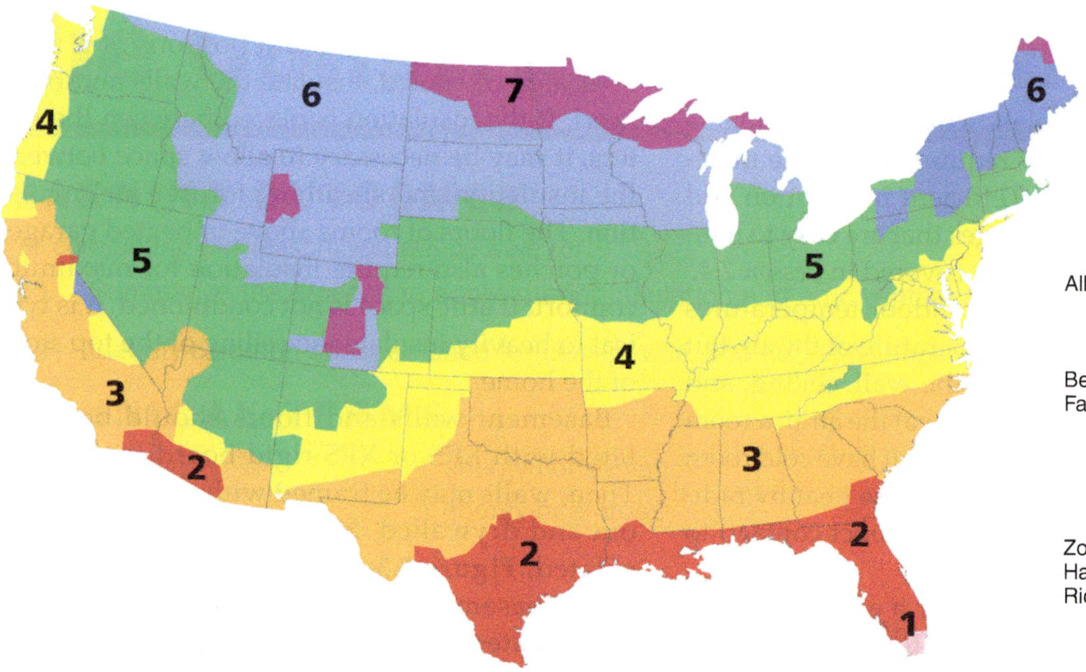

All of Alaska is in Zone 7, except for the following boroughs which are in Zone 8: Bethel, Dellingham, Fairbanks N. Star, Nome, North Slope, Northwest Arctic, Southeast Fairbanks, Wade, Hampton, and Yukon-Koyukuk
Zone 1 includes: Hawaii, Guam, Puerto Rico, and the Virgin Islands

New Wood-Framed Houses

Zone	Heating System	Attic	Cathedral Ceiling	Wall: Cavity	Wall: Insulation Sheathing	Floor
1	All	R30 to R49	R22 to R38	R13 to R15	None	R13
2	Gas, oil, heat pump	R30 to R60	R22 to R38	R13 to R15	None	R13
	Electric furnace					R19 to R25
3	Gas, oil, heat pump	R30 to R60	R22 to R38	R13 to R15	None	R25
	Electric furnace				R2.5 to R5	
4	Gas, oil, heat pump	R38 to R60	R30 to R38	R13 to R15	R2.5 to R6	R25 to R30
	Electric furnace				R5 to R6	
5	Gas, oil, heat pump	R38 to R60	R30 to R38	R13 to R15	R2.5 to R6	R25 to R30
	Electric furnace				R5 to R6	
6	All	R49 to R60	R30 to R38	R13 to R15	R5 to R6	R25 to R30
7	All	R49 to R60	R30 to R38	R13 to R15	R5 to R6	R25 to R30
8	All	R49 to R60	R30 to R38	R13 to R15	R5 to R6	R25 to R30

Existing Wood-Framed Houses

Zone	Uninsulated Attic	Wall: Existing 3 to 4 Inches of Insulation	Floor
1	R30 to R49	R25 to R30	R13
2	R30 to R60	R25 to R38	R3 to R19
3	R30 to R60	R25 to R38	R19 to R25
4	R30 to R60	R38	R25 to R30
5–8	R49 to R60	R38 to R49	R25 to R30

Wall Insulation Whenever Exterior Siding is Removed:

Uninsulated wood-frame wall:	Insulated wood-frame wall:
• Drill holes in the sheathing and blow insulation into the empty wall cavity before installing the new siding • Insulative sheathing may be used as necessary to meet R-value requirements	• For Zones 4-8, add R5 insulative sheathing before installing the new siding

Reference: Department of Energy/CE-01802008, Insulation Fact Sheet

US Department of Energy

Figure 7.3-2. North America is divided into zones according to the climate in each area. The R-values listed are recommended by the US Department of Energy.

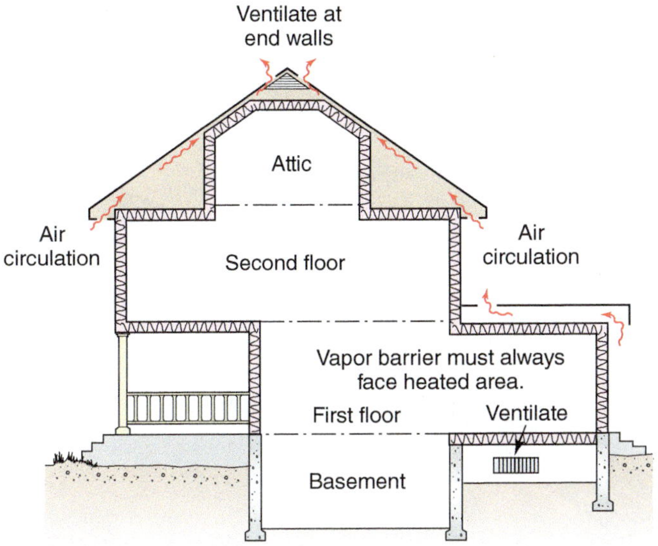

Goodheart-Willcox Publisher

Figure 7.3-3. Where to insulate residential structures. Basement walls and, in many cases, crawl space walls are also insulated.

Owens-Corning

Figure 7.3-4. Insulating a basement wall with R-11 unfaced fiberglass. A framework of 2×4 studs has been installed against the masonry wall and provides support for insulation and drywall or paneling.

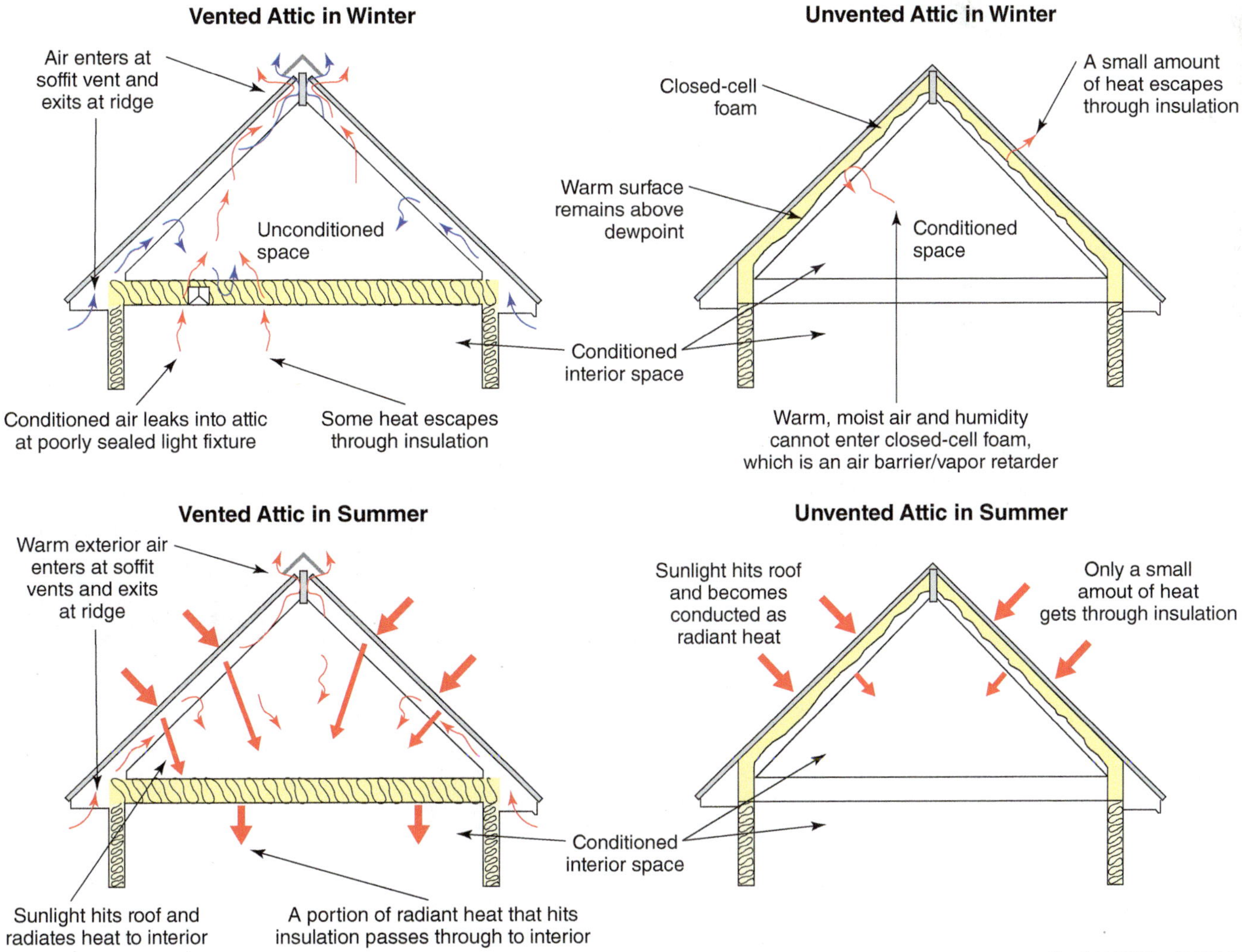

Goodheart-Willcox Publisher

Figure 7.3-5. Common roof systems. Unconditioned attic spaces must be ventilated.

covered with perforated vinyl, hardware cloth, corrosion-resistant wire cloth screening, or similar material.

A special roof truss, called an ***energy-heel truss***, has extra height at the eaves that allows more insulation over the plate, **Figure 7.3-6**. This can also be accomplished with a truss that adds an extended overhang to the building.

Older construction can present problems with heat loss and ice damming where there is not sufficient room between the roof and the wall plate. In such construction, use rigid foam insulation with the highest available R-value per inch of thickness. In any case, when installing thicker insulation at the eaves, do not block the airway from the soffit vent into the attic. Installation of baffles will keep airways open for adequate ventilation, **Figure 7.3-7**.

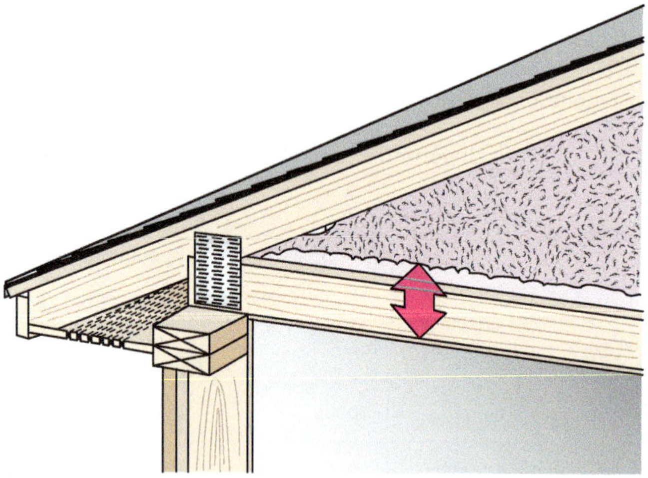

A

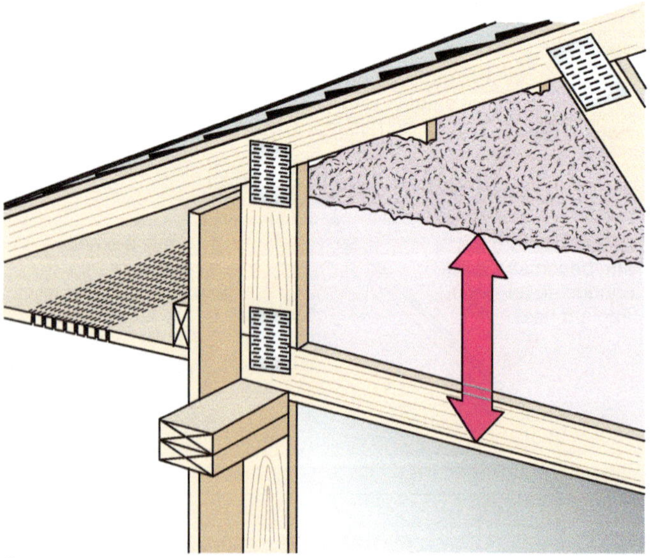

B

Goodheart-Willcox Publisher

Figure 7.3-6. A—Standard roof trusses restrict the amount of attic insulation that can be installed at the eaves of a building. B–Raised-heel trusses extend the wall surface to allow for additional insulation.

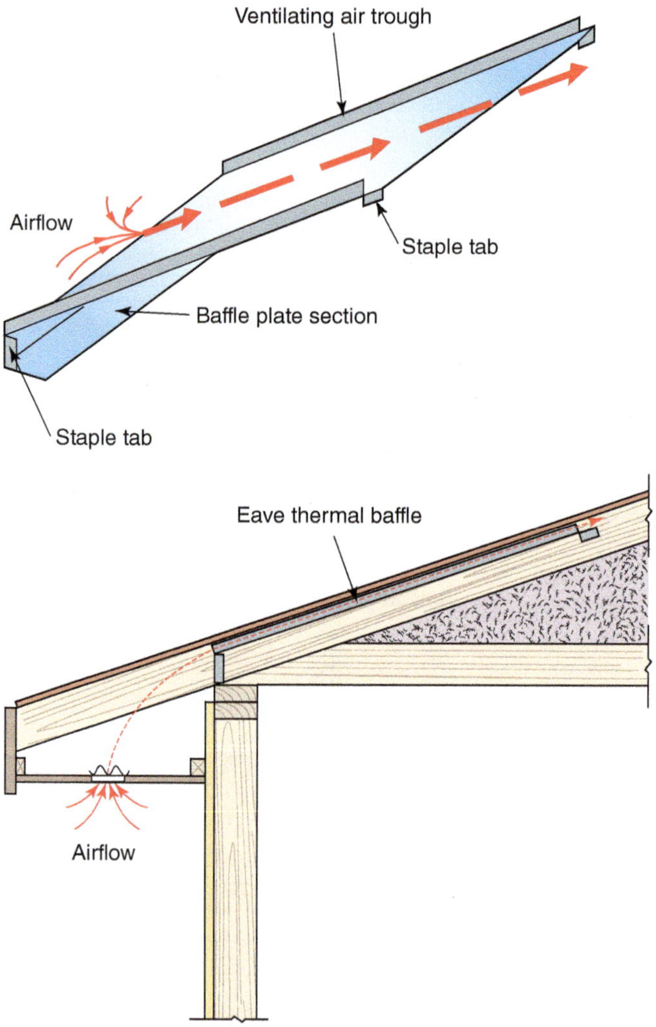

A

Pease Industries, Inc.

B

Owens-Corning

Figure 7.3-7. Maintaining an airway under the eaves for proper attic ventilation. A—These baffles are to be attached to the roof between rafters. They prevent insulation from shutting off airflow. B—This view of the attic shows the baffles installed at the eaves.

Ceiling, Wall, and Floor Coverings

Interior surfaces are crucial when it comes to the look, feel, and overall atmosphere inside a finished home, **Figure 7.3-8**. Considerations such as durability, maintenance requirements, budget, and personal preferences help determine which materials should be used. Because these materials make up the final fit and finish of a home, attention to detail and fine craftsmanship are critical.

BM_27/Shutterstock.com

Figure 7.3-8. The interior surfaces and millwork set the feel for the interior of a home.

Common interior surfaces used in new home construction are listed here.

- Ceilings
  - Drywall—smooth or textured
  - Wood planks
  - Faux wood planks
  - Surface mount ceiling tiles
  - Suspended ceilings
- Walls
  - Drywall—smooth or textured
  - Wood board paneling
  - Wood planks
  - Faux wood planks
  - Tile
- Floors
  - Laminate or vinyl plank flooring
  - Hardwood or softwood flooring
  - Carpet
  - Tile
  - Sheet vinyl

Many of these products have a different set of installation tools and techniques and require a unique skill set to make the job go smoothly. Trade specialists, such as drywall installers, tile installers, or carpet layers, are often brought in to maximize quality and efficiency when it comes to the installation of interior surfaces.

LESSON 7.3 Review

Answer the following questions based on information in this lesson.

1. *True or False?* It is possible to heat the air in a room to the correct comfort level and still have cold room surfaces. (7.3-1)
2. *True or False?* If the attic space in a home is not conditioned, it is typical to leave the ceiling of the top story of the home uninsulated. (7.3-1)
3. *True or False?* When installing insulation at the eaves of a home, it should block the airway from the soffit vent into the attic to reduce the amount of cold air that enters the home during colder months. (7.3-1)
4. Which type of covering is commonly used on floors? (7.3-2)
   A. Drywall
   B. Wood paneling
   C. Sheet vinyl
   D. None of these.

LESSON 7.4

Materials and Fasteners

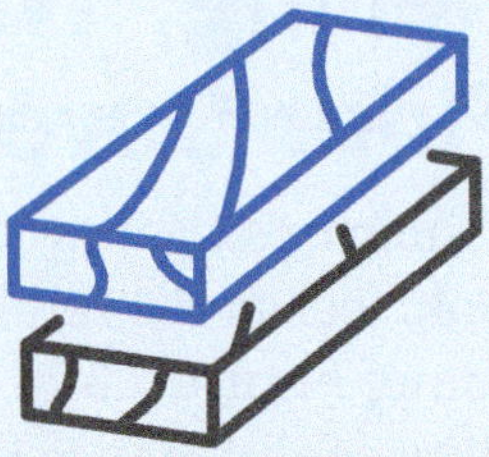

LEARNING OUTCOMES

After completing this lesson, you will be able to:

7.4-1 Classify materials used for interior insulation into five categories.

7.4-2 Describe the materials used for interior surfaces, including ceilings, walls, and floors.

7.4-3 Describe the bonding agents, fillers, surface coatings, and stains used in interior construction.

TECHNICAL TERMS

batt insulation
binder
blanket insulation
blemish
casein glue
cement board
contact cement
drywall
drywall compound
engineered wood flooring
epoxy
floor leveling compound
grout
gunnable
laminate flooring
loose-fill insulation
lumber defect
mastic
pigment
plank flooring
polyurethane adhesive
polyvinyl resin emulsion adhesive
reflective insulation
rigid insulation
side-and-end matched
spray foam insulation
strip flooring
suspended ceiling
thinset
trowelable
uncoupling membranes
underlayment
vinyl plank flooring

Insulation Materials

The variety of material choices available to use inside a home is vast and ever-expanding, **Figure 7.4-1**. As new product developments are made, outdated materials fall out of favor due to more durable or higher-quality materials becoming available, shifting design trends, or new developments in manufacturing. Some materials in this section are less utilized in new construction, but they remain relevant in the industry to match or repair materials on remodeling jobs or because they are a more economical choice.

Thermal insulation is made in many forms and may be grouped into five broad classifications: flexible, loose-fill, rigid, spray foam, and reflective, **Figure 7.4-2**.

Flexible Insulation

Flexible insulation is manufactured in two types: blanket and batt. ***Blanket insulation*** is generally furnished in rolls and in various widths suited to standard stud and joist spacing. It comes in thicknesses of 3/4"–12". The body of the blanket is made of loosely felted mats of mineral or vegetable fibers, such as glass fiber, rock wool, slag, wood fiber, and cotton. Organic fiber mats are usually chemically treated to make them resistant to fire, decay, insects, and vermin.

Cassiohabib/Shutterstock.com

Figure 7.4-1. The product selection for interior surface finishes is continually expanding and changing.

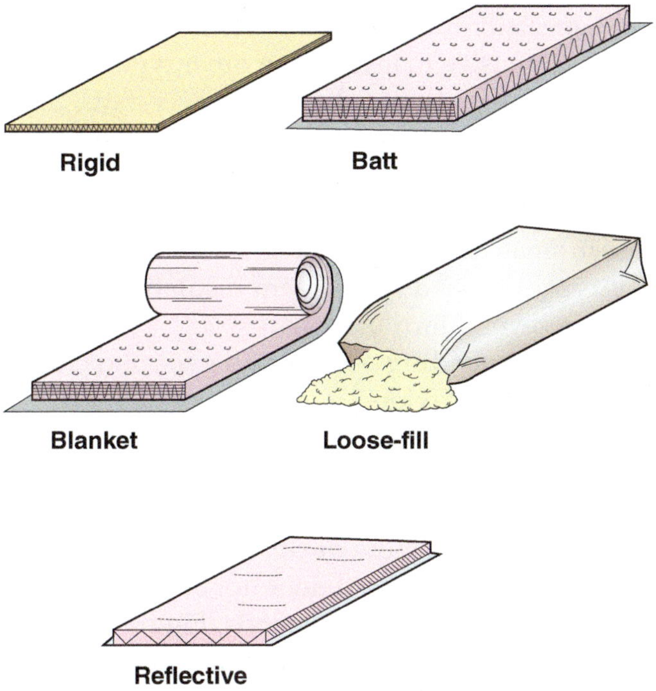

Goodheart-Willcox Publisher

Figure 7.4-2. Insulation is made in these basic forms plus as spray foam. Batt and blanket are both types of flexible insulation.

Safety Note

Follow these safety rules when working with insulation:

- Wear loose clothing. Long-sleeved shirts or blouses that are loose at the neck and wrists, caps, and long trousers will prevent most fibers from coming into contact with the skin. Loose clothing will prevent chafing where fibers do contact the skin. Gloves may be recommended in some circumstances.
- Protect eyes. Use goggles or safety glasses with side shields when applying fiberglass materials overhead or where loose particles or fibers may get into your eyes.
- Wear a mask covering your nose and mouth. This will prevent or reduce inhaling airborne fibers.
- Do not rub or scratch your skin. Instead, thoroughly but gently wash with warm water and soap. Barrier cream applied before working with fiberglass will minimize the effect of skin contact.
- Wash work clothes separately from other clothes. This practice will remove all possibility of fibers being transferred to other clothing. Rinse the washing machine thoroughly before reuse.

Batt insulation is made of the same fibrous material as blankets. Thickness can be greater in this form and may range from 3 1/2″–12″, **Figure 7.4-3**. Batts are generally available in widths of 15″ and 23″ and cut to 4′ and 8′ lengths.

Blankets and batts are available with or without a facing. The facing is a thin layer of material applied to one side of the batt. Typical facing

Goodheart-Willcox Publisher

Figure 7.4-3. Batts are sold in various thicknesses. Thicker batts are designed to give high insulation values for ceilings and attics.

materials include kraft paper, foil, or plastic film. Facing serves to provide vapor control properties to the insulation. It helps to provide additional durability to the insulation, adding to its tear and damage resistance. Foil facing also acts as a radiant barrier. Finally, kraft and foil facing provide a stapling flange as means to fasten the insulation.

Loose-Fill Insulation

Loose-fill insulation is made from such materials as rock wool, glass, slag wool, wood fibers, shredded redwood bark, shredded paper, granulated cork, ground or macerated (softened by soaking in a liquid) wood pulp products, vermiculite, perlite, powdered gypsum, sawdust, and wood shavings. Vermiculite and perlite are both naturally occurring materials that greatly expand, causing them to hold trapped air.

Loose-fill insulation is used in bulk form and supplied in bags or bales. Loose insulation may be poured or blown in using a machine, **Figure 7.4-4**. It is commonly used to insulate spaces between studs or to build up any desired thickness on a flat surface, such as above ceilings. One of the chief advantages of loose-fill insulation is that when insulating an older structure, only a few boards need to be removed in order to blow the material into the walls. Typical coverage of loose fiberglass insulation blown into walls and attics is shown in **Figure 7.4-5**.

Rigid Insulation

Rigid insulation is manufactured in sheets and other inflexible forms. These foam panels have higher R-values per inch than other forms of insulation, but they are more expensive. Their durability and rigidity allow them to be used in conditions where other types of insulation would not be suitable. For more information on rigid insulation types, properties, and R-values, refer to *Chapter 4, Drying In*.

Kurteev Gennadii/Shutterstock.com

Figure 7.4-4. Loose-fill insulation is often used over ceilings in attic areas. Coverage and R-values are listed on the bag.

Spray Foam Insulation

Spray foam insulation is a two-part liquid polyisocyanurate (PIR) material that rapidly expands and hardens after being pumped through a mixing spray nozzle. It is sprayed onto the surface to be insulated. It quickly foams up to several times its original volume and then cures into either semi-flexible foam or rigid foam, **Figure 7.4-6.**

This type of insulation requires special equipment and specially trained personnel, so it is generally more expensive to install than other types of insulation. It offers some advantages over batts, blankets, and loose-fill in that it does an excellent job of filling all voids and sealing any possible air leaks. It is also practical to fill voids in concrete blocks. For use in small quantities, it's also sold in aerosol cans.

Aerosol canned spray foam insulation is available in several formulations:

- **Low expanding**—Fills small gaps and cracks less than 1″ wide.
- **High expanding**—Fills large voids.
- **Window and door**—Insulates between window or door frames and the wall framing. It is more flexible than other spray foam products, so it does not bow the door or window frame as it expands, **Figure 7.4-7**.
- **Fireblock**—Seals around pipes and ducts to prevent flame spread during a fire.
- **Pestblock**—Openings are filled with bitter-tasting foam that repels rodents and other pests.

Reflective Insulation

Reflective insulation is usually a metal foil or foil-surfaced material. It differs from other insulating materials in that the number of reflecting surfaces, not the thickness of the material, determines its insulating value. In order to be effective, the metal foil must be exposed to an air space, preferably 3/4″ or more in depth.

Sidewall Coverage Information				
Nominal R-value	**Thickness (framing timber)**	**Density**	**Minimum weight per sq ft**	**Maximum coverage per bag****
To obtain a thermal resistance (R) of:	**Installed insulation should not be less than (inches):**	**Pounds per cubic foot:**	**Weight per sq ft of installed insulation should not be less than (lb):**	**Contents of bag should not cover more than (sq ft):**
R-14	3.5″ (2×4)	1.8	0.525	55
R-22	5.5″ (2×6)		0.825	35
R-29	7.25″ (2×8)		1.088	26
R-37	9.25″ (2×10)		1.387	21
R-15	3.5″ (2×4)	2.3	0.670	43
R-23	5.5″ (2×6)		1.054	27
R-31	7.25″ (2×8)		1.389	21
R-39	9.25″ (2×10)		1.773	16

Attic/Open Blow Coverage Information				
R-value	**Bags per 1000 sqft.***	**Maximum sq ft per bag***	**Minimum weight per sq ft**	**Minimum thickness**
To obtain a thermal resistance (R) of:	**Bags per 1000 sq ft of net area:**	**Contents of bag should not cover more than (sq ft):**	**Weight per sq ft of installed insulation should not be less than (lb):**	**Installed insulation should not be less than (inches):**
R-60	43.5	23.0	1.307	20.50″
R-50	35.2	28.4	1.057	17.50″
R-44	31.0	32.3	0.928	15.75″
R-38	26.7	37.4	0.803	14.00″
R-30	20.2	49.5	0.606	11.00″
R-26	18.2	54.8	0.547	10.00″
R-22	15.0	66.5	0.451	8.50″
R-19	13.1	76.1	0.394	7.50″
R-13	9.3	107.5	0.279	5.50″
R-11	7.9	126.6	0.237	4.75″

*Coverage per bag does not include framing members, which will increase coverage, depending on 16″ or 24″ OC.
**For sidewall applications—To compensate for wall framing, the net coverage per bag should be increased. To calculate for conventional wood stud framing, when framing is 16″ OC multiply coverage value shown by 1.14. When framing is 24″ OC multiply coverage value shown by 1.11.

Ark Seal, Inc.

Figure 7.4-5. These R-values and coverage charts are for a blown-in type of insulation.

Other Insulation Materials

There are insulation materials that do not fit the other classifications. Some examples include a confetti-like material mixed with adhesive and sprayed onto the surface to be insulated, multiple layers of corrugated paper, and lightweight aggregates like vermiculite and perlite used in plaster to reduce heat transmission.

Lightweight aggregates made from blast furnace slag, burned clay products, and cinders are commonly used in concrete and concrete blocks. They improve the insulating qualities of these materials.

Interior Surface Materials

Interior surface materials are the finished surfaces that cover the walls, ceilings, and floors of a home. Many of these materials, such as drywall and wood board paneling, have finishes that are progressively improved by the installer. Others,

CertainTeed Corporation

Figure 7.4-6. The installation of spray foam requires special training and equipment.

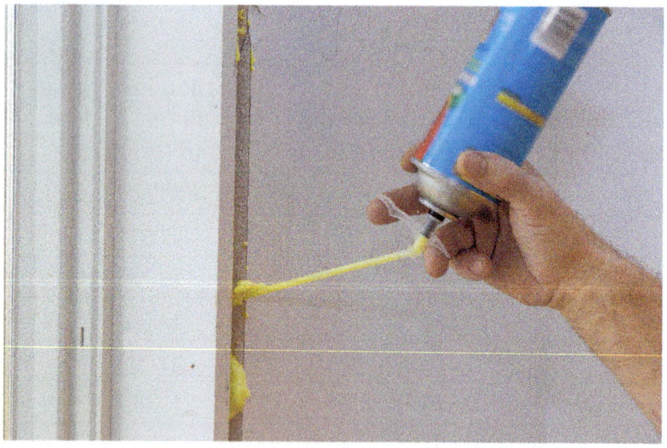

ungvar/Shutterstock.com

Figure 7.4-7. The space between door and window frames and the wall framing should be filled with flexible spray foam.

such as prefinished hardwood, laminate, or vinyl plank flooring, are a finished product as soon as they go in. These products require extra care when handling or installing them to keep them from being damaged.

Drywall

Drywall, also called *sheetrock, wallboard,* or *gypsum wallboard,* is the most common wall and ceiling covering used in construction. It consists of a core of gypsum, a naturally occurring mineral, sandwiched between two layers of paper or fiberglass mats. It is well suited to remodeling jobs, and for resurfacing damaged or cracked plaster walls. For both new construction and remodeling, 1/2″ or 5/8″ gypsum wallboard is used. Ceilings are covered first, then the walls, **Figure 7.4-8**.

Gypsum wallboard comes in several types, including regular, type X, and mold resistant. Regular wallboard is used where a special type is not specified. Type X has an improved, thicker core that does not crumble as quickly as other boards in a fire. It is specified for fire-rated construction, such as to separate an attached garage from living spaces or to separate attached dwelling units in an apartment building. A mold-resistant wallboard incorporates various methods of improving resistance to mold and mildew.

Drywall Installation

Drywall is installed by cutting the sheets to fit the dimensions of the area to be covered and then screwing them to the framing members.

When screwing drywall into framing members, there are two methods of arranging the sheets: parallel or perpendicular, **Figure 7.4-9**. A parallel arrangement has the long edges of panels run in the same direction as studs and joists. A perpendicular arrangement has the long edges of panels are at right angles (90°) to studs and joists. In either method, vertical wall joints must be centered on studs for proper fastening.

The perpendicular arrangement method is preferred because it saves time and reduces costs due to fewer feet of joints to be finished. This method has

Goodheart-Willcox Publisher

Figure 7.4-8. After ceilings are drywalled, apply drywall to walls. Try to get a close fit in the corners.

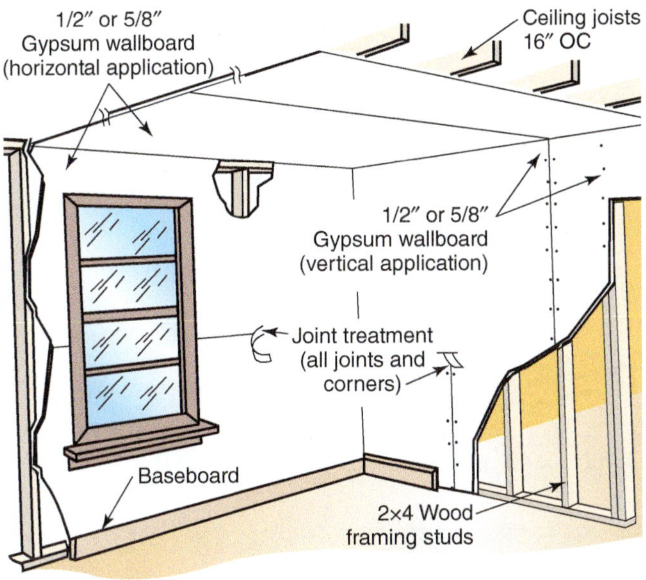

Goodheart-Willcox Publisher

Figure 7.4-9. The drywall on the left-hand wall and ceiling has a horizontal application. The right-hand wall has a vertical application.

the additional benefit of strengthening the building's frame. Panels bridge more studs and joists, the strongest dimension of the panel runs across the frame, and each piece is fastened to the greatest number of studs possible. This results in fewer problems with framing irregularities in alignment and spacing. Finally, horizontal joints are easier to treat because they are lower on the walls.

Wall panels are loosely butted against the ceiling panels, and the top wall panels are installed first. This is done so that any gaps or cut edges come at the floor where trim will cover them. Metal, plastic, or paper corner beads are applied to outside corners to achieve a crisp, straight edge, **Figure 7.4-10**. Installation continues with applying drywall compound and tape to create a seamless, consistent wall or ceiling surface. Tape adds strength to the wall and keeps the dried drywall compound from cracking. The process of achieving a smooth surface takes multiple coats of compound with a period of drying between each coat. After the final coat is applied, the surface is sanded. The drywall surface can be finished smooth or to have a texture. The surface is finally coated with primer and paint or wallpaper.

Suspended Ceilings

A ***suspended ceiling***, also called a *drop ceiling* or *false ceiling*, is a secondary ceiling installed below the primary ceiling framing of a building. When heating ducts and plumbing lines interfere with the application of a finished surface, a suspended ceiling is a practical option. In other instances, it provides a simple way of lowering high ceilings.

Goodheart-Willcox Publisher

Figure 7.4-10. Attaching metal corner bead to an outside corner.

Suspended ceilings are relatively simple to install and consist of a metal framework grid system designed to support tile or panels, **Figure 7.4-11A**. The framework is mainly supported by wires tied to the building structure directly above, **Figure 7.4-11B**. Ceiling tiles or panels are then placed within the grid, creating the ceiling surface. Tiles are either 2′ × 2′ or 2′ × 4′ and are available in a variety of styles.

Wood Board Paneling

Solid-wood paneling makes a durable and attractive interior wall surface. It may be used in nearly any type of room, and many species of hardwood and softwood are available. Softwood species most used for solid lumber paneling include pine, spruce, hemlock, and western red cedar. Boards range in widths from 4″ to 12″ (nominal size) and are dressed to a 3/4″ thickness.

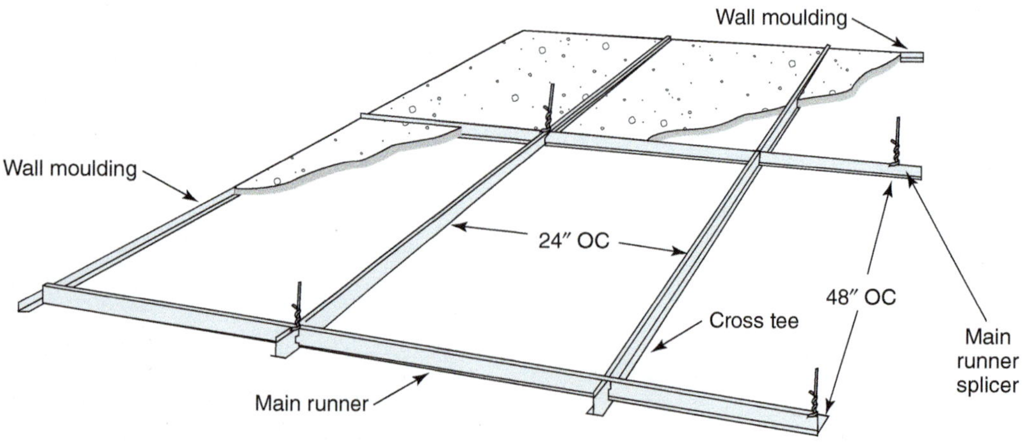

B Spectra-Physics Laserplane, Inc.

Figure 7.4-11. A—This sketch is a topside view, typical of the framework used for suspended ceilings. B—This installer is using a laser level to plumb the suspended ceiling framework.

Pro Tip

Interior solid-wood paneling may cause problems resulting from expansion and shrinkage. Be sure the material has a moisture content about equal to what it will attain in service. This should be about 8%–10% for most regions of the United States. Prior to installation, store the paneling in the room and allow room air to reach both sides of the wood.

Paneling patterns are often reversible, offering two choices in a single panel. Some patterns are smooth on one side and saw-textured on the other. **Figure 7.4-12** shows several patterns offered by one manufacturer. Dozens of variations are possible, varying by species of wood, texture, finish, and how the paneling is applied to a wall.

Tile

Ceramic tile and stone are popular, long-lasting, and versatile materials for use on walls and floors in homes. When properly prepared and installed, these materials can make a waterproof, long-wearing, attractive surface. The use of tile for backsplashes, tub surrounds, and showers is common, **Figure 7.4-13**.

Four kinds of tiles are widely used in both residential and commercial buildings:

- **Ceramic and porcelain**—Made of clay, minerals, and water, and then fired in a kiln to make a strong ceramic material. Firing porcelain at a higher temperature makes a denser, more water-resistant material. These tiles may be glazed, covered with a glass-like coating that is fired to make a water-repellant surface, or left unglazed for a more natural appearance. Ceramic and porcelain tiles are available in a wide range of patterns, colors, shapes, and sizes. Porcelain is generally more expensive.

- **Quarry**—Made from shale and natural clays, typically unglazed, and meant for high-traffic settings. Although reddish-brown is the most common color, it is also available in tan, grey, and deeper brown.
- **Glass**—Made of glass that is waterproof and generally translucent in color. This creates a different appearance from other tile products. Glass tile is often used for backsplashes and wall tile, but it is not suitable for floors.
- **Natural stone**—Many varieties of natural stone are available, but the most common are marble and granite. Each variety of stone has different properties. Marble, for example, is attractive but not as durable as other stone.

Proper tile installation is crucial to ensure longevity and prevent issues such as cracked tiles or water damage. The installation process includes preparing the surface, applying adhesive, laying the tiles, and grouting the joints. Tile setting requires strong attention to detail and knowledge of the specific type of tile being used.

Wood Flooring

Wood is popular as flooring in residential structures. Hardwood and softwood flooring is available in a variety of widths and thicknesses, as random-width planks, or in unit blocks. Wood flooring, especially hardwood, has the strength and durability to withstand wear, while providing an attractive appearance, **Figure 7.4-14**.

Pro Tip

The moisture content usually recommended for flooring at the time of installation is 6% for the dry southwestern states, 10% for the more-humid southern states, and about 7% or 8% for the remainder of the country.

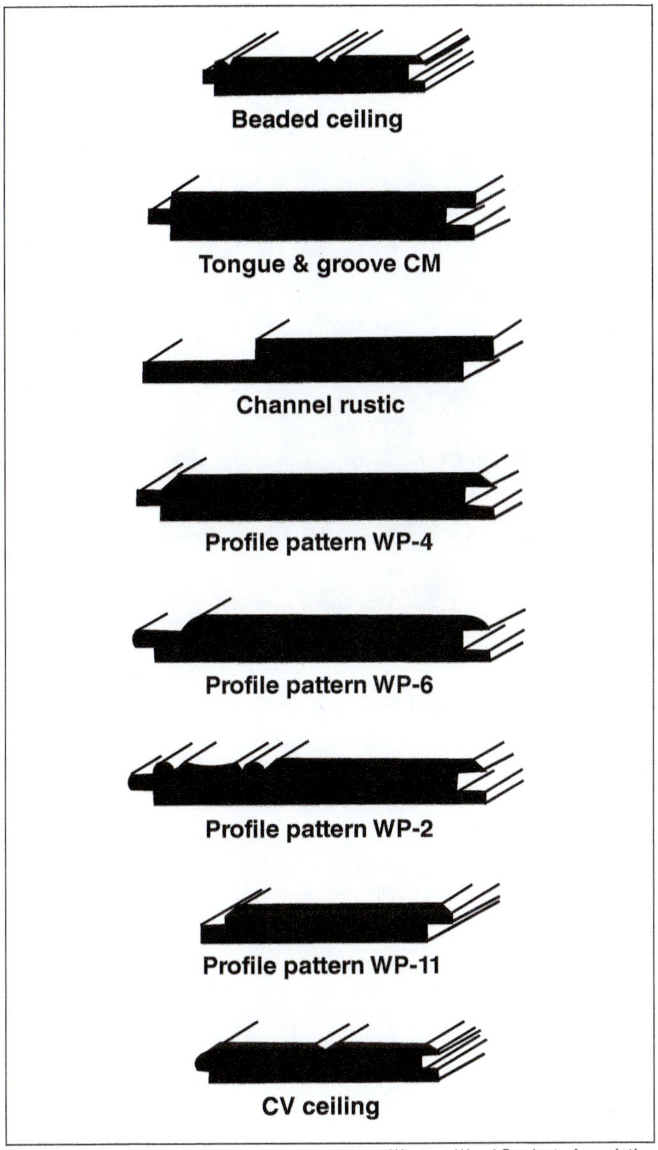

Western Wood Products Association

Figure 7.4-12. Smooth-surfaced solid-wood paneling is available in several pattern profiles.

Arsel Ozgurdal/Shutterstock.com

Figure 7.4-13. Ceramic tile makes an attractive, water-resistant, durable surface.

Artazum/Shutterstock.com

Figure 7.4-14. Strip hardwood flooring is durable and has a natural beauty brought out by a clear finish. A wide variety of hardwood flooring with a factory-applied finish is available.

Oak is a widely used species. However, maple, birch, beech, and other hardwoods also have desirable qualities. Softwood flooring includes such species as pine and fir. These are fairly durable when produced with an edge-grain surface.

These general types of wood flooring are commonly used in residential structures:

- Strip flooring
- Plank flooring
- Engineered flooring

Strip flooring consists of pieces of solid wood cut into narrow strips. It is the most widely used wood flooring and is laid in a random pattern of end joints. Most strip flooring is produced with tongue-and-groove on the sides and on the ends; this design is referred to as *side-and-end matched*. Another feature of strip flooring is the undercut, a wide groove on the bottom of each strip that enables it to lie flat and stable even when the subfloor surface is slightly uneven, **Figure 7.4-15**.

Unfinished strip flooring is manufactured with sharp, square corners on the face edges. After the floor is laid, the surface is sanded to remove unevenness. Prefinished strip flooring is sanded, finished, and waxed before it leaves the factory. Both edges of the face are chamfered. When the strips are laid, the chamfered edges form a vee joint. This makes any unevenness unnoticeable.

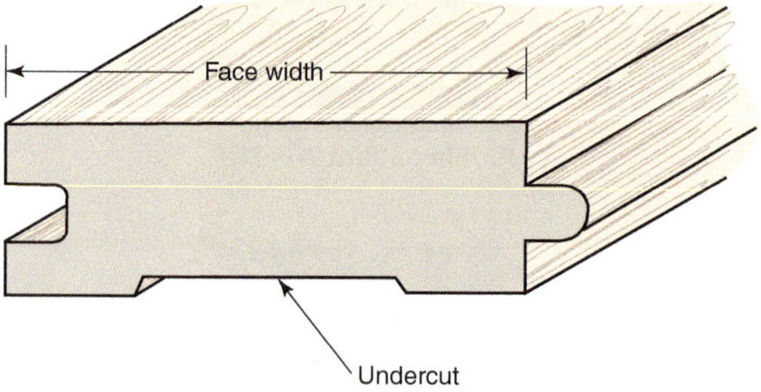

Goodheart-Willcox Publisher

Figure 7.4-15. Typical section of strip flooring. The ends are also tongue-and-groove. The undercut helps compensate for slightly uneven subfloors.

Plank flooring is solid wood that gives an informal appearance to the floor, **Figure 7.4-16**, and is particularly appropriate for colonial and ranch-style homes. Plank floors are usually laid in random widths. The pieces are bored and plugged in the millwork shop to simulate the wooden pegs traditionally used to fasten flooring in place. Today, this type of flooring has tongue-and-groove edges. It is laid in about the same manner as regular strip floors. Widths of the strips vary in different combinations.

Engineered wood flooring is designed to float on the subfloor, **Figure 7.4-17**. This flooring has a thin layer of hardwood glued to a stable base. The base may be made of hardboard, other composites, or veneer lumber. The top ply is prefinished. Individual pieces are 3/8″–3/4″ thick, 4″–8″ wide, and about 3′ long. The face veneer is made of two or three hardwood strips tightly edge joined to resemble solid strip flooring. The thickness of the veneer and the joining are so precise that chamfering edges is not necessary.

Robbins/Sykes, Tibbias Flooring Inc.

Figure 7.4-16. Softwood planks with walnut plugs give this floor an authentic colonial look.

Ingo Bartussek/Shutterstock.com

Figure 7.4-17. Engineered wood strip flooring is available in many wood grain patterns. The top layer is a veneer of real wood.

Lumber Defects

Sometimes, grades that contain blemishes or defects are used for interest, appearance, or dramatic effect. A *lumber defect* is an irregularity occurring in or on wood that reduces its strength, durability, or usefulness. A defect may also detract from or improve the wood's appearance. A *blemish* is an imperfection that impairs only the appearance of wood. For example, knots are commonly considered a defect and should not be used where strength is important, but they add to the appearance of wood paneling and flooring. Deep fissures in pecky cypress can also provide a dramatic effect.

Some common defects include the following:

- **Knots**—Caused by an embedded branch or limb and reduces the strength of the wood. The extent of damage depends on the type of knot, its size, and its location, **Figure 7.4-18**.
- **Splits and checks**—Separations of the wood fibers that run along the grain and across the annular growth rings. Splits usually occur at the ends of lumber that has been unevenly seasoned.

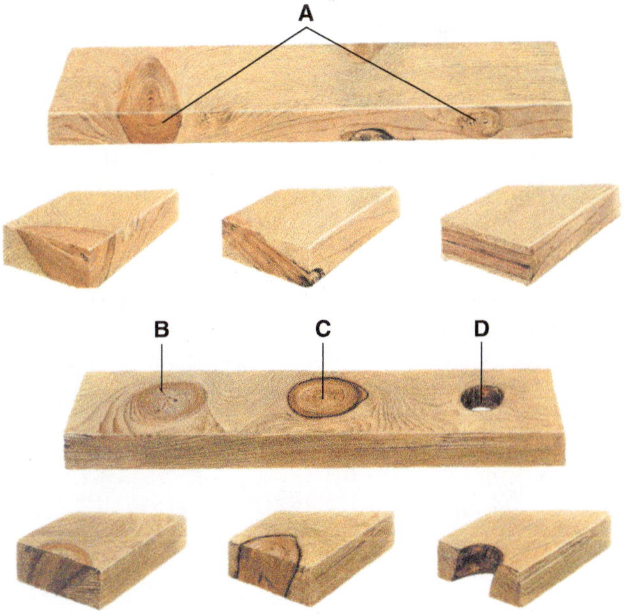

Western Wood Products Assn.

Figure 7.4-18. Types of knots. A—Knots may appear on one or both faces and on edges of sawn lumber. B—A sound (cannot be knocked loose), intergrown knot. C—A sound, encased knot. D—Unsound knots can fall out or be knocked out easily.

- **Shakes**—Separations along the grain and between the annular growth rings. Shakes are likely to occur only in species with abrupt change from spring to summer growth.
- **Pitch pockets**—Cavities that contain or have contained pitch in solid or liquid form.
- **Honeycombing**—Separation of the wood fibers inside the tree. Honeycombing may not be visible on the board's surface.
- **Wane**—The presence of bark or the absence of wood along the edge of a board; wane forms a bevel and/or reduces width.
- **Blue stain**—Caused by a mold-like fungus. Blue stain discoloration has little or no effect on the strength of the wood. However, it is objectionable for appearance in some grades of lumber.
- **Decay**—Disintegration of wood fibers due to fungi. Decay is often difficult to recognize in its early stages. In advanced stages, wood is soft, is spongy, and easily crumbles.
- **Holes**—Openings caused by handling equipment or boring insects and worms. These lower the lumber grade.
- **Warp**—Any variation from true or plane surface. Warp may include any one or combinations of the following: cup, crook, twist (or wind), and bow.

Laminate Flooring

Laminate flooring consists of a 1/4″ core of medium- or high-density fiberboard that is sandwiched between layers of melamine paper or some other base material and a protective topcoat. The top layer carries a synthetic wood-grained pattern. Sizes vary, but most planks are 46″–50″ long and 8″ wide.

The laminate is not fastened to the subfloor. Instead, like engineered wood flooring, it floats over the subfloor. An underlayment of foam takes care of cushioning and absorbs sound. Tongue-and-groove construction allows the planks to be snapped together as they are installed. Although the first types introduced required gluing of the joints, most manufacturers now offer flooring in both glue-less and glued joint types. Glued joints are used for installations where moisture could be present, such as bathrooms and kitchens.

Pro Tip

Follow the manufacturer's recommendations on allowing the laminated flooring to acclimate to the building's humidity level. Some types require storing the unopened cartons in the room for two days before installation, but others can be installed without this waiting period.

Vinyl Plank Flooring

Vinyl plank flooring, sometimes referred to as *luxury vinyl plank (LVP)*, is a versatile type of synthetic flooring material that mimics the appearance of hardwood flooring, **Figure 7.4-19**. It has gained popularity in recent years due to its durability, affordability, and ease of installation.

A Dmitry Kalinovsky/Shutterstock.com

B Dmitry Kalinovsky/Shutterstock.com

Figure 7.4-19. Vinyl plank flooring. A—Glue-down vinyl planks. B—Snap-together floating vinyl planks.

Vinyl plank flooring is made primarily of PVC and, depending on the design, can be installed as a floating floor system or glued to the subfloor using a troweled-on mastic. The result is a highly durable, waterproof floor surface that closely resembles the look of wood. This product is also available in a design that mimics the appearance of ceramic or stone tiles commonly called luxury vinyl tile (LVT).

Sheet Vinyl Flooring

Sheet vinyl flooring is extremely flexible, making installation much easier, **Figure 7.4-20**. Since sheets are available in 12′ widths, many installations can be made free of seams. This material usually can be installed over concrete, plywood underlayment, or noncushioned vinyl flooring.

The three types of sheet vinyl flooring are full-adhesive-bonded, perimeter bonded, and loose-lay. Full-adhesive-bonded flooring must be fastened to the substrate with a coat of adhesive covering the entire surface. Perimeter-bonded flooring is fastened down only around the edges and at seams. Loose-lay flooring requires no glue.

Flooring Underlayment

Underlayment is a layer of material that is installed between the subfloor and the finish flooring. Underlayment can vary in thickness, composition, and purpose. Common reasons for installing underlayment include cushioning and sound absorption, moisture protection, or subfloor imperfections.

Phovoir/Shutterstock.com

Figure 7.4-20. Installation of sheet vinyl flooring.

Materials used for underlayment include foam, cork, rubber, felt, polyethylene, plywood, luan, and fiberboard.

Tile requires a stable base (underlayment) for secure attachment and resistance to cracking. The framing under a tile floor must be sufficient to support the floor and prevent flexing. Plywood and OSB tend to expand and contract as their moisture content changes. To prevent these problems, cement board or other tile backing is applied to the surface to be tiled, **Figure 7.4-21**.

Cement board is a versatile fiber-reinforced cement panel material that is used for tile underlayment. It is fireproof, not damaged by water, and resists impact. Some cement board products are lightweight with fiberglass-reinforced matting and silicone-treated cores.

Effective alternatives to cement board are waterproofing and uncoupling membranes applied directly to gypsum and subflooring. *Uncoupling membranes* are designed to absorb movement and stress, preventing cracks in the tile and grout. They provide a flexible layer between the substrate and the tile, making them ideal for areas with potential movement or where there might be a difference in the expansion rate of different materials. Some uncoupling membranes are waterproof when installed, making them ideal for showers and tub surrounds. In dry areas, drywall can act as a suitable base for wall tile.

Pro Tip

Not all types of finish flooring require underlayment. Special care should be given to the manufacturer's instructions regarding when and what type of underlayment is required.

Fasteners, Bonding Agents, Fillers, and Coatings Used on Interior Surfaces

A variety of fasteners and adhesives are used during the installation of interior surfaces. Fillers, like drywall compound and floor leveler, are used

Casein glue is used for structural laminating and works well with wood that has high moisture content. It has good joint filling qualities and, therefore, is often used on materials that have not been carefully surfaced. Casein glue is used for gluing oily woods such as teak, padouk, and lemon wood. Its main disadvantages are that it stains the wood (especially such species as oak, maple, and redwood) and has an abrasive effect on tool edges.

A wide array of mastic products include waterproofing materials and adhesives. A *mastic* can be liquid and applied with a brush or roller, such as those used for foundation dampproofing, or heavy, paste-like adhesives applied with a notched trowel or a caulking gun. Usually designed for a specific type of material, mastics vary in their characteristics and application methods.

Application of Bonding Agents

Drywall installers often adhere the drywall to the framing of the home with a *gunnable* adhesive, meaning it can be applied with a caulk gun. Using adhesive on drywall enables installers to use fewer fasteners, making finishing easier. It also makes a stronger wall assembly, reduces cracking, and improves sound deadening.

Flooring installers use a wide variety of trowelable adhesives and mastics. Many of these products are designed to be used for a specific purpose, such as small ceramic tile, wood flooring, carpet, or vinyl. *Trowelable* adhesives are applied with a trowel. A notched trowel is used to apply mastics and adhesives in order to leave a predetermined amount of product on the surface, **Figure 7.4-24**. The size of the notch depends on the materials being used.

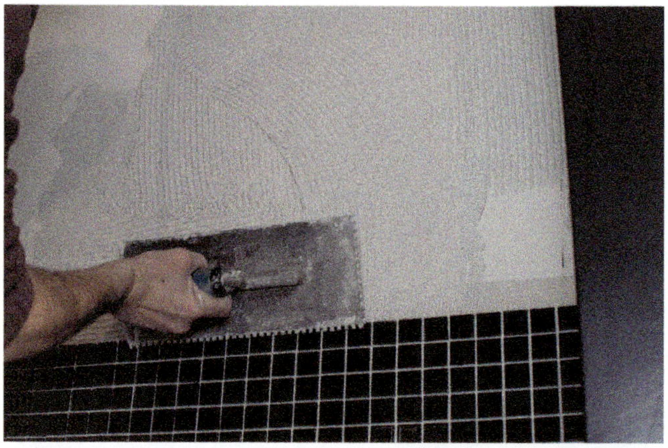

Denis Zhengal/Shutterstock.com

Figure 7.4-24. Adhesive being applied with a notched trowel.

Thinset is a trowelable, cement-based adhesive used for installing ceramic tile, porcelain tile, natural stone, and underlayment, **Figure 7.4-25**. It is composed of Portland cement, fine sand, and additives that enhance bond performance and workability.

ivan canavera/Shutterstock.com

Figure 7.4-25. Thinset is spread on the subfloor before applying the uncoupling membrane. Another layer will be applied to adhere the floor tile.

Fillers

Drywall compound is a sandable filler used to conceal joints and fasteners after installation of the wallboard. Drywall compound, when used in conjunction with drywall tape, adds strength to panel joints. The two primary types of drywall compound are setting compounds and drying compounds.

Setting compounds come in powder form that must be mixed with water. These compounds harden rather than dry on the wall, enabling the installer to apply an additional coat of compound sooner. They are available in various workability times, ranging from 5 to 90 minutes and higher. Workability refers to the amount of time the compound remains stable for use. The workability time is a subjective number and can vary depending on factors such as water temperature and humidity level.

Setting compounds shrink less than drying compounds as they harden and are more difficult to sand. Care must be taken to minimize sanding when using any drywall compound, but especially with setting compounds.

Drying compounds are available in powder form as well, though many drywall finishers prefer to use them premixed from a bucket. The consistency of the "drywall mud" as it is called can be adjusted by adding water and mixing it with a paddle mixer. Drying compounds are available in a variety of weights that can impact the strength and ease of sanding.

Floor leveling compound is another type of filler occasionally used during the installation of interior surfaces. Floor leveling compound is a cementitious product used to fill low spots and voids in floors before the finished floor surface is installed. It is available as a trowelable product that can be troweled smooth to a feathered edge, **Figure 7.4-26**, or as a self-leveling product that is poured into place.

Grout is a cement-based material used to fill the gaps between floor or wall tiles. Grout is available in a variety of colors. After it hardens, grout provides a finished look to the tile.

Coatings

After the drywall finishing process is complete, walls and ceilings are primed and painted. To protect finished products from paint splatters and drips, this process is completed before floors and tiles are installed. Wood paneling and floor stains and clear finishes are sometimes applied before these products are installed; if not, these coatings are applied shortly after installation.

Steve Heap/Shutterstock.com

Figure 7.4-26. In some cases, low spots or inconsistencies in subflooring need to be filled before the finish flooring is applied.

Green Note

Certain materials, such as paint, sealants, particleboards, and even carpeting, can off-gas chemicals, called volatile organic compounds (VOCs), and impact indoor air quality. High-VOC materials can be substituted with low-VOC and VOC-free alternative products.

Primer and Paint

Primers, paints, varnishes, and stains are adhesive coatings. They are spread in a thin, unbroken film that is designed to protect and beautify the surface. The term *paint* was once reserved for oil-based paints. However, it now includes coatings based on various resins such as acrylic latex. These are often referred to as water-based paints because they can be thinned or cleaned up with water.

The principal ingredients of paint are the ***pigment***, which provides color and opacity, and the vehicle. Paint stores and home improvement centers have small cards, called *paint swatches*, that show possible colors, **Figure 7.4-27**. These colors can be mixed by using different pigments. Vehicles are oils or resins that make the paint a fluid and allow it to be applied to, and spread over, a surface. The vehicle also includes a binder that remains behind after the oils or resins have dried. The ***binder*** is a material that holds the pigment particles together after the paint dries, forming a cohesive film on the surface. Binders serve to bind the pigment particles to each other and to the substrate (surface being painted), providing adhesion, durability, and protection.

Andy Dean Photography/Shutterstock.com

Figure 7.4-27. Paint stores are usually well stocked with paints and aids that help buyers in selection of colors.

Clear Finishes

Shellacs, lacquers, and varnishes are clear coatings applied over wood, **Figure 7.4-28**. They not only protect the wood surface but also enhance the beauty of the grain and the wood color.

Shellac has been used as a wood finish for centuries. It is made by dissolving the excretions of the lac bug in alcohol. The alcohol evaporates after it is applied and left exposed to the air, leaving a hard, attractive finish. Shellac is available as a liquid in an alcohol base or as dry flakes to be dissolved before use. In its natural state, it produces a warm, amber-colored finish called orange shellac. It is also available as white shellac, with the natural pigment removed, so it yields a nearly water-clear finish. Shellac produces an attractive finish, but it is easily damaged by small amounts of water and even more so by alcohol.

Lacquer is a synthetic finish that dries or cures by the evaporation of the solvent. Traditionally, that solvent is a volatile organic compound that evaporates quickly. In recent years, lacquers have been developed that use water as the solvent base. In either case, lacquer is usually sprayed, not brushed. It yields a hard finish and can achieve high gloss.

Green Note

Some finishes are safe to wash down the drain when cleaning tools. However, many finishes are harmful to the environment. Always read the cleanup instructions on the label to determine how to handle cleanup and disposal.

taurusphoto/Shutterstock.com

Figure 7.4-28. Clear finishes help to protect and beautify wood surfaces.

Colored lacquer is used to achieve smooth, glossy finishes on automobiles.

Traditional varnish is an oil-based product with solids similar to what is used to make oil-based paint. Traditional varnish is seldom used since the introduction of polyurethane varnish, or polyurethane. Polyurethane is essentially a plastic in liquid form until it cures through a chemical reaction. It does not dry. It is especially tough and can be formulated to produce gloss, semigloss, or satin finishes. Most polyurethane does not withstand the effects of the sun's ultraviolet rays well, so unless it is specifically marked for exterior use, such as spar polyurethane, it should not be used on building exteriors. Polyurethane is available as either water-based or oil-based. Water-based polyurethane is clear and does not change the color of the substrate (base material). Oil-based polyurethane is slightly amber in color and produces a warmer color on wood. It also withstands heat slightly better than does its water-based counterpart.

Stains

Many stains are designed for interior wood finishing, but stains for exterior uses on siding, decks, and fences are also popular. Interior stains are either penetrating or pigmented. Penetrating stains are brushed on and allowed to soak into the wood before the excess is removed with a dry cloth, **Figure 7.4-29**. Pigmented stains are designed to be applied with either a brush, roller, or spray. For deeper tones, the stain is brushed on and allowed to penetrate the wood and dry for a longer period of time before wiping.

MintImages/Shutterstock.com

Figure 7.4-29. Most stains are designed to be wiped off after application.

LESSON 7.4 Review

Answer the following questions based on information in this lesson.

1. Blanket and batt are types of ____ insulation. (7.4-1)
   A. flexible
   B. rigid
   C. spray foam
   D. reflective
2. Which statement is *true* about loose-fill insulation? (7.4-1)
   A. It is supplied in rigid sheets or panels that are highly durable.
   B. It has a higher R-value per inch than all other forms of insulation.
   C. Only a few boards must be removed when it is installed in older structures.
   D. When it is applied, it expands to fill in small spaces in the surrounding area.
3. *True or False?* A disadvantage of type X drywall is that it crumbles more quickly than other boards in a fire. (7.4-2)
4. *True or False?* Quarry tiles are specifically designed to be used in low-traffic areas. (7.4-2)
5. *True or False?* A blemish is an imperfection that impairs only the appearance of wood. (7.4-2)
6. *True or False?* If a board has honeycombing, it will always be visible on the surface. (7.4-2)
7. Vinyl plank flooring is made primarily of ____. (7.4-2)
   A. rubber
   B. PVC
   C. hardwood
   D. softwood
8. Which statement is *true* about casein glue? (7.4-3)
   A. It works well with wood that has high moisture content.
   B. It is made with a neoprene or latex rubber base.
   C. After mixing, it should be applied immediately.
   D. It sets without affecting the appearance of the wood.
9. A ____ is a material that holds the pigment particles together after paint dries. (7.4-3)
   A. thinset
   B. grout
   C. mastic
   D. binder
10. *True or False?* Shellac is easily damaged by small amounts of water. (7.4-3)

LESSON 7.5

Tools

LEARNING OUTCOMES

After completing this lesson, you will be able to:

7.5-1 Identify the equipment and tools used to apply interior insulation and surfaces.

TECHNICAL TERMS

drywall panel lifter
drywall router
flooring nailer
flooring roller
grout float
insulation blower
notched trowel
pull bar
tapping block
tile cutter
tile nippers
tile spacer
T-square
wet saw

Insulation Tools

Installation of insulation, drywall, and flooring surfaces of a home is often reduced to a separation of the trades. Subcontractors, each with their own specialized trade, are called upon to perform their part of the project. Many of these trades use a special set of tools that are not used for any other purpose. This section will feature an overview of the tools used during the installation of the interior surfaces of the home.

Batt insulation is installed using common hand tools. However, loose-fill insulation that is blown in requires the use of an ***insulation blower***. Bailed insulation is dumped into a hopper where an agitator breaks up the packed insulation before it drops into a blower, **Figure 7.5-1A**. The insulation is then blown through a hose to the location where the installer directs its output, **Figure 7.5-1B**. This machine can be used to blow insulation into attics, as well as wall or floor cavities.

A Kurteev Gennadii/Shutterstock.com

B Kurteev Gennadii/Shutterstock.com

Figure 7.5-1. A—Loading the insulation blower with bailed insulation. B—Cellulose insulation being installed in an attic space.

Green Note

Never throw away unused batt or cellulose insulation. Install excess insulation in interior walls and on top of attic insulation for extra protection, but do not force extra insulation in a confined space. Compressing insulation reduces its insulating value.

Some cellulose insulation blowers also feature a nozzle that mixes the insulation with water as it exits the hose, **Figure 7.5-2A**. The damp-spray process makes the insulation stick to the surface, allowing it to be sprayed into a rafter or wall cavity without falling. The cavity is overfilled and trimmed with a special insulation trimmer after drying, **Figure 7.5-2B**.

Spray foam insulation requires an extensive setup of equipment referred to as a spray foam rig. This equipment requires extensive training and practice to operate. The function of the rig is to deliver the two-part mix of resin and hardener to the nozzle where an operator carefully manages the speed and quantity of product as it reacts and expands.

Drywall Tools

During installation, drywallers use standard measuring, marking, and cutting tools along with a few specialty tools that can improve their speed or accuracy.

The utility knife is an important cutting tool for the drywaller. Gypsum, the core material of a sheet, is brittle. The paper surface gives the sheet strength. To make a straight cut, the installer scores one face of the sheet with a sharp utility knife, then flexes the sheet away from the scored face. This action snaps the sheet in a straight line leaving only the opposite paper face to be cut. This cut can be made easily and accurately by running the utility knife up the fold on the opposite side of the sheet, **Figure 7.5-3**.

To guide the cut on drywall, a T-square is used. The ***T-square*** enables a square mark or cut across the entire width of a sheet of drywall, **Figure 7.5-4**. This tool is also used to draw layout marks on the face of the sheet. It is important to mark the sheets to indicate where the framing members are located (for attachment) once the sheet is in place. The T-square is designed so 16″ or 24″ on center (OC) studs can be laid out quickly and easily.

Drywall screw guns, **Figure 7.5-5**, speed up drywall installation and provide better connections, all but eliminating "nail popping." These tools feature an adjustable nose cone that dimples the surface of the drywall around the screw. The screw gun is designed so that it does not operate until the screw and gun tip are pressed against the drywall. Most require drywall screws to be

A Kurteev Gennadii/Shutterstock.com

B Kurteev Gennadii/Shutterstock.com

Figure 7.5-2. A—Damp-spray cellulose sticks as it is applied, allowing for vertical and ceiling applications. B—Overfilled cavities are trimmed for wall surface installation.

Arturs Budkevics/Shutterstock.com

Figure 7.5-3. To cut drywall, score through the paper on one side, break the sheet away from the score, and cut through the paper on the other side. Cutting can be done with sheets laying down as shown, but it is more common to cut sheets while they are leaning against a wall for easier access to both sides of the sheet.

Goodheart-Willcox Publisher

Figure 7.5-4. A T-square can be used to mark a square cut line on the panel. Where only a portion of the width will be cut away, a pencil is used to first mark the cut line.

Hilti Tools

Figure 7.5-5. An electric screw gun with a special depth-adjusting clutch is used to drive drywall screws into steel studs. The clutch disengages when the nose strikes the panel surface.

inserted one at a time into the nose. Some drywall screw guns have a magazine that allows strips of drywall screws to be loaded for automatic screw advancement. Usually, a belt clip attached to the tool's housing allows the gun to be carried on a tool belt.

Irregular shapes and curves can be cut in drywall using a jab saw, **Figure 7.5-6**. A drywall router, also called a *drywall cut-out tool*, is used to cut around fixtures such as an electrical box. A ***drywall router*** is a small rotary power tool with a thin cutting bit

Klein Tools

Figure 7.5-6. A jab saw, just one type of drywall saw, is useful when cutting openings in drywall for electrical boxes. The large triangular teeth cut fast and do not easily clog.

to follow an object, allowing the cut to match the shape of the object, **Figure 7.5-7**.

A ***drywall panel lifter*** is a foot-operated lever device, **Figure 7.5-8**. Stepping on this device while one end is under the drywall will raise that panel against the sheet above or the ceiling. This is done so gaps or cut edges occur at the floor where the trim will cover them.

Drywall stilts are used by many drywall installers and finishers to reach high areas, **Figure 7.5-9**. Stilts allow continuous access to ceilings and other high areas.

Safety Note

Stilts require a clean work environment with no obstructions on the floor. The focus for a worker using stilts is at working height and they should be able to walk freely around the room without the distraction of stepping over or around debris, extension cords, or other obstructions.

RotoZip, a division of Robert Bosch Tool Corporation

Figure 7.5-7. A portable power tool makes cutting holes in drywall for electrical boxes easier and faster. The tool can also be used to cut circles or irregular shapes.

MARSHALLTOWN

Figure 7.5-8. Several designs of drywall panel lifters are available. This model is called a roll lifter.

Dmitry Kalinovsky/Shutterstock.com

Figure 7.5-9. Stilts are helpful when working on ceilings and upper walls. Many stilts are adjustable in height, making it possible to work on high ceilings and tall walls.

Finishing Tools

A mixing paddle is used in conjunction with a drill to mix or stir drywall compound, **Figure 7.5-10**.

Drywall taping and finishing knives, **Figure 7.5-11**, are used to apply drywall tape and joint compound. Taping knives have a 5″ or 6″ blade and are used to embed paper tape into drywall compound. Finishing knives are available with 4″ to 14″ blades and are used to apply a smooth layer of drywall compound to joints, corners, and fastener heads.

A variety of drywall sanders are available for sanding ceilings, walls, and corners, **Figure 7.5-12**.

Painting Tools

A variety of tools may be needed to tackle painting or finishing tasks. Brushes are sold in many sizes and grades and are designed for various

Nikita Rublev/Shutterstock.com

Figure 7.5-10. Mixing drywall mud smooths the consistency. Water may be added to thin it out.

A National Gypsum Co. B National Gypsum Co. C National Gypsum Co.

Figure 7.5-11. Taping wallboard joints with taping knives. A—First apply compound to the channel at the joint. B—Then, embed the tape. Be sure it is centered over the joint. C—Immediately apply a skim coat over the tape and smooth the edges. Use a broader knife for this step.

Figure 7.5-12. A variety of drywall sanders are available for different purposes and user preferences. A—Pole sanders help to sand tough to reach places without a ladder or stilts. B—Hand sanders use sanding screens that resist clogging better than sandpaper. C—Sanding sponges are available with square and angled edges. D—Power sanders are available with or without dustless systems.

purposes. The most common brushes are shown in **Figure 7.5-13**. The bristles used in brushes are either synthetic fibers or natural fibers (animal hair). It is important to match the brush with the task and type of coating you will be using.

Paint rollers are sold in various widths. They are designed to hold tube-shaped pads that are available in many different types, naps, and sizes. **Figure 7.5-14** shows a sample of rollers and roller covers. Rollers are ideal for painting large surfaces and can be used to apply many types of interior and exterior finishes, including oil-based paints, water-based paints, floor and deck paints, masonry paints, and aluminum coatings. Short-nap roller covers are used for smooth surfaces. The rougher the surface to be painted, the longer the nap should be.

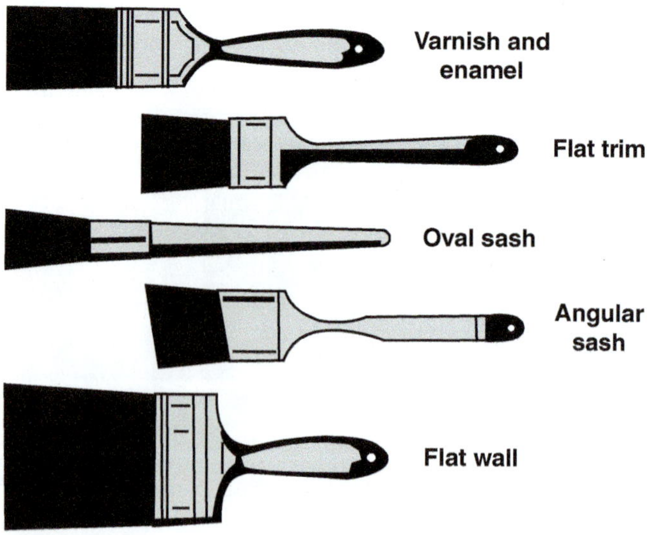

Dutcher Glass and Paint

Figure 7.5-13. Most painting or finishing tasks can be accomplished with these five kinds of brushes.

Radovan1/Shutterstock.com

Figure 7.5-14. Rollers and roller covers are made in different widths. Covers are designed in different naps for smooth to rough surfaces.

Pans for roller painting have a sloped bottom, with the lowest part designed to hold a supply of paint, **Figure 7.5-15**. The pan has a textured sloping surface for rolling off excess paint from the roller.

Nearly all types of paint, stain, and varnish can be applied with a sprayer. Spraying is the quickest and most practical method for painting large surfaces. Sprayers vary in function and purpose. The most common types used in construction are airless paint sprayers; high-volume, low-pressure (HVLP) sprayers; and pneumatic air sprayers.

Benoit Daoust/Shutterstock.com

Figure 7.5-15. A roller pan is designed for evenly spreading paint on a roller cover. The deepest section holds a supply of paint, while the sloping, upper part provides a surface for removing excess paint. This pan has a disposable plastic insert for easy cleanup.

Pro Tip

Great care must be taken to clean painting tools after each use. Paint that is allowed to dry in a brush, roller, or spray gun may be difficult or even impossible to remove. Of course, such a tool is useless unless the paint can be removed.

Tiling Tools

Installing both floor and wall tiles requires the same tools.

Cement board can be cut by scoring the face with a carbide-tipped scoring knife, **Figure 7.5-16**, and

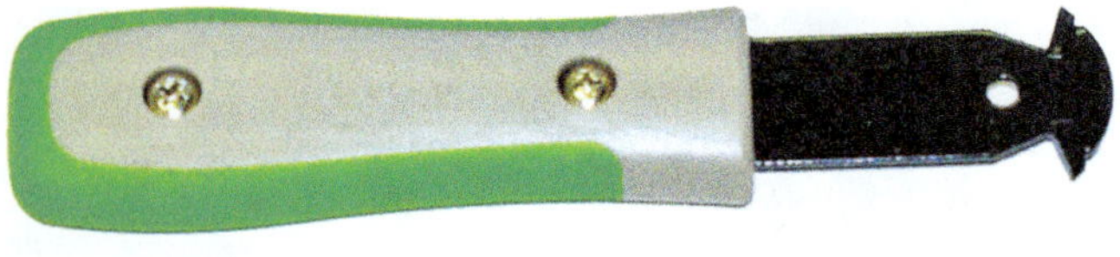

Goodheart-Willcox Publisher

Figure 7.5-16. A scoring knife has a carbide cutting edge. A utility knife can be used to score cement board, but it will dull quickly.

a straightedge and then snapping the board on the scored line.

A ***tile cutter*** is used to make straight cuts across tiles. Manual tile cutters use a track-mounted cutter to score the tile so that it breaks in a straight line, **Figure 7.5-17**.

A ***wet saw*** is a power tool with a circular diamond blade to cut tiles, **Figure 7.5-18**. The saw is outfitted with a small pump to circulate water to the blade for lubrication and dust control.

Tile nippers are a handheld cutting tool designed for making small cuts on tiles to chip away small amounts of material at a time. They are used to trim contours where a straight cut will not work, especially around obstacles like pipes, curves, or irregular edges. Tile nippers resemble a pair of pliers but have special jaws that allow for controlled chipping of the tile material, **Figure 7.5-19**.

A ***notched trowel*** is used to spread thinset mortar or mastic on the substrate before laying the tiles. The size and type of trowel depend on the tile size and the recommended adhesive coverage. Notched trowels can be square-notched, **Figure 7.5-20**, or v-notched.

A ***tile spacer*** is a small, cross-shaped plastic piece placed between tiles to maintain consistent grout lines, **Figure 7.5-21**.

A ***grout float*** is used to apply grout between the tiles after they are installed, **Figure 7.5-22**. Most grout floats are made with a rubber base that is used to direct and press grout into joints.

Flooring Tools

Standard carpentry tools are used for the installation of most flooring, with a few exceptions. Some specialty flooring tools are listed in this section.

Flooring installers spend a lot of time bent over, kneeling, and working while crouched on their

New Africa/Shutterstock.com

Figure 7.5-17. A tile cutter is a tool that requires no power source and keeps the work area clean, cutting without dust or spray.

Natallia Ploskaya/Shutterstock.com

Figure 7.5-18. Wet saws make precise cuts in tiles and are more versatile than tile cutters. They require setup in a place that can withstand some mess.

didden/shutterstock.com

Figure 7.5-19. Tile nippers remove small bits of tile at a time. Care must be taken not to break the tile.

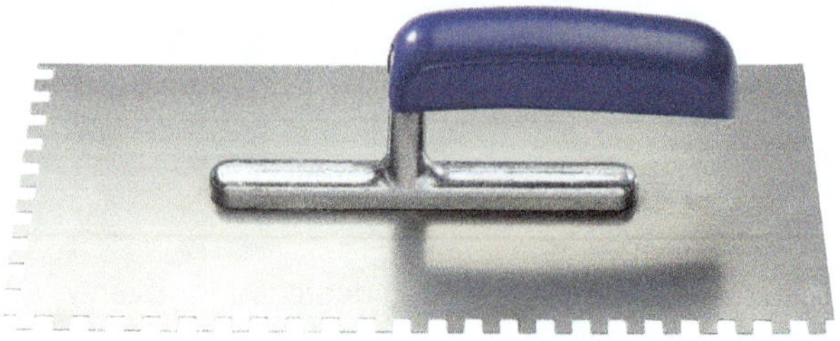

Gunter Kremer/Shutterstock.com

Figure 7.5-20. Notched trowels are available in a variety of notch sizes for different applications.

knees. Knee pads provide cushion and protection for the user's knees when working, **Figure 7.5-23**.

When flooring underlayment, such as luan or fiberboard, is required, it is vital to achieve a well-fastened product to ensure a squeak-free installation of the finish floor. Most of these products are fastened every 4″–6″. This high volume of fasteners is managed using an underlayment stapler. Underlayment staplers can be manual or pneumatic.

A ***flooring nailer*** is a specialized pneumatic nail gun designed specifically for the installation of wood strip flooring. A flooring nailer is essential for efficiently and securely attaching tongue-and-groove wood planks to the subfloor. The design of the nailer allows the user to operate the tool standing up, rather than kneeling, which allows for faster work and less user fatigue. To operate the tool, the user seats the base shoe of the flooring nailer on the subfloor and against the tongue of the wood plank. The installer then strikes the plunger with a mallet, **Figure 7.5-24**. This action triggers the tool while forcing the piece tight against the adjacent row of flooring. Because of the design of the base shoe, the nail is driven through the flooring at the ideal angle and placement, allowing the next row of flooring to cover the nail. This method is called blind nailing, **Figure 7.5-25**.

Tapping blocks and pull bars tools are commonly used for laminate wood floor installations. The ***tapping block***, **Figure 7.5-26**, is used to gently tap the flooring pieces together, while the ***pull bar*** is used to pull the pieces tight in areas where a tapping block cannot be used, **Figure 7.5-27**.

A ***flooring roller***, **Figure 7.5-28**, is a heavy rolling weight used to apply pressure to adhesive-backed flooring materials, such as sheet vinyl, ensuring a secure bond with the subfloor.

Carpet installers use specialized tools such as knee kickers and carpet stretchers, **Figure 7.5-29**, to properly stretch and tension carpet into place. They also use seaming tools, **Figure 7.5-30**, to splice carpets over wide rooms. Various carpet knives are used for cutting and trimming.

Oleksandr Lysenko/Shutterstock.com

Figure 7.5-21. Placing tile spacers is tedious and time consuming but essential for consistent grout lines.

ungvar/Shutterstock.com

Figure 7.5-22. After grout is worked into the joints, use a rough cloth or sponge to clean off grout from the tiles before it sets.

Lopolo/Shutterstock.com

Figure 7.5-23. Knee pads help reduce pain and fatigue when working on floors.

Michael Pettrigrew/Shutterstock.com

Figure 7.5-24. The force produced by the trigger blow helps to tighten flooring together as it is nailed.

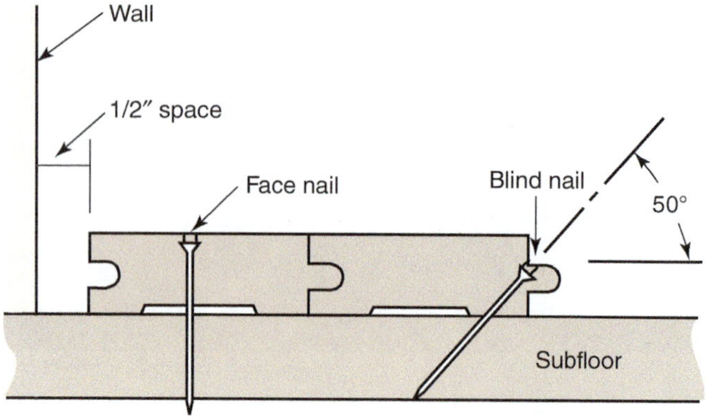

Goodheart-Willcox Publisher

Figure 7.5-25. Nail the starter strip through the face. Succeeding strips are blind nailed through the tongue.

Sharomka/Shutterstock.com

Figure 7.5-26. A tapping block protects the edge of the flooring as it is being installed.

Brenda Bakker/Shutterstock.com

Figure 7.5-27. A pull bar is used to draw flooring tight near a wall.

Levent Konuk/Shutterstock.com

Figure 7.5-28. A heavy flooring roller works out air bubbles and helps adhesive to bond under vinyl flooring.

septian intizom armedi/Shutterstock.com

Figure 7.5-29. A knee kicker has spikes on the bottom pad to grip the carpet while the installer strikes the pad with their knee. This action draws the carpet tight.

septian intizom armedi/Shutterstock.com

Figure 7.5-30. A carpet seamer is used to splice two pieces of carpet.

LESSON 7.5 Review

Answer the following questions based on information in this lesson.

1. Which tool is used to guide the cut on drywall? (7.5-1)
   A. Yardstick
   B. T-square
   C. Level
   D. Chalk line
2. *True or False?* Taping knives are used to embed paper tape into drywall compound. (7.5-1)
3. *True or False?* The rougher the surface to be painted, the shorter the nap on the roller cover should be. (7.5-1)
4. Which implement offers the quickest and most practical method for painting large surfaces? (7.5-1)
   A. Roller
   B. Natural fiber brush
   C. Synthetic fiber brush
   D. Sprayer
5. Tile ____ are handheld cutting tools designed for making small cuts on tiles to chip away small amounts of material at a time. (7.5-1)
   A. saws
   B. spacers
   C. nippers
   D. cutters
6. A ____ is a heavy weight used to apply pressure to adhesive-backed flooring materials to ensure a secure bond with the subfloor. (7.5-1)
   A. flooring roller
   B. tapping block
   C. knee kicker
   D. pull bar

LESSON 7.6

Safety

LEARNING OUTCOMES

After completing this lesson, you will be able to:

7.6-1 Explain key safety principles regarding material handling and storage.

7.6-2 Classify different types of fires and the extinguishers that can be used to fight them.

TECHNICAL TERMS

fire prevention

fire protection

Moving, Handling, and Storing Materials Safely

The weight and size of objects that must be moved on a construction site are significant factors that can contribute to injuries. Workplace accidents can easily be caused by bending, twisting, and turning motions, especially when handling heavy materials. Additional hazards, such as falling objects, improperly stacked materials, and various types of equipment, can also cause injuries, **Figure 7.6-1**.

Always be aware of the possibility of strains and sprains due to improper lifting techniques or handling loads that are too large or heavy. Fractures and bruises can result from being struck by materials or caught in pinch points. Falling materials due to improper storage or incorrect handling when cutting ties or securing devices may also result in cuts and bruises or worse.

Aleksei Ignatov/Shutterstock.com

Figure 7.6-1. Improperly stacked or stored materials can be a major hazard on the jobsite.

Moving and Handling Materials Manually

When moving heavy materials, OSHA recommends attaching handles or holders to loads whenever possible. Personal protective equipment such as proper eye and foot protection as well as gloves and long sleeves should also be worn when appropriate.

Recall from *Lesson 7.2* that improper lifting and carrying of heavy objects may cause injuries. When moving materials that have a large surface area, such as plywood, drywall, or steel roofing, be mindful of wind gusts that can throw off your balance or force you to twist and strain. If you have any doubt in your ability to manage it on your own, get help.

Whenever working with another person to perform a team lift, be mindful of your teammate's stability and grip. For instance, drywall sheets are carried on edge with the bottom of the sheet at about waist height with one hand supporting the load. The sheet is leaned against the shoulder with the top stabilized by the other hand. When carrying drywall with a partner, both workers should be on the same side of the sheet in this posture. If each worker is on opposite sides of the sheet, they will struggle against one another as they carry it, each trying to lean the sheet on their own shoulder. This is only one example of how a moment's forethought can be used to avoid discomfort and a potential injury.

Moving Materials with Equipment

Using mechanical equipment to move materials increases the potential for injuries, **Figure 7.6-2**. Workers must be aware of both manual handling safety concerns and safe equipment operation. Before operating any mechanical equipment, you must be trained by your employer.

Safety Note

It is important to have a pre-task plan prior to moving heavy materials to ensure safety. Follow these general safety rules when moving materials with equipment:

- Ensure that the equipment can handle the load. Equipment must meet the minimum load handling capacity, have the appropriate attachments for the load (forks, bucket, drum clamp, etc.), and have been inspected.
- Establish clear and effective communication between the equipment operator and ground worker. Use hand signals, radios, or other communication devices to coordinate movements.
- Ensure the equipment is on stable and level ground before operating. Avoid operating on soft, muddy, or uneven terrain that could cause instability.
- Keep the equipment's windows, mirrors, and cameras clean and unobstructed to maintain clear visibility of the surroundings.
- Operate the equipment at a safe and controlled speed.
- Distribute the load evenly and securely on the equipment's forks or bucket to maintain stability during transportation.
- Keep the load low whenever possible.
- Never walk or work under a suspended load. Stay clear of any area where the load could fall if it is upset.

Irene Miller/Shutterstock.comr

Figure 7.6-2. Moving materials with equipment is common on construction sites.

Material Storage and Disposal

Pay attention to how materials are stacked and stored on the jobsite. Materials stacked high or improperly can fall over, potentially injuring a worker. Imagine the force that a poorly stacked pile of CMU blocks could have if it were to fall over.

Overloading scaffolding, ladders, or other equipment can be hazardous. Storing materials overhead or on a sloped or uneven surface can be unsafe if materials are not properly secured. Ensure that material stacks are stable and self-supporting.

Do not store materials or tools on stairs. Maintain a clean and organized jobsite to avoid tripping hazards.

Consider the consequences of a mistake or miscommunication before attempting something potentially dangerous. For instance, drywall is often stored with the sheets upright, leaning against a wall. Attempting to pull one of the back sheets out of the pile (perhaps it is the size you need) without moving the rest could easily tip the entire pile over onto you or another worker.

Ensure flammable materials are stored properly. They should be in appropriate containers and kept far from any ignition source.

Dispose of debris, materials, and refuse appropriately, **Figure 7.6-3**. Remove all nails from used lumber before stacking. Keep trash and refuse in dumpsters or containers to avoid rodents. Rags that have been used on paints, stains, and solvents can spontaneously combust; follow manufacturers' disposal instructions for these products.

ungvar/Shutterstock.com

Figure 7.6-3. This overloaded dumpster is a hazard to workers. Unstable materials can easily come dislodged due to high winds or during pickup/transport of the dumpster. Do not stack materials above the sides of a dumpster.

Fire Protection

Construction workers should be aware of fire hazards; they must know the causes of fires and methods of controlling them. Employers are responsible to ensure that all workers are properly trained in fire protection and fire prevention, **Figure 7.6-4**. ***Fire protection*** involves measures taken to minimize the damage caused by a fire that has already started. ***Fire prevention*** involves taking measures to reduce the risk of fire occurring in the first place.

Fire is essentially a chemical reaction that requires the presence of three key elements to ignite and keep burning. These elements are commonly referred to as the "fire triangle," **Figure 7.6-5**, and they include:

1. Heat or some ignition source
2. Fuel, which serves as the combustible material
3. Oxygen, usually available in the air around us

Scientists introduced the concept of the fire triangle to better understand the causes of fires and how to prevent them or put them out. To start and sustain a fire, the right combination of heat, fuel, and oxygen must come together in a specific way. If any of these elements is missing or removed, a fire that's already burning will go out.

There are five different classes of fires:

- **Class A**—Result from burning wood and debris.
- **Class B**—Involve highly volatile materials such as gasoline, oil, paints, and oil-soaked rags.
- **Class C**—Result from faulty electrical wiring and equipment.
- **Class D**—Involve combustible metals, such as magnesium and sodium.
- **Class K**—Are associated with vegetable or cooking oils.

Any of the first three classes of fires can occur on a typical construction site. Combustible metals and cooking oils are less often found on construction

Yutthaphong/Shutterstock.com

Figure 7.6-4. Fire safety training often includes the local fire department. Here a fire pot is used to demonstrate proper use of a fire extinguisher.

Luciano Cosmo/Shutterstock.com

Figure 7.6-5. Fire is a chemical reaction between heat, fuel, and oxygen. If any one of the three elements of fire is taken away, the fire will go out.

sites but could be if the construction site is at an operational industrial center or restaurant.

Different types of fire extinguishers are designed to put out a specific class of fire, **Figure 7.6-6**. Using the correct fire extinguisher is an important consideration. Fire extinguishers should be available on the construction site. All fire extinguishers have a safety pin in the handle.

Fire Classifications

Class	Description	Requires	New Symbol	Old Symbol
A	Ordinary Combustibles (Materials such as wood, paper, and textiles)	Cooling/quenching		 A
B	Flammable Liquids (Liquids such as grease, gasoline, oils, and paints)	Blanketing or smothering		 B
C	Electrical Equipment (Wiring, computers, switches, and any other energized electrical equipment)	A nonconducting agent		 C
D	Combustible Metals (Flammable metals such as magnesium and lithium)	Blanketing or smothering		 D

Fire Extinguishers

Type	Description		Typically Approved for Use On	Not for Use On
Pressurized Water	Water under pressure		A	B C D
Carbon Dioxide (CO_2)	Carbon dioxide (CO_2) gas under pressure		B C	A D
Foam	Aqueous film-forming foam (AFFF) or film-forming fluoroprotein (FFFP)		A B	C D
Dry Chemical, Multipurpose Type	Typically contains ammonium phosphate		A B C	D
Dry Chemical, BC Type	May contain sodium bicarbonate or potassium bicarbonate		B C	A D
Dry Powder	May contain sodium chloride, sodium carbonate, copper, or graphite		D	A B C

Goodheart-Willcox Publisher

Figure 7.6-6. Not every fire extinguisher will put out every kind of fire. Check the label on the extinguisher. Using the wrong extinguisher could strengthen the fire, electrocute you, or produce toxic fumes.

To use the extinguisher, remember the PASS technique:

1. **P**ull the pin out.
2. **A**im the discharge nozzle at the base of the fire.
3. **S**queeze the handle.
4. **S**weep the extinguishing agent from side to side at the fire.

Stand about 6-8 feet away from the fire when you start to extinguish the fire, moving closer as it gets under control. Before using a fire extinguisher, make sure the fire is small and contained. If it's too large or spreading quickly, evacuate immediately and call emergency services.

Fire Prevention

Approved fire prevention practices should be followed throughout the construction project. Good housekeeping is an important aspect. Special precautions should be taken during the final stages of construction when heating equipment and wiring are being installed and when highly flammable surface finishes are being applied. Always keep containers of flammable materials closed when not in use. Proper storage and handling of flammable and combustible liquids must be followed. Dispose of oily rags and combustible materials promptly.

Code Note

Be careful when insulating around lights or exhaust fans recessed into the ceiling. As a safeguard against fire, the *National Electrical Code* requires a 3″ space around any heat-producing device, unless it is rated for insulation contact (IC).

To comply with standards regarding fire prevention, companies can implement administrative policies that include training and awareness, a good fire safety plan, smoking restrictions, and definitive procedures for welding and operating torches (hot work).

LESSON 7.6 Review

Answer the following questions based on information in this lesson.

1. *True or False?* Fire prevention involves measures taken to minimize the damage caused by a fire that has already started. (7.6-1)
2. What are the three components of the fire triangle? (7.6-1)
   A. Fire, fuel, and water
   B. Heat, fuel, and oxygen
   C. Solid, liquid, and vapor
   D. Wood, propellant, and hydrogen
3. Class _____ fires result from faulty electrical wiring and equipment. (7.6-2)
   A. A
   B. B
   C. C
   D. D
4. *True or False?* Fires involving vegetable or cooking oil are classified as Class E fires. (7.6-2)
5. What does the *A* in the acronym *PASS* stand for? (7.6-2)
   A. Aim
   B. Align
   C. Assess
   D. Act

LESSON 7.7

Math

LEARNING OUTCOMES

After completing this lesson, you will be able to:

7.7-1 Calculate volume for various construction estimating applications.

TECHNICAL TERMS

cube	rectangular solid	volume
cylinder	solid	
exponent	triangular prism	

Understanding Exponents

An ***exponent*** is the superscript value written to the right of the number to indicate how many times the number will be multiplied by itself. When a number is squared, for example 10^2, the exponent is 2, indicating that the number is 10 multiplied by itself. Another way of saying this is "10 to the second power." If the number were to be $10 \times 10 \times 10$, it could be written as 10^3 and called "10 cubed" or "10 to the third power." Exponents of 2 and 3 have names—squared and cubed, respectively—because they are the exponents used with area and volume measure. Higher exponents are only referred to as powers. For example, 10^5 is read as "10 to the 5th power." It is easier than saying "$10 \times 10 \times 10 \times 10 \times 10$." Both forms of that number equal 100,000.

Practice 7-1

1. What is 12 squared?
2. What is 8.5 cubed?
3. What is 4 to the 6th power?
4. What are two other ways to write $3 \times 3 \times 3 \times 3$?

Estimating Materials Using Volume

Understanding how to calculate ***volume*** is an important part of estimating materials that take up a three-dimensional space such as loose-fill insulation and concrete.

Just as area is always measured in square inches (in^2), square feet (ft^2), etc.; volume is always measured in cubic inches (in^3), cubic feet (ft^3), cubic yards (yd^3), etc. When a number is cubed, it is multiplied by itself three times. For example, 3 cubed ($3 \times 3 \times 3$) equals 27. Cubic units are written with a superscript 3, indicating that it is units × units × units.

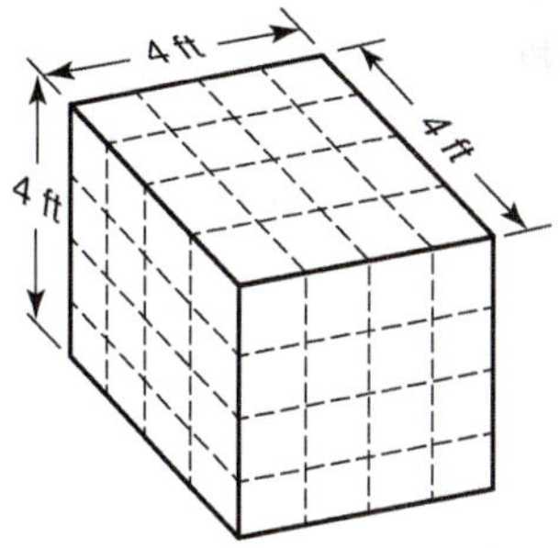

4 ft × 4 ft × 4 ft = 64 cubic ft or 64 ft^3

This cube is made up of 64 individual cubes, each measuring 1 foot by 1 foot by 1 foot.

As long as a ***solid*** (a three-dimensional shape) has the same size and cross-section shape throughout its depth, its volume can be found by multiplying the area of one surface by the depth from that surface. To find the volume of a ***cube*** (all edges are the same size) or a ***rectangular solid*** (a rectangle with a third dimension), multiply the length, width, and height.

For example:
Find the volume of a 3-inch cube.

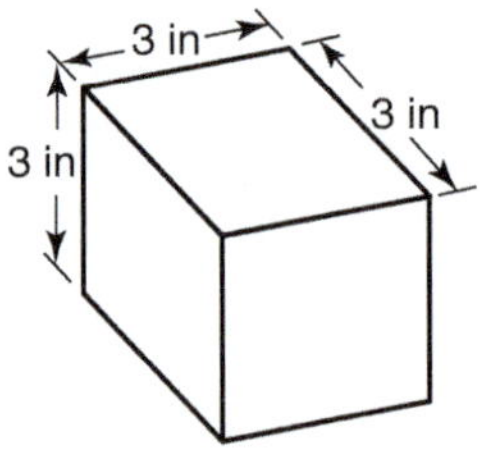

Goodheart-Willcox Publisher

Volume = 3 in × 3 in × 3 in = 27 in^3

For example:
Find the volume of the rectangular solid.

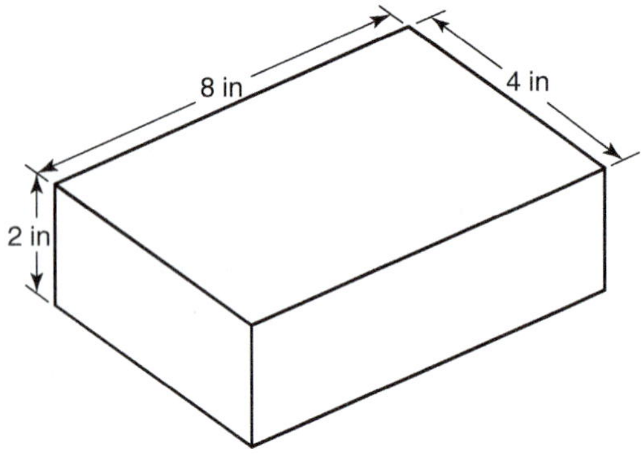

Goodheart-Willcox Publisher

Volume = 4 in × 2 in × 8 in = 64 in^3

A ***triangular prism*** is a solid with two opposite triangular faces. A ***cylinder*** is a solid with two circular faces. The volume of a triangular prism or a cylinder is found by multiplying the area of its face by its height.

For example:
Find the volume of the triangular prism.

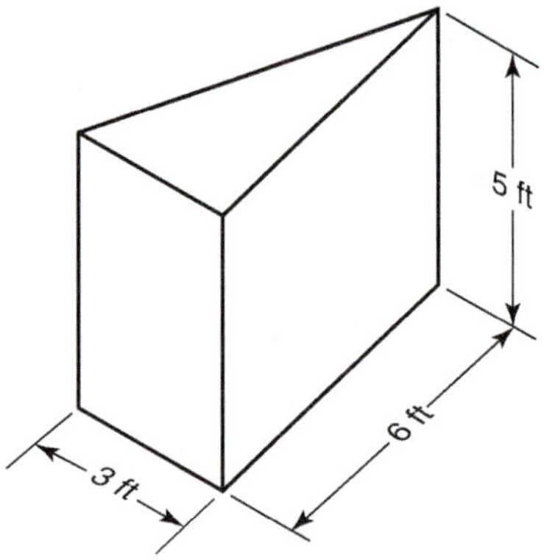

Goodheart-Willcox Publisher

Area of face = 1/2 × 3 ft × 6 ft = 9 ft^2
Volume = 9 ft^2 × 5 ft = 45 ft^3

For example:
Find the volume of the cylinder.

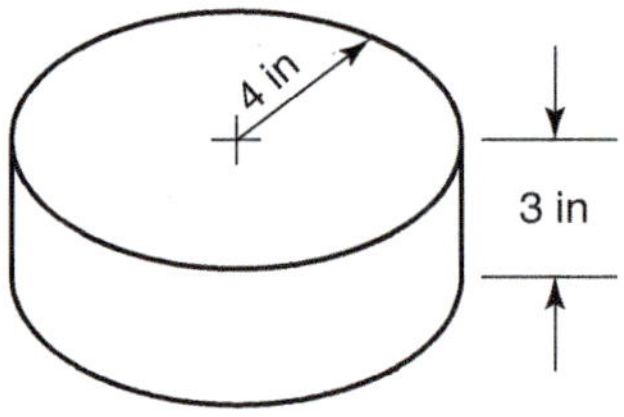

Goodheart-Willcox Publisher

Area of face = π × 4 in × 4 in = 50.2656 in^2
Round to 50.27 in^2
Volume = 50.27 in^2 × 3 in = 150.81 in^3

Practice 7-2

An insulation contractor needs to fill the floor of an attic space to a depth of 1.5′ of cellulose insulation to meet building code. The attic space is rectangular and measures 30′ by 50′. How many cubic feet are required to complete the job?

LESSON 7.7 Review

Answer the following questions based on information in this lesson.

1. 6^3 = _____. (7.7-1)
   A. 2 C. 36
   B. 18 D. 216
2. *True or False?* Volume is always measured in cubic terms, such as cubic inches. (7.7-1)
3. If a cube has one side that is 5 in long, the volume of the cube is _____. (7.7-1)
   A. 25 in^2 C. 125 in^2
   B. 25 in^3 D. 125 in^3
4. What is the volume of a triangular prism that is 6 ft long with a face that is 4 ft wide and 7 ft high? (7.7-1)
   A. 84 ft^2 C. 168 ft^2
   B. 84 ft^3 D. 168 ft^3
5. *True or False?* In the number 8^2, the exponent is 8. (7.7-1)

LESSON 7.8

Print Reading

LEARNING OUTCOMES

After completing this lesson, you will be able to:

7.8-1 Explain the importance of construction specifications.

7.8-2 Identify where specifications for common materials would be included in construction documents.

TECHNICAL TERMS

specifications

Specifications

When it comes to installing interior surfaces, the most important part of the working drawings are the specifications, commonly called *specs*. ***Specifications*** are detailed written documents that outline the specific requirements, materials, methods, and quality standards for construction, **Figure 7.8-1**.

Without specs, the drywall finishers would not have guidance on the level of quality their finish needs to meet; or a tile installer would have to rely on the word of the architect, general contractor, or homeowner to determine what tile to use in each room, what color grout to use, and what substrate to build it all on. This may be acceptable on certain jobs, but if there is a conflict of information, mistakes can be made and misunderstandings or disputes can occur between parties.

When a reliable set of specifications exists for a project, it reduces the risk of miscommunication. Specifications can contain a high level of detail, dictating procedures like how to store materials, how many coats of polyurethane or paint are applied to surfaces with instructions on sanding between coats, and so on.

Even in the pre-construction phase, it is critical to have the specifications worked out for fair competition when estimating and bidding on a project. Imagine being a contractor who takes pride in performing high-quality work, taking the time to do every job to the best standard, only to be outbid by a competitor who does substandard work. Specifications can even the playing field in these situations, as long as they are upheld and enforced throughout the project. This means that all supervisors, subcontractors, and project leaders should be aware of any specifications that impact their work. These documents are legally binding for the people responsible for carrying out the work.

Specifications for a residential structure generally include the following headings:

- General Requirements, Conditions, and Basic Information
- Excavating and Grading
- Masonry and Concrete Work
- Sheet Metal Work
- Rough Carpentry and Roofing
- Finish Carpentry and Millwork
- Insulation, Caulking, and Glazing
- Lath and Plaster or Drywall
- Schedule for Room Finishes
- Painting and Finishing
- Tile Work
- Electrical Work
- Plumbing
- Heating and Air Conditioning
- Landscaping

Under each heading, the content is usually divided into sections: scope of work, specifications of materials to be used, application methods and procedures, and guarantee of quality and

DIVISION 5 **PLAN NO. 12275**

CARPENTRY, MILLWORK, & HARDWARE

General Conditions:

This contractor shall read the General Conditions and Supplementary General Conditions which are a part of these Specifications.

Scope of Work:

Furnish and install all rough lumber, millwork, rough and finished hardware, including all grounds, furring, and blocking, frames and doors, etc., as shown on the drawings or hereinafter specified.

Rough Lumber:

All lumber used for grounds, furring, blocking, etc., shall be #1 Common Douglas Fir or Pine.

Grounds and Blocking:

Furnish and install suitable grounds wherever required around all openings for nailing metal or wood trim such as casings, wainscoting, cap, shelving, etc.; also building into masonry all blocking as required. All wood grounds must be perfectly true and level and securely fastened in place.

Doors:

Interior doors to be plain sliced red oak for stain finish. Solid core doors to be laminated 1 3/4" thick, or as called for in the door schedule. See plan for Label doors.

Temporary Doors and Enclosures:

Provide temporary board enclosures and batten doors for all entrance openings. Provide suitable hardware and locks to prevent access to work by unauthorized persons.

At any openings used for ingress and egress of material, provide and maintain protection at jambs and sills as long as openings are so used.

Rough Hardware:

All hardware such as nails, spikes, anchors, bolts, rods, etc., in connection with rough car-
... by this contractor. Use aluminum non-staining nails for all exterior wood.
around openings ...
edges of trim shall be lightly ...
smoothed and machine sanded at the ...
carpenters at the building in a first class manner before ...
applied.

Finishing Hardware:

An allowance of $1,000.00 shall be included for finishing hardware, which will be selected by the architect and installed by this contractor. Locksets shall match existing design and finish. Adjustment of cost above or below this allowance shall be made according to authentic invoice for hardware for this building from hardware supplier. Hardware supplier shall furnish schedule and templates to frame manufacturer.

Caulking:

Caulk all exterior windows, doors, etc., and openings throughout building with polysulifide based sealant, see specifications under cut stone.

Wood Paneling:

All wood paneling shall be Weyerhauser Forestglo, Orleans Oak 1/4" prefinished. Apply over 1/2" drywall on steel studs and furring.

Goodheart-Willcox Publisher

Figure 7.8-1. A portion of the specifications that are included with a set of house plans. Other topics included in specifications are plumbing, heating, and wiring.

performance. Carefully prepared specifications are valuable to the contractor, estimator, tradespeople, and building supply dealer. They protect the owner by helping to ensure quality work. In addition to the previously described items, specifications may include information and requirements regarding building permits, contract payment provisions, insurance and bonding, and provisions for making changes to the original plans.

LESSON 7.8 Review

Answer the following questions based on information in this lesson.

1. When it comes to installing interior surfaces, what is the most important part of the working drawings? (7.8-1)
   A. Specifications
   B. Plans
   C. Detail drawings
   D. Building elevations
2. *True or False?* The specifications are generally worked out as construction progresses. (7.8-1)
3. The "Insulation, Caulking, and Glazing" specifications usually include a section called _____. (7.8-2)
   A. Mechanical System Drawings
   B. Orthographic Specifications
   C. Specifications of Materials to Be Used
   D. General Requirements and Conditions
4. *True or False?* Carefully prepared specifications protect the owner by helping to ensure quality work. (7.8-2)

LESSON 7.9

Green Building and the Future of Construction

LEARNING OUTCOMES

After completing this lesson, you will be able to:

7.9-1 Describe methods for superinsulating homes.

7.9-2 Summarize how emerging technologies are being used in the construction industry.

TECHNICAL TERMS

deck-over roof insulation
double-stud wall construction
superinsulating
truss wall

Superinsulated Homes

Superinsulating is a building design strategy that minimizes the heat loss and gain of a building by using considerably higher levels of insulation and airtightness compared to standard practices and code requirements.

The main purpose of a superinsulated home is to reduce its heating requirements, potentially relying mostly on internal heat sources like waste heat from appliances and even the body heat of occupants, supplemented by a small backup heating system. This approach requires thorough attention to construction details and an efficient air filtration and ventilation system to manage indoor air quality.

On superinsulated homes, R-values often exceed R-40 in walls and R-60 overhead. These high R-values can be achieved through the extensive use of continuous insulation, spray foam, and alternative framing methods that allow more space for insulation. Advanced framing practices or SIP construction, covered in *Chapter 3, Framing Systems*, are good starting points to achieve better insulation performance. However, when superinsulating a home, a less traditional approach is often required.

These framing methods are not common in the United States and can seem quite foreign when compared with standard framing practices. However, examining these methods in detail can help develop a stronger understanding of the principles of heat transfer and construction methods in general.

Double-Stud Wall

Double-stud wall construction uses two stud-framed walls to form an extra thick wall cavity that can be filled with insulation. Along with the benefit of additional insulation, the interior and exterior framing are separated, reducing thermal bridging. Variations of double-stud wall construction exist, but the principles are the same.

One form of double-stud wall construction uses an exterior structural wall and a second nonstructural wall built to the inside, **Figure 7.9-1**. The studs of the two walls can be aligned or staggered to further reduce thermal bridging.

Another type of double-stud wall places the nonstructural wall on the outside of the structural wall, **Figure 7.9-2**. This method is more difficult to build but further reduces thermal bridging between floors at the rim joist. It also maintains more interior square footage.

Goodheart-Willcox Publisher

Figure 7.9-1. This double-stud wall uses a combination of spray foam and cellulose insulation to achieve a high R-value assembly.

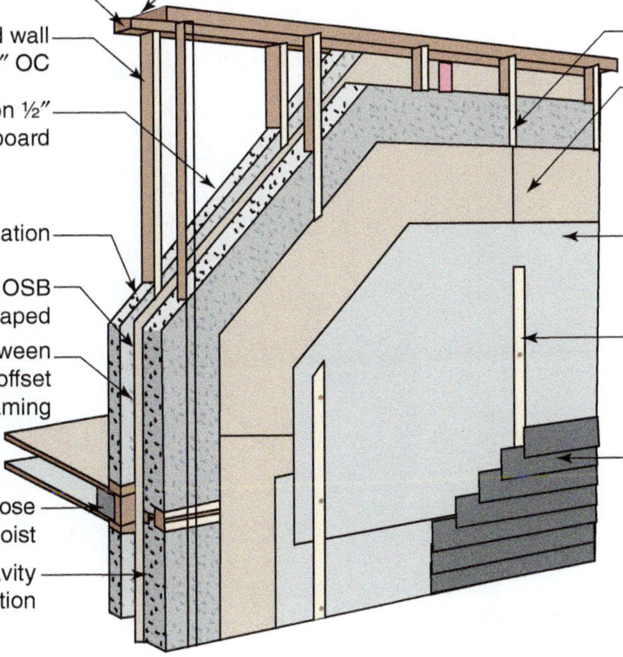

Goodheart-Willcox Publisher

Figure 7.9-2. Another double-stud wall design places the additional insulating studs on the outside of the structure to reduce thermal bridging at the floor joists.

Truss Wall Construction

A *truss wall* is similar to a double-stud wall in that it adds a cavity for wall insulation. However, rather than simply adding a stud cavity, the space for insulation is further increased through the use of wall trusses built from framing lumber and plywood gussets. Wall trusses are hung on the exterior of the structural wall and attached to the structural wall studs, **Figure 7.9-3**.

Deck-Over Roof Insulation

Although this method of insulating a roof can be used in designs that are simply meeting minimum code requirements and does not equate to superinsulating a structure, deck-over roof insulation is a method of insulating a roof that makes superinsulation easier when the thermal layer is at the roof level, rather than in the floor of an attic space.

Deck-over roof insulation uses a rigid insulation material over the rafter sheathing, **Figure 7.9-4**. This insulation reduces thermal bridging and combines with the insulation between the rafters. The deck-over insulation can be layered to suit the insulation needs for the job. An additional layer of sheathing is installed over the rigid insulation to provide a solid surface for roofing material installation.

Other Superinsulation Methods

There is no one prescribed way to superinsulate a home. Traditional framing methods, SIP construction, and ICF construction can all accommodate additional insulation if the details are thought through in advance. Builders have used many combinations to achieve insulation goals. The most common, and perhaps the simplest, way to meet these goals is by adding rigid insulation.

It is important to understand that the heat loss/gain benefit to adding a little insulation (if there were none) is far greater than if you have a lot of insulation and add more. A cost/benefit analysis should be done before settling on any design.

Finally, the performance of any insulation is highly dependent on the quality of the installation. Using best installation practices will yield a better return than simply adding more poorly installed insulation. Superinsulated homes are also

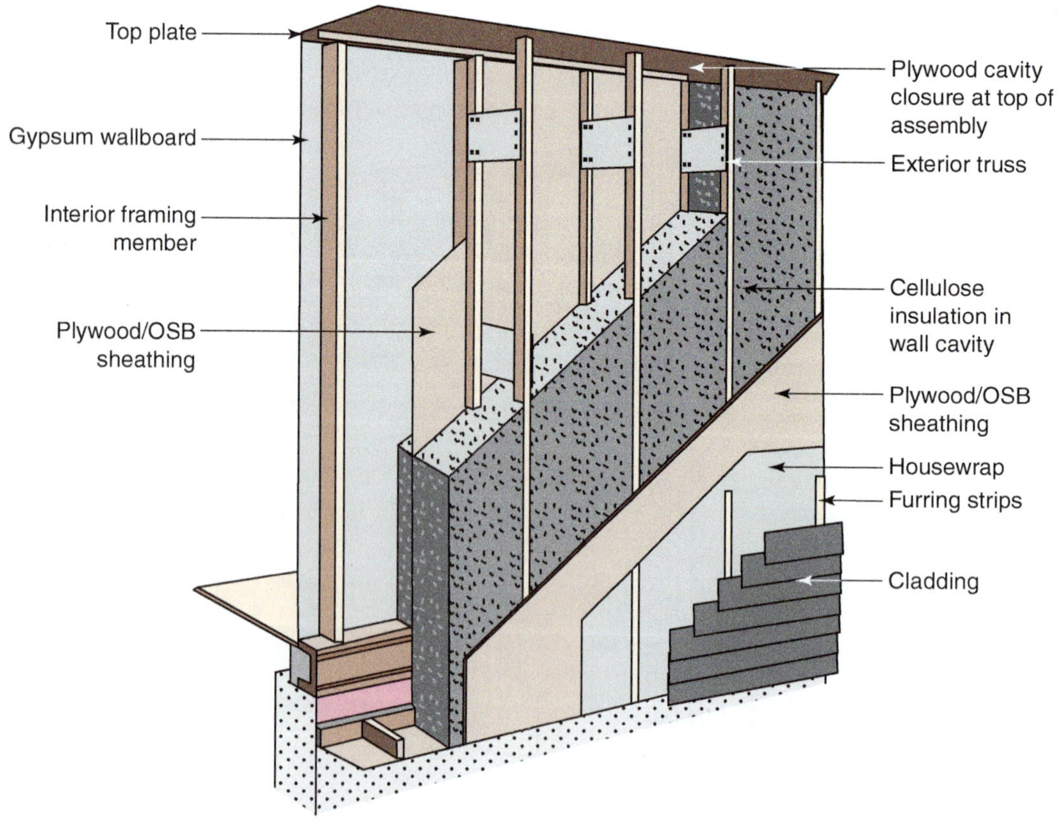

Goodheart-Willcox Publisher

Figure 7.9-3. Truss walls are similar to double-stud walls but offer the design flexibility to add even more insulation.

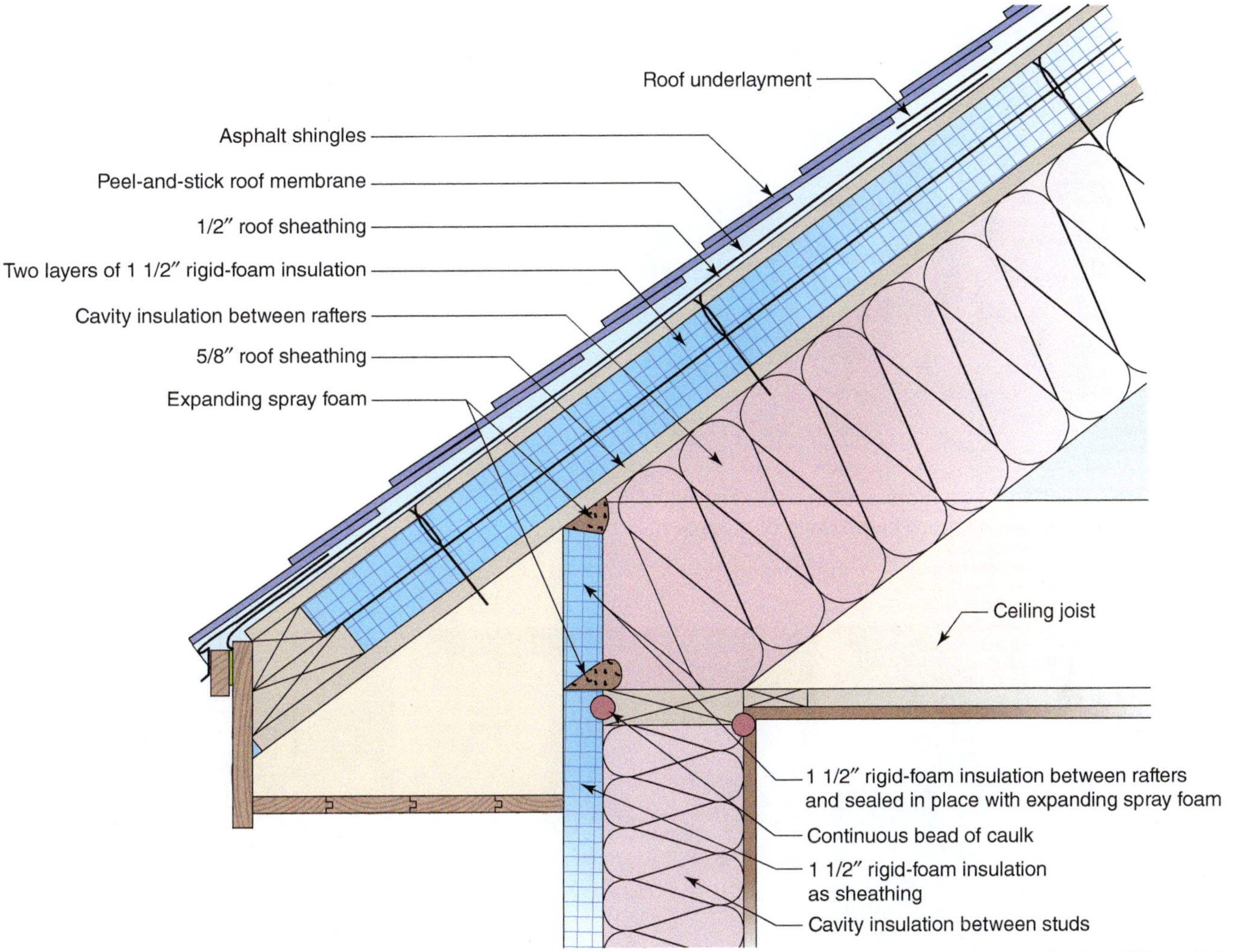

Goodheart-Willcox Publisher

Figure 7.9-4. Deck-over roof insulation helps to reduce thermal bridging and condensation issues in roof systems.

highly sensitive to issues related to air and water infiltration and condensation if the details are not carefully attended to.

Emerging Technologies: Construction Robots and Printed Walls

With companies facing persisting labor shortages and high production demands, the construction industry is turning to technology to help develop new ways to build homes in faster and more economical ways. Though it has not become a mainstream building practice, automated construction continues to make strides through robotics and 3-D printing. Robots can apply drywall joint compound to interior walls and sand surfaces to a smooth finish,

Canvas

Figure 7.9-5. Canvas is a robotic drywall finishing machine that can spray and sand smooth interior surfaces of gypsum wallboard.

Figure 7.9-5. Robots can also be used to lay bricks or CMU blocks, **Figure 7.9-6**. Giant 3-D printers, **Figure 7.9-7**, can construct the walls of an entire home in a single day.

FBR Limited

Figure 7.9-6. Hadrian X is a mobile robotic blocklaying machine capable of building the block walls of a home in a single day.

Each of these processes is designed to save time, reduce material waste, and protect human workers from jobsite hazards. Which of these amazing technologies will prove to be cost effective is hard to say. However, one thing is certain. Regardless of how many tasks become automated, the industry will continue to rely heavily on experienced tradespeople and professionals.

A Apis Cor B Apis Cor C futurewalk/Shutterstock.com

Figure 7.9-7. Developments in 3-D printing have led to more acceptance in the construction industry. A 3-D printer pumps a special blend of concrete through a nozzle that leaves a unique texture on the walls. Completed homes designed and built using this technology are leading to acceptance for production home builds.

LESSON 7.9 Review

Answer the following questions based on information in this lesson.

1. The main purpose of a superinsulated home is to _____. (7.9-1)
   A. reduce its heating and cooling requirements
   B. minimize air circulation
   C. reduce construction costs
   D. meet the basic code requirements
2. *True or False?* Using wall trusses eliminates the stud cavity. (7.9-1)
3. *True or False?* When deck-over roof insulation is used, an additional layer of sheathing is installed over the insulation for roofing material installation. (7.9-1)
4. *True or False?* Robots can be used to lay bricks. (7.9-2)

Chapter 7 Review

Summary

Lesson 7.1

- Becoming a stand-out candidate in the workforce involves crafting a professional résumé and cover letter, keeping an up-to-date portfolio, and mastering interview techniques, all to showcase your skills and suitability for the job. (7.1-1)

Lesson 7.2

- The construction field can be physically demanding. Key components to preventing injuries are understanding ergonomics and preparing for adverse weather conditions. (7.2-1)

Lesson 7.3

- A thorough check verifying proper completion of MEPs and framing, coupled with careful insulation placement, is critical for ensuring quality installation of interior surfaces. (7.3-1)
- Common interior surfaces used in new home construction for ceilings, walls, and floors include drywall, wood planks and board paneling, and tile. Common floor surfaces also include laminate and carpet. (7.3-2)

Lesson 7.4

- Thermal insulation is made in many forms and may be grouped into five broad classifications: flexible, loose-fill, rigid, spray foam, and reflective. (7.4-1)
- Interior surfaces play a pivotal role in shaping the look and feel of a home, with considerations such as durability, maintenance, and personal preference guiding the selection of materials. The selection and installation of interior surface materials, such as drywall, wood paneling, tile, and flooring determine the look and feel of a finished home. (7.4-2)
- A variety of fasteners, fillers, and adhesives are used during the installation of interior surfaces. Following manufacturers' recommendations regarding installation of products helps ensure the installation will meet the performance and warranty requirements. (7.4-3)

Lesson 7.5

- Installation of interior surfaces often requires specialized tools that are only used for one job. (7.5-1)

Lesson 7.6

- Key safety principles in moving, handling, and storing materials on construction sites include proper lifting techniques, clear communication, stable equipment operation, secure stacking to prevent collapses, maintaining a clean work area, and proper disposal of debris and flammable materials. (7.6-1)
- Construction workers need to understand fire hazards, causes, and control methods, with employers ensuring proper training on fire protection and prevention, including knowledge of fire classes, types of extinguishers, and adherence to fire safety protocols and administrative policies. (7.6-2)

Lesson 7.7

- Understanding how to calculate volume is an important part of estimating materials that take up a three-dimensional space. (7.7-1)

Lesson 7.8

- The specifications, the most important part of the working drawings with regard to interior surfaces, outline the specific requirements, materials, methods, and quality standards for construction. (7.8-1)
- Under each heading, the content of specifications is usually divided into sections: scope of work, specifications of materials to be used, application methods and procedures, and guarantee of quality and performance. Specifications may also include information and requirements regarding building permits, contract payment provisions, insurance and bonding, and provisions for making changes to the original plans. (7.8-2)

Lesson 7.9

- Superinsulating is a building design strategy that minimizes the heat loss and gain of a building by using considerably higher levels of insulation and airtightness. This is achieved through the extensive use of continuous insulation, spray foam, and alternative framing methods that allow more space for insulation. (7.9-1)
- Though it has not become a mainstream building practice, automated construction continues to make strides through robotics and 3-D printing. (7.9-2)

Apply and Analyze

1. After applying for a job, you receive a call to schedule an interview. Identify five things you will do before the day of your interview. (7.1-1)
2. The weather has turned cold, and today you will be working outside at the jobsite, assisting with the installation of rigid insulation sheets and sheathing. Should you select a pair of thick insulated gloves or a pair of thinner protective gloves, and why? (7.2-1)
3. When planning new construction, the homeowner tells you that he doesn't think the house will need much insulation because he doesn't often get cold. What types of information should you suggest the homeowner take into consideration when determining the type and amount of installation to use, and why? (7.3-1)
4. A design team examines the plans for a new home to see if there is a way to trim the budget to offset an unexpected increase in the price of lumber. One member of the team suggests eliminating the underlayment for the hardwood flooring in the upstairs bedrooms. Would you agree that this would be a good idea? Why or why not? (7.4-2)
5. A carpenter's apprentice needs to bring the carpenter a bonding agent to glue veneer strips to plywood edges. The apprentice brings epoxy. Is this a good choice for the intended purpose? Why or why not? (7.4-3)
6. You are installing 4′ × 8′ drywall sheets, but the distance from the floor to the ceiling in the room is 8′-1″. What type of equipment would help you install the drywall in this room? Why? (7.5-1)
7. On the first day at a jobsite, you and an apprentice are tasked with moving drywall sheets to a particular room. How should you instruct the apprentice to help you carry the drywall? (7.6-1)
8. If a 12′ pipe has a diameter of 6″, what is its volume rounded to the nearest foot? (7.7-1)
9. What would happen if the specifications for a project were poorly prepared? Why? (7.8-1)
10. Would the use of insulated double-stud walls help ensure a house is superinsulated? Why or why not? (7.9-1)

Critical Thinking

1. What documents and materials should be kept in a career portfolio? How does this help a potential employer better understand your skills and experience? (7.1-1)
2. What kind of impact can rodents, such as mice, rats, and squirrels, have on a building? Left unchecked, how can these animals damage the thermal performance of a home? (7.3-1)
3. Construction workers building a tile shower have several options for tile backing. In the past, unprotected gypsum wallboard was used as a backing (or base) for wall tile; this method is no longer approved. What are the issues with using unprotected gypsum wallboard as a tile backer in a shower? How do you think these issues manifest themselves over time? (7.4-2)
4. How might the lack of proper training and awareness regarding fire protection and prevention among construction workers impact overall safety on a construction site? What measures can employers take to mitigate such risks? (7.6-2)
5. Superinsulating a home is one way to improve energy performance. Research other green building practices and compare the benefits of each. What do you think is the most cost-effective green building practice? What method do you think is the most impactful and beneficial to our larger society? (7.9-1)

Communicating about Construction

1. **Art and Writing.** Draw a picture of a residence, including a typical wall section, showing key areas that require insulation for your climate zone. Include a comparison of a home in the same climate without insulation and describe the consequences of not having insulation protection. (7.3-1)
2. **Speaking and Writing.** Create a material list to cover all ceilings and walls with gypsum wallboard in the room where you are sitting. It may be helpful to draw a 2-D sketch of each wall and the ceiling complete with dimensions. Research the lengths and widths of gypsum board available locally (often 52″ widths are available). Justify what size and quantity of materials you would purchase for this job, considering factors such as the number of joints, minimizing waste, linear feet of joints to finish, delivery access, and material-handling restrictions. Present your list to the class. (7.4-2)
3. **Writing.** Write an essay describing the home you lived in while growing up. Be as specific as possible while trying to identify the different materials used in the construction of the home. Compare the house you grew up in with your dream house. Point out the differences, other than size, and include in your description the materials used and the design of both houses. Be detailed in your descriptions. (7.4-2)

Breadmaker/Shutterstock.com

CHAPTER 8 Interior Finishes

Chapter 8 Outline

Introduction

After the interior surfaces of a home have been installed and completed, the rough edges between material transitions, such as floors to walls, remain exposed. The interior finishes of a home cover these rough edges and add functionality and beauty to the interior. *Chapter 8, Interior Finishes*, highlights this process and explores the types of materials, methods, tools, and fasteners used during this phase of construction. We will consider the importance of working with a good team and the efforts needed to maintain teamwork. You will continue to build your knowledge of how to protect yourself and others as we investigate the health hazards associated with the construction trade. You will also continue honing your construction math skills by studying how averages and percentages are used in construction.

Determining the precise layout and assembly of interior finishes such as cabinetry and staircase assemblies requires careful attention to the interior elevation drawings. In *Lesson 8.8*, you will see examples of these drawings and how they relate to other drawings in a set.

As you have progressed through each chapter, you have seen how conservation is at the core of green building. In this chapter, you will learn how managing construction waste can contribute to or detract from a building's impact on the environment.

Finally, in this chapter, we consider what kind of employment benefits you might expect to receive (aside from a paycheck) with a career in residential construction.

LESSON 8.1

Industry and Careers

LEARNING OUTCOMES

After completing this lesson, you will be able to:

8.1-1 List various employment benefits that may be offered in the workplace.

TECHNICAL TERMS

employment benefit

insurance premium

Employment Benefits

Employment benefits are compensation and perks you receive from an employer in addition to your regular paycheck, **Figure 8.1-1**. These benefits are meant to improve the value of your career with a specific employer, create a better work–life balance, and promote a higher quality of life in both the short and long term. In the residential construction industry, employment benefits can vary depending on the company's size, location, and policies.

It is important to understand that not all employers offer benefits beyond a working wage (paycheck). Laws vary from state to state, and many smaller companies may not have the administrative or financial capability to provide employment benefits.

The following sections contain examples of possible benefits.

Traditional Benefits

- Health, dental, and vision insurance
- Short- and long-term disability
- Paid time off

Retirement & Financial Benefits

- 401(k) or 403(b)
- Tuition reimbursement
- Student debt relief

Nontraditional Benefits

- Accident insurance
- Pet insurance
- Identity theft protection

Trending Benefits

- Parental leave
- Mental health benefits
- Professional development
- Flexible work schedule

Goodheart-Willcox Publisher

Figure 8.1-1. Employment benefits vary from one employer to another. It is important to consider an entire benefit package and not just the salary when comparing jobs.

Health Insurance

Health insurance provides coverage for medical expenses and healthcare services, **Figure 8.1-2**. Employers may offer various types of health insurance plans to their employees.

fizkes/Shutterstock.com

Figure 8.1-2. Health insurance is a valuable employment benefit.

Typically, employers work with health insurance companies to provide coverage for a fee. This fee is called the health ***insurance premium***. The premium is the amount paid to the insurance company, usually taken directly out of your paycheck, to maintain the health insurance coverage. The premium must be paid to keep your health insurance policy active, even if you don't use any medical services.

Most health insurance plans require contributions from the employee to help offset the cost of the premium. How much the employee pays often depends on whether the employee receives only individual coverage or coverage for a family.

Most insurance plans do not pay all medical expenses. They may require the policyholder to pay deductibles and copayments, as well as for prescription medications. Some policies cover more of these expenses than others. These policy variations make health insurance an important benefit to understand when changing jobs or negotiating with a potential employer.

Retirement Plans

Many employers offer retirement plans to help their employees save money for the future and to help ensure their financial security after they stop working. Retirement plans come in a variety of structures with various tax advantages and investment options. Retirement plans include options such as 401(k), 403(b), traditional and Roth IRAs, and pension plans.

The most common employer-offered plans in the construction industry are 401(k) accounts. The 401(k) plan allows employees to choose to contribute to the account before taxes are paid on that income. This structure allows contributors to defer income tax that would have been paid in higher-earning years to later in life when the income generated by the investments is taxed.

Employer contributions to these plans can include an employer match, profit sharing, or fixed percentage contributions.

- Employer match—The company matches what the employee contributes to their retirement plan (typically about 2% to 5% of their pay).
- Profit sharing—The employer makes contributions based on the company's profitability. The amount and frequency of profit-sharing contributions can vary based on the company's performance. This type of plan is meant to encourage employees to hold a direct interest in the performance of the company.
- Fixed percentage contributions—The employer contributes a fixed percentage of the employee's pay, regardless of their own contributions.
- No contribution plans—An employer may choose to set up the structure for an employee to contribute to a retirement account without contributing themselves.

Paid Time Off

Some employers offer their employees paid time off (PTO) in the form of paid holidays, vacation days, personal days, sick leave, or maternity/paternity leave. PTO can add real material and quality-of-life value to employees, allowing them a break from work when they otherwise may not have been able to afford it.

Overtime Pay

The Fair Labor Standards Act (FLSA) is a federal law that governs overtime pay in the United States. It requires that employers (though there are some exemptions) must pay workers at least 1.5 times their regular rate of pay when they work more than 40 hours in a workweek. This law is a major factor in why most Americans have a 40-hour workweek.

Employers are not required to provide more than 40 hours of work to employees in a given week, but some offer or even may require it as a work condition. Some employees feel that overtime opportunities are a big benefit because of the extra pay, while others may not find them appealing.

Tool or Clothing Allowances

Employers may offer monetary credit or discount allowances for employees to purchase tools or clothing they are expected to use on the job.

Pro Tip

If an employer offers a tool/clothing allowance, don't try to pocket extra money by avoiding buying the required tools/apparel or by purchasing low-quality items. This can end up costing you more money in the long run.

This benefit can be beneficial to both employee and employer as it helps ensure employees have the tools and clothing needed to perform a job.

Transportation Benefits

Because vehicles, especially pickup trucks, are used extensively in residential construction, it is not uncommon for the employer to provide a truck and/or fuel card to some of their key employees for work purposes. Different companies have a variety of policies concerning work vehicles, but many employers allow company vehicles to be driven to and from personal residences to the jobsite. This can be a major financial benefit for the employee when the vehicle cost, maintenance, fuel, and insurance costs are factored in.

LESSON 8.1 Review

Answer the following questions based on information in this lesson.

1. *True or False?* Health insurance premiums are usually taken directly from a worker's paycheck. (8.1-1)
2. *True or False?* Most health insurance plans pay all medical expenses. (8.1-1)
3. Which statement about 401(k) accounts is true? (8.1-1)
   A. Income taxes are immediately deducted from contributions made to the account.
   B. The employer decides the amount that the employee will contribute to the account.
   C. They are the most common employer-offered plan in the construction industry.
   D. The employer is legally obligated to contribute matching funds to each employee's account.
4. *True or False?* A fixed percentage 401(k) plan is specifically designed to encourage employees to hold a direct interest in the performance of the company. (8.1-1)
5. Paid holidays are an example of _____. (8.1-1)
   A. a retirement plan
   B. paid time off
   C. overtime pay
   D. a tool or clothing allowance
6. The Fair Labor Standards Act requires that most workers be paid 1.5 times their regular rate of pay when they work _____ hours in a workweek. (8.1-1)
   A. less than 10
   B. more than 30
   C. less than 35
   D. more than 40

LESSON 8.2

Construction Fundamentals

LEARNING OUTCOMES

After completing this lesson, you will be able to:

8.2-1 Demonstrate how teamwork and collaboration are critical when working on construction projects.

TECHNICAL TERMS

collaboration

teamwork

Teamwork and Collaboration

Teamwork and collaboration contribute greatly to success when working in the residential construction industry. ***Teamwork*** refers to the combined efforts of a group of individuals working together toward a common goal through shared leadership or management. ***Collaboration*** involves individuals or groups working together, across different teams, crews, or trades, to achieve a shared objective. Collaboration is usually more flexible and can involve sharing knowledge, skills, and resources without a rigid structure. Collaboration often relies on mutual respect for achieving the overall outcome.

Pro Tip

Effective teamwork and collaboration mean actively listening, communicating openly, and respecting each team member's contributions. Practice these traits on a regular basis whether in school, at home, in sports, or at work.

The construction of a home is a complex and intricate process that requires coordination among various trade professionals, project participants, and stakeholders. Each team member brings their own expertise, experience, and personality into the mix. When faced with a challenge or crisis, these individual traits can become a powerful resource to produce creative solutions. To achieve project goals, individuals with diverse backgrounds, skills, and expertise must be able to come together to overcome personal and professional differences.

Coordinating tasks, time lines, and resources effectively is key to keeping projects on schedule and on budget and meeting quality standards, **Figure 8.2-1**. Often, this requires a give-and-take relationship across trades. The jobsite is a shared space where each trade and each worker must

BELL KA PANG/Shutterstock.com

Figure 8.2-1. Architects and builders need to have excellent communication skills to ensure project goals are met. This is just one example of the many team dynamics needed to complete a construction project.

recognize the need for others to use that space to get their work done, preventing delays in the overall schedule. Even though this principle sounds simple enough to achieve, an endless number of issues can arise on a job, raising stress levels and causing potential conflict between trades or between coworkers. Unfortunately, conflict on a job can escalate to a point of animosity, rivalry, and in some cases, sabotage of another's work. This type of behavior is never acceptable and can be grounds for disciplinary action or even termination.

Teamwork and collaboration are a major part of the entire industry; it is important to consider the long-term perspective. Often, crews of contractors, subcontractors, and tradespeople work together on multiple projects over time. In the residential construction industry, local professional circles tend to be small. These circumstances make being a team player even more important.

Building trust is also important and confidentiality is key to this. Throughout the course of a construction project sensitive information is shared between stakeholders in the project. This information can include personal and financial information about clients, proprietary information regarding designs and business strategies, financial information dealing with the project or clients, employee records, security information such as access codes for door locks, or even personal information that comes up in conversations. It is important not to share information or documents that could violate privacy, compromise intellectual property, or otherwise damage the trust held between coworkers, subcontractors, clients, and other parties involved.

One powerful tool construction workers use to enhance communication throughout the project is digital construction management software, featured in *Chapter 9, Finish MEP Systems and Punch List*. Tradesmen, architects, clients, vendors, and other stakeholders access the software through apps on their phone, tablet or a laptop. Good communication and coordination, respect, problem solving, patience, and adaptability are central to maintaining good teamwork and collaboration throughout a project. A good team leader, such as a superintendent, foreman, or lead tech, can be central to making sure these qualities are being practiced on the job. However, it is up to each member of the team to carry out the behaviors of a team player. Practicing these traits will help you to grow personally and professionally. Exhibiting this type of work ethic and responsibility on the job will build trust in relationships and can help ensure future projects will operate smoothly.

LESSON 8.2 Review

Answer the following questions based on information in this lesson.

1. Teamwork refers to _____. (8.2-1)
   A. the efforts of a group of individuals working together toward a common goal under shared leadership
   B. individuals working together, across different teams, to achieve a shared objective
   C. a competition between different individuals to achieve a goal in the least amount of time
   D. a group of self-managed individuals working independently of one another toward the same goal
2. *True or False?* Teamwork is usually more flexible than collaboration. (8.2-1)
3. *True or False?* In order to achieve project goals, individuals with different backgrounds, skills, and expertise must overcome their differences and work together. (8.2-1)

LESSON 8.3

The Construction Path

LEARNING OUTCOMES

After completing this lesson, you will be able to:

8.3-1 Identify which areas of a new home build require the application of interior finishes.

TECHNICAL TERMS

base shoe	platform	total rise
baseboard	prehung door unit	total run
casework	reveal	tread
crown molding	riser	unit rise
door casing	stair carriage	unit run
door frame	stairwell	winding stair
millwork	stool	
molding	straight run	

Interior Finishes

Interior finishes of a home include the installation of millwork, casework, and work surfaces, such as kitchen and bathroom countertops. ***Millwork*** refers to woodworking products produced in a mill or woodworking shop, such as stair railings (called the balustrade), interior doors and frames, and moldings to cap intersections between floors, walls, ceilings, windows, and doors. ***Casework*** is built-in cabinetry or shelving, such as kitchen or bathroom cabinetry, built-in bookshelves, and other storage units. Casework can be custom built or prefabricated.

Installing interior finishes is the job of a carpenter, requiring great skill and accuracy. These finishes have a major influence on the overall look and functionality of the home, **Figure 8.3-1**.

Stairs

Stair construction requires a high degree of skill. The quality of the work should compare with that found in fine furniture building, **Figure 8.3-2**. The parts for main stairways are usually made in millwork plants and then assembled on the job. Even so, the assembly work must be performed by a skillful carpenter who understands the basic principles of stair design and knows layout and construction procedures.

Stairs consist of ***risers***, the vertical stair members between consecutive stair treads, and ***treads***, the horizontal walking surface of a stair. Risers and treads are supported by carriages, **Figure 8.3-3**. A ***stair carriage***, sometimes referred to as a stringer, is

ungvar/Shutterstock.com

Figure 8.3-1. The installation of millwork and casework is an exciting time in a project.

sergiophoto/Shutterstock.com

Figure 8.3-2. An attractive main set of stairs is a desirable architectural feature in a residence. Stair design and construction has long been considered one of the highest forms of joinery.

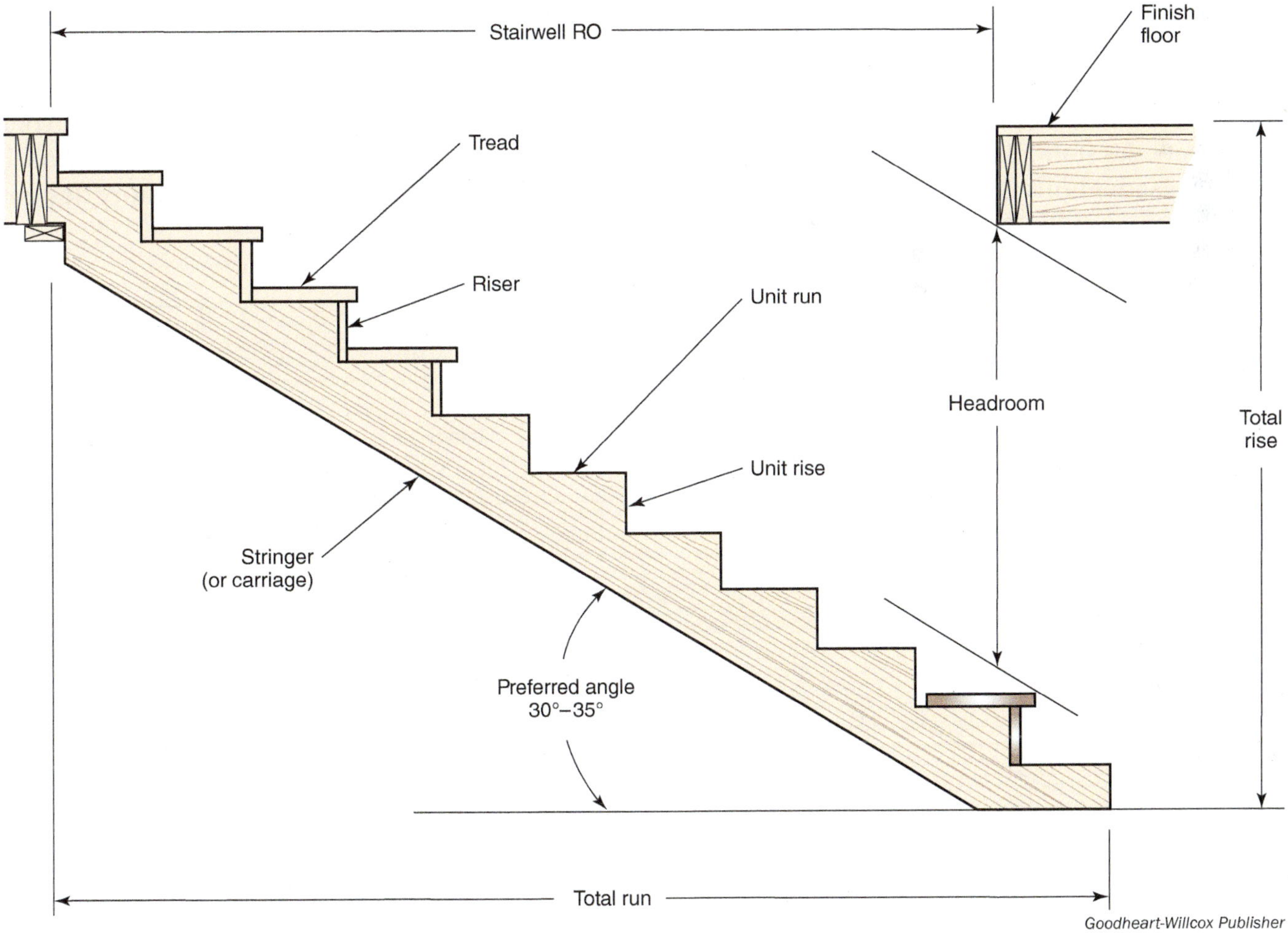

Goodheart-Willcox Publisher

Figure 8.3-3. Basic stair parts and terms. The total number of risers is always one greater than the total number of treads.

the main structural support component of the stair system. Stair carriages must be laid out and cut to evenly distribute the ***total rise***, or the overall vertical distance from one floor to the next. Additionally, the ***total run***, or the overall horizontal distance covered by the stairs, is used to calculate the length of the stair carriage. The relationship between the riser height, called ***unit rise***, and the tread width determines how easily the stairs may be negotiated. ***Unit run*** is the tread width minus the nosing. Research has indicated that the ideal riser height is 7″, while the ideal tread width is 11″.

In residential structures, treads (excluding nosing) are seldom more than 12″. Nosing is a small extension of the tread. In a given run of stairs, it is extremely important that all of the treads be the same size. The same is true of the risers. A person tends to subconsciously measure the first few steps and may trip if subsequent risers are not the same.

When the rise-run combination is wrong, climbing the stairs will be tiring and cause extra strain on leg muscles. Further, the toe may kick the riser if the tread is too narrow. A unit rise of 7″–7 5/8″ with an appropriate tread width provides both comfort and safety. As stair rise is increased, the run must be decreased, **Figure 8.3-4**.

A stair should be wide enough to allow two people to pass without contact. Further, it should provide space so furniture can be moved up or down.

These rules are basic principles that make for a comfortable and safe staircase, but each job has its own limitations. Adjustments can be made to suit these limitations; however, stair building requires a full understanding with respect to the building codes.

Carpenters build temporary stairs from framing lumber to provide access until the permanent stairs are installed. These are often designed as a detachable unit so they can be moved from one project to another. Sometimes, permanent stair carriages are installed during the rough framing and temporary treads are attached.

Main stairways are usually not built or installed until after interior wall surfaces are complete and finish flooring or underlayment has been laid. There are two basic stair categories: service stairs and main stairs. These may be closed, open, or a combination of open and closed, **Figure 8.3-5**. In addition, the type of stairs may be straight run, platform, or winding.

The ***straight run*** stairway is continuous from one floor level to another without landings or turns. It is the easiest to build. Standard multistory designs

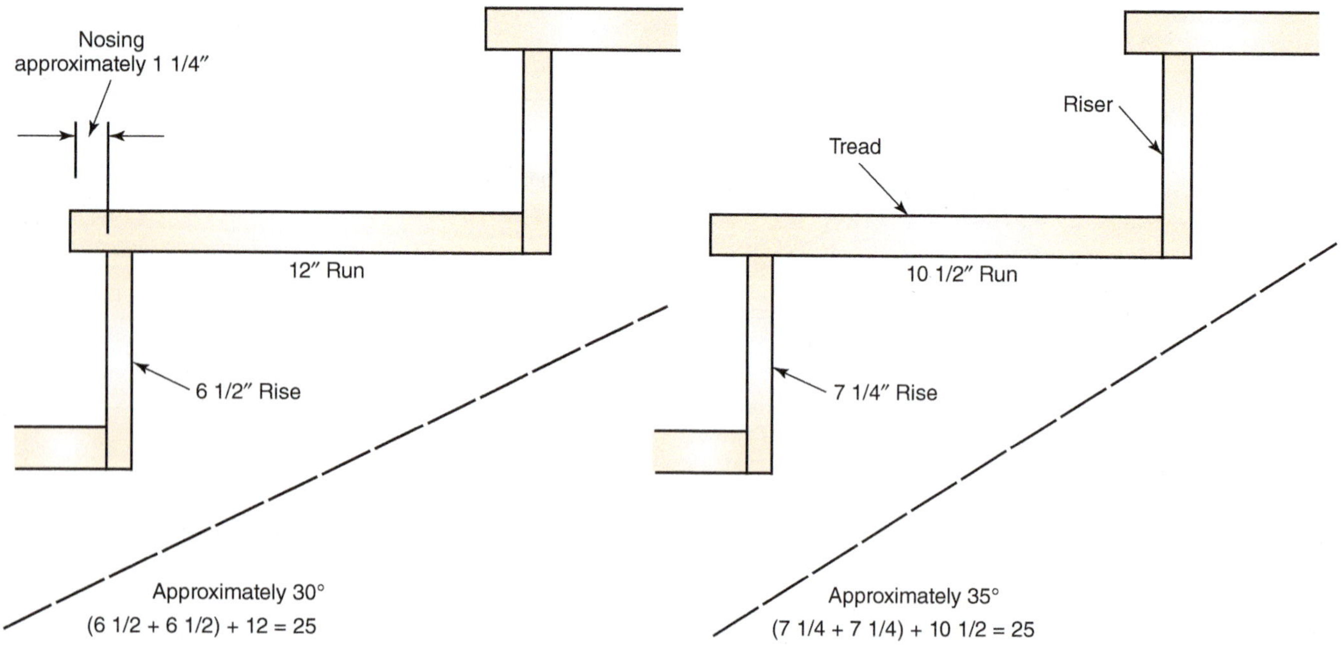

Goodheart-Willcox Publisher

Figure 8.3-4. The relationship between rise and run must be considered when building stairs. An increase in unit rise generally coincides with a reduction in unit run.

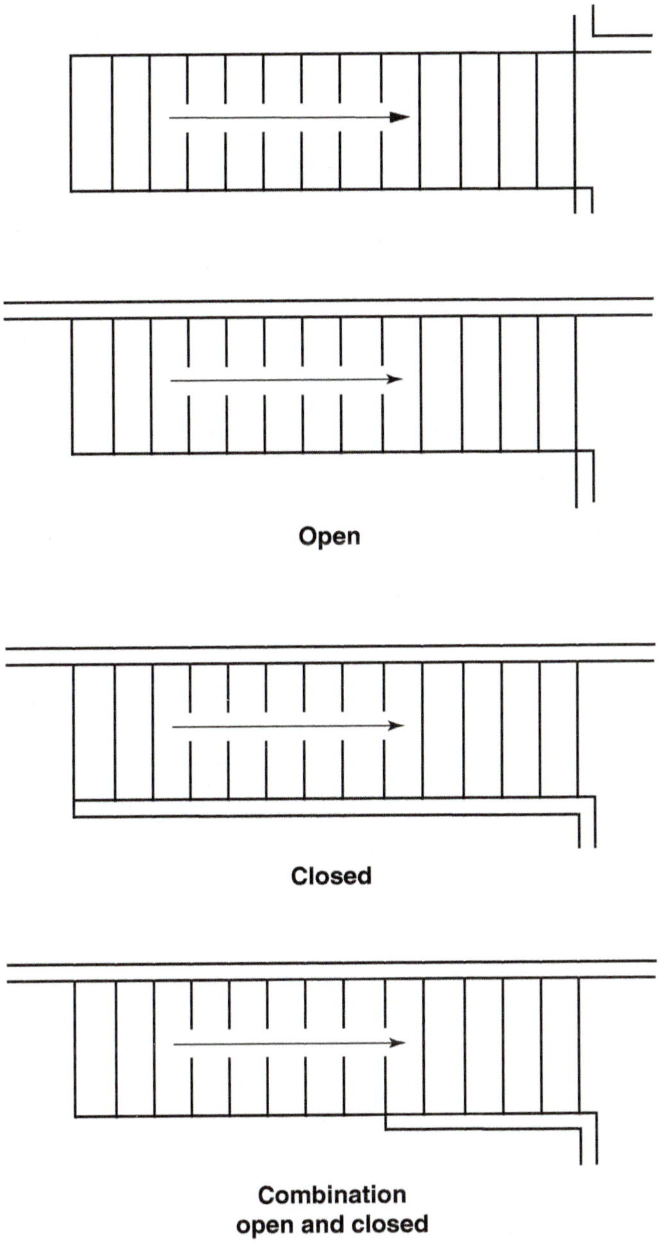

Goodheart-Willcox Publisher

Figure 8.3-5. Open and closed stairways. A stair is called open even if one side is enclosed by a wall.

require a long ***stairwell*** in the floor above to provide headroom. This often presents a problem in smaller structures. A long run of 12–16 steps also has the disadvantage of being tiring. It offers no chance for a rest during ascent.

The ***platform*** type includes one or more landings where the direction of the stair run is usually changed. Such descriptive terms as L-type (long L and wide L), double L-type, and U-type (wide U and narrow U) are commonly used, **Figure 8.3-6**.

Winding stairs, also called *geometrical stairs*, are circular or elliptical. They gradually change directions as they ascend from one level to another. These often require curved wall surfaces that are difficult to build. Because of their expense, winding stairs are usually only found in high-end homes.

Interior Doors and Trim

The interior ***door frame*** is an assembly of wood parts that form an enclosure and support for a door and its hardware. The door frame covers the unfinished edges of the door opening. It also supports the trim pieces that are attached after installation.

The frame consists of two side jambs and a head jamb, **Figure 8.3-7**. Door frames are usually cut, sanded, and fitted (prehung) in millwork plants. This allows quick assembly on the job. Door frames should receive the same care in storage and handling as other finished woodwork.

A ***prehung door unit*** consists of a door frame with the door panel already installed. See **Figure 8.3-8**. Trim, or ***door casing***, is applied to each side of the door frame to cover the space between the jambs and the wall surface. This secures the frame to the wall structure and stiffens the jambs so they can carry the door. **Figure 8.3-9** shows a section view through a doorjamb. The casing covers the shims and is attached to both the jamb and wall surface. It is usually installed with a ***reveal***, a setback from the inside edge of the jamb of about 1/4″ or 3/8″. The reveal allows room for hinges and striker plates, while improving the appearance of the trim.

Pro Tip

After a prehung door unit is installed, some builders remove the door from its hinges and carefully store it to avoid damage during continuing construction.

Window Trim

Interior window trim consists of the casing, ***stool*** (forms the interior sill cap of a window), apron, and jamb extensions, **Figure 8.3-10**.

Sometimes, the stool and apron are eliminated. Instead, a piece of beveled sill liner is installed to match the window jamb. Regular casing

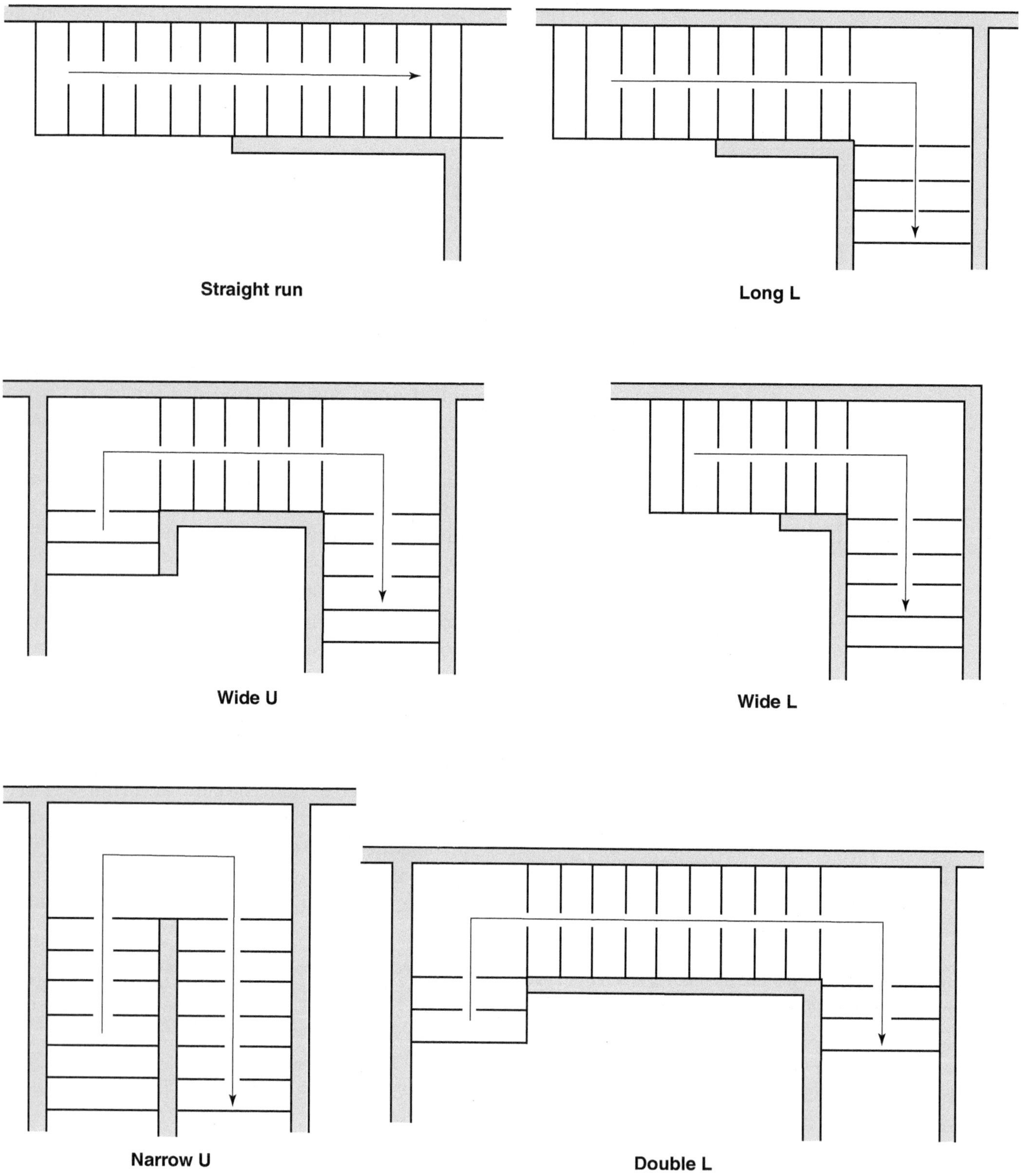

Goodheart-Willcox Publisher

Figure 8.3-6. Terms used to define various stair types.

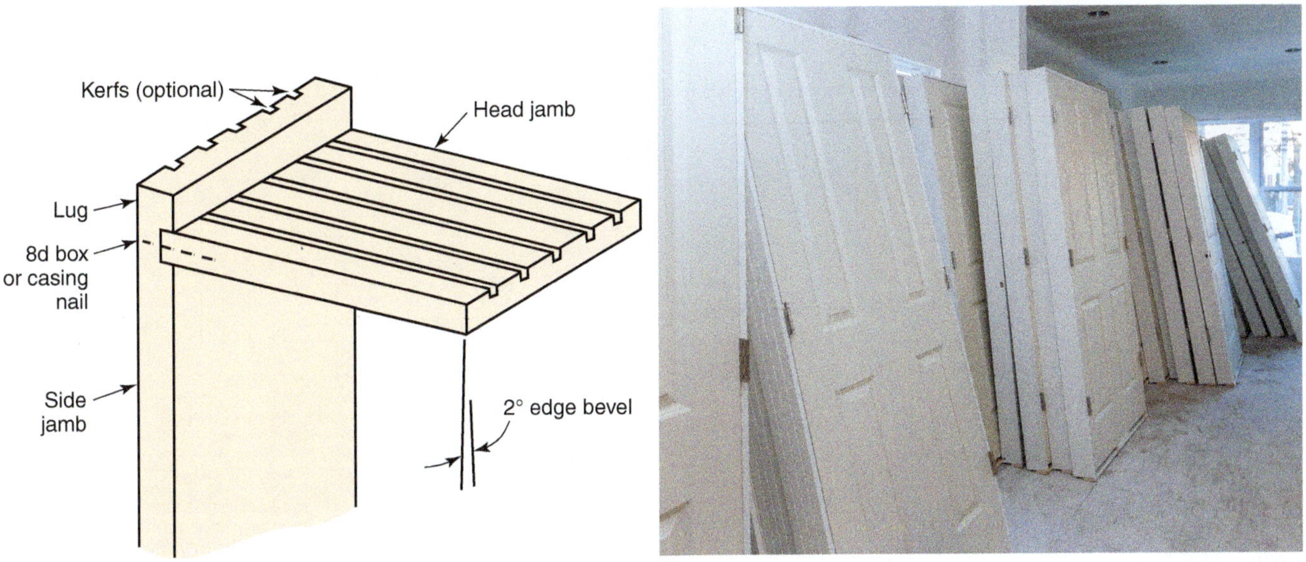

A Goodheart-Willcox Publisher B ungvar/Shutterstock.com

Figure 8.3-7. A—Section view of an interior door frame. Parts are listed. Note that edges of all jambs may be beveled slightly so trim will fit snugly. B—Prehung door frames should be stored with care to avoid damage.

General Procedure for Installing a Prehung Interior Door		

1. Remove door from packing and check for damage. Place it on sawhorses with the exposed hinge barrels up.	2. Install door casing to the door frame on the side facing up. Slide the door unit into the rough opening. Side jambs should rest on the finished floor or spacer blocks.	3. Carefully plumb door frame and nail casing to the wall structure.

4. Move to the other side of the wall and install shims between the side jambs and the rough opening at locations indicated by the door manufacturer. Nail through jambs and shims into the rough framing.	5. Install casing on the other side of the door unit.	6. Check door operation and fitment. Make any required adjustments. Install lock set.

Goodheart-Willcox Publisher

Figure 8.3-8. General procedure for installing a prehung interior door.

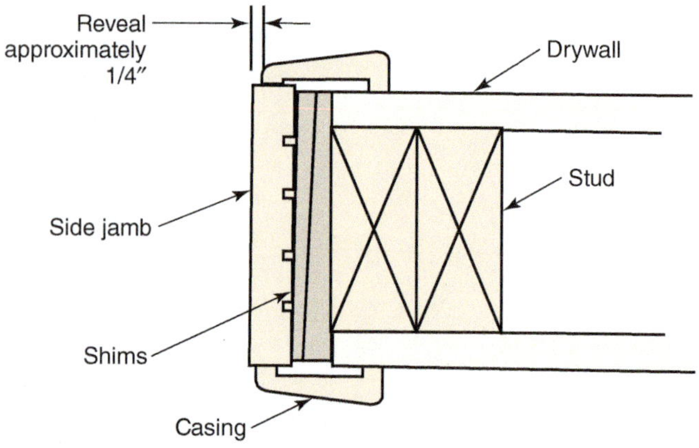

Goodheart-Willcox Publisher

Figure 8.3-9. This plan-view section drawing shows the position of casing, which is attached to the doorjamb.

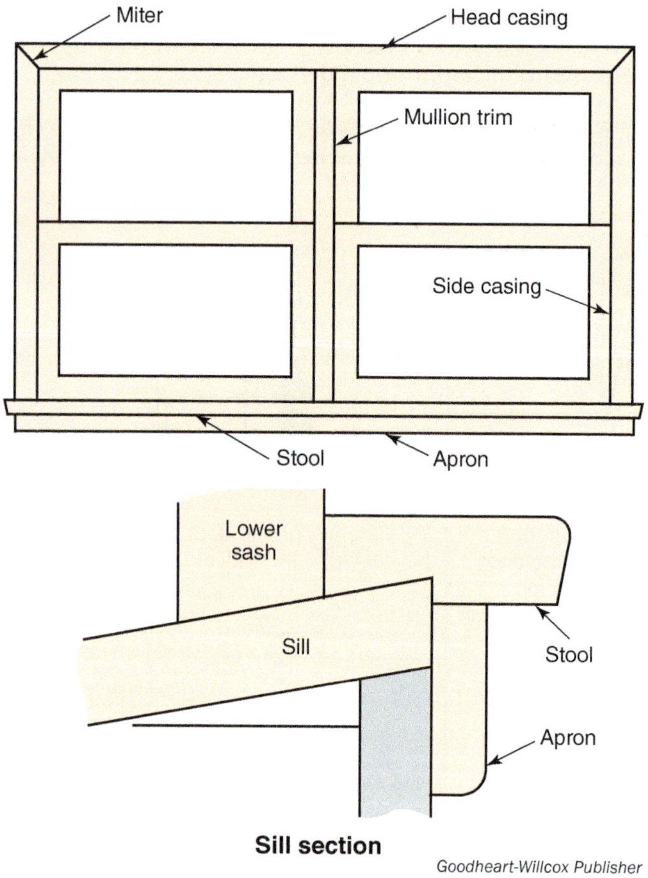

Goodheart-Willcox Publisher

Figure 8.3-10. Trim members used for a traditional wood double-hung window.

is then applied around the entire window. See **Figure 8.3-11**. This is known as picture-frame trimming.

Moldings

Moldings are decorative wood, plastic, or composite trim pieces used on doors, windows, ceilings,

Artazum/Shutterstock.com

Figure 8.3-11. Notice how casing can be used to replace the stool and apron.

and walls. Window and door casings, mentioned earlier, are examples of molding.

Crown molding is a decorative piece applied to the top of a wall to make the transition from the wall to the ceiling. **Figure 8.3-12A** shows one style of crown molding. There are many profiles (shapes), but all crown molding is installed at an angle, leaving an open space behind the molding, **Figure 8.3-12B.**

The ***baseboard*** is a molding that covers the joint where the wall intersects the finish flooring, **Figure 8.3-13**. It is among the last of the interior trim members to be installed, since it must be fitted to the door casings and cabinetwork and over the finish flooring. Baseboards run around the room between door openings, cabinets, and built-ins.

Base shoe is narrow molding that is sometimes used around the perimeter of a room where the baseboard meets the finish flooring. It is used to seal the joint between the baseboard and the floor. It is usually fitted at the time the baseboard is installed, but not nailed in place until after any surface finishes have been applied. Base shoe is often used to cover the edge of tile or carpet.

Casework and Countertops

In the kitchen, and in other areas of the home, storage units should be designed for the items that they will contain. Space must be carefully allocated. Drawers, shelves, and other elements should be

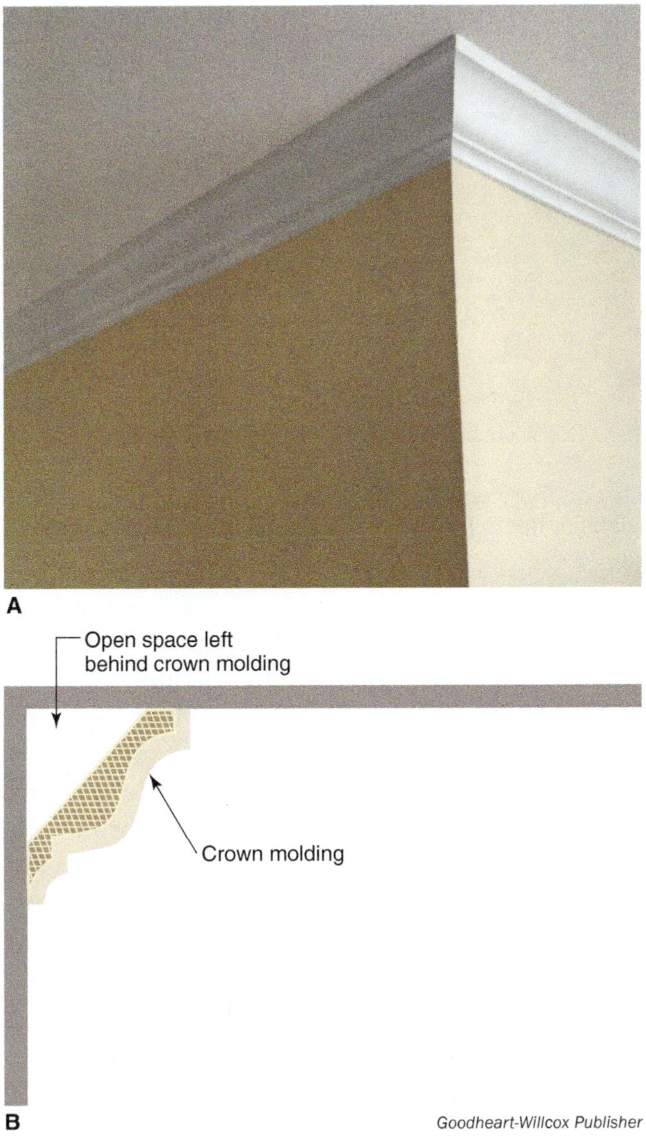

Goodheart-Willcox Publisher

Figure 8.3-12. A—Crown molding is available in many patterns and sizes. B—Crown molding is installed at an angle between the wall and ceiling.

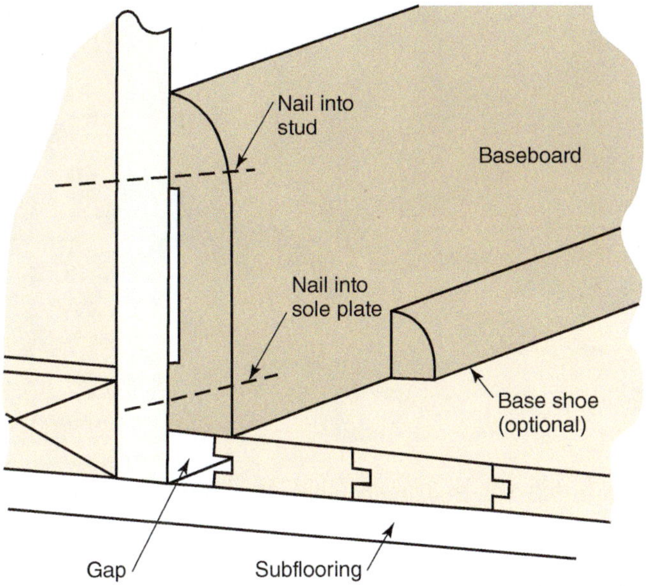

Goodheart-Willcox Publisher

Figure 8.3-13. Cutaway of baseboard and base shoe. They conceal the gap between the flooring and finished wall.

proportioned to satisfy specific needs. Three types of cabinetwork are most often used in homes:

- Units built on the job by a carpenter—In many places in a home, a carpenter may be required to build shelving or casework. Even when most of the cabinets are factory produced, a carpenter is responsible for the installation. This task requires skill and careful attention to detail.
- Custom-built units constructed in local cabinet shops or millwork plants—Custom cabinets in high-end homes are generally built in a cabinet shop and brought to the building site ready to be installed.
- Mass-produced cabinets from factories that specialize in this area of manufacturing—Except in high-end custom-built homes, most cabinets are made in a cabinet factory and shipped to the site ready for installation.

Most of the cabinetwork for residential and commercial buildings is constructed in factories that specialize in cabinetry. Modern production machines and tools save time and produce high-quality cabinetry. Mass-produced parts are assembled with the aid of jigs and fixtures.

Manufacturers offer a variety of shades and colors that are applied at the factory by experts, **Figure 8.3-14**. Because of the controlled conditions and special equipment, finishing materials with high resistance to moisture, acids, and abrasion can be applied. Once the finishing process is complete and hardware has been installed, the units are carefully packaged and shipped to a distributor or directly to a construction site.

Cabinet Design

Fabricators of casework rely on universal design principles for comfort, convenience, and usability. Standardization of some key dimensions also helps to ensure compatibility across a range of products and manufacturers for appliances, fixtures, and accessories, **Figure 8.3-15**. Kitchen base cabinets are typically 34 1/2″ high. When the countertop

A il21/Shutterstock.com

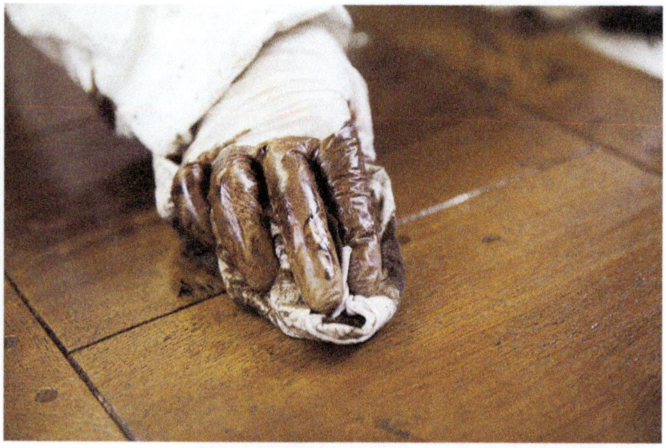

B MintImages/Shutterstock.com

Figure 8.3-14. Building cabinets in a factory can improve quality at a more economical price. A—Cabinet finishes are sprayed on for a consistent, quality finish. B—Premium-quality cabinet doors receive a coat of hand-rubbed stain from highly skilled wood finishers.

is added, they are typically 36″ high. The countertop extends about 1″ beyond the base cabinets. The vertical distance between the top of the base unit and the bottom of the wall unit is normally 18″, but it can vary to suit special circumstances. Above a sink or cooktop, this distance is usually at least 24″, although it is common to have a window over a kitchen sink. Wall cabinets above a refrigerator are usually 72″ above the finished floor.

The standard depth for base cabinets is 24″. The countertop usually overhangs the base cabinets by about 1″. If the cabinets were deeper than 24″, it would be difficult to reach items in the back of the cabinets. Wall cabinets are usually 12″ deep, but other depths are available for other locations, such as over the top of a refrigerator.

Kitchen cabinets are generally available in widths from 9″ (narrow spice drawers) to 36″ or more in 3″ increments. Where the cabinets cannot be arranged to exactly fit the available space with these 3″ increments, a filler piece (usually 3″ wide) can be made to fill any remaining space. The filler strip has the same finish as the cabinets, so it blends in with the rest of the kitchen.

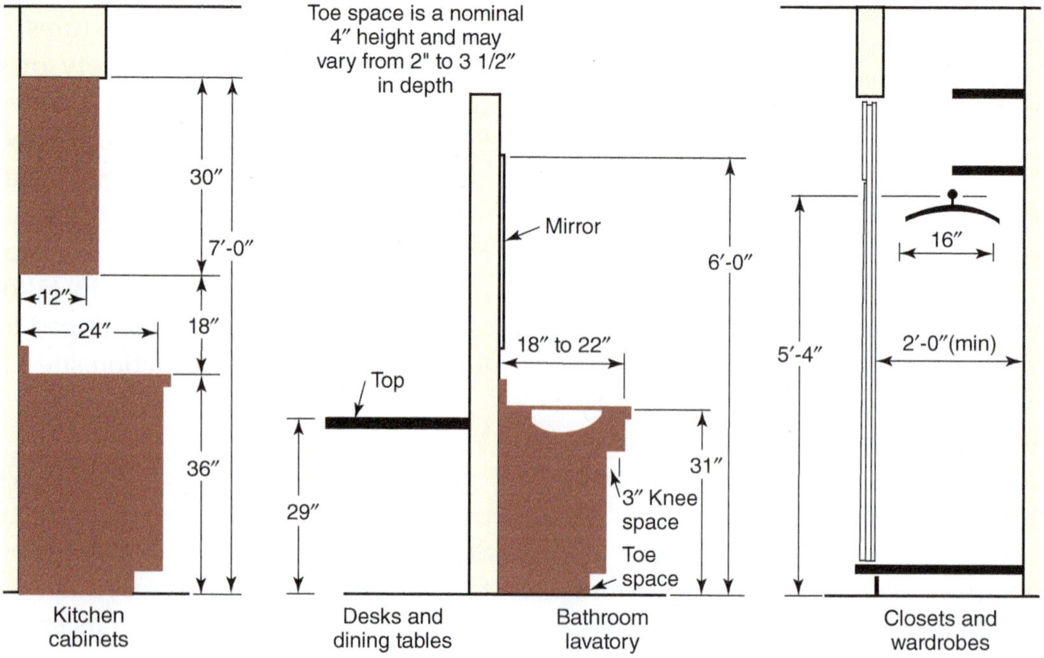

Goodheart-Willcox Publisher

Figure 8.3-15. Cabinet dimensions are standardized.

A variety of storage features can be added to standard cabinet units. Examples include revolving shelves (lazy Susan), special compartments and dividers in drawers, slide-out breadboards, and slide-out shelves. Storage units for canned goods provide extra convenience with swing-out wood shelving, **Figure 8.3-16**. Other available storage features include the following:

- Slide-out or tilt-out wire racks for under-sink storage
- Wire racks for special lid storage
- Swing-out storage trays
- File cabinets
- Wall tambour storage
- Pull-out ironing boards
- Swing-out, multi-storage wire shelves for base cabinets

In bathrooms, vanity cabinets with built-in lavatories are normally 31″ high, **Figure 8.3-17**. The depth depends on the type of fixture. The counter surface may be plastic laminate, stone, synthetic stone, or tile. Some lavatories provide knee room; others do not.

Built-in units provide drawers and cabinets that save space. They improve efficiency and convenience. For example, many homes are being equipped with closet organizer systems for more efficient use of storage space, **Figure 8.3-18**. These systems consist of modular cabinet and shelving units that can be installed in various combinations, depending on the space available and specific storage needs. In other rooms, carefully planned built-in cabinets provide efficient storage and display for items used in work, recreation, and hobby activities. Their custom-built appearance usually adds to room decor. They may even eliminate the need for some movable pieces of furniture.

Countertops

Work surfaces for kitchen, bathroom, home office, laundry room, and other areas of the home must provide functionality and durability. These

Rev-A-Shelf, Inc.

Figure 8.3-16. Cabinets can be ordered with special drawers, bins, and roll-out shelves.

Breadmaker/shutterstock.com

Figure 8.3-17. Bathroom lavatories are often built into a cabinet called a vanity.

Goodheart-Willcox Publisher

Figure 8.3-18. A combination of wall- and floor-mounted cabinetry is used in this walk-in closet to provide various types of storage.

surfaces play a significant role in the aesthetics of a room. As one of the primary touch points for various activities in a home, they are a major design feature of the living space. A variety of natural and manufactured materials are used for work surfaces. Refer to *Lesson 8.4* to see commonly used materials.

LESSON 8.3 Review

Answer the following questions based on information in this lesson.

1. The balustrade or _____ is an example of millwork. (8.3-1)
   A. bathroom vanity
   B. stair railing
   C. built-in bookshelf
   D. kitchen countertop
2. The _____ is the main structural support component of the stair system. (8.3-1)
   A. riser
   B. tread
   C. carriage
   D. run
3. The unit run is the _____. (8.3-1)
   A. tread width minus the nosing
   B. riser height
   C. distance from the lower floor to the upper floor
   D. distance from the foundation to the upper floor
4. What is the ideal tread width? (8.3-1)
   A. 7″
   B. 9″
   C. 11″
   D. 14″
5. *True or False?* As stair rise is increased, the run must be increased. (8.3-1)
6. Which type of stairway is continuous from one floor level to another without landings or turns? (8.3-1)
   A. Platform stairs
   B. Winding stairs
   C. Geometrical stairs
   D. Straight run stairs
7. *True or False?* All crown molding is installed at an angle. (8.3-1)
8. When the countertop is added to a kitchen base cabinet, the total unit is typically _____ high. (8.3-1)
   A. 24″
   B. 25 1/2″
   C. 34 1/2″
   D. 36″
9. The standard depth for base kitchen cabinets is _____. (8.3-1)
   A. 12″
   B. 24″
   C. 30″
   D. 36″

LESSON 8.4

Materials and Fasteners

LEARNING OUTCOMES

After completing this lesson, you will be able to:

8.4-1 Choose appropriate materials for various interior finishing applications.

8.4-2 Describe the fasteners and adhesives used during the installation of interior finishes.

TECHNICAL TERMS

core	hollow wall anchor	pocket hole fastening
engineered quartz	nosing	rail
FAS (firsts and seconds)	panel door	solid-surface
flush door	plastic laminate	stile
granite	pocket door	

Interior Finish Materials

Apart from flooring, the interior finishes of a home see more daily use than most of the other materials in the building. Consider how often you open and close the doors and cabinets where you live. Think of how often the countertop surface is used throughout the day. When the floor is swept, mopped, or vacuumed, how many times is the baseboard trim bumped into or brushed across? The fact that these features are so often touched, bumped, slammed, leaned on, and generally used means they need to be made of durable materials that are easily cleaned and maintained without losing any of their beauty or functionality.

Though durable for everyday use, finish materials need to be protected before, during, and after installation to maintain their beauty throughout the construction phases. Therefore, attention to detail and careful work practices must be strictly maintained. A hammer dropped from a ladder onto a countertop could cost the builder thousands of dollars in damage. Many of the materials featured in this section are manufactured as end-use products, or at least close to end-use. Cabinets usually arrive on the job assembled with their final coat paint or sealer already applied. Milled materials are shaped to their final profiles. Unlike framing lumber or drywall, damage to these materials cannot easily be repaired or covered up. Always use care when handling and working with finish materials.

Hardwood Grades

Hardwood used in building construction is primarily used for flooring, cabinets, and some interior trim. The National Hardwood Lumber Association establishes grades for hardwood lumber. ***FAS (firsts and seconds)*** is the highest grade. It specifies that pieces be no less than 6″ wide by 8′ long and yield at least 83.33% clear cuttings. The next lower grade, selects, permits pieces 4″ wide by 6′ long. A still lower grade is No. 1 common. Lumber in this group is expected to yield 66.66% clear cuttings.

Staircase Materials

At the most basic level, the finishes of a staircase are made up of treads, risers, newel posts, handrail, and balusters, **Figure 8.4-1**. While many parts

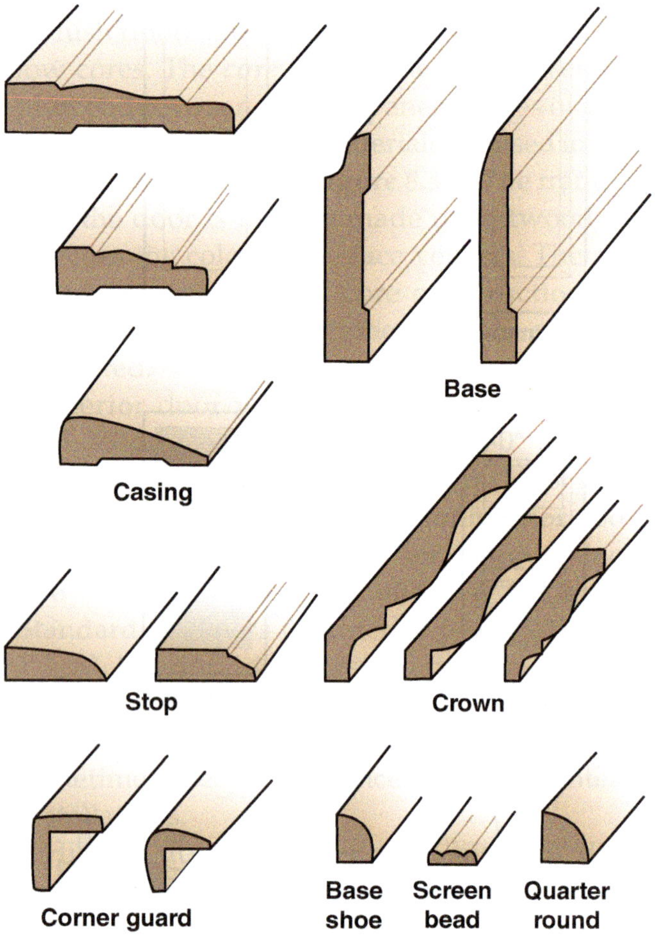

Figure 8.4-10. Profiles of some typical molding patterns.

- Baseboard, base shoe, and base cap are installed where the wall and floor meet.
- Quarter round used as a base cap dresses the top of the base when the base's top edge is squared off.

Names of other moldings come from their use. Examples include casing, chair rail, and panel molding.

Casework

Factory-built cabinets are made of a wide range of materials. Low-priced cabinets are usually made from panels of particleboard with a vinyl film applied to exposed surfaces. The vinyl is printed with a wood grain pattern.

High-quality cabinets can be made from veneers and solid hardwoods such as oak, birch, ash, and hickory. Hardboard, particleboard, and waferboard may be used for certain interior panels and drawer bottoms. Frames are assembled with accurately made joints. Dovetail joints are generally used in drawer assemblies. See **Figure 8.4-11. Figure 8.4-12** shows six basic styles of cabinet doors. Some manufacturers offer panel doors with glass inserts for displaying items inside the cabinets. A completed kitchen installation using hardwood cabinetry is shown in **Figure 8.4-13**.

Countertops

Countertops can be made from a wide variety of materials:

- Ceramic tile
- Concrete
- Copper
- Glass
- Granite
- Marble
- Plastic laminate
- Soapstone
- Solid-surface (synthetic resin)
- Wood
- Zinc

Each has its own characteristics, benefits, and drawbacks. As with cabinets, most countertops are prefabricated in a shop that specializes in that work and then brought to the jobsite to be installed.

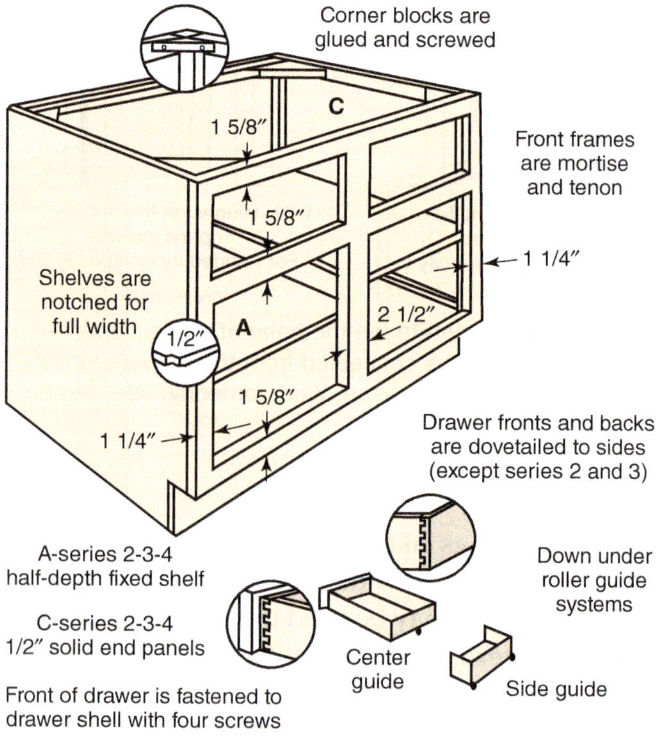

Figure 8.4-11. Construction details of a factory-built cabinet.

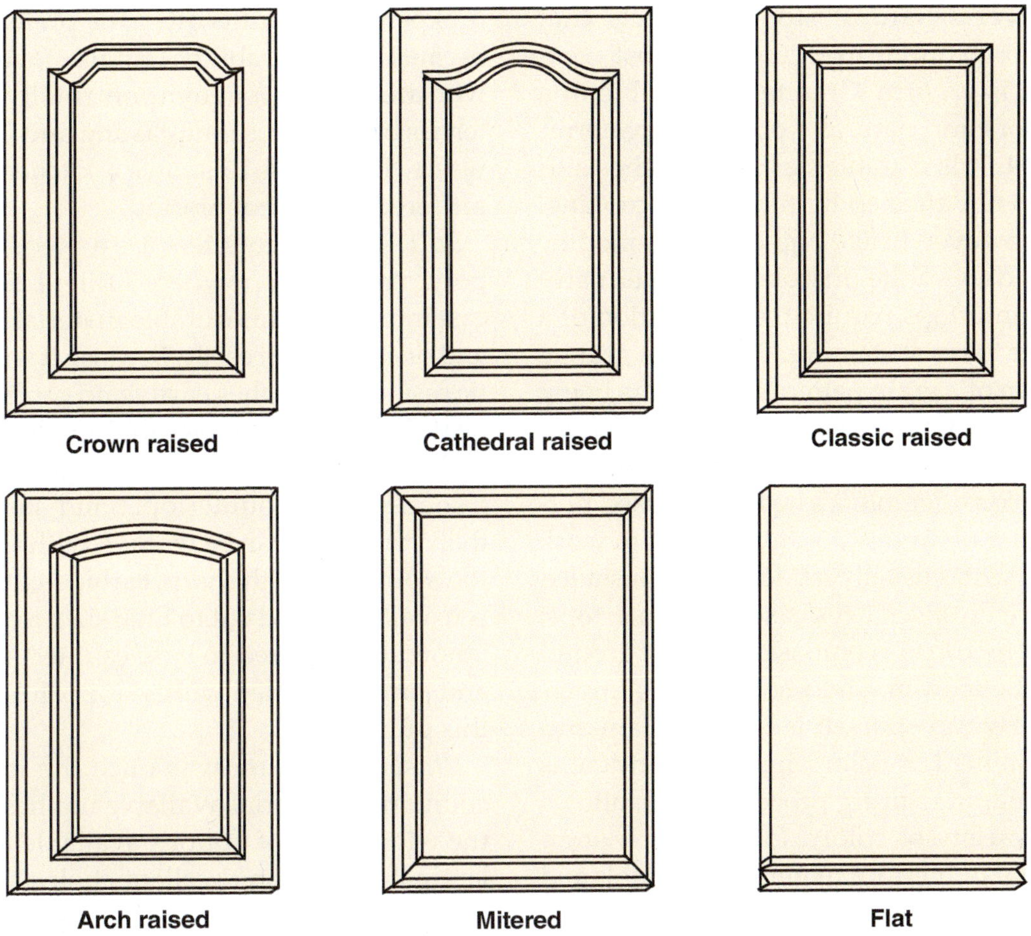

Merillat Industries, Inc.

Figure 8.4-12. Six basic styles used for cabinet doors.

Artazum/Shutterstock.com

Figure 8.4-13. This modern kitchen includes a large number of factory-built cabinets.

Of the several natural stone products that make excellent countertops, the one most commonly used is *granite*. Granite is available in a range of colors and naturally occurring patterns, **Figure 8.4-14**. It is durable, resists scratches and dents, and is not affected by the heat of cooking. Granite is, however, heavy and can be stained by colored liquids. Granite, marble, and other natural stone countertops require the application of a special sealer to resist staining. The cost of granite varies, with exotic grain patterns and colors being most expensive. Granite suppliers often require that you purchase the entire slab, which means there can be considerable waste from unused portions. Special tools and skills are required to work with granite, so it is usually fabricated and installed by someone who specializes in stone countertops.

Engineered quartz is a composite made of ground quartz, synthetic resin, and pigment. It is impervious to stains and spills. It is heat resistant and scratch resistant. The color pigment is added as part of the manufacturing process, which allows for a wide variety of colors. Like granite, engineered quartz countertops require special tools and skills to fabricate, so they are usually installed by someone who specializes in stone countertops. It is expensive, but it is often the preferred material because of its durability. It is also favored by those who are striving for environmentally friendly materials because it is long-lasting and involves little waste. **Figure 8.4-15** shows a kitchen with an engineered quartz countertop.

Solid-surface counters are a blend of acrylic or polyester resins, powdered fillers, and pigments cast into slabs. This durable material is nonporous and does not stain easily. Because it is made of plastic resins, it scratches easily and can be affected by heat. It can be manufactured in almost any color and made to look like stone. **Figure 8.4-16** shows a solid-surface countertop. Solid-surface is softer than stone, so it can be worked with woodworking tools outfitted with sharp, carbide-cutting edges. It is usually ordered cut to final size, ready for installation. It is adhered to base cabinets with silicone, and joints are glued with seam adhesive made for this purpose.

Plastic laminate was once the most popular countertop material available but has lost favor to the wide range of options available today. Plastic laminate is still an excellent choice and attractive for a variety of applications. It is easily worked with carpentry tools. However, most plastic laminate

Arizona Tile

Figure 8.4-14. Granite from different locations around the world has different colors and grain patterns.

Goodheart-Willcox Publisher

Figure 8.4-15. Engineered quartz is durable, resistant to stains, and available in many colors and patterns.

Goodheart-Willcox Publisher

Figure 8.4-16. Solid-surface countertops can be manufactured in almost any color.

countertops are prefabricated and delivered to the site ready for installation. Plastic laminate for countertops is usually approximately 1/16″ thick, offers high resistance to wear, and is unharmed by boiling water, alcohol, oil, grease, and ordinary household chemicals, **Figure 8.4-17**. Although the laminate is hard, it does not possess great strength and is serviceable only when bonded to plywood or particleboard. This base or core material must be smooth and dimensionally stable.

Concrete countertops have gained popularity in recent years. Concrete is a durable material that is highly versatile and well-suited for countertop usage. These countertops can be site-built or fabricated in a mold and brought to the site. A wide range of finishing techniques, aggregates, and colors are available for a completely custom look. One

Formica Group North America

Figure 8.4-17. Plastic laminate countertops are hard and durable, resisting stains and wear. Laminates are available in a wide variety of colors and patterns.

challenge with concrete countertops is that their porosity is higher than that of other countertop materials. Various types of food-safe concrete sealers are available, but each comes with potential drawbacks such as frequent reapplication, food stains, or surface scratching.

Fasteners and Adhesives Used for Interior Finishes

Many of the fasteners and adhesives used for interior finishes have been discussed elsewhere in this text. Various trim nails, including finish nails and brads, are used to secure interior doors and trim work. Wood glue is used for joinery and cabinetmaking.

Additional fasteners and fastening systems used during this phase of construction include the use of pocket hole fastening, cabinet screws, and hollow wall fasteners.

Pocket Hole Fastening

The ***pocket hole fastening*** system is a form of joinery that creates a strong, tight wood joint using screws rather than traditional woodworking techniques like dowels, biscuits, or mortise and tenon joinery. Using pocket hole fastening is quicker and easier than these traditional methods. The system works by creating pocket holes, angled holes drilled into one piece of wood using a pocket hole jig. The pocket allows a screw to be driven into the adjoining piece at the correct angle. The screws are installed on the back side of the joint, concealing the fastener.

The jig has a drill guide with a metal sleeve that helps drill a precisely angled hole into the wood, **Figure 8.4-18**. A special drill bit with a pilot end is used to set the exact angle for the screw to follow. You can adjust the drill guide to set the desired depth for the hole based on the wood thickness and the type of joint you want to create. Once the pocket holes are drilled, specialty pocket screws are inserted into the holes and driven into the adjoining piece of wood. The screws are specially designed with a self-tapping tip and a coarse thread to provide strong holding power in the pocket holes, **Figure 8.4-19**. Using the recommended screw length is critical to keep the tip from piercing the face of the material being joined. Glue can be applied to the joint before the screws are installed to increase the strength of the joint. If glue is not used, the screws can be removed for disassembly.

kasarp studio/Shutterstock.com

Figure 8.4-18. A pocket hole jig aligns and clamps the stock for correct drilling of pocket holes.

Vjacheslav_Kozyrev/Shutterstock.com

Figure 8.4-19. A basic pocket hole kit includes a simple jig, a drill bit and depth collar, an extended drive bit, and a variety of screws.

Cabinet Screws

Cabinet screws typically have a washer head that is flat with a built-in washer underneath, **Figure 8.4-20**. They have a coarse thread design to provide excellent holding power. They are available in various lengths.

Cabinet screws are specialized screws designed for assembling cabinets and securing them to walls. When fastening cabinets together, pilot holes must be drilled to avoid splitting and alignment issues. The pilot hole should be slightly smaller than the screw diameter but large enough to allow the screw to pass through without binding.

Hollow Wall Anchors

Stud walls finished in drywall and other nonstructural surfaces are considered hollow walls. To securely screw or nail an object to a hollow wall, the fastener must penetrate into a stud or wood blocking. Heavy objects, like cabinetry, should be securely fastened to walls into studs, backing, and blocking that is installed during the framing phase, **Figure 8.4-21**. Ideally, any item that is secured to a hollow wall would have backing installed prior to installation of drywall. However, it is sometimes difficult to predict what or where certain objects will need to be attached during the framing phase of home construction. Often, homeowners will decide on a specific fixture, appliance, grab bar, decoration, or other item after walls are covered up.

Attachment of lightweight to medium-weight objects to drywall and other hollow wall surfaces is made possible with the use of ***hollow wall anchors***. These fasteners secure to or through gypsum and other wall materials to provide improved holding strength.

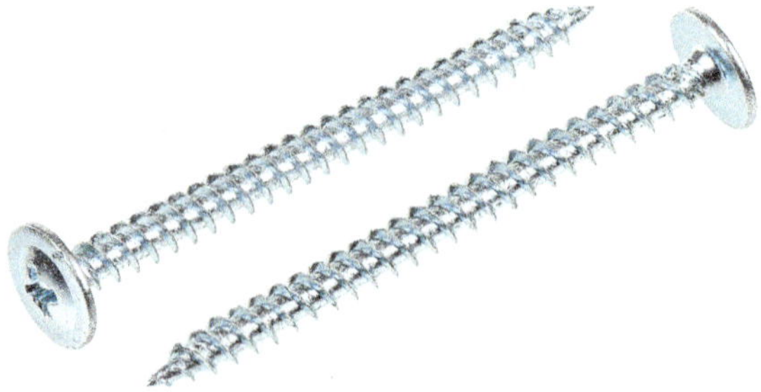

Nikita Rublev/Shutterstock.com

Figure 8.4-20. The washer head on cabinet screws gives the head more surface area for better holding power.

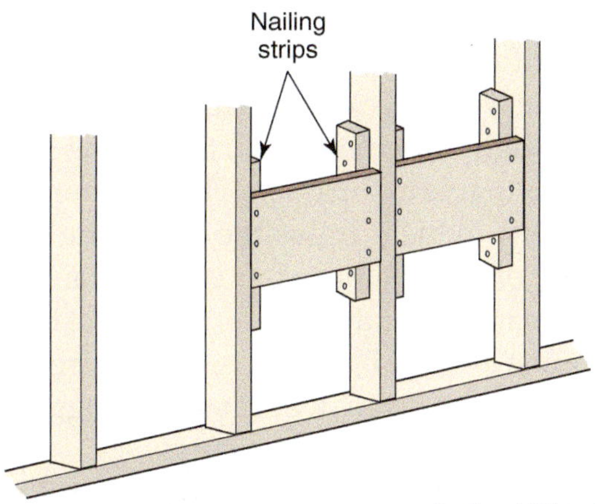

Goodheart-Willcox Publisher

Figure 8.4-21. Backing is installed during the framing phase for attachment of fixtures that are known or anticipated before drywall is applied.

Pro Tip

Choose the correct type of hollow wall anchor for your load requirements and wall material. Ensure proper installing by predrilling the right size hole and tightening the anchor to provide a secure hold.

Light-Duty Hollow Wall Anchors

Various plastic and aluminum screw anchors are used for lightweight applications. They are available in a variety of styles that provide holding power through a course threaded casing, **Figure 8.4-22A**, or clamping force that grips the back of the wallboard as a screw is tightened, **Figure 8.4-22B**. Plastic anchors are ideal for items like towel racks, picture frames, or small shelves.

Medium-Duty Hollow Wall Anchors

Molly bolts, also called hollow wall expansion anchors, are a medium-duty hollow wall fastener that uses a metal sleeve as a compression device to distribute the load to a wide area on the back of the wallboard, **Figure 8.4-23**. To use these fasteners, a hole is predrilled corresponding with the size of the anchor. The sleeve is inserted into the hole, then the bolt is tightened to secure the sleeve. As the bolt is tightened, the sleeve flares against the back of the wallboard, creating tension to provide a secure hold. The bolt can then be removed from the sleeve and replaced, adding the object to be secured to the wall.

A Cico/Shutterstock.com

B wabeno/Shutterstock.com

Figure 8.4-22. A—Aluminum screw anchors, such as these, are threaded into drywall with a screwdriver as a sleeve for a mounting screw. B—The expanded surface area and clamping force of this anchor provides its holding power. The screw is installed through whatever fixture is being secured.

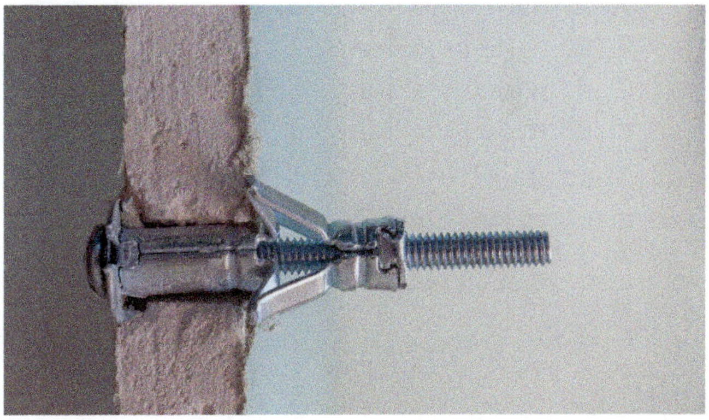

ANDY RELY/Shutterstock.com

Figure 8.4-23. These anchors expand against the back of the drywall as the screw is tightened.

Heavy-Duty Hollow Wall Anchors

Toggle bolts are used for medium and medium-heavy weight objects, **Figure 8.4-24**. They are the strongest hollow wall anchors because they expand to use more surface area for holding power inside the wall. Toggle bolts consist of two parts: the toggle, a spring-loaded mechanism with two arms that

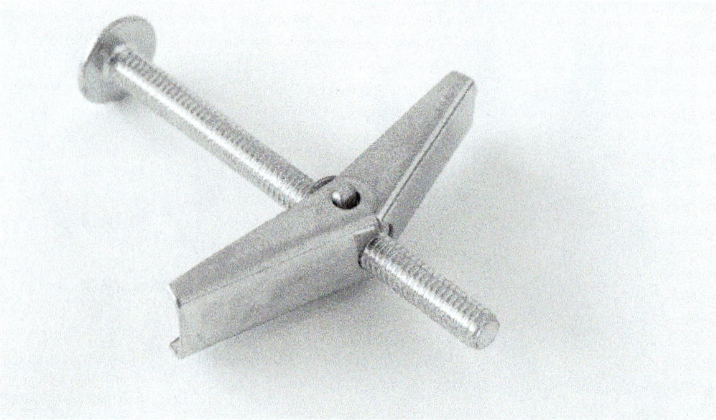

Michelle Lee Photography/Shutterstock.com

Figure 8.4-24. The spring-loaded wings of a toggle bolt fold around the bolt as it is inserted through a predrilled hole. Once inside, they flip open and tighten against the back of the drywall.

open into the wall cavity, and a machine screw that threads into the toggle. To use a toggle bolt, drill a hole in the wall large enough for the folded toggle to fit through. When inserted, the spring pressure causes the toggle to flip open. Then tighten the machine screw, which draws the toggle tight to the back side of the wallboard. It is important to note that when inserting the toggle, the object being secured must already be in place. Otherwise, removing the machine screw will cause the toggle to fall into the bottom of the wall cavity.

Variations to these light-, medium-, and heavy-duty anchors are available, but the same principles apply.

LESSON 8.4 Review

Answer the following questions based on information in this lesson.

1. What is the highest grade of hardwood lumber? (8.4-1)
   A. No. 1 common
   B. FAS (first and seconds)
   C. Selects
   D. Premium common
2. Selects lumber permits pieces _____. (8.4-1)
   A. 2″ wide by 4′ long
   B. 2″ wide by 6′ long
   C. 4″ wide by 4′ long
   D. 4″ wide by 6′ long
3. Main stair treads are commonly _____ thick. (8.4-1)
   A. 1 1/12″
   B. 1 1/8″
   C. 1 1/4″
   D. 1 1/2″
4. *True or False?* The risers and treads on a staircase should be made of the same type of lumber. (8.4-1)
5. The standard thickness for interior passage doors is _____. (8.4-1)
   A. 1 1/8″
   B. 1 1/4″
   C. 1 3/8″
   D. 1 3/4″
6. *True or False?* If the hinges on a door are on the left of the viewer, opening away from you and into the room, it is a right-hand door. (8.4-1)
7. Which type of molding is used to cover the raw edges of shelving? (8.4-1)
   A. Half round
   B. Quarter round
   C. Full round
   D. Baseboard
8. What is a disadvantage of granite countertops? (8.4-1)
   A. They are easily scratched and dented.
   B. They are available in just a few colors.
   C. They are affected by the heat of cooking.
   D. They require a special sealer to resist staining.
9. *True or False?* The pocket hole fastening system uses traditional woodworking techniques to create a strong, tight wood joint. (8.4-2)
10. Which statement about toggle bolts is true? (8.4-2)
    A. They expand inside the wall to provide holding power.
    B. They are recommended for items such as towel racks and picture frames.
    C. They have a sleeve that is inserted into the hole during installation.
    D. They can be installed in a wall without predrilling.

LESSON 8.5

Tools

LEARNING OUTCOMES

After completing this lesson, you will be able to:

8.5-1 Model the correct operation of tools commonly used for installing interior finishes.

TECHNICAL TERMS

backsaw	finishing sander	saber saw
belt sander	kerf	T-bevel
block plane	oscillating multi-tool	try square
combination square	power plane	wood chisel
coping saw	pull saw	
crosscut saw	ripsaw	

Tools Used for Interior Finishing

The nature of interior finish work requires some tools that provide an additional level of precision and refinement. For fine craftsmanship, it is important to keep cutting tools sharp and protect their blades and cutting edges from abuse.

Handsaws

Handsaws are available for a variety of purposes in various lengths and tooth sizes (given as teeth per inch or points per inch). The number of teeth per inch determines how fine the blade cuts. Saws with a high number of teeth per inch generally have smaller teeth, removing less material and creating a smoother cut.

Safety Note

Before using a handsaw, ensure the workpiece is securely clamped to prevent movement, reducing the risk of accident or injury.

The most common types of handsaws are the crosscut saw and the ripsaw. *Crosscut saws*, as the name implies, are designed to cut across the wood's grain. Their teeth are shaped so they come to a point on alternating sides, **Figure 8.5-1A**. Every carpenter needs a good crosscut saw with a tooth size ranging from 8 to 11 points. A crosscut saw for general use has 8 teeth per inch. A finishing saw used for fine cutting has 10 or 11 teeth per inch. Using the thumb as a guide will ensure getting a crosscut started on the line, as shown in **Figure 8.5-2**.

Ripsaws have chisel-shaped teeth that cut best along the grain. Note that the teeth are set (bent) alternately from side to side, **Figure 8.5-1B**. This is so the *kerf* (cut in the wood) will be large enough for the blade to freely run. Some saws have a taper ground from the toothed edge to the back edge, eliminating the need for a large set on the teeth. Most ripping operations are now performed with power saws, making the ripsaw an optional tool in a carpenter's arsenal. In fact, most handsaws sold today use teeth machined for both cross cutting and rip cutting. These saws are not for fine woodworking, but instead are used for convenience. Unlike crosscut and ripsaws, the teeth on these combination saws are not meant to be resharpened.

A saw occasionally used for cutting curves is the *coping saw*, **Figure 8.5-3**. This saw has a thin, flexible blade pulled tight by the saw frame. With a

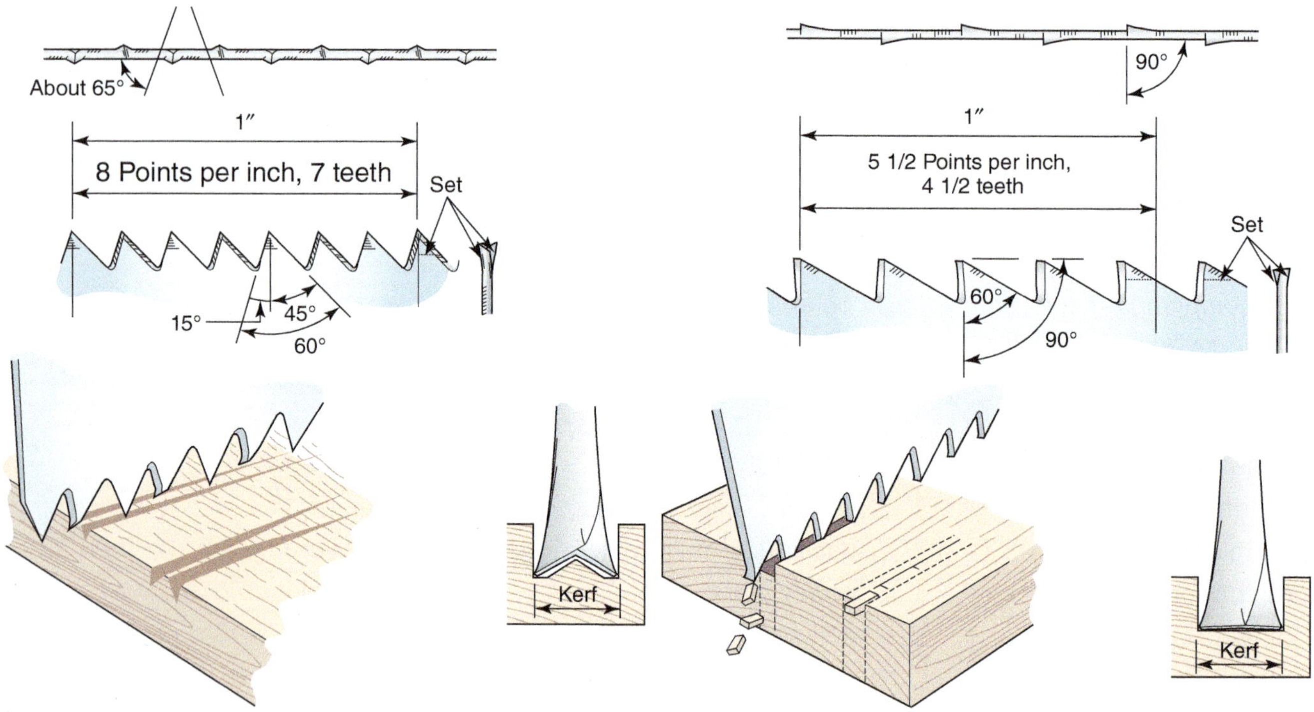

A B

Goodheart-Willcox Publisher

Figure 8.5-1. The teeth of crosscut saws and ripsaws are quite different. A—Crosscut teeth cut like a knife. The top illustration shows the shape and angle of teeth. The bottom illustration shows how teeth cut. B—Ripsaw teeth cut like a chisel. The top illustration shows shape and angle of teeth. The angle is often increased from 90° to give a negative rake. The bottom illustration shows how ripsaw teeth cut.

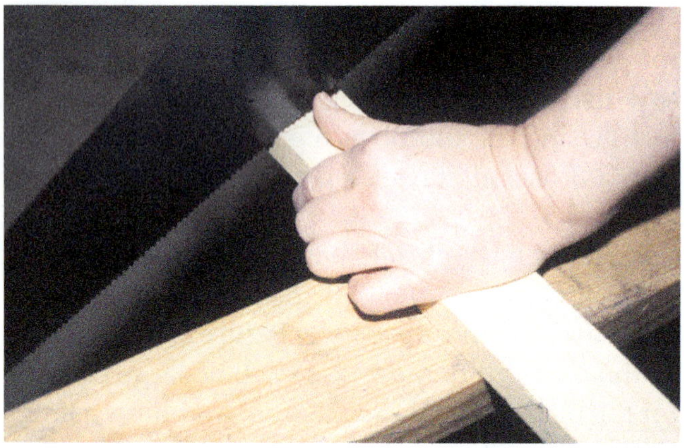

Goodheart-Willcox Publisher

Figure 8.5-2. Use your thumb as a guide against the saw blade when starting a cut with a handsaw.

Scott J. Carson/Shutterstock.com

Figure 8.5-3. A coping saw is designed to cut along curves. Its thin, flexible blade is supported under tension by the saw's frame. The blade has about 15 teeth per inch.

blade that has 15 teeth per inch, the coping saw can make fine cuts.

The ***backsaw*** has a thin blade reinforced with a heavier steel strip along the back edge. The teeth are small (14–16 teeth per inch). Thus, the cuts produced are fine. It is used mostly for interior finish work.

Pull saws come in many shapes and sizes, but generally have a handle that is in a straight line with the blade. The blade is thin and has fine teeth that come to a sharp point. The saw cuts as it is pulled *toward* the user.

Surfacing Planes and Chisels

Surfacing planes are used to shave the surface of wood. Standard surfacing planes include the smooth plane (8″–9″ long), jack plane (14″ long), and fore and jointer plane (18″–24″ long). **Figure 8.5-4** illustrates the parts of a plane and how they fit together. The jack plane is usually selected for general-purpose work.

Another surfacing plane, the ***block plane***, is especially useful. The block plane is a small surfacing plane (6″–7″ long) that can be used with one hand. It produces a fine, smooth cut, making it suitable for fitting and trimming work and can be used with one hand. The blade is mounted at a low angle, and the bevel of the cutter is turned up. This plane produces a fine, smooth cut, making it suitable for fitting and trimming work.

Wood chisels are used to trim and cut away wood or composition materials to form joints or recesses, **Figure 8.5-5**. They are also helpful in paring and smoothing small, interior surfaces that are inaccessible for other edge tools. Chisel widths range from 1/8″ to 2″. For general work, a carpenter usually selects 3/8″, 1/2″, 3/4″, and 1 1/4″ sizes. A soft-face hammer or mallet should be used to drive the chisel when making deep cuts.

Safety Note

When using a chisel, maintain a firm grip and always use controlled force, directing the tool away from your body to prevent slips and potential injuries.

Squares

The ***try square*** is the most basic version of a marking square. Try squares are available with fixed blades that are 6″–12″ long. Handles are made of wood or metal. These are used to check the squareness of surfaces and edges. They are also used to lay out lines perpendicular to an edge, **Figure 8.5-6**.

The ***combination square*** serves a similar purpose as a try square but can also be used to lay out miter joints. Its adjustable, sliding blade allows it to be used as a gauging tool. It can be used as a marking gauge to make parallel lines and as a square for 90° and 45° angles. See **Figure 8.5-7**. Although the combination square has multiple uses, it is difficult to carry in a tool belt and it is not as durable as a speed square.

The ***T-bevel*** has an adjustable blade, making it possible to transfer an angle from one place to another. It is useful in laying out cuts for hip and valley rafters. **Figure 8.5-8** illustrates how the framing square is used to set the T-bevel at a 45° angle.

Poppy Pix/Shutterstock.com

Figure 8.5-5. The chisel should be kept sharp enough to cut easily with hand pressure or light mallet blows. It should never be struck with a hammer.

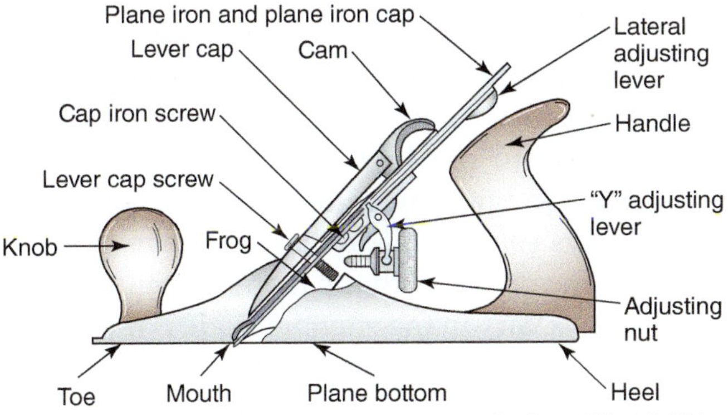

Goodheart-Willcox Publisher

Figure 8.5-4. A standard plane is made up of many parts.

Anna Berdnik/Shutterstock.com

Figure 8.5-6. The try square is handy for laying out lines perpendicular to an edge. This student is able to lay out a precise mark to square off this board.

pirke/Shutterstock.com

Figure 8.5-7. A combination square is a versatile marking and layout tool.

Kzenon/Shutterstock.com

Figure 8.5-9. A saber saw cuts with an action much like a handsaw.

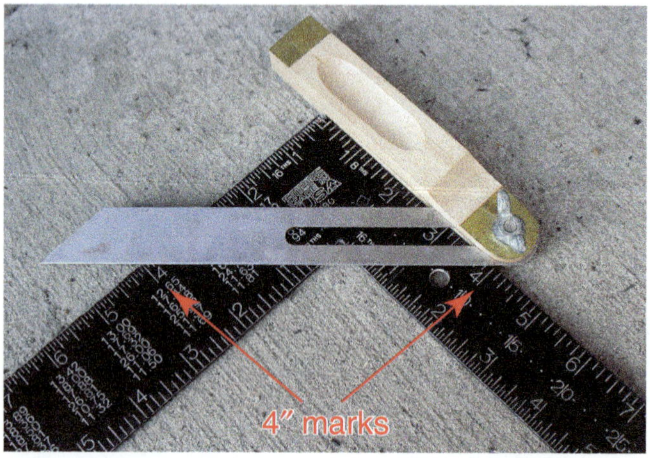

Goodheart-Willcox Publisher

Figure 8.5-8. T-bevel squares have an adjustable blade that can be used to match angles up to 180°. Resting the handle of the sliding T-bevel square firmly against one side of the framing square and aligning the adjustable blade to the same number on both edges of the framing square (in this case 4") sets the angle at 45°.

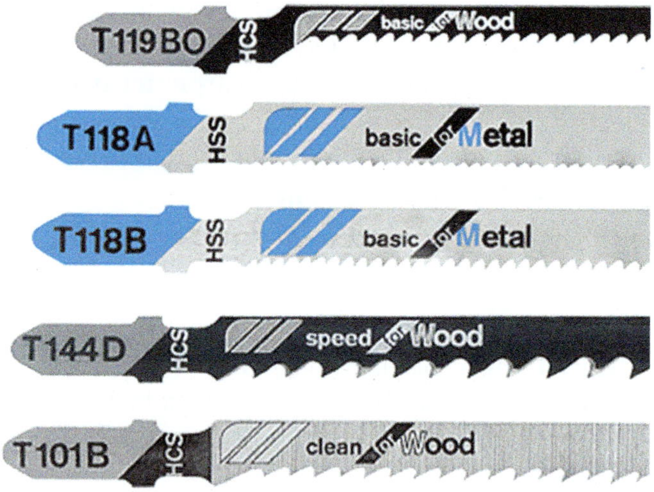

Photo courtesy of Robert Bosch Tool Company®

Figure 8.5-10. Saber saw blades come in a variety of configurations to suit different tasks. Narrow blades are used for curved cuts in wood or plastic. Some blades have small teeth for fine cuts in wood and plywood or metal. Blades with larger teeth are used for faster wood cutting.

Powered Cutting Tools

The ***saber saw***, also called a portable jigsaw, is a portable saw useful for a wide range of light work. See **Figure 8.5-9**. Carpenters, cabinetmakers, electricians, and home craftspeople use it. The stroke of the blade is about 1/2". The saw operates at a speed of approximately 2,500 strokes per minute, but some have a variable speed control.

A wide variety of blade types are available for saber saws, **Figure 8.5-10**. In general, narrower blades can cut tighter curves in wood or plastic. Blades with more teeth per inch make smoother cuts but cut more slowly. Blades for wood cutting have 6–12 teeth per inch. For general-purpose work, a blade with 10 teeth per inch is satisfactory. Always select a blade that will have at least two teeth in contact with the material being cut.

Saber saws vary in the way the blade is mounted in the chuck. Follow the directions and lubrication schedule in the manufacturer's manual.

The saber saw can be used to make straight or bevel cuts. Curves are usually cut by guiding the saw along a layout line. However, circular cuts may be more accurately made with a special guide or attachment.

Because the blade cuts on the upstroke, splintering takes place on the top side of the work. This consideration is important when making finished cuts, especially in fine hardwood plywood. Always firmly hold the base of the saw against the surface of the material being cut.

When cutting internal openings, a starting hole can be drilled in the waste stock or the saw can be held on end so the blade cuts its own opening. See **Figure 8.5-11**. This approach is called plunge cutting and must be done with considerable care. Firmly rest the toe of the base on the work and turn on the motor. Then, slowly lower the blade into the stock.

The blade on an ***oscillating multi-tool*** is offset, which makes it useful for cutting, scraping, sanding, and grinding in difficult to reach places. See **Figure 8.5-12**. These tools are available in both corded and cordless versions and can be equipped with a wide variety of attachments. The multi-tool is designed for accuracy rather than speed. As its name suggests, the tool's blade oscillates at a rapid pace, allowing it to make fine cuts or remove material when using an abrasive attachment.

Cutting blades are available for wood, plastic, metal, and carpet. The blades come in a variety of shapes to accommodate different tasks. Scraping blades are used to remove old paint or caulk.

Goodheart-Willcox Publisher

Figure 8.5-11. On internal cuts, a saber saw can make its own opening. Be sure to rest the base on the workpiece before starting the cut. Plunge the blade in slowly while holding the trigger.

Photo courtesy of Robert Bosch Tool Company®

Figure 8.5-12. A multi-tool is useful for shortening door jams when installing a new floor. The offset blade allows for a flush cut.

They have a sharpened straightedge, rather than cutting teeth. Sanding pads are available in a variety of shapes that can be used for profile sanding. Carbide rasps and grout removal attachments are used for tile restoration work, **Figure 8.5-13**.

The ***power plane*** produces finished wood surfaces with speed and accuracy, **Figure 8.5-14**. The motor, which operates at about 20,000 rpm, drives a spiral cutter. The depth of cut is adjusted by raising or lowering the front shoe. The rear shoe (main bed) must be kept level with the cutting edge of

Photo courtesy of Robert Bosch Tool Company®

Figure 8.5-13. Manufacturers produce a variety of useful multi-tool attachments.

Photo courtesy of Robert Bosch Tool Company®

Figure 8.5-14. Portable power planes are sometimes used instead of hand planes because they are faster and more accurate.

MDV Edwards/Shutterstock.com

Figure 8.5-15. Belt sanders are the most aggressive type of portable sander.

Photo courtesy of Robert Bosch Tool Company™

Figure 8.5-16. Using an orbital sander to perform finish sanding on cabinetwork. Some orbital finishing sanders are equipped with a dust bag.

the cutter head. The power plane is equipped with an adjustable fence for planing bevels and chamfers. For surfacing operations, the fence is removed.

Portable Sanders

A ***belt sander*** is a portable tool that consists of a belt of sandpaper that is rotated by a motor, **Figure 8.5-15**. The size of a belt sander is determined by the width and length of the belt. Belt sanders aggressively sand large surfaces. Using this sander takes some practice and skill.

Finishing sanders are used for final sanding, where only a small amount of material needs to be removed. They are also used for cutting down and rubbing finishing coats. There are two general types of finish sanders, orbital and oscillating. An orbital sander is shown in **Figure 8.5-16**.

Also called a pad sander, the finishing sander has a pad backing up the sanding paper. Unlike belt sanders, where the sanding belt moves, the paper is held stationary on the pad by mechanical grippers or is attached to the pad with either a strong adhesive or hook-and-loop surfaces. The pad moves, carrying the paper.

LESSON 8.5 Review

Answer the following questions based on information in this lesson.

1. A finishing saw used for fine cutting has ____ teeth per inch. (8.5-1)
   A. 8 or 9
   B. 10 or 11
   C. 12 or 13
   D. 14 or 16
2. Which type of saw has small teeth and a thin blade reinforced with a heavier steel strip along the back edge and is used mostly for interior finish work? (8.5-1)
   A. Coping saw
   B. Crosscut saw
   C. Ripsaw
   D. Backsaw
3. *True or False?* A pull saw cuts as it is pulled toward the user. (8.5-1)
4. Which statement about the block plane is true? (8.5-1)
   A. It produces a large, rough cut.
   B. Its blade is mounted at a low angle.
   C. The bevel of its cutter is turned down.
   D. It is about 10″ to 12″ long.
5. *True or False?* When using a saber saw, splintering takes place on the top side of the wood. (8.5-1)
6. The ____ has a motor that operates at a speed of about 20,000 rpm and a spiral cutter that produces finished wood surfaces with speed and accuracy. (8.5-1)
   A. power plane
   B. portable jigsaw
   C. jack plane
   D. oscillating multi-tool

LESSON 8.6 Review

Answer the following questions based on information in this lesson.

1. *True or False?* A chronic health condition is characterized by a health issue that develops quickly and has a relatively short duration. (8.6-1)
2. Which group of people is most susceptible to the harmful effects of lead? (8.6-1)
   A. Children 6 years and younger
   B. Adolescents
   C. Adults 18 to 40 years
   D. Adults 65 years and older
3. *True or False?* The VOCs in paint can continue to be released even after the paint has dried. (8.6-1)
4. *True or False?* The information found on a safety data sheet applies to all types of chemicals. (8.6-1)
5. Biological hazards exist any time a worker encounters the potential for exposure to _____. (8.6-1)
   A. explosive materials
   B. fall hazards
   C. mold, fungi, or viruses
   D. dangerously sharp objects

LESSON 8.7

Math

LEARNING OUTCOMES

After completing this lesson, you will be able to:

8.7-1 Perform construction math calculations using averages and percentages.

TECHNICAL TERMS

average

percent

Averages and Percentages

Computing averages and percentages is a helpful skill in construction. Among other things, averages can be used effectively to estimate costs by looking at cost data from previous jobs that are similar in nature. For instance, when generating an estimate for replacing the roof on a house, a builder may choose to look at their past reroofing jobs to calculate their average cost per square foot and use that knowledge to inform the estimating process. Armed with this information, they may want to determine how much of the cost could be attributed to labor and how much to material. For this, it is useful to calculate the percentage of the cost that makes up each category. They can also use averages and percentages to track how costs have changed over time, answering questions such as Have material costs remained consistent over the past three years? and How have labor costs impacted the profit margins? and other important financial statistics. These and other business operations such as interest rates, taxes, discounts, and other functions regularly use percentages and averages.

Averages and percentages are not just helpful in making good business decisions, but they are also used on the jobsite for determining things like how much artificial or natural light or ventilation is required in a home, efficiency calculations for heating and cooling equipment, moisture content in materials, waste factors for materials, and much more.

Code Note

Building codes and zoning laws related to percentages and averages often pertain to aspects like maximum occupancy, allowable floor area ratios, or requirements for minimum green space as a percentage of the total lot area. For instance, zoning regulations might dictate that a building's footprint cannot exceed a certain percentage of the total lot size, or fire codes might specify the maximum occupancy load for a given space based on square footage.

Computing Averages

An ***average*** is a typical value of one unit in a group of units. For example, if 4 windows have areas of 12.0 sq ft, 11.2 sq ft, 11.2 sq ft, and 14.5 sq ft; the average size of one of those windows is 12.2 sq ft. The average is computed by adding all the unit values and dividing that sum by the number of units in the group.

Example:

$$\begin{array}{r} 12.0 \text{ sq ft} \\ 11.2 \text{ sq ft} \\ 11.2 \text{ sq ft} \\ +\ 14.5 \text{ sq ft} \\ \hline 48.9 \text{ sq ft} \end{array}$$

Divide the sum by the total number of units:

$$48.9 \text{ sq ft} \div 4 = 12.225$$

The quotient should be rounded off to the same number of decimal places as is used in the problem. The average window size is 12.2 sq ft.

Practice 8-1

Compute the averages of these groups.

1. 80 lb, 83 lb, 88 lb, 79.5 lb, 81.6 lb, 84 lb
2. 11 cu yd, 13 cu yd, 11.5 cu yd, 12 cu yd, 12.8 cu yd

Percentages

A *percent* is one part in a hundred. One penny is 1% of a dollar. Twenty-five cents is 25% of a dollar. Twenty-five cents is also 50% of a half-dollar, because if the half dollar were divided into 100 parts of 1/2 cent each, 50 of those 1/2 -cent increments would be 25.

Think of percent as hundredths. 100% is equal to 1.0, 50% is equal to 0.50, 25% is equal to 0.25, and so on. Therefore, to convert a decimal part of a number to a percentage, simply multiply that decimal number by 100, or move the decimal point two places to the right and add the percentage sign.

Example:

0.62

Multiply by 100:

$0.62 \times 100 = 62$

Add the percent sign:

62%

The process of converting a percentage to a decimal is the opposite of what was done to convert a decimal to a percent: simply remove the percent sign and divide by 100, or move the decimal point two places to the left.

Example:

43%

Remove the percent sign:

43

Divide by 100:

$43 \div 100 = 0.43$

To find a given percentage of an amount, multiply the amount times the desired number of hundredths.

Example:

To find 12% of \$4.40, we can think of 12% as 0.12 and multiply by \$4.40.

$$\$4.40 \times 0.12 = \$0.528 \text{ or } 53 \text{ cents}$$

Using what we have learned, we can calculate the whole if we know the percentage.

Example:

What was the total spent on tools if \$22.00 was spent on sales tax and the tax rate is 8%? We know that \$22.00 was spent on taxes, which is the total multiplied by the tax rate of 8%.

$$\begin{aligned} \$22.00 &= \text{Total} \times 8\% \\ &= \text{Total} \times 0.08 \\ \$22.00 \div 0.08 &= \text{Total} \\ \$275.00 &= \text{Total} \end{aligned}$$

Practice 8-2

1. What is 10% of 225?
2. What is 6.5% of \$350.50?
3. If merchandise cost \$22.25 and the bill comes to \$23.14, what percentage was added for sales tax?
4. If a 500-gallon tank is 60% full of water, how much water does it contain?
5. A kitchen remodel costs a total of \$65,000 for material, labor, and profit. If the kitchen cabinets cost \$26,650, what percentage of the cost are they?

LESSON 8.7 Review

Answer the following questions based on information in this lesson.

1. If five windows have areas of 10.0 sq ft, 11.5 sq ft, 12.0 sq ft, 9.5 sq ft, and 9.5 sq ft, the average size of the windows is _____. (8.7-1)
   A. 9.5 sq ft
   B. 10.5 sq ft
   C. 12 sq ft
   D. 52.5 sq ft
2. *True or False?* A percent is one part in ten. (8.7-1)
3. 0.85 = _____ (8.7-1)
   A. 15%
   B. 1.5%
   C. 85%
   D. 8.5%
4. 36% = _____ (8.7-1)
   A. 0.18
   B. 0.36
   C. 1.8
   D. 3.6
5. 28% of $225.00 is _____. (8.7-1)
   A. $8.04
   B. $22.50
   C. $56.26
   D. $63.00

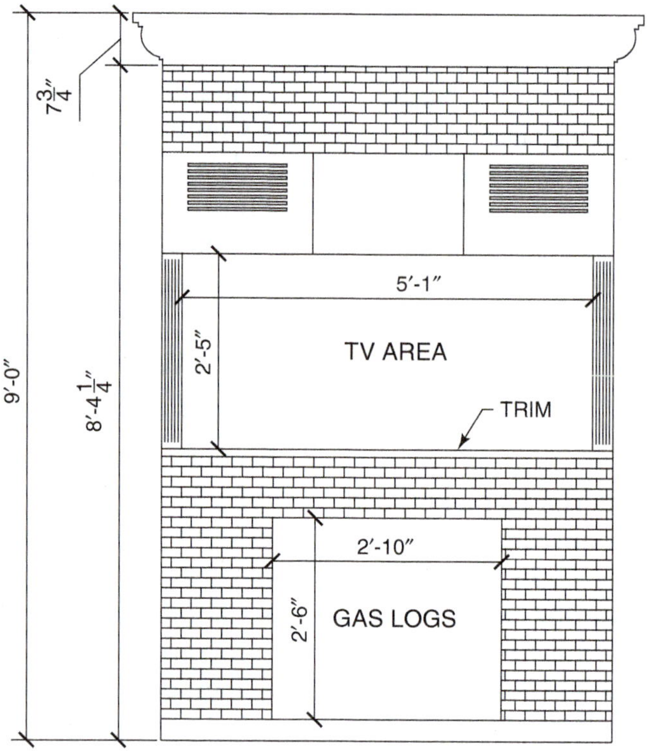

Goodheart-Willcox Publisher

Figure 8.8-3. Interior elevation drawings may be included in blueprints to give details regarding fireplaces, bathrooms, utility rooms, staircases, and other detailed areas.

LESSON 8.8 Review

Answer the following questions based on information in this lesson.

1. Where would the builder look to find details pertaining to the type of hardware to use when installing kitchen cabinets? (8.8-1)
   A. Specifications
   B. Finish schedules
   C. Interior elevations
   D. Floor plans
2. The _____ provide(s) details such as the type of tile that will be installed on the laundry room floor. (8.8-1)
   A. framing plan
   B. floor plan
   C. finish schedule
   D. specifications
3. The builder would check the interior elevations to find details such as the _____. (8.8-1)
   A. color code for ceiling paint
   B. material used for a countertop
   C. type of stain used for a cabinet
   D. placement of wainscoting

LESSON 8.9

Green Building and the Future of Construction

LEARNING OUTCOMES

After completing this lesson, you will be able to:

8.9-1 Give examples of how reducing waste relates to green building.

TECHNICAL TERMS

deconstruction

source reduction

Reducing Waste on the Job

It is estimated that the amount of debris generated by the construction industry is more than twice the amount of municipal waste generated, including household garbage, recyclables, and yard waste. Much of this construction waste is generated from demolition and renovations; however, a significant amount also comes from new construction where excess materials, scraps, and product packaging all add up to millions of tons of refuse every year, **Figure 8.9-1**.

ungvar/Shutterstock.com

Figure 8.9-1. Strategies to reduce excessive construction waste can both save money and reduce the volume entering landfills.

Incorporating strategies to reduce the amount of waste entering landfills is an important part of building green. Reducing waste, reusing and repurposing material, and recycling are all effective practices for cutting back the amount of construction and demolition materials entering landfills. Not only can conservation efforts save landfill space and processing resources, but they can also create employment and business opportunities in the recycling industry and reduce overall project costs through better material management and lower disposal fees.

Source Reduction

As briefly discussed in *Chapter 5, Exterior Finishes*, material sourcing is a contributing factor with regard to green building. Source reduction is a top priority when it comes to managing waste. ***Source reduction*** is about the process of decreasing how much we use, the energy we consume, and the waste we produce throughout the entire life of materials. Instead of dealing with waste after it's made, source reduction stops it from being created in the first place.

Preserving existing structures is one effective way to apply source reduction. Renovating, rather than demolishing and building new, reduces waste, **Figure 8.9-2**. When new buildings are built, we can make sure they are just the right size. Durable design helps ensure buildings will last a long time and prevent waste from entering landfills in the future.

hanohiki/Shutterstock.com

Figure 8.9-2. Preserving existing structures saves resources and can retain historical character.

Source reduction isn't just about changing how we design and build. It also involves smart purchasing decisions. By purchasing only what is really needed and using efficient building methods, excess material and packaging waste can further be reduced.

Salvaging and Reusing Materials

Building demolition is not a resource-efficient practice when debris is simply hauled away to a landfill. Existing buildings are full of materials that are valuable. Recovering used materials can be an effective way to save money while protecting natural resources.

Deconstruction is the process of dismantling buildings to salvage components that can be reused or recycled. Deconstruction can be done in various ways to save materials and significantly reduce waste. Many deconstructed materials are not used on the same jobsite but are instead donated or sold to a local resale outlet to be resold, **Figure 8.9-3.**

Pro Tip

To salvage interior trim for reuse, do not attempt to remove the nails by driving them back through the face of the trim. Instead, use a small wrecking bar or nippers to pull them through the back. This avoids splintering that would spoil the exposed face.

INTREEGUE Photography/Shutterstock.com

Figure 8.9-3. Construction material recycling centers are becoming more popular as material prices rise and awareness of resource limitations increases.

Recycling

In many cases, building materials can be recycled. Recycling involves processing the materials for reuse. For example, materials, such as concrete, brick, and gypsum wallboard, that cannot be easily deconstructed can be taken to a processing center and repurposed for use as fill, soil amendment, subbase, or aggregate for new concrete, **Figure 8.9-4**. Wood can be used in the production of engineered-wood materials, or it can be turned into mulch, wood pellets, and other useful products. Metals, such as steel, copper, and brass, are also valuable when recycled. Even cardboard packaging from construction sites can be recycled.

Recycling is an effective way to reduce waste and the resources put into extracting new raw materials for production of new goods.

Peter Horrox/Shutterstock.com

Figure 8.9-4. Used brick and masonry materials are separated and salvaged so they can be crushed and repurposed.

Green Note

When recycling material on or off the jobsite, reuse it as directly as possible. When building materials are 'recycled,' they are often 'down-cycled' into lower-quality products. Lumber becomes wood chips and mulch. Concrete is ground down into low-value aggregate. When a material is reused directly, without recycling, no quality is lost.

LESSON 8.9 Review

Answer the following questions based on information in this lesson.

1. *True or False?* Source reduction is about the process of decreasing how much we use. (8.9-1)
2. *True or False?* The process of demolishing a building and building new produces less waste than renovating. (8.9-1)
3. *True or False?* Deconstructed materials can only be used on the same jobsite from which they were salvaged. (8.9-1)

Chapter 8 Review

Summary

Lesson 8.1

- Employment benefits are additional perks and compensation provided by employers beyond regular paychecks, designed to enhance work–life balance and overall quality of life. These benefits, which may vary by company size and location, can include health insurance, retirement plans, paid time off, overtime pay, tool or clothing allowances, and transportation benefits. (8.1-1)

Lesson 8.2

- Teamwork and collaboration are vital in the residential construction industry, requiring individuals and groups to work together effectively toward common goals. (8.2-1)

Lesson 8.3

- Interior finishes in a home include the installation of millwork, such as stair railings, doors, and moldings, and casework, such as built-in cabinetry and shelving. (8.3-1)

Lesson 8.4

- Interior finish materials, including doors, cabinets, countertops, and moldings, see frequent use and must be durable and easy to maintain. Many of these materials are manufactured off-site and may come to the jobsite prefinished. Extra care must be taken during handling, storage, and installation and after installation to protect them from damage. (8.4-1)
- Fasteners and adhesives used for interior finishes include trim nails, wood glue, pocket hole screws, cabinet screws, and hollow wall anchors. Pocket hole fastening creates strong joints with screws, while cabinet screws are used for assembling cabinets and securing them to walls. Hollow wall anchors, such as light-duty plastic anchors, medium-duty molly bolts, and heavy-duty toggle bolts, provide secure attachment to drywall and other nonstructural surfaces. (8.4-2)

Lesson 8.5

- Interior finishing tools prioritize precision and refinement, with sharp cutting tools such as various handsaws, surfacing planes, and chisels essential for fine craftsmanship. Additionally, powered tools—saber saws, oscillating multi-tools, power planes, and portable sanders—are used for more efficient and accurate cutting, shaping, and sanding tasks. (8.5-1)

Lesson 8.6

- Construction workers face various health hazards, including chemical, physical, and biological risks, which can lead to acute or chronic health conditions. (8.6-1)

Lesson 8.7

- Averages and percentages are crucial in construction for estimating costs, tracking financial trends, and making everyday jobsite decisions. These computations are used to calculate factors that impact code compliance, energy efficiency, and material layout and placement. (8.7-1)

Lesson 8.8

- During construction, determining what interior finishes to use in a home involves consulting specifications, finish schedules, interior elevations, and section and detail drawings to understand details such as trim, cabinetry, countertops, and paint colors. (8.8-1)

Lesson 8.9

- Reducing waste on construction sites is crucial for environmental sustainability. Strategies that include source reduction, salvaging and reusing materials through deconstruction, and recycling play key roles in minimizing the millions of tons of construction debris sent to landfills annually. These approaches also offer cost savings, resource conservation, and employment opportunities in the recycling industry. (8.9-1)

Apply and Analyze

1. A construction worker applies to several companies for employment and receives two offers. The first job pays $25 per hour and offers a comprehensive benefits package. The second job pays $30 per hour without benefits. Which job should the construction worker take, and why? (8.1-1)
2. You are considering taking a job working on a large development project that will last several years and involve working in a fast-paced environment with tradespeople and subcontractors across a variety of fields. Other than your skills working with tools, what personal skills would you need to work as an effective member of the team and collaborate well with others? (8.2-1)
3. A company is building a large multimillion-dollar home for a wealthy client in an exclusive neighborhood. Although the client is spending a lot of money on this project, they want to keep the project within their budget without sacrificing quality. When it comes time to select cabinets for the kitchen, they ask for some options. Which type of cabinets should the builder recommend—mass-produced cabinets from a cabinet factory, custom-built units from a local cabinet shop, or custom-built units from an on-site carpenter? Why? (8.3-1)
4. A client has an alcove that is 40″ wide in their kitchen and wants to install a base cabinet that fills the space from side to side. A custom-built unit would be too expensive, but the closest-sized cabinets available are 39″ and 42″. What would be the easiest and least expensive solution to this problem? (8.3-1)
5. You are renovating a kitchen for a client who will be using the home as a rental property when it is finished. When you ask what type of countertops they want, they say, "I've been hearing a lot about concrete countertops. That sounds like a nice strong material that I won't have to worry about maintaining once it's installed." Would you recommend concrete countertops for this client? Why or why not? (8.4-1)
6. A construction worker will be working today as a floating assistant, helping the crew doing interior finishing work throughout a large home. The tasks will involve checking the squareness of surfaces and edges, laying out miter joints, and using a marking gauge to make parallel lines. When selecting the necessary tools for the day, the worker chooses a combination square for these tasks. A more experienced worker suggests that they select a different tool instead. Which tool does the more experienced worker recommend, and why? (8.5-1)
7. Two construction workers will be applying a chemical sealant to a wood floor today. While putting on their PPE, one worker asks the other why they aren't wearing a mask. The second worker tells the first worker that as long as they're wearing protective overalls and gloves, they'll be fine because they won't be touching the sealant with their skin. Should the workers also be wearing protective masks? Why or why not? (8.6-1)
8. When a builder purchased materials for a job, they spent $86.00 on sales tax. If the sales tax rate is 8%, what was the original cost of materials before taxes? Explain how you arrived at your answer. (8.7-1)

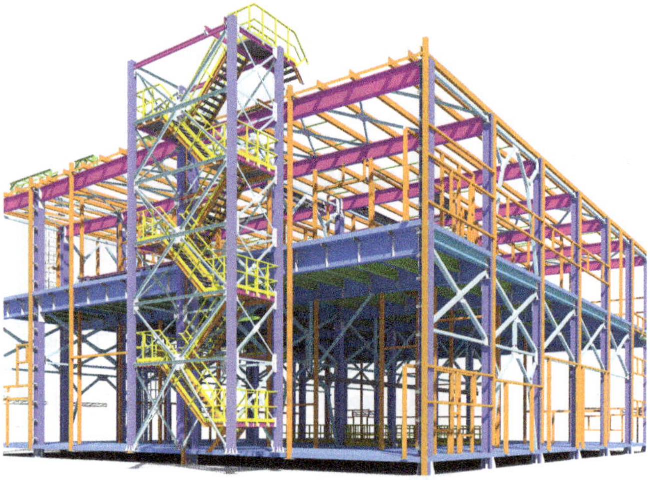

KRAUCHANKA HENADZ/Shutterstock.com

Figure 9.1-3. This BIM screenshot shows the structural layer of a commercial building.

Gorodenkoff/Shutterstock.com

Figure 9.1-4. Drones continue to gain popularity for use on construction sites.

Drones are now being used on construction sites for numerous purposes:

- **Site surveying and mapping**—Drones equipped with high-resolution cameras and LiDAR technology can survey construction sites efficiently. LiDAR, which stands for light detection and ranging, is a remote sensing technology that uses laser light to measure distances and create detailed, high-resolution maps and 3-D models of environments. These 3-D maps and models help in initial site assessments, land surveying, and planning.
- **Inspections**—Drones can be quickly and easily deployed to access areas that are difficult to see or get to safely, such as rooftops, chimneys, confined spaces, and other potentially dangerous or challenging spaces.
- **Monitoring progress**—Drones can capture aerial photos and video of a project at regular intervals, documenting or highlighting progress.
- **Marketing**—Drone footage can provide unique perspectives useful for marketing a construction business's services or the newly built home when it comes time to sell.

Construction Equipment

Construction technology does not have to be digital, or even "high-tech," to be helpful. All kinds of equipment used on construction sites help to reduce labor, improve efficiency or quality, and make jobs safer or easier, **Figure 9.1-5**. Material handling equipment, cordless tools, motorized concrete screeds, excavation equipment, nail guns, power saws, and hundreds of other examples can be cited for improving work practices in construction over the years.

Some examples of newer equipment technology that is becoming more accessible and gaining widespread use by many residential contractors include items such as the following:

- **Thermal imaging cameras**—Used to evaluate efficiency, search for leaks, assess HVAC and electrical systems, or find hidden pipes.
- **Wearable tech/equipment**—Exoskeletons, discussed in *Chapter 7, Interior Insulation and Surfaces*, reduce muscle strain and fatigue; vests and watches monitor vitals for health,

Leszek Glasner/Shutterstock.com

Figure 9.1-5. Technological improvements continually help to make construction jobs easier, safer, and more efficient. Technology is applied to all levels of construction, from personal protective equipment (PPE) and tools to equipment and materials.

air quality, temperature, and a variety of other hazards; safety harnesses, helmets, and other PPE continue to improve in comfort, performance, and features.

- **Inspection cameras**—Equipment outfitted with cameras can be used for inspection of inaccessible areas such as inside walls, ducts, tanks, and drainpipes, preventing unnecessary demolition for exploratory purposes, **Figure 9.1-6**.
- **Lasers**—A variety of laser tools are available for tasks such as measuring distances, setting or determining a grade or slope, or providing references for plumb, straight, and level.

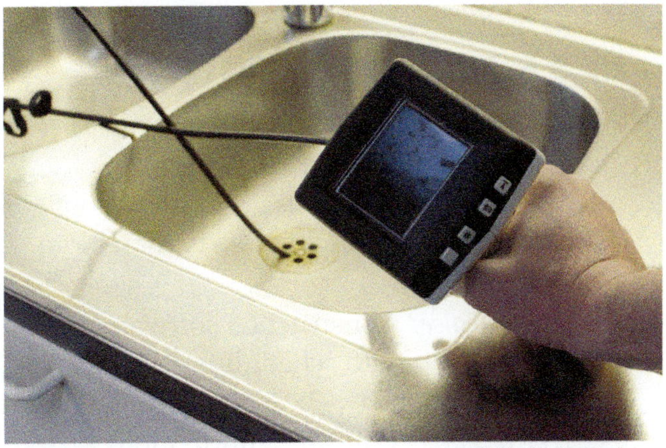

JRJfin/Shutterstock.com

Figure 9.1-6. Inspection cameras, sometimes called borescopes, are available for a variety of applications in construction. Cameras like this one are useful for troubleshooting problems while minimizing the need for deconstruction.

Concluding Thoughts on Industry and Careers

Throughout this text, we explored the residential construction industry; highlighted some of the many career options available to those interested in construction, from laborers to management to even more niche career options; reviewed types of business structures that are used in the United States; considered ways to develop experience and credentials as a professional builder or remodeler; and discussed organizations and associations that support and influence the residential construction industry. We discussed the job search, application, and hiring process, and the many possible employment benefits one might need to understand when comparing employment from one company to another.

In this chapter, we considered some of the technologies you may encounter when working in the industry. Becoming familiar with this technology may help you become a more desirable candidate in the job market. If any of these specific developments interest you, it would be worth learning more so you can pursue the type of career that aligns with your interests and passions.

With this final Industry and Careers lesson, reflect on what the next steps may be for you and your career in construction. How do you envision your future in residential construction, and what preliminary steps can you take to achieve those goals? What career options most interest you and why? How can you begin to establish network connections within the construction trades? What further training and credentials would you benefit from as a starting point?

Considering these questions and others like them can help direct you toward a rewarding career in the residential industry. May you be successful in whatever direction you go!

LESSON 9.1 Review

Answer the following questions based on information in this lesson.

1. *True or False?* Smaller residential construction companies typically adopt new technology more quickly than large companies. (9.1-1)
2. *True or False?* Construction software can be used to help with scheduling and managing subcontractors. (9.1-1)
3. *True or False?* Building Information Modeling software allows the user to view a structure as a three-dimensional model. (9.1-1)
4. LiDAR technology is used to _____. (9.1-1)
   A. manage documentation for complex projects
   B. take thermal images to identify leaks
   C. market a construction company's services
   D. create high-resolution maps of environments
5. Wearable technology used in the construction industry includes vests and watches that _____. (9.1-1)
   A. take photos of inaccessible areas
   B. monitor the wearer's health vitals
   C. determine a grade or slope
   D. replace traditional PPE

LESSON 9.2

Construction Fundamentals

LEARNING OUTCOMES

After completing this lesson, you will be able to:

9.2-1 Summarize specific challenges associated with residential remodeling projects.

TECHNICAL TERMS

remodeling

renovating

Remodeling

Even though much of this text has covered new construction, nearly all new construction principles apply to renovating and remodeling a home. However, working on an existing building poses unique challenges. ***Renovating*** involves updating or restoring a space to a good state of repair, while ***remodeling*** involves construction that changes the form or function of a space. Renovating can be as simple as repainting walls or replacing flooring. Remodeling is more transformative and includes modifications such as removing or adding walls to alter a floor plan, adding on to a building, or converting a space for a different function.

Pro Tip

Often, building permits must be obtained before remodeling and renovation work can begin. Building inspectors may need to inspect the work at various stages of the project. Always check with your local municipality to determine if permits and inspections are required. The permit and inspection requirements will be based on the scope of the project.

Remodeling, renovating, and repairing existing buildings account for approximately 30% to 40% of all residential construction. This segment of the construction industry includes work ranging from freshening up an exterior or a room, adding an extension, or repairing a structural issue, to a complete overhaul of the home. In this lesson, we end our survey of construction fundamentals by considering some of the specific challenges associated with residential remodeling.

Remodeling and renovating are well worth the effort, **Figure 9.2-1**. Remodeling brings greater comfort and usability for the owner. Further, remodeling and repairing can forestall the structure's deterioration, enhance its value, or have other personal, aesthetic, or environmental benefits. Additionally, as stated in *Chapter 8, Interior Finishes*, remodeling is a green building practice that keeps a tremendous amount of waste out of the landfill and saves the expenditure of resources that would have been consumed during the production of new materials for reconstruction on the building site.

Critical Thinking and Problem Solving

Remodeling and renovating projects generally take longer and require more flexibility than new construction. Oftentimes, remodeling jobs do not go exactly as planned because of unforeseen challenges that arise as the existing building components are dismantled, **Figure 9.2-2**. Such challenges require effective critical thinking and can be severe. For instance, a performance failure can cause serious damage to a home if a flashing, siding, WRB, or roofing product fails. Part of successful material selection is researching products and following the manufacturer's instructions during installation. As we conclude the Materials lessons, take a moment to consider a home where you spent an extended amount of time when you were younger; it could be a childhood home, a relative's home, or even a close friend's house. Can you list the materials used on the inside and outside of that home? Why do you suppose those specific products were chosen? Can you identify the type of the home's heating system? As you go about your daily activities in public spaces, notice the materials used in the buildings. Can you identify them? Do you notice the quality of the craftsmanship? With exercises such as these, you continue to develop a broader understanding of materials and their purposes. It can also give you a better understanding of the value of fine craftsmanship and attention to detail.

Fasteners

The fasteners used while finishing the mechanical systems and punch list items are an extension of what this text has already covered apart from electrical wire nuts.

Wire Nuts

Aside from connecting wires to devices, wires must often be connected to one another to complete a circuit. A *wire nut* is a connector made of an insulating plastic shell with an internal threaded metal insert that is used to connect two or more wires. Wire nuts join electrical wires when stripped wire ends are twisted together and the nut is threaded over them to cover any exposed conductors, **Figure 9.4-12**.

Pro Tip

Ensure wire nuts are tightly twisted onto stripped wires. Tug slightly on each wire to verify they are securely connected and won't easily pull out.

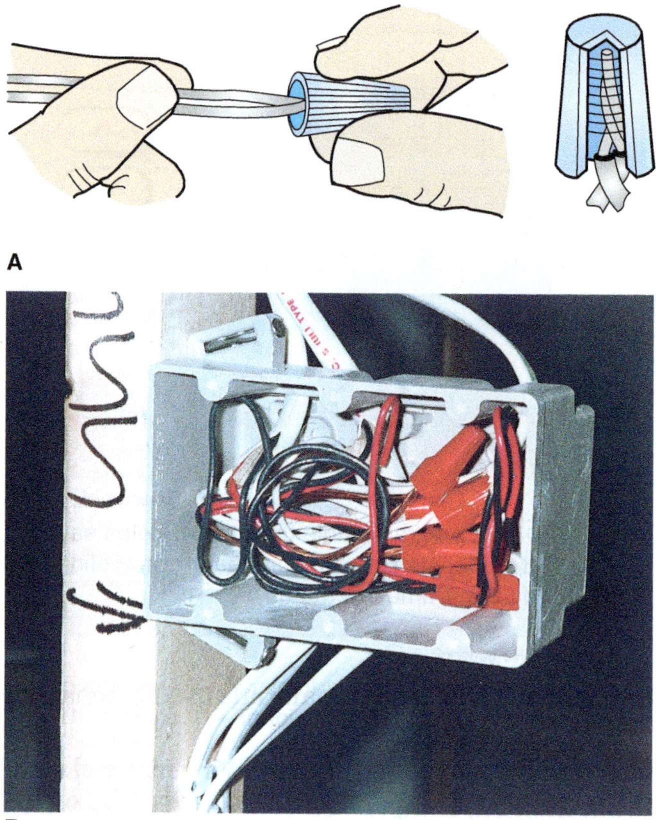

A

B

Goodheart-Willcox Publisher

Figure 9.4-12. The proper use of wire nuts to connect bared ends of conductors. A—Twist the conductors together, insert the bare conductors into the wire nut, and twist the nut clockwise until it is tight. B—Wire nuts insulate connected wires against accidental shorting.

Concluding Thoughts on Construction Fasteners

Thousands of different fasteners are used when building a home. Though the variety of fasteners used for construction is extensive, the knowledge you have gained throughout this book provides the basis for understanding how different fastening systems work for specific purposes. Understanding these fundamental principles should enable you to properly select and use nearly any construction fastener.

Next time you visit a building supply store, browse the fasteners and hardware aisles. Identify some fasteners that you have never used or seen. See if you can figure out what they are used for and how they work. This type of exercise can be useful in the future when you encounter a unique situation where a more creative solution is needed. Can you think of any areas in your home where a problematic item could be made better with a different kind of fastener? Take on the challenge and fix it! Armed with knowledge of materials and fasteners gained throughout this text, along with the instructions provided by the manufacturer, you will be able to grasp the purpose and usage of many building products with minimal effort.

LESSON 9.4 Review

Answer the following questions based on information in this lesson.

1. When connecting wires, you must strip several inches of the jacket away and then strip the insulation from about _____ of each conductor. (9.4-1)
   A. 1/8″
   B. 5/8″
   C. 7/8″
   D. 1″
2. When installing an electrical receptacle, you attach the _____. (9.4-1)
   A. hot wire to the silver-colored terminal
   B. neutral wire to the copper-colored terminal
   C. bare copper wire to the copper-colored terminal
   D. grounding wire to the green terminal
3. A(n) _____ is a gasket used on faucets to produce leakproof seals between the faucet body and internal parts. (9.4-1)
   A. slip-joint washer
   B. wax ring
   C. O-ring
   D. lock washer
4. *True or False?* When using Teflon® tape on a threaded plumbing connection, it should be wrapped around the male threads. (9.4-1)
5. A(n) _____ is used to form a seal between a toilet and the closet flange. (9.4-1)
   A. O-ring
   B. wax ring
   C. slip-joint washer
   D. lock washer
6. One example of a hardscape material is _____. (9.4-1)
   A. shrubbery
   B. mulch
   C. landscaping fabric
   D. concrete pavers
7. Which of the following statements about split face retaining wall block is *true*? (9.4-1)
   A. It is designed for creating walls at elevation changes.
   B. It features a smooth, even surface on all sides.
   C. It is shaped like a standard brick and must be installed with mortar.
   D. Although attractive, it tends to produce unstable structures.

LESSON 9.5

Tools

LEARNING OUTCOMES

After completing this lesson, you will be able to:

9.5-1 Identify the tools used to install electrical, HVAC, and plumbing finishes.

TECHNICAL TERMS

basin wrench
circuit tester
flexible copper tubing
impact wrench
multimeter
nibbler
powered metal shears
ratchet
socket
spanner wrench

Tools Used During the Finishing Phase

Mechanical contractors use certain specialty tools during the finishing phase of their work that have not yet been featured in the text. We briefly identify these tools in the following sections.

Electrical Tools

A ***circuit tester*** is used to determine whether there is voltage at any point in a circuit, **Figure 9.5-1A**. A simple tester may have only a pair of contact leads and a neon bulb that lights when voltage is detected. Most electricians also carry a ***multimeter***, a device that not only detects and measures voltage,

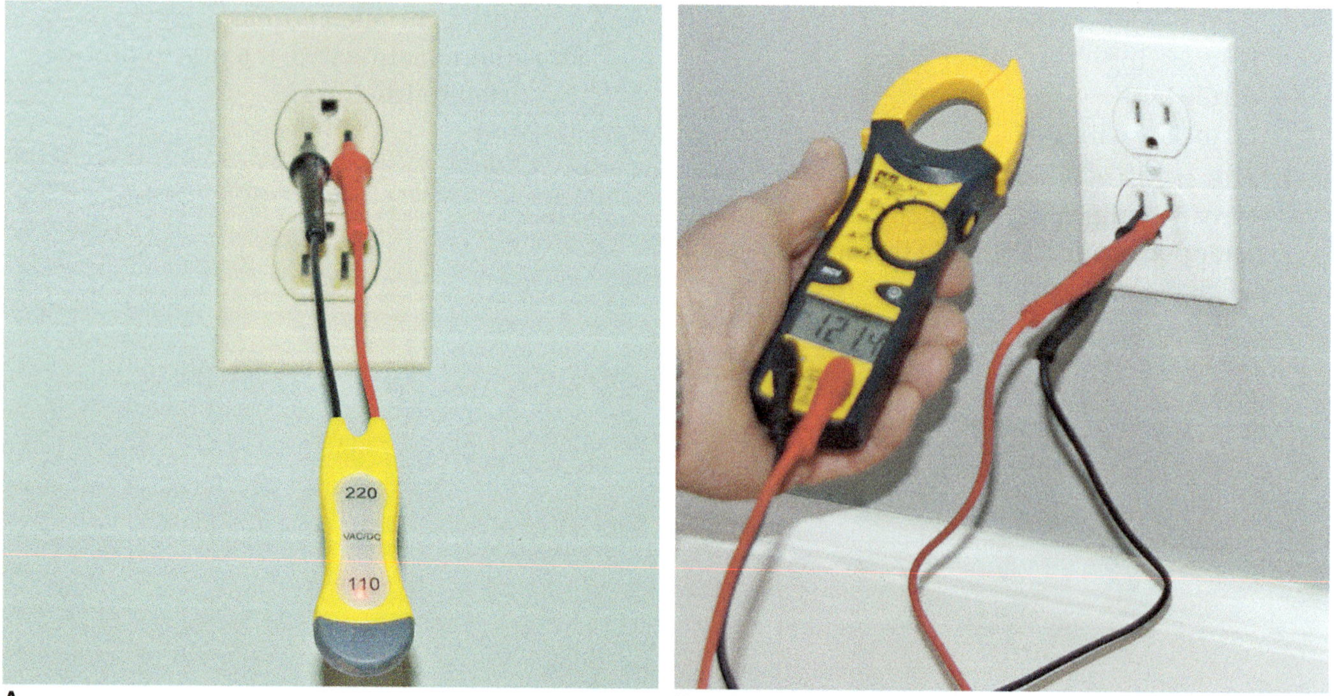

A

B

Bosch Tools

Figure 9.5-1. Electrical test equipment. A—A neon circuit tester determines only if voltage is present. B—A multimeter can be used to measure AC and DC circuits for voltage, amperage, and resistance.

but also measures current flow (amps) and resistance (ohms), **Figure 9.5-1B**.

Various other tools that complete an electrician's tool kit include a trouble light, pipe cutter, conduit bender, conductor bender, and a set of socket wrenches.

Safety Note

For electrical work, use only wood or fiberglass ladders because neither are good conductors of electricity.

HVAC Tools

Much of the ductwork for the HVAC system was likely installed during the rough-in phase. However, if alterations or additional ductwork needs to be installed, powered metal shears or nibblers may be used for cutting. A handheld electric tool, ***powered metal shears*** can quickly and easily cut through sheet metal faster than aviation snips or tin snips. Most are about the size and shape of an electric drill, **Figure 9.5-2**. Shearing attachments that clamp onto a battery drill are also available. A ***nibbler*** is a handheld cutting tool that shears off small bits of sheet metal with a mechanized punch. Its action is similar to that of a sewing machine. Nibblers are more versatile than shears and can cut around bends and profiles easier than shears, **Figure 9.5-3**.

Plumbing Tools

When installing sinks and drains, a plumber may use various wrenches, such as a ***basin wrench***, **Figure 9.5-4**, to reach into areas that other tools

Milwaukee Tools

Figure 9.5-2. This cordless metal shear has a double-cutting action that removes a swath of metal for ease of cutting.

Milwaukee Tools

Figure 9.5-3. Metal cutting nibblers work well for cutting irregular shapes and profiles.

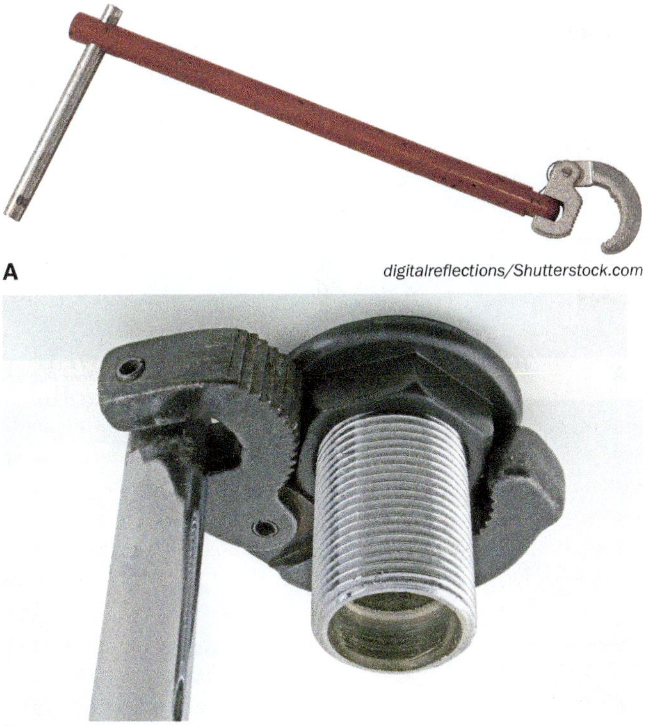

A digitalreflections/Shutterstock.com

B Richard z/shutterstock.com

Figure 9.5-4. A—A basin wrench has a unique design. B—This design allows a basin wrench to reach around a threaded drainpipe where it attaches to a sink basin.

cannot. Various designs and shapes of wrenches called spanner wrenches are also available. A ***spanner wrench*** is used for installing and removing specific drain components.

The American Heart Association and American Red Cross recommend a hands-only CPR without mouth-to-mouth resuscitation in most cases. Studies have shown hands-only CPR to be more effective in saving lives and that onlookers are more likely to engage in lifesaving CPR using this simpler, more effective approach.

Emergency Action Plan

Accidents on a construction site can occur without warning or predictability. Some accidents can cause unsafe circumstances for uninvolved employees, even after the event has occurred. Consider what could happen if a scaffold were to collapse, trapping employees in the wreckage, or the possibility of a toxic gas like carbon monoxide in a confined space asphyxiating another worker, or a fire breaking out on site. These circumstances are all examples where a decision to try and help another worker could place you in danger. Some incidents require immediate evacuation of the jobsite.

To help instruct during an emergency, employers in the construction industry are required to have an emergency action plan in place. An ***emergency action plan (EAP)*** is a written document mandated by OSHA regulations. Its primary purpose is to streamline and coordinate the actions of employers and employees during workplace emergencies. Effective emergency plans, combined with thorough employee training that clarifies individual roles and responsibilities, can reduce the number of employee injuries and the severity of injuries that may occur.

Safety Note

OSHA requires all buildings on the worksite to maintain unobstructed exits from all areas and at all times the building is occupied.

To produce an effective EAP, employers first evaluate the workplace, identifying possible emergency situations and assessing requirements for exit routes, fire extinguishers, and detection systems, and reviewing their fire prevention plan. The employer then outlines how employees will react to emergencies within the specific worksite layout, structural features, and emergency systems. At a minimum, OSHA requires an EAP to include the following:

- Means of reporting fires and other emergencies
- Evacuation procedures and emergency escape route assignments
- Procedures for employees who must remain to operate critical operations before they can evacuate
- Accounting for all employees after an emergency evacuation has been completed
- Rescue and medical duties for employees performing them
- Names or job titles of persons who can be contacted

Simply having an EAP in place does not ensure any level of safety for the workers on the job. The employer must also ensure individuals are trained in the elements of the plan. Responsible people must be assigned to supervise and coordinate an evacuation in the event of an emergency. The plan must also be kept up to date for each jobsite.

Emergency Evacuation Coordinator

When creating an emergency action plan, some employers appoint a responsible individual to lead and coordinate the emergency plan in the event of an evacuation. It is crucial for workers to know who this coordinator is and to understand that the coordinator has the authority to make decisions during emergencies.

The coordinator's responsibilities should include:

- Assessing the situation to determine if an emergency exists and, if so, initiating emergency procedures
- Supervising all emergency efforts in the area, including personnel evacuation
- Ensuring external emergency services, such as the local fire department or emergency medical service, are notified and coordinating with them upon arrival
- Directing the shutdown of operations when necessary
- Ensuring that only trained workers use portable fire extinguishers

- Keeping routes for emergency vehicles and paths for emergency responder access clear
- Informing arriving emergency responders about the incident location, conditions, and status of any remaining occupants
- Having knowledgeable workers available to assist emergency responders

Concluding Thoughts on Construction Safety

In *Chapter 1, Preparing for Construction*, you began a journey learning about the safety hazards of working on a construction site and how they are managed. The knowledge you gained throughout these chapters is consistent with information given in a typical OSHA 10 training course. These subjects have been identified by OSHA as particularly important for someone entering the industry. Understanding, practicing, and communicating about hazards encourage a safety mindset and safety culture within an organization, which can be a major factor in keeping workers from harm on the job. Take a moment for some self-reflection. Consider the importance of your own personal safety on the job. Imagine a scenario where you or someone you love was injured permanently. How would that impact your future career?

Practicing a safety mindset is so valuable because an accident on the jobsite could have lasting and permanent consequences. Following the standards that OSHA requires for employees and employers will go a long way in protecting you.

Pro Tip

To learn more about workplace safety, OSHA, or OSHA training programs such as OSHA 10 and OSHA 30, visit *OSHA.gov*.

LESSON 9.6 Review

Answer the following questions based on information in this lesson.

1. A first aid kit should contain _____. (9.6-1)
   A. a scalpel
   B. a stitch kit
   C. prescription medications
   D. antiseptic wipes
2. *True or False?* CPR should be administered to any person who has lost consciousness due to an injury. (9.6-1)
3. *True or False?* Studies have shown that CPR used in conjunction with mouth-to-mouth resuscitation is more effective in saving lives than hands-only CPR. (9.6-1)
4. The first step in producing an effective emergency action plan is for the employer to _____. (9.6-2)
   A. outline a response to a specific emergency
   B. train employees in the elements of the plan
   C. evaluate the workplace for possible emergencies
   D. designate an emergency evacuation coordinator
5. A primary responsibility of the emergency evacuation coordinator is to _____. (9.6-2)
   A. ensure external emergency services are notified
   B. assemble first aid kits for on-site use
   C. develop the emergency action plan
   D. select the appropriate PPE for on-site tasks

LESSON 9.7

Math

LEARNING OUTCOMES

After completing this lesson, you will be able to:

9.7-1 Calculate material quantities required for various construction projects.

TECHNICAL TERMS

estimating

Estimating Materials

Throughout this text, you have studied various math functions that are used in construction. One of the most essential times a construction worker uses math is to estimate materials for a job, **Figure 9.7-1**. *Estimating* is the process of determining accurate quantities of equipment, labor, and materials for a job. When estimating, even a small mistake, oversight, or miscalculation can have a dramatic impact on the profitability and outcome of a project. It is vital to be able to recognize errors and double-check critical conclusions. Experienced contractors can often pick out mistakes by recognizing inconsistent patterns in material quantities.

BearFotos/Shutterstock.com

Figure 9.7-1. Whether estimating materials on the job, or behind a desk, accurate calculations are critical for ensuring the correct material quantities.

Estimating Concrete and Masonry Materials

Concrete is measured and sold by the cubic yard. A cubic yard is 3′ wide, by 3′ deep, by 3′ high. It contains 27 cu. ft. ($3 \times 3 \times 3 = 27$). To determine the number of cubic yards needed for any square or rectangular area, use the following formula. All dimensions should be converted to feet and fractions of feet.

$$\text{Cubic yards} = \frac{\text{Width} \times \text{Length} \times \text{Thickness}}{27}$$

When using this formula, you are first calculating the volume of concrete in cubic feet (width × length × thickness) and then converting the cubic feet to cubic yards by dividing by 27.

For example:

To find the concrete needed to pour a basement floor that is 30′ × 42′ × 4″, complete the following steps:

$$\text{Cubic yards} = \frac{30 \times 42 \times 1/3}{27}$$

$$= \frac{\overset{10}{\cancel{30}} \times \overset{14}{\cancel{42}} \times 1/3}{\underset{9}{\cancel{27}} \quad \underset{1}{\cancel{3}}} = \frac{140}{9}$$

$$= 15.56 \text{ cubic yards}$$

Account for waste or slight variation in the cross-sections of the concrete form. When placing an order, an additional 5%–10% of concrete is usually added. For the preceding example, you would order 17 cubic yards of concrete.

The number of concrete masonry units needed can be estimated by first determining the number of units required in each course, and then multiplying that by the number of courses between the footing and sill.

$$\text{Total number} = \frac{\text{Perimeter}}{\text{Unit length}} \times \text{Number of courses}$$

For example:

Find the number of 8×8×16 blocks required to construct a foundation wall with a perimeter (total distance around the outside) of 144′. The wall will be laid 11 courses high.

$$= \frac{144'}{16''} \times 11$$

$$= \frac{144'}{(4/3)'} \times 11$$

$$= \frac{144 \times 3}{4} \times 11$$

$$= \frac{\overset{36}{\cancel{144}} \times 3}{\underset{1}{\cancel{4}}} \times 11$$

$$= 108 \times 11$$

$$= 1188 \text{ blocks}$$

Another method of estimating block and mortar quantities is based on the amount of material needed for 100 sq ft of wall area.

The table in **Figure 9.7-2** shows that 100 sq ft of wall area contains 112.5 8×8×16 blocks. Thus, to estimate the number of blocks needed for a wall, you can determine the area of the wall in square feet and divide by 100. (This determines the number of 100 sq ft "units" in the wall.) Then multiply this value by 112.5 to calculate the number of blocks required, or multiply by 2.6 to calculate the cubic feet of mortar required.

Quantities of Concrete Block and Mortar

Wall Thickness	For 100 Sq Ft of Wall: Number of Blocks	For 100 Sq Ft of Wall: Mortar*	Mortar* for 100 Concrete Blocks
8″	112.5	2.6 cu ft	2.3 cu ft
12″	112.5	3.9 cu ft	3.5 cu ft

Based on block having an exposed face of 7 5/8″ × 15 5/8″ and laid up with 3/8″ mortar joints.
*With face shell mortar bedding – 10% waste included.

Portland Cement Association

Figure 9.7-2. Needed quantities of concrete block and mortar can be calculated with this chart.

Practice 9-1

Calculate the amount of concrete needed for a sidewalk measuring 4″ thick × 4′ wide × 36′ long.

Estimating Framing Materials

When estimating quantities for framing materials, a thorough understanding of requirements for the job is critical. Framing member spacing [16″ on center (OC), 24″ OC, etc.] will determine the number of each joist, stud, and rafter you will need. Sizing of framing members based on plans and specifications, or span tables, also determines how many of each size component will be required.

Estimating Floor/Ceiling Joists and Subfloor

To estimate the number and size of floor joists on a job, first check the plan and determine the lengths that are needed, **Figure 9.7-3**. Be sure to allow sufficient length for full bearing on girders and partitions. Average residential structures require several different lengths.

To determine the number of joists needed for spacing 16″ OC, multiply the length of the wall that carries the joists by 3/4 and then add 1.

For 24″ OC spacing, multiply the length of the wall by 1/2. Also, add extra pieces for doubled joists under partitions and trimmer joists at openings. If solid wood joists are being used, add extra pieces for headers at openings. If the floor is to be

Goodheart-Willcox Publisher

Figure 9.7-3. Floor joists being installed on foundation walls.

framed with I-joists, add LVL according to the plans for headers at openings.

No. of joists = (Wall length × 3/4) + 1 + Extras

The sizes and numbers of joist headers will be the lengths of the foundation sections on which they rest. The depth of the band joists will be the same as the adjoining floor joists. Your estimate should include the cross-sectional size and the number of pieces of each length.

A complete estimate might appear as one of the following:

Required floor joists:

- 20 pcs—1 3/4″ × 9 1/2″ × 28′-0″ I-joists
- 12 pcs—1 3/4″ × 9 1/2″ × 32′-0″ I-joists
- 2 pcs—1 3/4″ × 9 1/2″ × 14′-0″ I-joists

Required joist headers:

- 2 pcs—1 1/4″ × 9 1/2″ × 20′-0″ LVL rim board
- 2 pcs—1 1/4″ × 9 1/2″ × 12′-0″ LVL rim board
- 1 pc—1 1/4″ × 9 1/2″ × 8′-0″ LVL rim board for header

For solid wood:

- 40 pcs—2×10 × 16′-0″
- 36 pcs—2×10 × 14′-0″

Required band joists:

- 4 pcs—2×10 × 16′-0″
- 2 pcs—2×10 × 12′-0″

Estimating ceiling joist requirements uses the same method as estimating floor joists. Because ceilings are relatively free of openings, no extras or waste need to be included. For this reason, the short method should not be applied to ceiling joists.

Procedures for estimating the subflooring vary, depending on the type of material used. Usually, the area is calculated by multiplying the overall length and width, and then subtracting major areas that are not to be covered, including breaks in the wall line and openings for stairs, fireplaces, and other items. This calculation will determine the net area and the basic amount of material needed. Waste and other extras should be considered and added to the total.

Little waste is involved when using sheet materials. The net area is divided by 32 (number of square feet in a 4×8 sheet) and rounded out to the next whole number to arrive at the required number of pieces of OSB or plywood. Be sure to specify the type of sheet material, its thickness, and its span rating.

The same estimating principles are used to determine materials required for walls and roofs, with some adjustments made for differences in components.

Estimating Wall and Ceiling Framing

To estimate wall and ceiling framing materials, first determine the total lineal feet of the walls. Add the length of each wall and partition, **Figure 9.7-4**. The plans include the dimensions of outside walls. These amounts can be added together. Short partitions may not be dimensioned. In this case, use a scale on the drawing. It is a good idea to place a colored pencil check mark on each wall and partition as its length is added to the list.

For plates, multiply the total figure by 3 (one sole plate plus two top plates) if you are using a double top plate or by 2 if you are using a single top plate. Add about 10% for waste. Order this number of linear feet of lumber in random lengths or convert to the number of pieces of a specific length. Each lineal foot of 2×4 equals 2/3 bd ft. Each lineal foot of 2×6 equals 1 bd ft. Thus, the total lineal feet of walls and partitions can be quickly converted to board measure if required.

For example:

Wall and partition length = 240
Total plate material = (Length × 3) + 10%
= (240 × 3) + 10%
= 792 lineal ft
= or 57 pcs

Goodheart-Willcox Publisher

Figure 9.7-4. Students in a construction program panelized the walls of this home to be transported to a jobsite. Before assembly, they learned how to estimate the materials required.

Estimating Studs

The total length of all walls and partitions is also used to estimate the number of studs required. As with floor joists, when studs are spaced 16″ OC, multiply the total length (in feet) by 3/4. If the framing is 24″ OC, divide the total length by 2. Next, add two more studs for each corner, intersection, and opening. Using 240′ for the wall length, assume 12 corners, 10 intersections, 20 openings, and 16″ OC framing.

For example:

Find the number of studs needed, including outside walls and partitions.

$$\text{Total studs} = (\text{Total length} \times 3/4) + 2\,(\text{Corners} + \text{Intersection} + \text{Openings})$$

$$= \left(\frac{240 \times 3}{4}\right) + 2(12 + 10 + 20)$$

$$= \left(\frac{\overset{60}{\cancel{240}} \times 3}{\cancel{4}}\right) + 2(42)$$

$$= (60 \times 3) + 84$$

$$= 264$$

Many carpenters estimate the number of studs by simply counting one stud (spaced 16″ OC) for each lineal foot of wall space. The overrun on spacing provides the extras needed for corners and openings. This method is rapid and fairly accurate. However, it does not estimate enough material for a small house divided into many rooms. On the other hand, too many studs will likely be estimated for a large house with wide windows and open interiors.

Pro Tip

Carefully check over your rough opening layouts for errors. Do the math before cutting and framing.

Estimating Headers

To estimate the amount of header material required, begin by identifying which headers, if any, will be doubled or built up. If header hangers are being used without trimmer studs, the header length is the actual rough opening width. If trimmers are being used, add 3″ (1 1/2″ for the trimmer at each side) of the opening. If LVL or other engineered lumber is to be used, add all header lengths together, plus several inches for saw kerfs. If dimensional lumber will be used, add the lengths and divide by 10 to determine the number of 10′ pieces required. The same calculation can be made to use 8′ pieces (divide by 8).

Code Note

The table in **Figure 9.7-5** gives the allowable spans of headers according to the *International Residential Code (IRC)* for various load conditions.

Allowable Header Spans

Header Supporting	Size	Span with 30 lb per sq ft Snow Load		Span with 50 lb per sq ft Snow Load		Span with 70 lb per sq ft Snow Load	
		20′	28′	20′	28′	20′	28′
Roof and Ceiling	2-2×4	3′-6″	3′-2″	3′-2″	3′-6″	2′-10″	2′-6″
	2-2×10	8′-5″	7′-3″	7′-3″	6′-3″	6′-6″	5′-7″
	3-2×8	8′-4″	7′-5″	7′-5″	6′-5″	6′-8″	5′-9″
Roof, Ceiling, and One Clear-Span Floor	2-2×4	3′-1″	2′-9″	2′-9″	2′-5″	2′-7″	2′-3″
	2-2×10	7′-0″	6′-2″	6′-4″	5′-6″	5′-9″	5′-1″
	3-2×8	7′-2″	6′-3″	6′-5″	5′-8″	5′-11″	5′-2″
Roof, Ceiling, and Two Clear-Span Floors	2-2×4	2′-1″	1′-8″	2′-0″	1′-8″	2′-0″	1′-8″
	2-2×10	4′-9″	4′-1″	4′-8″	4′-0″	4′-7″	4′-0″
	3-2×8	4′-10″	4′-2″	4′-9″	4′-1″	5′-5″	4′-8″

Goodheart-Willcox Publisher

Figure 9.7-5. The allowable header spans for exterior walls under various loads according to the *IRC*.

plan view of the roof is helpful in adding together these materials. All ridges and eave lines are seen as true length and can simply be measured with a scale to find their length. Hips and valleys are not seen as true length so a small amount must be added. Using the percentage listed for converting ground area to roof area usually provides sufficient accuracy.

Practice 9-4

How many squares of shingles would you order to shingle a simple gable roof with a ground area of 1850 sq ft with a 6-in-12 roof slope? Add 10% as a waste factor.

Estimating Insulation Materials

The amounts of insulating materials are figured on the basis of area (square feet). The thickness is then specified as separate data. The size of packages varies, depending on the type and thickness. For example, one manufacturer packages 1 1/2″ blankets in rolls of 140 sq ft. A 3″ blanket in the same width comes in rolls of 70 sq ft. Batts are furnished in packages (sometimes called tubes) that contain as much as 100 sq ft. A 6″ batt usually contains 50 sq ft, **Figure 9.7-7**.

Serhii Krot/Shutterstock.com

Figure 9.7-7. Fiberglass insulation is available in rolls or blankets.

Green Note

In addition to advanced framing, other alternative green framing methods are sometimes used. Double stud walls and truss walls use an additional layer of framing and insulation on the outside of the structure to eliminate thermal bridges and add insulation to a structure. These systems can be labor intensive to build and can substantially add to the overall cost of construction. However, the owner can often recover those costs over time through energy savings.

To determine the amount of insulation needed for exterior walls, first calculate the total perimeter of the structure. Then, multiply that number by the ceiling height. Deduct from the total the area of doors and windows. Many carpenters deduct only large windows or window walls and disregard doors and smaller openings. This extra allowance makes up for loss in cutting and fitting and also provides for additional material needed around plumbing pipes, recessed lighting boxes, and other items.

For example:

$$\begin{aligned} \text{Perimeter} &= 30' + 40' + 36' + 20' + 6' + 20' \\ &= 152' \\ \text{Area} &= 152' \times 8' \\ &= 1216 \text{ sq ft} \\ \text{Window wall} &= 12' \times 8' \\ &= 96 \text{ sq ft} \\ \text{Net area} &= 1216 - 96 \\ &= 1120 \text{ sq ft} \end{aligned}$$

The same procedure can be used to estimate reflective insulation. Make a greater allowance for cutting and waste, especially when using the accordion type. This type cannot be effectively spliced. Rolls of reflective insulation hold from 250 to 500 sq ft.

Rigid insulation is also estimated on the basis of area. It can be calculated from dimensions shown on the working drawings.

When estimating the amount of insulation for floors and ceilings, use the same figures calculated for the subfloor area. Stairwell openings and openings for large fireplaces can be deducted. However, not deducting these amounts may provide the extra insulation needed for waste and special packing around fixtures.

For a perimeter insulation strip in a concrete slab floor, multiply the perimeter of the building by the width of the strip. There is little waste on such an installation, since even small pieces can be used.

For example:

Fill insulation comes in bags that usually contain 3 or 4 cu ft. Calculate the number of bags needed.

$$\begin{aligned}\text{Area} &= 1200 \text{ sq ft}\\ \text{Thickness} &= 4''\\ &= 1/3'\\ \text{Cu ft required} &= 1200 \times 1/3\\ &= 400\\ \text{Less 10\% allowance for joists 16'' OC} &= 400 - 40\\ \text{Net amount} &= 360 \text{ cu ft}\\ \text{Number of bags (4 cu ft per bag)} &= 90\end{aligned}$$

Manufacturer's directions and specifications usually have tables that provide a direct reading of the number of bags required for a certain thickness of application. These are especially helpful in estimating the amount needed for such items as filling the cores of concrete blocks.

Practice 9-5

Loose-fill insulation is available in 4 cubic foot bags from your local supplier. How many bags of loose-fill insulation are required to insulate the attic of the 1850 sq ft home used in the roofing estimate? The attic must be insulated to a thickness of 16″. Do not subtract an allowance for joists.

Estimating Interior Finish Materials

The amount of material required to cover walls or ceilings is determined by first calculating the total number of square feet of wall or ceiling area. Regular-size door and window openings are disregarded. This provides an allowance for waste. However, larger openings, such as window walls or picture windows, should be subtracted.

The ceiling area is usually the same as the floor area. It is much easier to take floor dimensions and multiply the length times the width. To find wall area, add all the wall lengths and multiply by the wall height.

Estimating Sheet Materials

When ordering sheet materials such as wallboard or paneling, be sure to specify the length. With horizontal applications, always plan to use the longest practical sheet. This will keep the number of butt joints to a minimum or may even eliminate them. Divide the total length of the walls by the width of the sheets to find the number of sheets needed for a vertical application.

Separately estimate each room. Directly take dimensions from the walls or carefully scale the plan. Consult tables and charts of suppliers for estimates of joint compound, adhesives, and nails.

When working with expensive hardwood panels, it is usually advisable to make a scaled layout. Refer to **Figure 9.7-8** for typical horizontal and vertical arrangements.

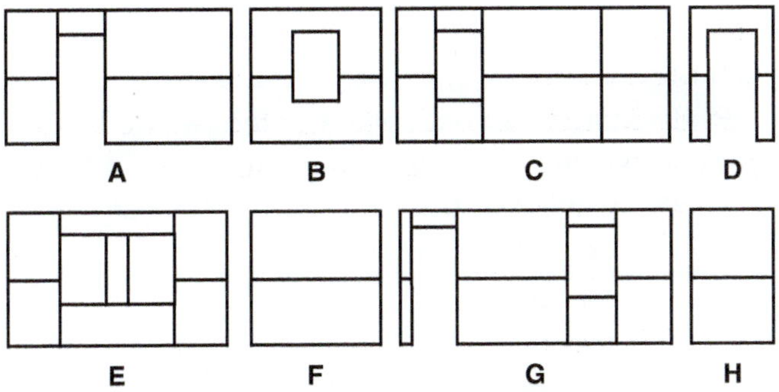

In design 1, a simple two-panel horizontal arrangement has the single continuous horizontal joint placed midway betwen the floor and ceiling. The panel design is defined by the vertical joints at openings, such as those in elevations A, C, E, and G. However, key panels may be omitted where the panel and length exceed the wall element width, as in elevations B and D.

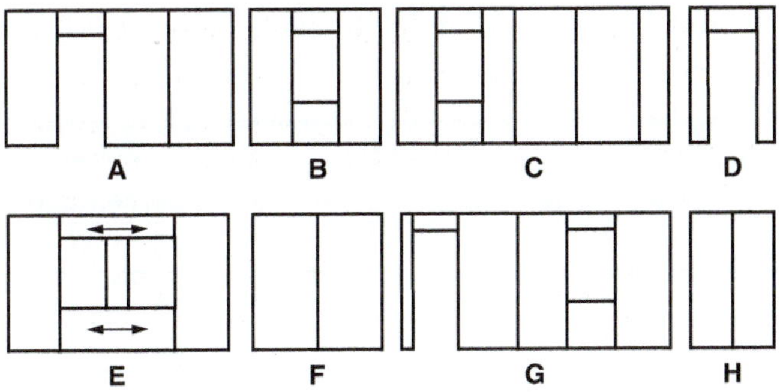

In design 2, a vertical-arrangement panel design involves lining up vertical joints with wall openings. Then, the plain wall space is vertically divided in widths proportionate to the openings. When the width of a door or window opening exceeds the panel width, panels may be placed horizontally, as shown by arrows in elevation E.

Goodheart-Willcox Publisher

Figure 9.7-8. Two commonly used arrangements for laying out wall panels. The basic rule is to "work from the openings." First, line up vertical joints above doors and above and below windows. Divide the remaining plain wall space into an orderly pattern as stud location allows.

Estimating Solid Wood Paneling

Estimates for solid paneling are based on its nominal (unfinished) size. Seasoning and planing bring it to its dressed size. Forming of joints may further reduce the actual face widths. For example, a 1×4 tongue-and-groove board has a face width of 3 1/8″.

First, calculate the square footage of the wall to be covered. Then, multiply by one of the following factors:

- For 1×4 tongue-and-groove boards, use 1.28
- For 1×6 tongue-and-groove boards, use 1.17

On standard vertical applications, add 5% for waste when required lengths can be selected or when the lumber is end-matched.

Estimating Ceiling Tile

Quantities of ceiling tile are estimated by figuring the area (square footage) to be covered. Round off any fractional parts of a foot in width and length to the next larger full unit when making this calculation.

Add extra units when it is necessary to balance the installation pattern with border tile along each wall. When using 12″ × 12″ tile, the number required equals the square footage plus the extra allowance described. Standard 12″ × 12″ ceiling tiles are packaged 64 sq ft to a carton.

Estimating Strip Flooring

To determine the number of board feet of strip flooring needed to cover a given area, first calculate the area in square feet. Then, add the percentage listed in **Figure 9.7-9** for the size being used. The figures listed are based on laying flooring straight across the room. They provide an allowance for side matching plus 5% for end matching.

For example:

Total area = 900 sq ft
Flooring size = 3/4″ × 2 1/4″
Bd ft of flooring = 900 + 38% of 900
= 900 + 342
= 1242
Number of bundles = 51.75 (round up to 52)
(24 bd ft to bundle)

There are typically 24 bd ft in a standard bundle of wood strip flooring. This amount can differ with other types of strip flooring.

Where there are many breaks and projections in the wall line, allow additional amounts.

Estimating Resilient Tile Flooring and Adhesive

To estimate the number of tiles needed, first find the area of the room. Multiply the length in feet and inches by the width in feet and inches. If the room has an irregular shape, section the room into squares or rectangles. Then, find each area and add the results.

To determine the allowance for waste, multiply the square footage by the factor shown in **Figure 9.7-10**. For example, suppose that the area of a room to be floored is 88 sq ft. The chart indicates that 10% should be added for waste. Thus, 88 × 1.10 = 96.8 or 97 sq ft. If one tile is 12″ × 12″, you

Estimating Strip Flooring

Flooring board size	Additional percentage
3/4″ × 1 1/2″	55%
3/4″ × 2″	43%
3/4″ × 2 1/4″	38%
3/4″ × 3 1/4″	29%
3/8″ × 1 1/2″	38%
3/8″ × 2″	30%
1/2″ × 1 1/2″	38%
1/2″ × 2″	30%

Goodheart-Willcox Publisher

Figure 9.7-9. When estimating strip flooring, add the listed percentage to total area to allow for waste.

Percentage of Tile Waste for Various Areas

Area (square feet)	Percentage
Up to 50	14
50–100	10
100–200	8
200–300	7
300–1000	5
1000–5000	3
5000–10,000	2–3
Over 10,000	1–2

Goodheart-Willcox Publisher

Figure 9.7-10. Tile flooring waste percentages for 12″ × 12″ floor tile.

will need about 97 tiles to complete the job. Add 1/3 more when laying down 9″ × 9″ tiles.

Estimate adhesive for laying resilient tile according to square yardage or square footage. Do not include the extra added for the waste. Divide the square yardage or footage by the per-gallon spreading capacity of the adhesive, which is shown on the container.

Estimating Porcelain, Ceramic, or Quarry Tile

To determine the quantity of tile needed, first calculate the square footage of the area to be covered. Measure the length and width and multiply the two figures. Add an additional 10% to compensate for breakage, cuts, and extras for future repairs. Manufacturer's coverage charts are available where tile is sold. A sample is shown in **Figure 9.7-11**.

Square Feet of Coverage		
	Tile size	
	4 1/4″ × 4 1/4″ 6″ × 6″	4′ × 4′ 8′ × 8′ 10′ × 10′ 12′ × 12′
	Carton size	
Number of cartons	**12.5 square feet per carton**	**11.1 square feet per carton**
1	12	11
2	25	22
3	37	33
4	50	44
5	62	55
6	75	66
7	87	77
8	100	88
9	112	99
10	125	111

American Olean Tile Co.

Figure 9.7-11. A coverage chart aids in determining the amount of tile needed.

Concluding Thoughts on Construction Math

The math skills you have practiced throughout this book have a significant bearing on your ability to accurately estimate and order materials for construction projects. The estimating techniques featured in this chapter are not the only methods construction estimators use when considering material quantities; however, the math principles you have studied are typical for any method of material estimating. Concepts such as working with fractions, figuring square footage (area) and volume, and working with percentages are used regardless of estimating methods.

Hopefully, you have gained an understanding of the importance of math in residential construction. Can you think of examples of how a misstep in basic calculations may lead to a serious error for a builder? Who ultimately pays the price for these kinds of mistakes? What impact would a laborer's math mistake have on a job? What about a tradesperson's error? Would the consequences likely be more costly or less costly if a math mistake were made by a foreman, estimator, or superintendent?

Pro Tip

Problems in estimating, although based on simple formulas, usually contain so many variables that good judgment must be applied to their solution. This judgment is acquired through experience.

Remember, anytime you make calculations for a job, first, step back and ask if the solution makes logical sense; then, double-check your math. It could save thousands of dollars in labor and materials costs.

LESSON 9.7 Review

Answer the following questions based on information in this lesson.

1. How much concrete should you order to pour a basement floor that is 25′ × 12′ × 6″? (9.7-1)
   A. 1 yd^3
   B. 5.56 yd^3
   C. 6 yd^3
   D. 66.67 yd^3
2. If spacing 16″ OC, how many studs are needed for a wall that is 30′ long? (9.7-1)
   A. 22
   B. 23
   C. 24
   D. 25
3. *True or False?* You should use the short method of determining the number of joists needed when estimating the number of ceiling joists required for a job. (9.7-1)
4. How many 2″ × 4″ × 14′ pieces of lumber will you need to order for wall plates if you are using a double top plate and the total wall and partition length is 320′? (9.7-1)
   A. 68
   B. 69
   C. 75
   D. 76
5. How many studs are required if the total length of all walls and partitions is 400′; there are 10 corners, 8 intersections, and 16 openings; and the framing is 24″ OC? (9.7-1)
   A. 168
   B. 268
   C. 368
   D. 468
6. *True or False?* When estimating headers, if the header hangers are being used without trimmer studs, the header length is the actual rough opening width. (9.7-1)
7. How many sheets of 4′ × 8′ wall sheathing will be needed if the net area is 2400 sq ft? (9.7-1)
   A. 38
   B. 57
   C. 75
   D. 113
8. When estimating the number of rafters required for a plain gable roof with 16″ OC spacing, you must _____, then round up and add 1 more. (9.7-1)
   A. multiply the length of the building by 3/4
   B. multiply the length of the building by 1/2
   C. divide the length of the building by 3/4
   D. divide the length of the building by 1/2
9. If the roof slope is 5-in-12, and the ground area is 1680′, how many squares will be needed for asphalt shingles? Add 10% for waste. (9.7-1)
   A. 17
   B. 18
   C. 19
   D. 20
10. If the total perimeter of a home is 175′, the ceilings are 8′ high, and the total area of windows and doors is 80 sq ft, what is the net area that needs to be insulated? (9.7-1)
    A. 1120 sq ft
    B. 1320 sq ft
    C. 1410 sq ft
    D. 2040 sq ft
11. *True or False?* To determine the number of sheet materials for a vertical application, you must multiply the total length of the walls by the width of the sheets. (9.7-1)
12. If you need to order 9″ × 9″ tiles for a rectangular room that is 18′ long and 12′ wide, and 7% should be added for waste, how many tiles will you need? (9.7-1)
    A. 216
    B. 232
    C. 287
    D. 309

LESSON 9.8

Print Reading

LEARNING OUTCOMES

After completing this lesson, you will be able to:

9.8-1 Summarize why a request for information (RFI) is used on a job.

9.8-2 Describe a scenario where a change order is necessary for a job.

TECHNICAL TERMS

change order

material takeoff

request for information (RFI)

Project Conceptualization

Chapter 1, Preparing for Construction, introduced construction drawings and how a set of plans is used to communicate the information needed to build a home to the architect's specifications. Throughout your study, you saw how different drawing types and perspectives work together to create a complete guide to the construction of that home. This lesson concludes our survey of print reading by summarizing how prints are used throughout the job.

Establishing Design

Before an architect can begin designing a home, some basic considerations about the home's purpose need to be established. What are the needs of the owner? What size home meets those needs? What style home would best suit the environment and the owner's preferences? How will the home be used? What is the lifestyle of the owner? If the lot has been purchased, what are the site limitations? These and many more questions must be answered before a design concept for the home can be developed.

Once the architect and owner have established a design, a full set of working plans can be drawn up. As they are drawn, engineering challenges, system designs, and material choices must be addressed in order to establish a project budget.

Drawing Distribution and Use

When a full set of drawings is complete, builders can perform material takeoffs and estimate labor. A ***material takeoff*** is the process of determining the quantities of specific materials that will be required for a given project, **Figure 9.8-1**. Individual builders use their material takeoffs and labor estimates to produce a bid or estimate for consideration. When a builder is chosen, pre-construction planning begins.

In the pre-construction phase of a project, multiple sets of prints are produced and shared. Before the digital age, physical sets of working drawings would be printed and distributed to subcontractors,

epiximages/Shutterstock.com

Figure 9.8-1. Prints allow estimators and contractors to perform accurate material estimates.

code officials, building departments, owners, and even material suppliers. Generally, a clean master copy would be kept for documentation. Today, some tradespersons still like to work from paper prints; however, the same goals can often be met with digital copies of the working drawings. Files can be shared through email and construction management software, allowing for instant access and collaboration among the construction team. It is still common to have a scale drawing printed and used on the jobsite. This drawing allows trades workers to easily meet and collaborate around a common table, pointing out and working through challenges and planning their work accordingly.

Requests for Information (RFIs)

Throughout the course of a construction project, it is common for a builder to have a question that is not answered by studying the plans. New or incomplete information, minor oversights, and real-world challenges often cause questions and problems that need to be addressed by the architect. A *request for information (RFI)* is a formal inquiry used in the construction industry to attain clarification or details about various aspects of a project. RFIs are used to document such interactions between the design team and the construction team. They usually contain details such as the requester's information, the nature of the request, specific questions or concerns, and an appeal for a response within a specified timeframe. The party responsible for providing the information typically responds formally, addressing the questions or concerns raised in the RFI.

Pro Tip

Submit detailed and specific RFIs as early as possible to avoid project delays, including clear references to drawings or specifications and suggested solutions when appropriate.

Change Orders

Once a builder is contracted to perform a job, the working drawings become part of the legally binding contract between the owner and builder. To fulfill the agreement, the home must be built according to the plans and specifications set forth by the owner and architect, **Figure 9.8-2**. This agreement does not prohibit changes; it simply means that any change or variance to the original plan must be approved by all involved parties. Changes to the original plan usually mean an adjustment to the contracted cost of the build. All terms of changes must be documented using a change order. A *change order* refers to any alteration or modification made to the original scope of work outlined in a construction contract. This change can involve adjustments to the project's design, specifications, materials, timelines, or other aspects.

Once approved, change orders become part of the contract and modify the original agreement. They are essential for documenting changes to ensure clarity and prevent disputes over changes made during the construction process. Without the use of change orders, builders run the risk of financial and design disputes with clients and architects.

Pro Tip

Change orders that clearly outline scope changes, cost adjustments, and revised timelines should be signed by all parties before proceeding to avoid disputes and ensure smooth project management.

worawit_j/Shutterstock.com

Figure 9.8-2. Construction drawings are part of the legal contract between the owner and the builder. The home must be built according to the expectations set by the architect through the drawings. Changes to the original plan must be documented using change orders.

Concluding Thoughts on Print Reading

Construction drawings are an essential part of residential construction. Proficiency in reading and interpreting prints is important to any home build, **Figure 9.8-3**. The principles you have learned in this book about print reading must be practiced and applied in order to fully appreciate the depth of information communicated through a full set of working drawings. Site plans, foundation plans, floor plans, framing plans, MEP plans, section drawings, details, and specifications all work together to provide a complete picture of what is to be built. The more print reading experience a novice construction worker can get, the quicker they will be on their way to making critical decisions and running jobs. If these are goals you embrace, take every opportunity available to study and use construction drawings in a real-world setting.

wichayada suwanachun/Shutterstock.com

Figure 9.8-3. Joining coworkers in planning and discussing construction drawings can help build your proficiency in print reading.

As you complete this lesson, consider how important construction drawings are to a builder. If you were a contractor and were asked to build a custom home for someone who did not want to hire an architect or use a set of construction drawings, what would be your response? Imagine the client simply describing to you what they wanted you to build for them. What complications do you think would arise? Do you think the finished product would meet their expectations and match the original vision they had for their home?

LESSON 9.8 Review

Answer the following questions based on information in this lesson.

1. *True or False?* Digital copies of the working drawings can be used for instant collaboration among the construction team. (9.8-1)
2. A request for information is a formal inquiry used in the construction industry to _____. (9.8-1)
   A. determine the quantities of materials required in order to estimate the project
   B. modify the original scope of work outlined in the construction contract
   C. identify the sections of the working drawings that apply to a specific task
   D. obtain clarification about one or more aspects of a project
3. *True or False?* A change order might be necessary if a problem such as a broken septic pipe was discovered during the remodeling process. (9.8-2)

LESSON 9.9

Green Building and the Future of Construction

LEARNING OUTCOMES

After completing this lesson, you will be able to:

9.9-1 List possible sources of indoor air pollution.

9.9-2 Describe how a builder would resolve poor indoor air quality.

TECHNICAL TERMS

energy recovery ventilator (ERV) heat recovery ventilator (HRV) radon

Indoor Air Quality

Green building principles focus not only on energy efficiency and sustainability, but also on creating a healthy environment for the building's occupants. Indoor air quality is a major concern with homes that are built using green building principles because of their tight building envelope, **Figure 9.9-1**. This concern grows as building codes continue to require tighter construction of all buildings. Without the addition of mechanical ventilation, contaminants and pollutants can become trapped inside a home at unhealthy levels.

Passive House Institute US, Inc.

Figure 9.9-1. Tight building envelopes like the one used on this passive house severely limit the exchange of indoor air and, without the use of mechanical ventilation, lead to poor air quality.

Sources of Indoor Air Pollution

Minimizing and managing potential sources of indoor air pollution are necessary to maintain a healthy atmosphere inside a home. It is easy to overlook or fail to consider the potential sources of air pollution that exist, **Figure 9.9-2**.

Common sources include:

- **Combustion sources**—Appliances that burn fuels, such as gas stoves, fireplaces, wood stoves, and furnaces, can release carbon monoxide, nitrogen dioxide, and other harmful pollutants into the air if not properly vented or maintained.
- **Building materials and furnishings**—New construction materials, paints, stains, varnishes, glues, solvents, carpets, and furniture often release volatile organic compounds (VOCs) such as formaldehyde and benzene, contributing to indoor air pollution.
- **Household products**—Many household products like air fresheners, pesticides, cleaning agents, and personal care products contain chemicals that can emit pollutants into the air, impacting indoor air quality.

bsd studio/Shutterstock.com

Figure 9.9-2. Pollutants can come from various sources and can build up to high concentrations in a home when indoor air quality is not properly managed.

- **Tobacco smoke**—Smoking indoors leads to high levels of indoor air pollution, including secondhand smoke, which contains numerous toxic chemicals harmful to health.
- **Pets**—Pets produce dander, fur, odors, and allergens that can contribute to poor indoor air quality.
- **Radon**—*Radon* is a naturally occurring colorless, odorless gas that can enter homes through cracks or unsealed seams in the foundation and accumulate to harmful levels. Long-term exposure to radon is linked to lung cancer.
- **Mold and moisture**—Damp areas in homes can promote mold growth, which releases spores and other contaminants and can lead to respiratory issues and allergies.

Managing Indoor Air Quality

Eliminating pollutants is the most effective way to prevent indoor air contamination. In some cases, this can be done using a substitution or alternative product, such as paint or carpet that emits low VOCs, **Figure 9.9-3**. Some air quality issues can be addressed by simply altering construction practices. For instance, certain products may only emit high VOCs the first week after they are applied. Allowing enough time for VOC reduction with proper ventilation can reduce the risk of poor air quality. However, this is not the case for all sources of pollution.

Goodheart-Willcox Publisher

Figure 9.9-3. Low VOC and no VOC paints and other products are available as an alternative to traditional materials.

Ventilation

Exhausting indoor air from the building enclosure and replacing it with fresh air reduces the level of pollutants to a point where they are no longer harmful. Ventilation can be accomplished through nonmechanical means, such as through windows, doors, and breaches in the building envelope. However, nonmechanical ventilation is unpredictable and unreliable and is highly subject to changes in conditions, such as wind gusts and temperature fluctuations. Additionally, it is important that the air being replaced in a home is not contaminated. Air quality outside during seasons of high pollen or high humidity can also be the cause of health issues, as can general outdoor air pollution.

To solve these issues, builders use an air exchanger to exhaust air from a building while drawing in fresh air, **Figure 9.9-4**. These units manage the ventilation in a way that addresses the concerns of nonmechanical ventilation. Two main types of air exchangers can be used. When used during the heating season, a ***heat recovery ventilator (HRV)*** uses the heat in the air that is being removed from the building to heat the incoming fresh air. During the cooling season, the HRV transfers incoming heat to the outgoing stale air, maintaining the cool temperatures inside the building, **Figure 9.9-5**. HRVs recapture 60%–80% of the conditioned air temperature that would otherwise be lost to straight ventilation. An ***energy recovery ventilator (ERV)*** performs the same functions as an HRV, but it also removes humidity from the incoming air to maintain more comfortable humidity levels inside the building, **Figure 9.9-6**. Both types of air exchangers filter the incoming air to remove pollen, mold spores, and dust particles. HRVs and ERVs are typically designed as ducted, whole-house systems.

Through mechanical ventilation, the rate of air coming into and out of the building is controlled and conditioned, rather than leaking through the building envelope. Testing indicates that indoor air should be exchanged with fresh outdoor air every 2–3 hours.

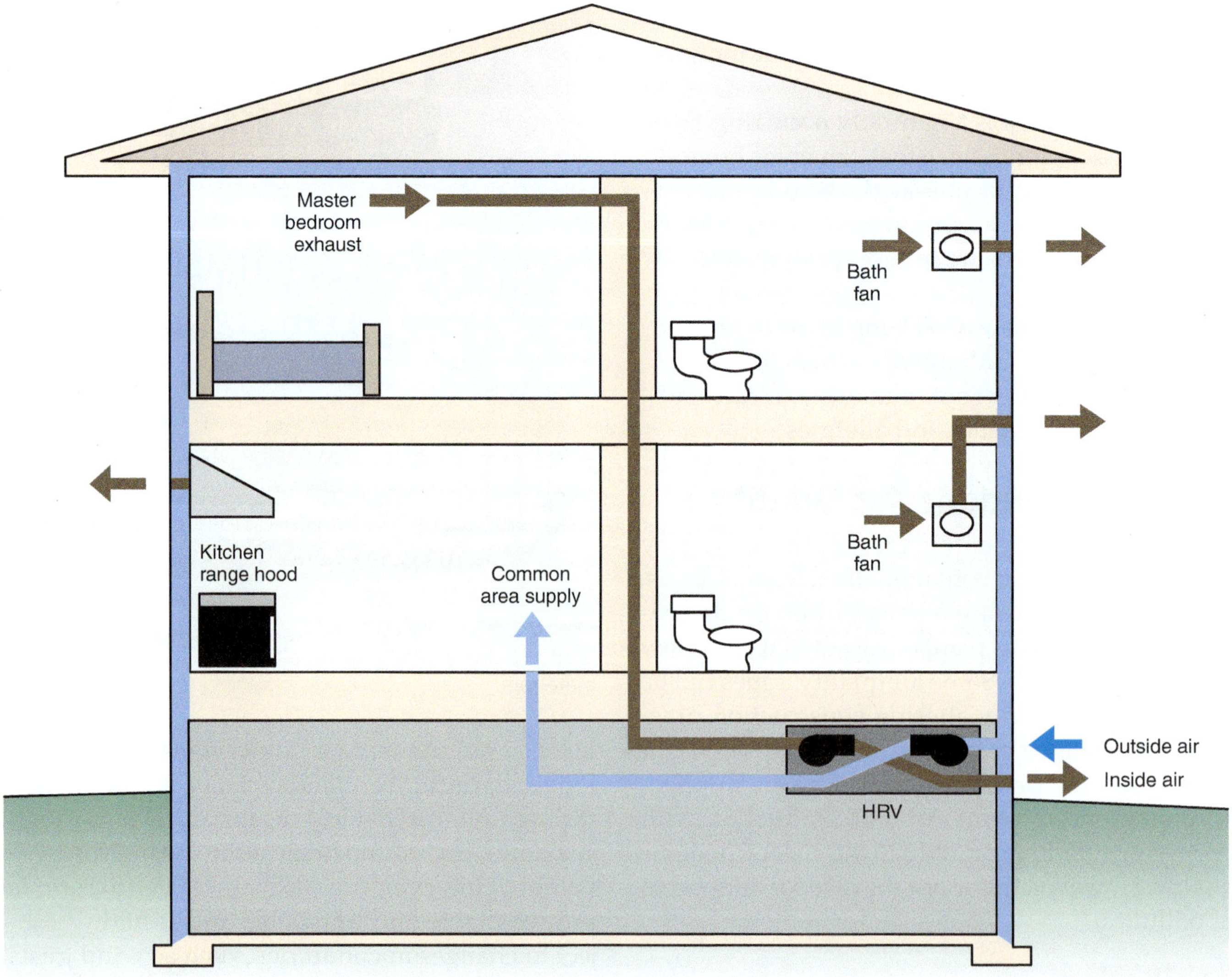

Goodheart-Willcox Publisher

Figure 9.9-4. Heat recovery ventilators and energy recovery ventilators scavenge heat from outgoing air, or vice versa in a cooling climate, while bringing fresh, filtered air into the building.

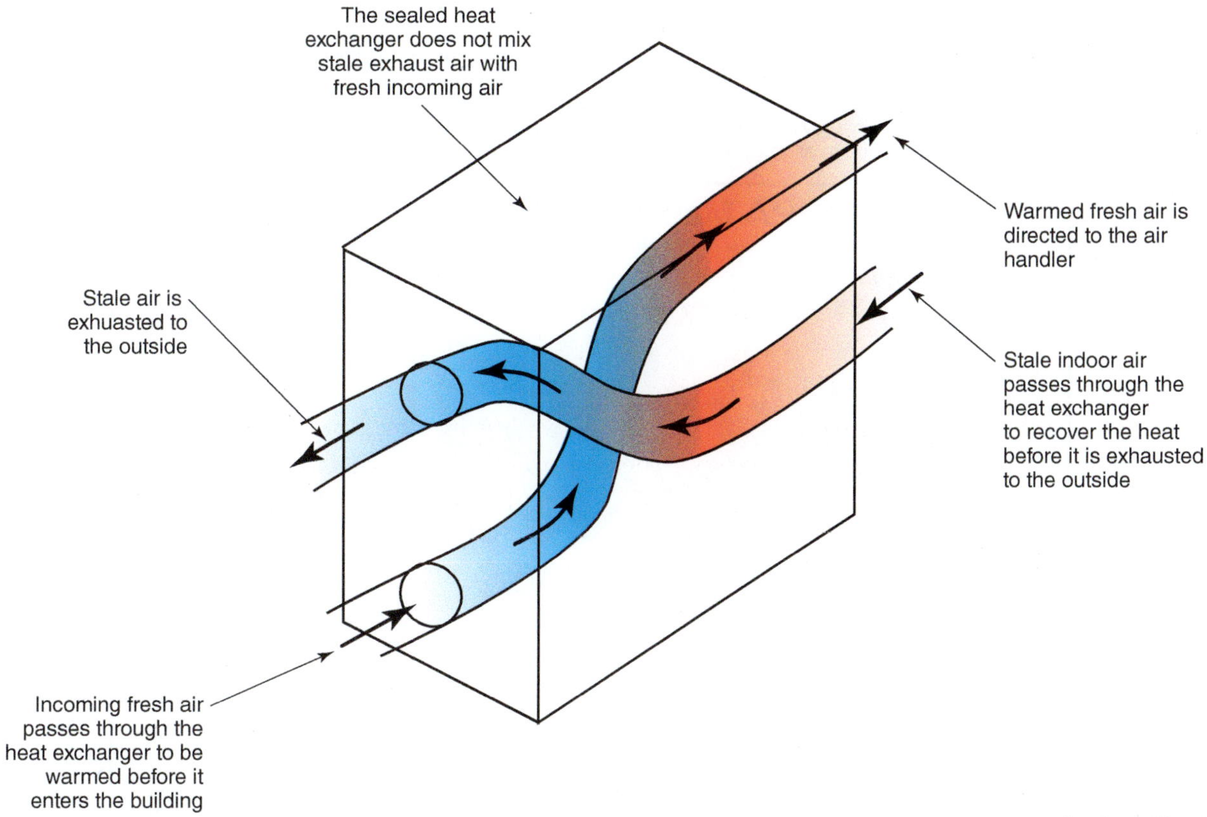

Goodheart-Willcox Publisher

Figure 9.9-5. An HRV exchanges the heat in the incoming fresh air with the outgoing stale air to maintain the air temperature inside the building without using additional heating or cooling energy.

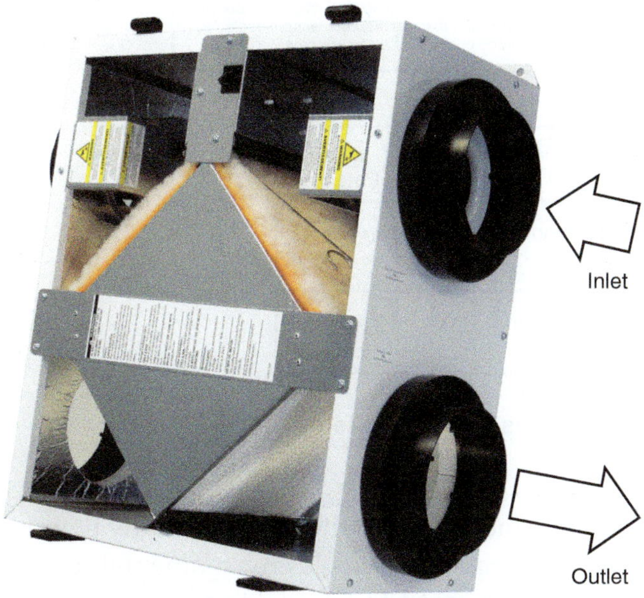

RenewAire

Figure 9.9-6. This ERV maintains both the existing air temperature and humidity level inside the building.

Radon Gas

In some areas, radon gas in the soil can be at levels high enough to compromise indoor air quality, **Figure 9.9-7**. New houses should be designed and built in a way that minimizes the ability for radon gas to enter the living space. If built in a region where radon gas is a potential concern, homes should be constructed in such a way that will facilitate post-construction radon control. This includes ensuring the HVAC systems do not depressurize or draw intake air from the basement. It may also include the installation of a sub-slab depressurization system, **Figure 9.9-8**.

Pro Tip

Applying radon prevention methods in the construction of a new home can prevent radon issues throughout the life of the home.

Operation, Maintenance, and Homeowner Education

Even if the builder achieves a high standard for indoor air quality, the systems in a green home must be operated and maintained properly. Homeowners must understand that the things they bring

Radon Zones Map

The U.S. Environmental Protection Agency (EPA) created this map to identify areas with the potential for elevated indoor radon levels. The EPA Map of Radon Zones helps national, state, and local organizations implement radon-resistant building codes. The map should not be used to determine if a home in a given zone should be tested for radon. Homes with elevated levels of radon have been found in all three zones. All homes should be tested for radon.

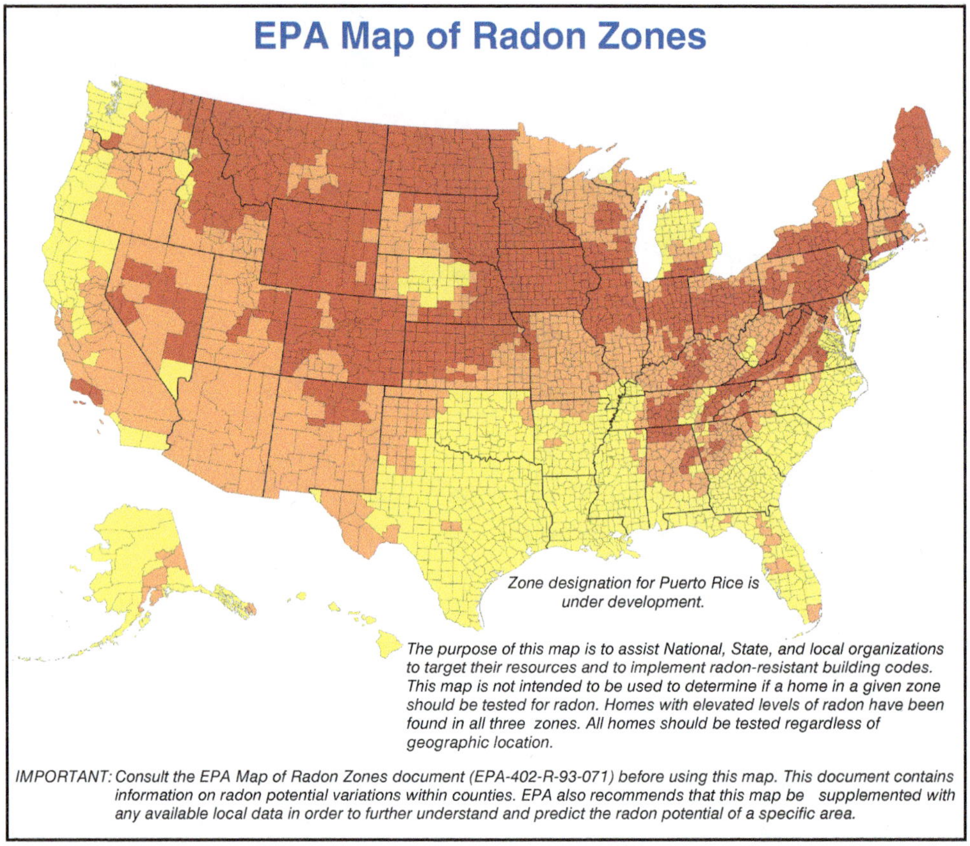

What do the colors mean?

Zone 1 counties have a predicted average indoor radon screening level greater than 4 pCi/L (pico curles per liter)	Highest Potential
Zone 2 counties have a predicted average indoor radon screening level between 2 and 4 pCi/L	Moderate Potential
Zone 3 counties have a predicted average indoor radon screening level less than 2 pCi/L	Low Potential

U.S. Environmental Protection Agency

Figure 9.9-7. Radon is prevalent in different concentrations in soil across the United States.

into a house can pose a threat to their health. Armed with knowledge of alternatives, homeowners can make choices for furnishings, clothing, and cleaning products that minimize pollutants inside the house and can enjoy the benefits of high indoor air quality.

Homeowners play the biggest role in how a green home performs over time and how much energy it uses. Before homeowners move into a new green home, the building team must teach them how to operate and maintain the home and how their behavior affects the home's environmental impact. The owners need to know, for example, how to program the HVAC system controls and clean or replace the filter to get the highest efficiency from the equipment. Each trade has a hand in educating homeowners either directly by leading a training session or indirectly by teaching the general contractor or designer, who then trains the owners, **Figure 9.9-9**.

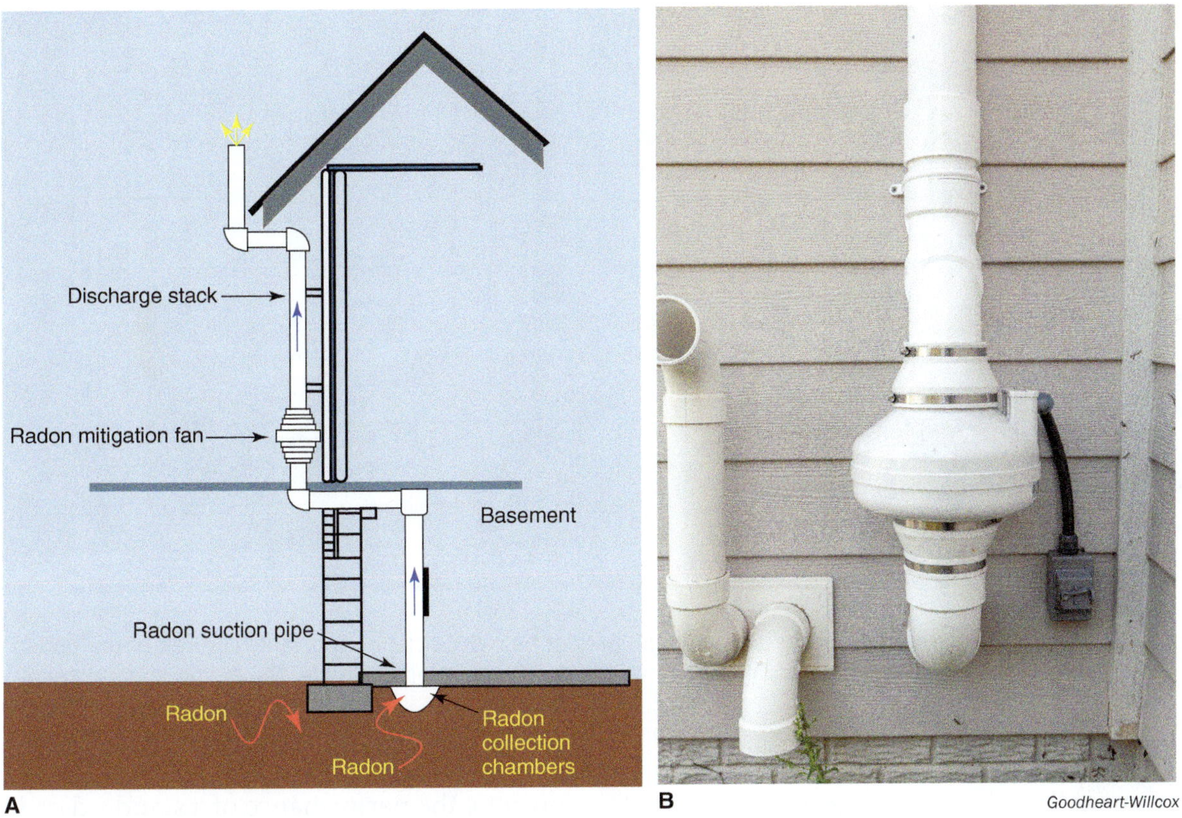

Figure 9.9-8. Typical sub-slab depressurization system to remove radon before it enters the home.

Jacob Lund/Shutterstock.com

Figure 9.9-9. An important step in the finalization of new home construction is ensuring the homeowner is trained in the proper maintenance and operation of the home and its systems.

Safety Note

Homeowners should regularly check and maintain HVAC systems to reduce indoor air pollution and protect occupants from harmful pollutants and allergens.

Because of the large amount of information about the systems and materials in a house, owners may not be able to absorb it all during a few hours of training. Therefore, the building team must compile a reference manual for every green home. The manual details how to operate and optimally set the various systems and appliances. The manual also contains valuable information for future owners or tradespeople who may work on the house during routine maintenance and renovation work, such as a list of all the materials and products used to build the house, as well as manufacturer printed literature. It often includes photographs of key elements of the house taken during installation, **Figure 9.9-10A**, and before the wall finishes were applied, **Figure 9.9-10B**. This visual documentation enables future workers to see what is inside the walls, making their work easier and safer. The manual also includes lifestyle impact information about using each of the home's systems, such as how the amount of time taken for showers affects utility costs. Without this knowledge and understanding, the owners can defeat many of the green features built into a house without realizing it.

A Goodheart-Willcox Publisher

B ungvar/Shutterstock.com

Figure 9.9-10. Photos taken throughout the construction project help document important features of the house and should be shared with the owner in the reference manual. A—Foundation drainage. B—Piping and electrical configuration.

Today, many builders use project management software for estimating, bidding, scheduling, and documenting all aspects of the build. Many of these software programs allow the builder to share select portions of these data with the home buyer. They also have the capability of creating the reference manual digitally, opening up another avenue for communication between the builder and the client after the home is complete. When used effectively, these programs compile a more complete picture of the project from start to finish.

Concluding Thoughts on Green Building

The construction of homes and their use after construction are significant factors in the environmental impact humans have on our planet. Builders can affect this impact through the incorporation of green building techniques and participation in green building programs. Throughout this book, you have learned strategies that builders and architects incorporate into the design and construction of green homes. Using any one of these strategies does not make a home "green." However, any number of green building practices can be used to improve the performance of a standard code-built home or remodeling project.

Green building can range from the construction of an individual home to the consideration of meta-factors when designing an entire housing development. In prior chapters, we looked at the factors impacting individual homes. However, builders, developers, architects, and city planners also consider factors such as efficient use of city infrastructure, impacts on traffic patterns when residents commute to work or school, availability of public services, recreational opportunities, the walkability of neighborhoods, usability and availability of public transportation, and many other related factors when considering green building on a larger scale.

On an individual home, you learned how green building begins with the design of the home and its orientation on the site. Material choices and how they are selected, sourced, and assembled make an impact from the foundation all the way to the finishes and appliances. The building envelope must be designed and installed properly to manage water, air, vapor, and thermal variables. Durability and longevity of the home and its components are major factors in reducing emissions caused from the sourcing, manufacturing, and installation of replacement materials. Even construction debris and the future deconstruction of the home should be considered as builders look at strategies for green building.

Green building practices and materials will continue to gain traction in the construction industry as they have in the past. This progression is natural as more and more manufacturers, government agencies, builders, and regulators experience the pressure to reduce global emissions and combat environmental challenges of the future.

After learning green building practices and strategies for new home construction, you now have knowledge that can help you to make any home more energy-efficient. What are some steps you can take to improve the efficiency of the home you live in? Can any areas be better sealed? What efficient practices with regard to thermostat settings, water conservation, or other behavioral changes could you make to improve your resource usage? Next time you perform a renovation or remodel, what products could you chose that are green rated or more sustainably sourced? Minor changes in each of our choices can add up and begin to make a real difference in the overall impact.

Development of Residential Materials and Building Systems

Many of the components and systems we use in residential construction have been developed or adapted from those used in commercial and industrial construction. Commercial applications need to be strong and durable enough to endure demanding environments produced by the high use of many people. Industrial applications go a step further because they are designed for heavy-duty operations and manufacturing processes. Industrial buildings commonly house specialized equipment requiring specific methods and materials for construction. Components, materials, and hardware used in commercial and industrial buildings are designed, engineered, and built to high standards of durability, safety, and efficiency. Though you may not use these components in residential construction today, it is common for some commercial and industrial applications to make their way into residential construction as manufacturing costs decrease or regulations for residential construction become more stringent. **Figure 9.9-11** shows typical components and systems used in commercial and industrial construction. Specific information

Commercial and Industrial Construction Systems		
	Commercial Construction	**Industrial Construction**
Structural Components/ Systems	▪ Structural steel beams and columns that improve strength, flexibility, and fire resistance. ▪ Commercial buildings typically have flat roofs framed with bar joists.	▪ Deep foundations that are highly reinforced concrete. ▪ Concrete tilt-up panels are often poured on site and lifted into place. These systems are combined with structural steel framework. ▪ Post-tensioned or pre-tensioned concrete slabs for floors and roofs ▪ High-bay construction may require high ceilings and large spaces to accommodate equipment.
Exterior Components/ Systems	▪ Walls are typically masonry, EIFS systems, panelized cladding, curtain walls, or glass systems. ▪ Windows often feature specialty glazing to meet performance or cosmetic requirements.	▪ Concrete; metal paneling; steel cladding. ▪ Heavy-duty rollup and loading dock doors; reinforced doors; high security access control
Insulation Systems	▪ Exterior continuous insulation systems. ▪ Extensive use of interior insulation to reduce noise transmission.	▪ Specialty insulation systems for industrial applications where high-heat or ultra-low temperature requirements must be met.
MEP Components/ Systems	▪ Cast iron and copper plumbing, roof drains for flat roof systems. ▪ Complex, highly automated HVAC systems which often use roof-mounted units. ▪ Stormwater management systems for parking lots. ▪ Armored cable and conduit for electrical wiring.	▪ Equipment to support operations such as dust collection systems and large-scale ventilation for extraction or filtration of caustic fumes/gases. ▪ Some applications may require high-performance filtration such as in a cleanroom environment. ▪ Specialized refrigeration equipment for cold storage may be required for climate-controlled applications.

(continued)

Goodheart-Willcox Publisher

Figure 9.9-11. Commercial and Industrial building applications differ from those in residential construction.

Commercial and Industrial Construction Systems		
	Commercial Construction	**Industrial Construction**
Fire Protection Systems	▪ Advanced fire alarm systems, sprinkler systems, exit signs, smoke seals on automatic doors, fire dampers in HVAC ductwork to slow the spread of fire. ▪ Finishes and surfaces are typically fire-resistant materials.	▪ Advanced fire suppression systems. ▪ Explosion-proof construction accommodations. ▪ Specialized hazardous material storage accommodations.
Flooring and Ceiling Systems	▪ Dropped Ceilings are used to access concealed ductwork, wiring, plumbing, and other mechanical equipment. ▪ Raised access flooring may be used for data cabling and HVAC systems. ▪ Durable flooring surfaces such as commercial vinyl tile, epoxy coatings, and polished concrete.	▪ Reinforced flooring systems to accommodate forklifts and other heavy equipment. ▪ Chemical-resistant epoxy-coated floors.
Hardware, Fasteners, and Other Features	▪ Stronger door hardware, door closers, panic bars. ▪ Specialty fasteners for use in cladding systems, drop ceilings, steel framing, etc.	▪ Epoxy anchors for fastening/securing equipment to masonry walls and floors. ▪ Advanced safety systems such as noise or vibration controls, railing and fall protection systems, and environmental monitoring systems.

Goodheart-Willcox Publisher

Figure 9.9-11. *(Continued)* Commercial and Industrial building applications differ from those in residential construction.

about these examples can be found by performing an internet search. Materials, systems, and technologies that have qualities which prove themselves valuable to a home owner or builder may begin to be used in residential construction when they are readily available and become reasonably affordable to purchase.

In addition to more durable components, commercial buildings must be built to the standards set forth in the Americans with Disabilities Act (ADA). The ADA is a federal law that prohibits discrimination against people with disabilities. The law requires that all new construction and alterations to facilities used by the public must be accessible to people with disabilities. ADA compliant buildings must include provisions for people with vision or hearing impairments, and for those who use a wheelchair or walking assistance. Provisions include ramps and elevators to access entrances and various levels of a building, signage that is easy to read and includes braille, handicapped parking, and provisions made for drinking fountains and restrooms to accommodate those using a wheelchair.

LESSON 9.9 Review

Answer the following questions based on information in this lesson.

1. New furniture often releases VOCs such as _____. (9.9-1)
   A. radon
   B. nitrogen dioxide
   C. formaldehyde
   D. carbon monoxide
2. A benefit of using an energy recovery ventilator to exhaust air from a building is that it also removes _____. (9.9-2)
   A. humidity
   B. radon
   C. mold spores
   D. asbestos
3. A(n) _____ plays the biggest role in how a green home performs over time. (9.9-2)
   A. electrician
   B. homeowner
   C. head contractor
   D. safety manager
4. *True or False?* The green home reference manual is simply a set of instructions for how to operate the green systems and appliances that have been installed. (9.9-2)
5. *True or False?* The components and systems used in residential construction are just as durable as those used in commercial and industrial construction. (9.9-2)

Chapter 9 Review

Summary

Lesson 9.1

- Residential construction is continually evolving as technological advancements in material science, software, tools, and equipment are made and adopted by industry professionals. (9.1-1)

Lesson 9.2

- Remodeling is an important part of the residential construction industry, and it often requires complex problem-solving and critical thinking skills. (9.2-1)

Lesson 9.3

- Finish electrical, plumbing, HVAC, and sitework are the last major steps of a home build before securing the certificate of occupancy and closing out the project. (9.3-1)
- Without a certificate of occupancy, the construction project is not complete and cannot be legally occupied. (9.3-2)

Lesson 9.4

- Materials required to complete the finish MEP systems include control devices such as switches, thermostats, and faucets, along with all associated covers and assemblies to make them aesthetically pleasing and functional. (9.4-1)

Lesson 9.5

- Several specialty hand tools and testers not used in other phases of construction are used during the installation of the finish MEP systems. (9.5-1)

Lesson 9.6

- First aid and first aid training are important parts of construction safety. The OSHA requirements for first aid save lives when properly implemented. (9.6-1)
- An effective emergency action plan (EAP), combined with thorough employee training and clear roles, can significantly reduce the number and severity of injuries during such events. (9.6-2)

Lesson 9.7

- Accurately estimating materials and labor for construction requires a good understanding and application of the math principles featured in this book. (9.7-1)

Lesson 9.8

- Navigating construction drawings with competence helps builders bring a home from concept to completion with accuracy and contractual compliance. (9.8-1)
- A change order refers to any alteration or modification made to the original scope of work outlined in a construction contract. This change can involve adjustments to the project's design, specifications, materials, timelines, or other aspects. (9.8-2)

Lesson 9.9

- Indoor air quality is a major factor in the health of a home's occupants and is especially important in homes constructed with green building principles. (9.9-1)
- Managing indoor air quality involves eliminating pollutants, using low-VOC products, and employing proper ventilation. Mechanical ventilation systems like HRVs and ERVs are effective in maintaining air quality by exchanging indoor air with filtered, fresh outdoor air while managing temperature and humidity. (9.9-2)

casement window. A window with a sash that is hinged on the side and swings outward. (4.4)
casework. The built-in cabinetry or shelving in a home, which can be custom built or prefabricated. (8.3)
cat's paw. A type of pry bar used to pull a nail head back above the surface of the wood to be grasped by a claw hammer or a bigger pry bar. (3.5)
ceiling frame. The system of support for all components of the ceiling. (3.3)
ceiling joist. The main ceiling framing members. (3.3)
cement board. A fireproof, moisture-resistant, fiber-reinforced cement panel used as a base for finishing materials on walls, floors, and countertops. (7.4)
center line. A line on a construction drawing that indicates the center of a symmetrical object, circular feature, or column. (3.8)
central heating. A system in which a single large heating plant produces heat, which is then circulated throughout the structure. (6.3)
certificate of occupancy. A document issued by the building department that closes out the building permit and signifies an end to construction and commissioning the building for use as a residence. (9.3)
chain dimension. The dimension lines linked across a drawing linearly to form a continuous series of dimensions. (3.8)
chalk line. A tool used to mark long straight lines; the case contains a spool of string and is filled with colored chalk powder and the chalk sticks to the string as it is unwound and stretched tightly between two marks. (2.5)
change order. A document that identifies any alteration or modification made to the original scope of work outlined in a construction contract. (9.8)
chop saw. A variation of the power miter saw used to cut light-gauge metal. (3.5)
chronic health condition. An illness that persists over a long period of time, often for months, years, or even a lifetime. (8.6)
circuit. A continuous path from the source of electric power, through various loads or current-using devices, and then back to source. (6.3)
circuit breaker. An overcurrent-protection device that opens a circuit, preventing the flow of electricity when an overload or short is detected so that the circuit will not be damaged. (6.4)
circuit tester. A tool used to determine if voltage is present at any point in a circuit. (9.5)
circular saw. A power tool with a round blade used to cut framing lumber and other wood products. (1.5)
circumference. The perimeter of a circle. (6.7)
coil stock. A roll of sheet metal, typically 24″ wide by 50′ long, that can be cut and bent to form a desired shape. (5.4)
collaboration. The act of individuals or groups working together, across different teams, crews, or trades, to achieve a shared objective. (8.2)
color-match trim nail. A fastener made of aluminum or stainless steel and painted to match coil stock colors. This nail is used to secure aluminum trim pieces such as fascia trim, soffit, and other exterior trim. (5.4)
combination square. A marking square that can be also used to lay out miter joints. Its adjustable, sliding blade allows it to be used as a gauging tool. (8.5)
common denominator. The same bottom number for two fractions being calculated. (2.7)
competent person. A worker who is on-site, making sure work is proceeding in a way that complies with the OSHA standards. (2.6)
composite decking. A durable, low-maintenance material made from a combination of wood fibers, recycled plastics, and binding agents. (5.4)
composite wood siding. A type of manufactured exterior cladding material that is designed to closely mimic the look of traditional wood siding. It is made by combining wood strands with resin and compressing them under high pressure and heat. Also called *engineered wood siding*. (5.4)
compression fitting. A connector that seals joints by compressing a ring of soft material, such as brass. (6.4)
compressive strength. A material's ability to resist deforming when being pushed together. (2.4)

concrete. A building material that begins as a flowable, formable blend of ingredients and then hardens through a chemical reaction to form a strong, durable product that can withstand tremendous loads. (2.4)

concrete broom. A specialty broom used to create a rough texture on the surface of concrete after it has been troweled. (2.5)

concrete edger. A handheld concrete tool used to produce a smooth and consistent radius edge on concrete. (2.5)

concrete float. A handheld concrete tool used to level and smooth concrete surfaces for finish troweling. (2.5)

concrete masonry unit (CMU). A block made from a mixture of Portland cement and aggregates such as sand, fine gravel, or crushed stone. (2.4)

concrete paver. A durable paving material made from molded, compressed concrete. (9.4)

concrete placer. A tool used to spread and move wet concrete as it is being poured. (2.5)

concrete saw. A saw that uses a large, circular blade to cut through concrete, CMU block, or steel rebar. Also called a *cutoff saw* or *demolition saw*. (2.5)

condenser. The part of an air-conditioning unit that receives hot refrigerant and allows it to release heat to the atmosphere and return to a liquid state. (6.3)

condensing unit. The air-conditioning component that houses the compressor, condenser, and evaporator. (6.4)

conditioned. The areas within the home that are actively heated, cooled, or otherwise controlled to maintain comfortable indoor temperatures and humidity levels. (7.3)

conduction. The movement of heat through substances from one end to the other. (6.3)

conductor. A material that carries an electric current. (6.4)

conduit. A tube of metal or plastic in which wiring is installed. (6.4)

construction drawing. A drawing that enables architects, designers, and engineers to communicate what is to be built and how it is to be built. (1.8)

construction expeditor. The person who ensures that the project runs smoothly and efficiently by managing logistics, paperwork, and communication among different stakeholders involved in the project. (2.1)

construction manager. An administrator who works to ensure a construction company meets its organizational goals by keeping the construction projects profitable. (2.1)

contact cement. An adhesive made with a neoprene or latex rubber base that is excellent for applying plastic laminates or joining parts that cannot be easily clamped together. (7.4)

control layer. A material, or assembly of materials that work together to mitigate the impact that environmental factors have on a building. (4.9)

convection. The movement of heat by way of a carrier, such as air or water. (6.3)

coping saw. A saw with a thin, flexible blade used for cutting curves. (8.5)

core. The innermost layer(s) of plywood between the face panels of a flush door. (8.4)

cornice. The parts that enclose or finish the overhang of a roof. It usually includes the fascia board, soffit, and gutter, along with any decorative moldings. (5.3)

corporation. A type of business structure in which the business and its liabilities are separate from the owner or owners of a company. (3.1)

counterbore. A pilot hole that has sides perpendicular to the top. (1.4)

countersink. A pilot hole that is angled at the top. (1.4)

cover letter. A letter sent to an employer to provide contextual details about why an individual is the most effective candidate for an open position. (7.1)

crawl space foundation. A foundation system with walls only a few feet high. (2.3)

credential. A type of documentation that recognizes the completion of, qualification for, or experience in a specific area of expertise. (6.2)

cripple stud. A stud used above or below a wall opening. Extends from the header to the top plate or from the sole plate to the rough sill. (3.3)

critical thinking. The ability to analyze an issue to form a judgment. (1.2)

crosscut saw. A handsaw designed to cut across the wood's grain. (8.5)

crown molding. A decorative piece applied to the top of a wall to make the transition from the wall to the ceiling. (8.3)

crushed stone. The material used to promote stability and drainage under concrete slabs and against foundation walls. Made of rock that has been quarried and run through a rock crusher. (1.4)

cube. A solid with all edges the same size. (7.7)

cultured stone. A material made of concrete, designed to resemble natural stone through the use of authentic shapes, textures, and pigments. Also called *manufactured stone*. (5.4)

custom home. A home that is designed and built to meet the specific needs of an individual client. (2.2)

cut line. A drawing line type that indicates the orientation and location referenced in a section drawing. (5.8)

cut-in point. The temperature setting that causes the thermostat to send a signal to the furnace or air conditioner. (9.4)

cut-out point. The temperature setting that causes the thermostat to stop signaling. (9.4)

cylinder. A solid with two circular faces. (7.7)

D

deceleration device. A PFAS component that is integrated into a lanyard that reduces the impact forces of a fall. (4.6)

decimal foot system. A system that uses dimensions that are graduated in feet, tenths of a foot, and hundredths of a foot rather than feet and inches. (2.7)

decimal fraction. A fraction whose denominator is a multiple of ten, such as 1/10. (4.7)

deck. An outdoor living space that is fully or nearly independent of the design of the home and tends to be built around the landscape of the site, considering privacy and usage needs. (5.3)

deck-over roof insulation. The use of rigid insulation material over the rafter sheathing to reduce thermal bridging. (7.9)

deconstruction. The process of dismantling buildings to salvage components that can be reused or recycled. (8.9)

dehydration. A condition that occurs when the body loses more fluids than it takes in, leading to an imbalance in the body's electrolytes and impairing its ability to function properly. (7.2)

denominator. The bottom number of a fraction. (2.7)

design-build firm. A custom home building firm that has a staff with architects who can do the design work in-house. (2.2)

detail drawing. A drawing that shows the builder any important and complex construction that cannot be included in plan and elevation drawings. It is large scale and shows how various parts are to be located and connected. (4.8)

diameter. The total distance across a circle through its center point. (6.7)

differential. The difference in temperature between the cut-in and cut-out points. (9.4)

diffusion. The tendency for water vapor to migrate slowly from a higher to a lower humidity level through building materials. (4.9)

dimension line. The lines on a drawing that show distance and size. (2.8)

direct heating system. A heating system that distributes heat directly to objects without benefit of ducts or a piping system. (6.3)

dividend. The number being divided in a division problem. (1.7)

divisor. The number the dividend is divided by in a division problem. (1.7)

door casing. A type of trim that is applied to each side of a door frame to cover the space between the jambs and the wall surface. (8.3)

door frame. An assembly of wood parts that form an enclosure and support for a door and its hardware. (8.3)

double-hung window. A window with two sashes that slide up and down past each other in the window frame. (4.4)

double-stud wall construction. A framing technique using two stud-framed walls to form an extra thick wall cavity that can be filled with insulation. (7.9)

downspout. A decorative pipe or tube through which water from the roof is directed from the gutter toward the ground and away from the building. (5.3)

drainage, waste, and venting (DWV) system. A plumbing subsystem that carries away wastewater and solid waste from bathrooms, kitchens, and laundries. (6.3)

dressed dimension. The actual dimensions of lumber after it has been surfaced. (1.4)

drying in. The process of making a home watertight and protected from the elements during the construction phase. (4.3)

drywall. A building material that consists of a core of gypsum, a naturally occurring mineral, sandwiched between two layers of paper or fiberglass mats. Also called *sheetrock*, *gypsum wallboard*, or *wallboard*. (7.4)

drywall compound. A sandable filler used to conceal joints and fasteners after installation of the wallboard. (7.4)

drywall panel lifter. A foot-operated lever device used to raise drywall panels against the sheet above or the ceiling so gaps or cut edges occur at the floor where the trim will cover them. (7.5)

drywall router. A small rotary power tool with a thin cutting bit to follow an object, which allows the cut to match the shape of the object. (7.5)

duct seal compound. A putty that is used to fill gaps and seal joints in ductwork. (6.4)

ductless mini-split. An HVAC system that provides both heating and cooling without the need for ductwork. (6.4)

E

electric radiant heat. A system that transfers heat by moving it through solid matter, such as a concrete floor. (6.3)

electric resistance baseboard heating. A heating system that consists of a heating element wrapped in a tubular casing. Metal fins spaced along its length help radiate the heat to the room air. (6.3)

electric shock. An injury that occurs when the body becomes a part of a circuit. (6.6)

electrical current. The flow of electrons. (6.3)

electrician. A worker who installs electrical systems in new or existing residential, commercial, and industrial structures. (2.1)

electrocution. A type of death that occurs as a result of a severe electrical shock. (6.6)

elevation drawing. A construction drawing showing the outside walls of the structure, providing a visual representation of its height, shape, and architectural details. (5.8)

elevation view. A perspective that shows the front, rear, and sides of a building. (1.8)

emergency action plan (EAP). A written document mandated by OSHA regulations that specify actions to be taken by employers and employees during workplace emergencies. (9.6)

employment benefit. A type of compensation or perk you receive from an employer in addition to a regular paycheck. (8.1)

endpoint. The tick marks, dots, or arrowheads at the end of dimension lines indicating the extent of the dimension noted. (2.8)

energy conservation. The practice of reducing the consumption of energy resources while still achieving desired outcomes such as comfortable temperatures and high indoor air quality. (7.3)

energy recovery ventilator (ERV). An appliance that performs the same functions as an HRV but also removes humidity from the incoming air to maintain more comfortable humidity levels inside the building. (9.9)

energy-heel truss. A framing member with extra height at the eaves to allow more insulation over the top plate. (7.3)

engineer. A construction designer who utilizes math and science to make a home safe and durable. (2.1)

engineer's rod. A leveling rod graduated in feet, tenths of a foot, and hundredths of a foot. (2.7)

engineered lumber product (ELP). A wood structural unit that has been altered through manufacturing processes to make it stronger, straighter, and more dimensionally stable than sawn lumber. (3.4)

engineered quartz. A composite material made of ground quartz, synthetic resin, and pigment, commonly used for countertops. (8.4)

engineered wood flooring. A type of flooring manufactured by laminating several plies and prefinishing the top ply. It is designed to float on the subfloor. (7.4)

granite. A natural stone product commonly used for countertops. (8.4)

green building. A system of planning and construction where all parts of the home work together to lessen the impact on the environment, reduce energy consumption, and extend the usable life of the building with minimum maintenance. (1.9)

ground fault. An unintentional electrical discharge to ground. (6.6)

ground-fault circuit interrupter (GFCI). An electrical device designed to detect and interrupt the flow of electric current when a ground fault is detected. (6.6)

grout. A cement-based material used to fill the gaps between floor or wall tiles. (7.4)

grout float. A hand tool with a rubber base that is used to direct and press grout into joints. (7.5)

growth mindset. A willingness to take on new challenges to improve your skills. (1.2)

gunnable. An application method using a caulk gun. (7.4)

gutter. A trough that collects the water runoff from the roof and discharges it into a downspout. (5.3)

H

hammer drill. A rotary power tool that combines the rotational motion of a standard drill with a rapid hammering or pulsating action to drill through masonry materials. (6.5)

hammer tacker. A stapler that uses a compression tip to drive a staple when the tool is swung like a hammer and contacts the surface. (4.5)

handrail. A hand grab that runs the length of a stairway. (4.6)

hard skill. An ability you develop through the technical knowledge of your trade. (1.2)

Hazard Communication Program (HazCom). A mandatory "right to know" program, required by OSHA and designed to ensure employees know and understand the chemical hazards they face on the job. (8.6)

health hazard. A substance, situation, or condition that has the potential to cause harm to workers through physical, chemical, biological, ergonomic, or psychological means. (8.5)

heat exchanger. A component of an HVAC system used to transfer heat from a source to the air. (6.3)

heat exhaustion. A heat-related illness that can occur with prolonged exposure to high temperatures, combined with physical exertion. (7.2)

heat pump. An HVAC component that combines heating and cooling in one. (6.4)

heat recovery ventilator (HRV). An air exchanging appliance that uses the heat in the air that is being removed from the building to heat the incoming fresh air. (9.9)

heat stroke. A severe and life-threatening heat-related illness that occurs when the body's temperature regulation system fails, leading to a dangerously high body temperature. (7.2)

hierarchy of controls. A system used to mitigate safety hazards that categorizes and ranks the effectiveness of safeguards used on a jobsite. (3.6)

hollow wall anchor. A fastener that secures to or through gypsum and other wall materials to provide improved holding strength. (8.4)

hopper window. A window with a sash hinged along the bottom that swings inward. (4.4)

horizontal sliding window. A window with two or more sashes where at least one of them can horizontally move within the window frame. (4.4)

house flipping. The process of buying a home at a bargain price, renovating it, and reselling it to generate a profit. (4.1)

housewrap. A plastic sheet material designed to protect from bulk water infiltration on walls while allowing water vapor to pass through. (4.4)

housewrap tape. An adhesive tape specifically designed to be used on housewrap or other select materials. (4.4)

HVAC (heating, ventilation, and air conditioning) technician. The worker responsible for installing, diagnosing, and repairing a building's climate control systems. (2.1)

hydronic perimeter system. A central heating system that uses water as a medium to move heat from the heating unit throughout the home. (6.3)

hydronic radiant system. A central heating system that uses water to transfer heat energy from a heat source to tubing located in floors or other structural parts of a building. (6.3)

hypotenuse. The diagonal line in a right triangle that connects the other two sides. (5.7)

hypothermia. A condition that occurs when the body loses heat faster than it can produce it, resulting in a dangerously low body temperature. (7.2)

I

ice dam. A buildup of ice on the eves of a roof that prevents water from draining. (4.4)

ice-and-water barrier. A self-adhering membrane used to protect against water infiltration caused by ice dams. (4.4)

I-joist. A beam consisting of flanges (also called *chords*) of structural composite lumber, such as laminated veneer lumber, or sawn lumber, and a vertical web of oriented strand board, 3/8″ or 7/16″ thick. (3.4)

impact wrench. A power tool with a drive mechanism that allows sockets to be attached. (9.5)

improper fraction. A fraction where the numerator is larger than the denominator. (2.7)

inspector. An official who ensures that a building and its systems conform to codes, standards, and local regulations. (4.1)

insulated concrete form (ICF). A system made of light polystyrene foam or other types of rigid insulating material into which concrete is poured; material used for both foundation walls and exterior walls of a building. (2.9)

insulation. A material that does not readily transmit energy in the form of heat, electricity, or sound. (7.3)

insulation blower. A piece of equipment used to install blown insulation. (7.5)

insurance premium. The amount paid to the insurance company, usually on a monthly basis, to maintain the insurance coverage. (8.1)

interior elevation. A section drawing that shows complex details on a vertical surface inside the home. (8.8)

J

jointer. A tool used to shape mortar joints in brick or block work before the mortar hardens. This tool gives the joint a smooth, concave shape. (2.5)

joist. The horizontal member of a floor frame. (3.3)

K

kerf. A cut in the wood left from the blade of a saw. (8.5)

king stud. The full-length stud at the side of a rough opening, which provides support to the trimmer stud. (3.3)

L

laborer. A worker who does manual labor, working with their hands to assist tradespeople. (2.1)

laminate flooring. A type of snap-together flooring that consists of a fiberboard core with a top layer of synthetic wood-grained pattern. (7.4)

laminated-strand lumber (LSL). An engineered lumber product made from strands of wood bonded together for use in door and window headers, rim joists in floors, and core stock for flush doors with veneer overlays. (3.4)

laminated-veneer lumber (LVL). A lumber constructed by gluing together a number of thin veneer panels with all of their grain running in the same direction. (3.4)

lanyard. A length of rope or strap made of synthetic material with locking snaprings on each end. (4.6)

laser level. A leveling instrument that emits a level laser beam over 360° without being attended. (2.5)

level. The state of being perfectly horizontal. (1.5)

leveling instrument. An instrument used for determining and identifying differences in elevation. (2.5)

leveling rod. A long rod marked with numbered graduations that is used, along with a builder's level or transit, to sight differences in elevation over long distances. (2.5)

lifeline. A cable, rope, or other device that acts as a movable anchorage to improve the mobility of workers using PFAS. (4.6)

lighting schedule. A list that gives details about the light fixtures to be used. (6.8)

limited liability company (LLC). A type of business structure that offers the owner(s) some level of protection from the debts and liabilities of the business. (3.1)

line laser. A leveling instrument that casts stationary lines, used to establish a plumb, level, and/or square reference. (2.5)

line of sight. A straight line that projects from a leveling instrument and does not dip, sag, or curve. (2.5)

lockout (LO). The practice of locking a mechanism or an electrical switch in the open position so that maintenance or service work can be performed safely. (6.6)

lockout/tagout (LOTO). A safety procedure designed to protect workers from the unexpected release of energy by using locks and tags to isolate energy sources such as electrical panels, cords, switches, circuit breakers, or mechanical components to prevent accidental start-up of equipment or energizing a circuit when it is not safe to do so. (6.6)

loose-fill insulation. An insulation used in bulk form and supplied in bags or bales. (7.4)

low-impact material. A material that has a low environmental impact throughout its entire life cycle. (5.9)

lumber defect. An irregularity occurring in or on wood that reduces its strength, durability, or usefulness. (7.4)

M

manifold. A pipe with several outlets that distribute heated water to different areas. (6.3)

mason. A worker who builds walls, floors, walkways, or other structures from brick, concrete block, or stone. (2.1)

mason's trowel. A handheld tool used to spread and shape mortar for brick and block laying. The tool's tapered blade and shape allow experienced masons to apply mortar quickly and accurately. (2.5)

mastic. An adhesive that can be a liquid, applied with a brush or roller, such as those used for foundation dampproofing, or a heavy, paste-like adhesive, applied with a notched trowel or a caulking gun. (7.4)

material takeoff. The process of determining the quantities of specific materials that will be required for a given project. (9.8)

measuring tape. A measuring tool available in a variety of lengths ranging from 50′–300′. It is bulky and has a crank to wind the tape back into the case. (1.5)

MEP system. The mechanical, electrical, and plumbing components of a home that work together to provide comfort and convenience to a building's occupants. (6.3)

metal framing connector. A stamped bracket or strapping designed to make wood-to-wood, wood-to-masonry, or wood-to-concrete connections. (3.4)

millwork. The woodworking products produced in a mill or woodworking shop. (8.3)

mixed number. A number made up of a whole number and a fraction. (2.7)

model code. A set of building codes that is generally accepted across the industry and can apply to any geographic location. (3.2)

modulation. The ability of the heating system to adjust its heat output based on the desired indoor temperature, effectively matching the required heating load. (6.9)

molding. A decorative wood, plastic, or composite trim piece used on doors, windows, ceilings, and walls on a home or building. (8.3)

monolithic slab-on-grade foundation. A single, continuous concrete slab that serves as both the floor structure and the foundation of the building. (2.3)

mortar. A mixture of Portland cement, lime, and masonry sand. (2.4)

mortar hoe. A tool that has two holes in the blade and is used to mix mortar in a wheelbarrow or mortar box. (2.5)

multimeter. A tool used to measure voltage, current flow, and resistance in electrical circuits. (9.5)

N

nail gun. A power tool used for driving nails. (3.5)

National Electrical Code (NEC). A collection of rules and standards developed by experts to ensure the proper installation of electrical conductors and related devices. (6.3)

nibbler. A handheld cutting tool used to shear off small bits of sheet metal with a mechanized punch in a manner similar to the action of a sewing machine. (9.5)

nominal dimension. The rough, unfinished dimensions (width and thickness) of lumber. Lumber is typically referred to in these dimensions. (1.4)

nosing. The part of the stair tread that overhangs the riser. (8.4)

notched trowel. A hand tool with a gapped blade used to spread thinset mortar or mastic on the substrate before laying the tiles. (7.5)

numerator. The top number of a fraction. (2.7)

O

Occupational Safety and Health Administration (OSHA). A government agency that assures safe and healthful conditions for workers. (1.6)

operating engineer. A person who is qualified to operate a number of different kinds of heavy equipment. (2.1)

oriented strand board (OSB). An engineered wood panel made of individual flakes of wood that are laid down like a blanket and adhered to each other with suitable resins and glues. (3.4)

O-ring. A gasket used on faucets to produce leakproof seals between the faucet body and internal parts. (9.4)

orthographic projection. A method of showing a three-dimensional object in two dimensions by displaying various views. (1.8)

oscillating multi-tool. A hand-held tool with an offset blade that oscillates at a rapid pace, allowing it to make fine cuts or remove material, if using an abrasive attachment. (8.5)

OSHA standards. A set of rules that describe the methods employers must use to protect employees from health and safety risks. (1.6)

P

panel door. A cabinet door that has a frame with separate panels of plywood, hardboard, or steel set into the frame. (8.4)

panel schedule. A list that gives information associated with the distribution panel, including the number and sizes of circuit breakers and the devices protected by each one. (6.8)

parallel-strand lumber (PSL). An engineered lumber made up of strands of wood up to 8′ long that are bonded together using adhesives, heat, and pressure for use in structural posts and beams. (3.4)

partnership. A type of business structure in which two or more people own a business together. (3.1)

percent. The name given to one part in a hundred. (8.7)

perm rating. A measure of how much water vapor can pass through a material. A lower number indicates that the material is more resistant to the flow of water vapor. (4.9)

personal fall arrest system (PFAS). A combination of gear that works together to stop a falling worker. It includes a body harness, a lanyard or other connection device, and an anchorage. (4.6)

personal protective equipment (PPE). A piece of clothing or equipment worn to provide protection against safety hazards. (3.6)

phantom power consumption. The energy use that occurs when electrical devices are in standby mode. (6.9)

pigment. A principal ingredient of paint that provides color and opacity. (7.4)

pipe compound. A paste sealant that is brushed onto male threads to seal threaded connections between pipe and other parts. (9.4)

pipe dope. A paste-like substance applied to the male threads of a pipe connection to form a tight seal. (6.4)

plan view. A perspective that shows the horizontal layout of a part of the structure viewed from above. (1.8)

plank flooring. A type of flooring similar to strip flooring, but bored and plugged at the factory to simulate the wooden pegs used to fasten the strips in colonial times. (7.4)

plastic laminate. A hard, thin material that is bonded to a plywood or particleboard substrate to make countertops. (8.4)

platform. A type of staircase that includes landings where the direction of the stair run is usually changed. (8.3)

platform framing. A system of framing a building where the floor joists of each story rest on the top plates of the story below, and the bearing walls and partitions rest on the subfloor of each story. Also called *western framing*. (3.3)

plumb. The state of being perfectly vertical. (1.5)

plumber. A worker who installs and repairs water supply, waste disposal, drainage, gas systems, and their associate fixtures. (2.1)

plumber's putty. A thick, doughy putty used to produce a seal on sink rims, strainers, and faucets. (9.4)

plumbing fixture. A water-using device, such as a sink, bathtub, bidet, urinal, toilet, or shower, that receives water from the supply system and delivers wastewater to the DWV system. (9.4)

plywood. An engineered wood panel constructed by gluing together a number of thin layers (plies) of wood with the grain direction turned at right angles in each successive layer. (3.4)

pneumatic. A device that is operated by compressed air. (1.6)

pocket door. A door suspended on a track that slides into a recess or pocket in a partition wall when opened. (8.4)

pocket hole fastening. A form of joinery that creates a strong, tight wood joint using screws rather than traditional woodworking techniques. (8.4)

point laser. A leveling instrument that casts stationary beams, used to establish a plumb, level, and/or square reference. (2.5)

polyurethane adhesive. A single-component, moisture-catalyzed adhesive that uses moisture from the air or wood to set. As it sets, it foams and spreads to fill gaps. (7.4)

polyvinyl chloride (PVC) trim. A low-maintenance exterior trim material that can be used as an alternative to exterior wood trim. (5.4)

polyvinyl resin emulsion adhesive. An adhesive that sets up rapidly, does not dull tools or stain wood, and securely holds wood parts. It comes ready to use in plastic squeeze bottles. Also called *yellow glue*, *white glue*, and *wood glue*. (7.4)

porch. An outdoor living space of a home, typically attached to or built into the building as an extension of the main structure. (5.3)

portable band saw. A power tool with a toothed blade made from a continuous band of metal stretched between two wheels. (6.5)

portable field brake. A piece of equipment commonly used by siding contractors to put sharp, precise bends in sheet metal, forming it into flashing or finish trim profiles for the exterior of a home. (5.5)

portfolio. A collection of photos, documentation, or samples of projects an individual has participated in or completed to showcase their experience, skills, and achievements to potential employers. (7.1)

post-and-beam framing. A framing method in which loads are carried by a frame of posts connected with beams, eliminating the need for loadbearing walls. (3.3)

potable. The state of being safe to drink. (6.3)

powder-actuated. A device that is operated by a powder charge. (1.6)

power miter saw. A semi-portable saw that rests on a stand or work platform used to cut wood. (3.5)

power plane. A cutting and surfacing tool that produces finished wood surfaces with speed and accuracy. The motor, which operates at a speed of about 20,000 rpm, drives a spiral cutter. (8.5)

power stapler. An electrically or pneumatically powered tool that drives staples. (3.5)

powered metal shears. A handheld electric tool about the size and shape of an electric drill that is able to quickly and easily cut through sheet metal. (9.5)

precast concrete wall. The panels that are formed and poured in a factory, delivered to the construction site, and placed with a crane. (2.9)

prehung door unit. A door frame with the door panel already installed. (8.3)

pressure differential. A difference in air pressure between the inside and outside of a building envelope. (4.9)

print. A copy of an architect's original drawings that are used on the jobsite. (1.8)

problem solving. The process of identifying an issue and finding a solution. (1.2)

product. The result of multiplication. (1.7)

production home. A house that is built, usually in a development, as cost-effectively as possible with highly specialized crews who focus on one aspect of the process. (2.2)

professional development. The ongoing process of improving the knowledge, skills, and abilities required for your career to progress. (5.1)

pry bar. A tool specifically designed to pull nails and dismantle assemblies. (3.5)

pull bar. A bar is used to pull laminate flooring pieces tight in areas where a tapping block cannot be used. (7.5)

pull saw. A saw with a thin blade with fine teeth that cuts as it is pulled toward the user. (8.5)

punch list. A to-do list created at the end of a job to indicate work that still needs to be completed. (9.3)

Pythagorean theorem. A geometric principle that relates to right triangles. It states that the sum of the squares of the sides of a right triangle is equal to the square of the hypotenuse. (5.7)

Q

quotient. The result of a division problem. (1.7)

R

radiation. The movement of heat by wave motion. (6.3)

radius. The distance from a circle's center to its outer edge. (6.7)

radon. A naturally occurring colorless, odorless gas that can enter homes through cracks or unsealed seams in the foundation and accumulate to harmful levels. (9.9)

rafter. A sloped framing member that runs downward from the peak of the roof to the plates of the outside walls. (3.3)

rail. The horizontal member of a door. (8.4)

rainscreen. A gap between the cladding and WRB that allows water to drain freely and for ventilation. (5.3)

ratchet. A handheld tool with a handle and a mechanism that applies force in one direction but not the other, enabling it to turn a socket mated with a bolt or nut without requiring the user to lift and reposition the tool after each turn. (9.5)

rebar. A steel bar commonly used to reinforce concrete against tensile forces. (2.4)

receiving channel. A slot on vinyl or aluminum siding trim pieces that accepts and conceals the cut end of a siding or soffit panel. (5.4)

reciprocating saw. A portable power tool that has linear (back-and-forth) cutting action. (3.5)

rectangular solid. A rectangle with a third dimension. (7.7)

reference. An individual who will provide important information about you to a prospective employer. (7.1)

reference mark. The symbol used in conjunction with a cut line to note where the section drawing can be found within the set of working drawings. (5.8)

reference symbol. A symbol that links two drawings to show an alternative perspective, a cutaway view, or a detail drawing. (1.8)

reflective insulation. A metal foil or foil-surfaced material. The number of reflecting surfaces, not the thickness of the material, determines its insulating value. (7.4)

register. A grille installed over duct openings that have shutoffs for controlling airflow. Some have adjustable vanes that control the direction of the airflow as it exits the duct. (6.3)

remodeling. The construction process that changes the form or function of a space. (9.2)

renewable material. The products made from raw materials that are managed and harvested responsibly and can be replenished in a relatively short timeframe. (5.9)

renovating. The construction process that involves updating or restoring a space to a good state of repair. (9.2)

request for information (RFI). A formal inquiry used in the construction industry to attain clarification or details about various aspects of a project. (9.8)

residential construction. The process of building or renovating single and multifamily homes. (1.1)

résumé. A brief outline of your education, work experience, and other qualifications for work. (7.1)

reveal. A setback from the inside edge of the jamb that allows for hinges and striker plates. (8.3)

rigging. The slings, cables, and chains used with a crane to lift heavy objects into place. (3.6)

right angle. A 90° angle. (5.7)

right triangle. A triangle with one angle that equals 90°. (5.7)

rigid insulation. An insulation manufactured in rigid panels. (7.4)

ripsaw. A handsaw designed to cut best along the grain. (8.5)

riser. The vertical stair members between consecutive stair treads. (8.3)

roof truss. An engineered and prefabricated assembly that combines rafter and ceiling joist in one. (3.3)

roofing nail gun. A pneumatic or battery-operated tool that uses a coil of wire collated roofing nails to attach shingles. (4.5)

roofing underlayment. A secondary layer of protection against weather that offers temporary protection before the finished roofing surface is installed. (4.4)

rope grab. A PFAS connecting device that attaches the lanyard to the lifeline. (4.6)

router. A power tool used to cut irregular shapes and to form various contours on edges. (5.5)

R-value. A measure of a material's resistance to the passage of heat. (4.9)

S

saber saw. A portable saw useful for a wide range of light work, also called a *jig saw*. (8.5)

safety data sheet (SDS). A standardized document that provides detailed information about a specific chemical's properties, hazards, handling procedures, storage, and emergency procedures. (8.6)

safety manager. A worker who identifies hazards and works with project leaders to reduce or eliminate those hazards. (2.1)

safety mindset. A thinking process of concerning yourself with your personal safety and welfare, and the safety and welfare of others. (1.6)

sash. The individual glass panel assembly of a window, including the frame around the glass panel. (4.4)

scaffolding. A temporary structure erected to provide a safe working platform for construction, maintenance, repair, or renovation work. (5.6)

scale. The measuring system used to divide the spaces on a measuring device. (1.7)

screed. A straight board used for striking off concrete to the proper and consistent level after it is placed. (2.5)

section drawing. A construction drawing that shows how a part of the structure looks when cut by a vertical plane. (4.8)

semi-custom home. A custom home built from a stock plan. (2.2)

service entrance. The conductors that bring power from a transformer to the home. (6.3)

set of plans. A comprehensive collection of drawings that communicates all necessary information to build a home. (1.8)

setback. The distance a building must be from the street and the minimum clearances from adjoining properties. (5.2)

sheathing. The boards or prefabricated panels attached to the exterior studding or rafters of a structure. (3.3)

sheet metal hand bender. A tool used to bend and shape thin metal flashing and trim. (5.5)

side-and-end matched. A strip flooring that is tongue-and-groove on the sides and ends. (7.4)

sidelight. A narrow window on one or both sides of a door. (4.4)

sill. The lowest member of the frame of a structure that rests on the foundation and supports the uprights of the frame. (3.3)

site plan. A drawing that shows the entire property or lot to be developed. (1.8)

site preparation. Any demolition work that needs to occur, clearing the land of obstacles or brush, stripping the top layer of vegetation and soil, and laying out where the building will be located. (1.3)

skill set. The range of abilities that help you perform a job. (1.2)

slab-on-grade foundation. A concrete foundation system that is placed directly on the ground. (2.3)

sledgehammer. A hand tool with a heavy head and long handle to maximize driving force. (1.5)

slip-joint washer. A gasket used to seal P-trap connections beneath sinks. (9.4)

snap lock punch. A hand tool used to form a locking tab on vinyl siding panels that mates into a corresponding trim piece. (5.5)

socket. An attachment that fits onto the end of a ratchet and comes in various shapes and sizes to fit different fasteners, such as bolts and nuts. (9.5)

soffit. The member of cornice trim that closes off the space between the wall and the fascia board. (5.3)

soft skills. The competencies used in productively interacting with others. (1.2)

sole plate. The horizontal member at the bottom of the wall frame, supported by a wood subfloor, concrete slab, or other closed surface. (3.3)

sole proprietorship. A type of business structure with only one owner who is responsible for all aspects of the business. (3.1)

solid. A three-dimensional shape. (7.7)

solid-surface. A blend of acrylic or polyester resins, powdered fillers, and pigments cast into slabs for use as a countertop material. (8.4)

source reduction. The process of decreasing material used, energy consumed, and waste produced. (8.9)

span. The distance between structural supports for a horizontal member. (3.3)

spanner wrench. A tool used specifically for installing and removing drain components. (9.5)

spec home. A home that the builder designs and constructs *speculating* that it will sell for a profit upon completion. (2.2)

specialty fastener. A screw or nail designed to be used for a specific purpose. (5.4)

specifications. A detailed, written document that outlines the specific requirements, materials, methods, and quality standards for construction. Also called *specs*. (1.8; 7.8)

split face retaining wall block. A specific type of interlocking CMU block that undergoes a splitting process to achieve a textured, uneven surface on one or more sides. They are designed for creating decorative walls at elevation changes. (9.4)

spray foam insulation. A two-part liquid polyisocyanurate (PIR) material that rapidly expands and hardens after being pumped through a mixing spray nozzle. (7.4)

spread footing. A type of footing that transmits the load from the walls, pilasters, columns, or piers to the soil beneath. (2.3)

square foot. A unit of measurement that represents an area equal to $1' \times 1'$. (6.7)

stair carriage. The main structural support component of the stair system, sometimes called *stringers*. (8.3)

stair rail. A guardrail system that protects an open side of the stairway. (4.6)

stairwell. The opening in a floor that allows adequate headroom for ascent of the stairs from the floor below. (8.3)

standards organization. An establishment whose primary function is to create uniformity across products and procedures by establishing criteria for industry participants. (6.1)

staple gun. A tool used to shoot a staple into various materials. (4.5)

stem wall foundation. A type of slab-on-grade that incorporates a footing placed below the frost line. (2.3)

stepladder. A self-supporting, A-frame ladder that folds open and rests on four legs. They are used for tasks that require a relatively low reach. (4.6)

stile. The outside vertical member of a door. (8.4)

stock plan. A home design that is not unique to a single build. (2.2)

stool. A type of window trim that forms the interior sill cap of a window. (8.3)

straight run. A type of staircase that is continuous from one floor level to another without landings or turns. (8.3)

strip flooring. A floor consisting of wood cut into narrow strips, tongued and grooved, and laid in a random pattern of end joints. (7.4)

vinyl siding. An inexpensive, low maintenance, durable siding material manufactured from rigid PVC resin. (5.4)

vinyl siding removal tool. A small hand tool with a hooked blade that allows the user to unhook an installed piece of vinyl siding for removal without removing the course above it. Also called a *siding unlocking tool*. (5.5)

volatile organic compound (VOC). A carbon-containing chemical that readily evaporates into the air. When it enters the air, it reacts with other elements to produce ozone, which causes air pollution and a host of health issues including breathing problems, headache, nausea, and burning, watery eyes. (5.9; 8.6)

volume. A measure of the space taken up by an object or material. (7.7)

W

water-resistive barrier (WRB). A material or system used to provide a primary layer of protection against water intrusion on the exterior walls of a home. (4.3)

water supply system. A plumbing subsystem that distributes water under pressure throughout the structure for drinking, cooking, bathing, and laundry. (6.3)

water vapor. A gaseous form of water present in the air. (4.9)

wet saw. A tile cutter that uses a circular diamond blade to cut tiles. The saw is outfitted with a small pump to circulate water to the blade for lubrication and dust control. (7.5)

wheelbarrow. A piece of construction equipment consisting of a tub, wheel(s), and handles that is used to move materials such as sand, gravel, concrete, and mortar around the jobsite. (2.5)

whole number. A number without a fraction or decimal point. (1.7)

winding stair. A type of staircase that gradually changes directions as it ascends from one level to another. (8.3)

window and door schedule. A reference table that includes, among other things, the manufacturer's numbers and rough opening sizes for each unit or combination. (5.8)

wire nut. A type of connector made of an insulating plastic shell with an internal threaded metal insert used to connect two or more wires. (9.4)

wood chisel. A hand tool used to trim and cut away wood or composition materials to form joints or recesses. (8.5)

work ethic. A desire and attitude that shows a determination to do what is right in the workplace. (1.2)

wrecking bar. A type of pry bar used to strip concrete forms and disassemble scaffolding and for other rough work involving prying, scraping, and nail pulling. (3.5)

Z

zoning laws. The local ordinances that designate and regulate the use of specific land for residential, commercial, industrial, agricultural, or mixed use. (1.3)

zoning ordinance. A law that defines and dictates how property can be used. (5.2)

INDEX

A

B

C

D

E

F

G

H

I

J

K

L

M

N

O

P

Q

R

T

U

V

W

X

Y

Z